T5-BCG-553

THE NEUROSCIENCES
AND MUSIC II

From Perception to Performance

**Fondazione
Pierfranco e Luisa Mariani**
ONLUS
neurologia infantile

The Mariani Foundation for Pediatric Neurology

The Pierfranco and Luisa Mariani Foundation was created in 1984 as a permanent organization dedicated to many aspects of the study of neurology in children. Its activities include services, training, research, and international cooperation.

Service is the Fondazione Mariani's primary concern: it provides help to those affected by diseases that damage the quality of life of both children and their families. The Foundation also gives assistance and subsidies for effective and innovative complementary programs. Pioneering work has been carried out in the area of metabolic and genetic childhood diseases and on the prevention of neurologic handicaps.

Education and training encompass annual updating courses on child neurology and annual residency courses on clinical genetics, developmental neurorehabilitation, and the Feuerstein method. Educational initiatives are accredited by the Italian Ministry of Health and certified according to ISO 9001. Innovative e-learning projects have been initiated and are developed through the integration of current practices, allowing a high degree of collective interaction. Abundant publishing activities also complement this wide-ranging educational work.

Research funding has also increased in importance and is also ISO 9001–certified. The Fondazione Mariani supports selected projects in keeping with a comprehensive approach, favoring research paths that analyze the biological as well as the psychological–cognitive and behavioral aspects of childhood. Most projects fall in the general areas of animal models, epilepsy, genetics and metabolic diseases, cerebral palsy, and cognition and learning. Calls for proposals are issued once a year on the Foundation's website www.fondazione-mariani.org.

International cooperation for development is part of the Foundation's most recent history. The objective of every activity is aimed at independent development and self-help through the transfer of expertise, protocols, and technologies, as well as through prevention and information.

For the Mariani Foundation helping those who suffer is the primary form of service. In concrete terms this derives from the fruits of training and research, which in turn are a point of departure for world-wide cooperation.

ANNALS OF THE NEW YORK ACADEMY OF SCIENCES
Volume 1060

THE NEUROSCIENCES AND MUSIC II

From Perception to Performance

*Edited by Giuliano Avanzini, Stefan Koelsch, Luisa Lopez,
and Maria Majno*

The New York Academy of Sciences
New York, New York
2005

Copyright © 2005 by the New York Academy of Sciences. All rights reserved. Under the provisions of the United States Copyright Act of 1976, individual readers of the Annals *are permitted to make fair use of the material in them for teaching or research. Permission is granted to quote from the* Annals *provided that the customary acknowledgment is made of the source. Material in the* Annals *may be republished only by permission of the Academy. Address inquiries to the Permissions Department (editorial@nyas.org) at the New York Academy of Sciences.*

Copying fees: *For each copy of an article made beyond the free copying permitted under Section 107 or 108 of the 1976 Copyright Act, a fee should be paid through the Copyright Clearance Center, Inc., 222 Rosewood Drive, Danvers, MA 01923 (www.copyright.com).*

⊛ *The paper used in this publication meets the minimum requirements of the American National Standard for Information Sciences—Permanence of Paper for Printed Library Materials, ANSI Z39.48-1984.*

Library of Congress Cataloging-in-Publication Data

The neurosciences and music II : from perception to performance / edited by Giuliano Avanzini ... [et al.].
    p. cm. — (Annals of the New York Academy of Sciences ; v. 1060)
    "This volume is the result of a conference entitled The Neurosciences and music II: from perception to performance, held on May 5–8, 2005 in Leipzig, Germany."
    Includes bibliographical references and index.
    ISBN 1-57331-610-5 (cloth : alk. paper) — ISBN 1-57331-611-3 (pbk. : alk. paper)
    1. Neurosciences—Congresses. 2. Music—Physiological aspects—Congresses. 3. Music—Psychological aspects—Congresses. 4. Musical perception—Congresses. I. Title: Neurosciences and music two. II. Title: Neurosciences and music 2. III. Avanzini, G. IV. Series.
    Q11.N5 vol. 1060
    [RC321]
    500 s—dc22
    [612.8/

          2005034233

GYAT / PCP
Printed in the United States of America
ISBN 1-57331-610-5 (cloth)
ISBN 1-57331-611-3 (paper)
ISSN 0077-8923

ANNALS OF THE NEW YORK ACADEMY OF SCIENCES

Volume 1060
December 2005

THE NEUROSCIENCES AND MUSIC II

From Perception to Performance

Editors
GIULIANO AVANZINI, STEFAN KOELSCH, LUISA LOPEZ, AND MARIA MAJNO

Scientific Advisors
ISABELLE PERETZ AND ROBERT J. ZATORRE

This volume is the result of a conference entitled **The Neurosciences and Music II: From Perception to Performance**, held by the Fondazione Pierfranco e Luisa Mariani in partnership with the Max Planck Institute for Human Cognitive and Brain Sciences Leipzig, and in cooperation with the New York Academy of Sciences; the European Society for the Cognitive Sciences of Music (ESCOM), Belgium; the Society for Music Perception and Cognition (SMPC), USA; the International Foundation for Music Research (IFMR), USA; the 8th International Conference on Music Perception and Cognition (ICMPC 2004, Northwestern University School of Music), USA; the International Society for Music in Medicine (ISMM), Germany; and with promotion partner, Society for Education, Music and Psychology Research (SEMPRE), UK, and held on May 5–8, 2005 in Leipzig, Germany.

CONTENTS

The New York Academy of Sciences believes it has a responsibility to provide an open forum for discussion of scientific questions. The positions taken by the participants in the reported conferences are their own and not necessarily those of the Academy. The Academy has no intent to influence legislation by providing such forums.

Preface

GIULIANO AVANZINI,[a] LUISA LOPEZ,[b] STEFAN KOELSCH,[c] AND MARIA MAJNO[d]

[a]Department of Experimental Research and Diagnostics, National Neurologic Institute "C. Besta," Milan, Italy

[b]"Eugenio Litta" Rehabilitation Center, Grottaferrata and Child Neurology Unit, University of Rome "Tor Vergata," Rome, Italy

[c]Max Planck Institute for Human Cognitive and Brain Sciences, Leipzig, Germany

[d]Fondazione Pierfranco e Luisa Mariani, Milan, Italy

This volume contains the distilled results of the Mariani Foundation's second conference on a theme that has elicited our interest and stimulated our efforts for more than six years. As its inaugural venture in this field, the Foundation participated in the conference entitled **The Biological Foundations of Music,** held in May 2000 and sponsored by the New York Academy of Sciences. The Academy also published the proceedings, as Volume 930 of the *Annals* with the support of the Foundation. Here, contacts were established for future steps leading to the **Neurosciences and Music** gathering in Venice in October 2002 on the little island of San Servolo in cooperation with the International School of Neurological Sciences.

Volume 999 of the *Annals* ensued from that meeting as a collection of papers that reflected the complementary perspectives of neurologists interested in music and those of musicians involved in and attracted by the neurosciences. In keeping with the Foundation's main commitment to child neurology, the Venice program was subtitled **Mutual Interactions and Implications on Developmental Functions;** the conference and publication were highly regarded by participants and readers.

Given the success of these conferences in a field where both general interest and specific research have been experiencing a veritable "boom," our aim is now to follow this favorable trend and further spin the thread through the relationship among music, neurosciences, and development. In the future, the Mariani Foundation plans to hold meetings approximately every two and a half years, with variable frameworks, depending on the specific outlook, inasmuch as progress in research suggests different angles, as more results are achieved and other issues and questions emerge.

For the 2005 conference, Leipzig was chosen as the most appropriate site, following a brilliant and encouraging suggestion expressed in Venice by the director of the Max Planck Institute for Human Cognitive and Brain Sciences; the Institute happily became the main partner in this endeavor and its staff a major actor in its production. Leipzig hosts one of Europe's oldest universities (founded in 1409) and remains a musical city par excellence: here Johann Sebastian Bach lived and worked as Thomaskantor for most of his life. The conference coincided with the Bach Festival, which enlivens the city's mood every spring. More than 400 participants enlisted in the conference.

Ann. N.Y. Acad. Sci. 1060: xi–xii (2005). © 2005 New York Academy of Sciences.
doi: 10.1196/annals.1360.099

The program, subtitled **From Perception to Performance**, was put together by the organizing committee, who later undertook the editing of the book, with the valuable assistance of two scientific advisors, Isabelle Peretz and Robert Zatorre. With other leading specialists, including Christo Pantev and Laurel Trainor, Drs. Perez and Zatorre, as two "guardian angels," led introductory workshops, the purpose of which was to offer a hands-on understanding of the current status in four main areas of interest: the relevant materials are available on the Foundation's web site (<http://www.fondazione-mariani.org/>).

Seven sessions were scheduled, as reflected in this volume, and the contributions show a thorough process of presenting the current interests and results of first-line scientists in the field. In order to properly encourage research efforts, the most representative and established figures were asked to participate as chairs or discussants; in turn upcoming younger scientists in their laboratories were asked for presentations of their ongoing work. We are also grateful to these (relatively) senior figures for the enlightening introductions and discussions they have provided to the volume's sections.

An ample poster session elicited the breathtaking submission of over 140 posters. A difficult selection process, necessary because of space and production constraints in publication, yielded the choice of the fraction published here; but we are certainly proud of the complete list, which indicates an explosion of consistently high-level work. These short presentations are listed at the end of this volume and the authors of each poster can be contacted for details of their work.

As a further sign of attention to developmental topics, the issue of music therapy, deliberately set aside in the Venice meeting, was included, with the purpose of starting the validation of evidence-based results of methods where music practice is effectively used in rehabilitation techniques: the presentations from the roundtable are a stepping stone in this direction.

In a territory where scientific contributions are multiplying at a speed so impressive that it has become a major commitment to track the literature day by day, we hope to offer a tool that more than fulfills its promise to provide robust information until the next encounter of this tightly woven, yet open, community. This volume aims to crystallize the impact of work whose value is unanimously appreciated for its capacity to open new avenues for a future deeper understanding of the unique multifaceted relationship among music, the nervous system, and the human arts.

Part I: Ethology/Evolution—Do Animals Have Music or Something Else?

Introduction

CAROL L. KRUMHANSL

Department of Psychology, Uris Hall, Cornell University, Ithaca, New York 14853, USA

This section contains three responses to the question, Ethology/evolution: do animals have music or something else? This question probes a relatively undeveloped area in the biology of music, so that it is important that the three contributions are careful to lay out certain general theoretical considerations that must be borne in mind. At the same time, each contribution brings a unique perspective to the question and, perhaps most importantly, offers suggestions for future research that would advance our understanding of the evolutionary history of human music and its relationship to the music-like behaviors of nonhuman species.

McDermott and Hauser's approach is to consider different aspects of human music perception and to ask three questions: (1) Is the feature in question innate in humans? (2) Is it unique to humans? and (3) Is it specific to music? If the answer to all three questions is in the affirmative, then the results support the argument that this particular aspect of music perception is a human adaptation for music. These researchers' empirical study with New World monkeys, the cotton-top tamarins, points to such an adaptation: the preference for consonant over dissonant intervals. In a series of carefully designed experiments, they find that the monkeys will place themselves closer to a preferred sound source than to a nonpreferred sound source (preferring, for example, monkey chirps to monkey screams). However, they show no preference when one sound source is playing consonant intervals and the other sound source is playing dissonant intervals. This is true even though other research has shown that monkeys can discriminate consonant from dissonant intervals. Thus, it is the preference for one over the other that these studies suggest may be a human music adaptation.

Merker's contribution focuses our attention on another aspect of human music, what he calls the "ritual stratum" of human culture. These are characterized as nonfunctional behaviors, in the sense that they are not directed at a specific goal or purpose. Moreover, ritual behaviors are performed with literal duplication. The ability to produce literal duplications requires an extended period of vocal learning. To understand the nature of this process, Merker points us to the large literature on song learning in birds. Birds move through a series of stages with increasing precision, finally arriving at the ability to faithfully reproduce the crystallized song. Humans,

Address for correspondence: Carol Krumhansl, Department of Psychology, Uris Hall, Cornell University, Ithaca, NY 14853. Voice: 607-255-6351; fax: 607-255-8433.
clk4@cornell.edu

Ann. N.Y. Acad. Sci. 1060: 1–2 (2005). © 2005 New York Academy of Sciences.
doi: 10.1196/annals.1360.059

however, surpass all other species in the extent of ritual behaviors, which, according to Merker, not only characterize the arts of music, dance, and poetry, but pervade most all aspects of human life.

In the final contribution, Fitch provides a broad survey of issues arising from comparative data. The first provides an intraspecies comparison between human language and human music. Using the framework of Charles Hockett's features for language, he provides a table showing which apply also to instrumental and vocal music and to innate calls of nonhuman primates. In a complementary way, he then provides a table of design features for music, showing which ones of these also apply to spoken language and to innate calls. From this, he concludes that the two communication systems share not only physical and physiological features, but also deeper formal similarities. He then considers interspecies comparisons, pointing to what might be an interesting homology (similarities between species that are derived from their common ancestor): "drumming" in African great apes. Finally, he considers the possible adaptive functions of music, from sexual selection, to social cohesion, to infant caregiving, concluding that currently the last is most firmly grounded empirically. To this, he adds Charles Darwin's hypothesis that music may have been a protolanguage, a scaffold for the evolution of intentionally meaningful speech.

These three contributions provide a survey of the fundamental issues in the biology of music, useful citations to a wide-ranging literature, and provocative insights that will encourage new empirical work on the ethology and evolution of music.

Comments on Music, Ethology, and Evolution

IAN CROSS

Centre for Music and Science, Faculty of Music, University of Cambridge, Cambridge, CB3 9DP, United Kingdom

The three papers in this section adopt different approaches to the issue of music in evolution and its relationship to behaviors of species other than humans. McDermott and Hauser present the results of an empirical study that demonstrates that tamarins exhibit neither a preference for consonant over dissonant musical sounds nor an aversion to a sound that humans find extremely unpleasant, concluding that a preference for consonant over dissonant sounds may be a species-specific—and music-specific—trait in humans. Merker gives a general account of the developmental trajectory of vocal learning in birds in the context of a discussion of ritual—as opposed to instrumental—behavior, rooting an account of the origins of music in the human capacity for highly developed conformal or imitative behavior. Fitch provides a broad and original treatment of the evolutionary origins of music that makes insightful use of comparative data; in the course of his discussion, he identifies a crucial and as yet virtually unexplored research question, that of the origins of human rhythmic behaviors, suggesting that "African ape percussive behavior...may...indicate a potential precursor of human instrumental music."

McDermott and Hauser explore continuities in "musical" perception between nonhuman species and humans by investigating whether tamarins are sensitive to a distinction between consonant and dissonant sounds. In part, this is motivated by findings that human infants reliably discriminate between consonant and dissonant sounds, and that this capacity might well be innate. However, from the fact that human infants are capable of distinguishing between types of sounds, it does not follow that such discriminations necessarily reflect mature musical (or, for that matter, linguistic) usages that may in part be the consequence of processes of enculturation. As has been demonstrated,[1] infants may exhibit sensitivities to aspects of musical structures, to which adults who have extensive (though informal) musical experience in particular cultures may not be sensitive. In order to investigate cross-specific perceptual continuities that would bear on the issue of the roots of the human faculty for music, it would seem necessary to identify classes of perceptual sensitivities that are *essentially* musical within a wide range of cultures. Alternatively, one might seek to distinguish contexts for the exercise of perceptual judgment that reflect commonalities of cross-cultural musical practices and investigate the existence of analogues of such contexts in the behavioral ecologies of nonhuman species.

Address for correspondence: Ian Cross, Centre for Music and Science, Faculty of Music, University of Cambridge, West Road, Cambridge, CB3 9DP, UK. Voice: +44-1223-335185; fax: +44-1223-335067.

ic108@cus.cam.ac.uk

Ann. N.Y. Acad. Sci. 1060: 3–5 (2005). © 2005 New York Academy of Sciences.
doi: 10.1196/annals.1360.060

Merker notes the existence of similarities in processes of vocal learning in humans and in the rather distantly related avian taxonomic class. These similarities must be regarded as homoplasies, products of processes of convergent evolution, arising independently as a result of some common selection pressures (rather than constituting behavioral capacities that are a consequence of a common genetic heritage). He suggests the requirement for faithful duplication that lies at the heart of "ritual performance" (as opposed to instrumental behavior, which is causal with respect to an evident goal) shapes processes of vocal learning in both birdsong and human music. Despite the attractiveness of this proposal, it can be argued that "transformation" rather than duplication might better represent the processes involved in the transmission of human behaviors (indeed, as has been argued,[2] with respect to all human cultural behaviors). Recent work by Nicholas Magriel (personal communication) on the learning of North Indian music within a hereditary musician caste indicates that when music-learning processes are primarily oral and performative, children's earlier musical productions (at around age 5 or 6) embody global and generic musical features, with vocal performance becoming precise and controlled, though generative rather than duplicative, from about age 11. Similar processes of transformative generation rather than duplication seem to underpin the emergence of children's musical repertoires in other societies,[3] though all too little research has been conducted on vocal and musical learning outside western traditions.

Fitch clearly sets out a broad agenda for exploring music in comparative and evolutionary contexts, and most explicitly seeks to address functional, cross-specific commonalities in the use of sound and action that may bear on an understanding of music. While many of his conclusions appear both substantive and substantial, his paper reflects the nascent state of research in raising more questions than answers. However, one specific point in this paper, concerning the question of music's semanticity, requires investigation.

In Fitch's initial discussion of the relationship between language and music, he suggests that while language possesses semanticity, music does not. This claim can be contested, not least on the grounds that music is reported as meaningful by those who engage with and in it;[4] moreover, a recent neuroscientific study cited by Fitch[5] seems to indicate that both the time course and the neural substrates, implicated in the experience of semantic incongruity, are common for both music and language.

From the perspective of conventional formal semantics where the unit of meaning is the word, Fitch is undoubtedly correct; however, recent approaches to semantic issues (such as functional semantics[6]) broaden the notion of the semantic in ways that would certainly allow music a large measure of semanticity. Indeed, it can be argued that music and language are differentiable not by music's lack of language's semantic capacities, but by the degree of consensual semantic specificity that is achievable in language and music.[7] Music's meanings may from time to time appear precise and specific for any individual listener, but listeners are unlikely to experience a common, specific, and precise meaning with respect to any particular musical experience; music's meanings appear to embody a fundamental ambiguity. Fitch suggests that "however one attempts to characterize musical meaning, the meaning of music and that of language are clearly distinct, and this difference is crucial to understanding these two related faculties." It can be claimed instead, however, that language and music are both communicative media that have profoundly semantic dimensions yet are polar opposites with respect to their capacity to articulate specificity of meaning. Hence not only differences, but similarities, in meaning processes in language and in music require consideration in assessing the human communicative faculties and their evolutionary roots.

REFERENCES

1. HANNON, E.E. & S.E. TREHU. 2005. Metrical categories in infancy and adulthood. Psychol. Sci. **16:** 48–55.
2. SPERBER, D. 1996. Explaining Culture. Blackwell. Oxford.
3. BLACKING, J. 1967. Venda Children's Songs: A Study in Ethnomusicological Analysis. Witwatersrand University Press. Johannesburg.
4. BOHLMAN, S. 1999. Ontologies of music. *In* Rethinking Music. N. Cook & M. Everist, Eds.: 17–34. Oxford University Press. Oxford.
5. KOELSCH, S. *et al.* 2004. Music, language and meaning: brain signatures of semantic processing. Nat. Neurosci. **7:** 302–307.
6. MILLIKAN, R.G. 2004. Varieties of Meaning. The Jean Nicod Lectures. F. Recanati, Ed.: 228. MIT Press. Cambridge, MA.
7. CROSS, I. 2005. Music and meaning, ambiguity and evolution. *In* Musical Communication. D. Miell, R. MacDonald & D. Hargreaves, Eds.: 27–43. Oxford University Press. Oxford.

Probing the Evolutionary Origins of Music Perception

JOSH McDERMOTT[a] AND MARC D. HAUSER[b]

[a]*Perceptual Science Group, Department of Brain and Cognitive Sciences, Massachusetts Institute of Technology, Cambridge, Massachusetts 02139, USA*

[b]*Cognitive Evolution Laboratory, Department of Psychology, Harvard University, Cambridge, Massachusetts 02138*

ABSTRACT: Empirical data have recently begun to inform debates on the evolutionary origins of music. In this paper we discuss some of our recent findings and related theoretical issues. We claim that theories of the origins of music will be usefully constrained if we can determine which aspects of music perception are innate, and, of those, which are uniquely human and specific to music. Comparative research in nonhuman animals, particularly nonhuman primates, is thus critical to the debate. In this paper we focus on the preferences that characterize most humans' experience of music, testing whether similar preferences exist in nonhuman primates. Our research suggests that many rudimentary acoustic preferences, such as those for consonant over dissonant intervals, may be unique to humans. If these preferences prove to be innate in humans, they may be candidates for music-specific adaptations. To establish whether such preferences are innate in humans, one important avenue for future research will be the collection of data from different cultures. This may be facilitated by studies conducted over the internet.

KEYWORDS: music; preferences; monkey; consonance; evolution; adaptation

INTRODUCTION

From the standpoint of evolutionary theory, music is among the most puzzling things that people do. As far as we know, music is universal, playing a significant role in every human culture that has ever been documented. People everywhere love music and expend valuable resources in order to produce and listen to it. Yet despite its central role in human culture, the evolutionary origins of music remain a great mystery. Unlike many other things that humans enjoy (e.g., food, sex, and sleep), music confers no obvious value to an organism, and for this reason music has puzzled evolutionary theorists since the time of Darwin.[1]

Although the adaptive function of music, if any, remains unknown, there is no shortage of proposals for how it might have evolved. Some have noted that music

Address for correspondence: Josh McDermott, Perceptual Science Group, Department of Brain and Cognitive Sciences, Massachusetts Institute of Technology, NE20-444, 3 Cambridge Center, Cambridge, MA 02139. Voice: 617-258-9412; fax: 617-253-8335.
 jhm@mit.edu

Ann. N.Y. Acad. Sci. 1060: 6–16 (2005). © 2005 New York Academy of Sciences.
doi: 10.1196/annals.1360.002

might promote social cohesion in group activities like war or religion; others have proposed a sexually selected role in courtship.[1–6] Developmental psychologists have drawn attention to the pacifying effect music has on infant listeners, which could constitute an adaptive function.[7] Still others suggest that music was not a product of natural selection and, instead, is a side effect of mechanisms that evolved for other functions.[8] Despite the longstanding interest in music's origins, there has thusfar been little empirical data with which to decide between these and other theories (see McDermott and Hauser[9] for a review).

Rather than continue to speculate on putative adaptive functions, we have focused on gathering further empirical constraints on music's origins. Our approach is to examine aspects of human music perception, and for each of them attempt to answer three questions: (1) Is the feature in question innate in humans? (2) Is it unique to humans? and (3) Is it specific to music?

Each of these questions plays an important role in thinking about the evolution of music. Capacities that are innate, that is, determined from properties present in an organism at birth, are potential targets for evolutionary explanations, unlike capacities that are learned. The question of uniqueness plays an equally important role, particularly for music, because music is something that only humans do (see recent reviews[10,29] for a discussion of animal song). If some feature of human music perception is found to be shared by a nonhuman animal, and that feature is assumed to be homologous to the human feature, then the feature in question must not have evolved for the purpose of making music. Testing for aspects of human music perception (e.g., octave equivalence,[11–13] or relative pitch perception[11,12,14]) can thus place useful constraints on music's origins. The third question of music specificity is most relevant for features of music perception that have been found to be uniquely human. If some aspect of music perception in humans is found to be innate and uniquely human, the possibility remains that it evolved to serve some uniquely human function other than music, such as language or mathematics. In contrast, perceptual capacities that are innate, unique, and specific to music are strong candidates for adaptations for music. We thus suggest that evolutionary theories of music perception would be well served by posing these three questions about different aspects of music perception.

PREFERENCES

In this paper we will discuss one particular aspect of music perception—preferences—framed by the three questions about innateness, uniqueness, and specificity. Clearly, many preferences that humans have for music are culture specific, as humans tend to prefer the music of their own culture. Preferences for entire pieces or genres of music may, however, be built on more elementary preferences that could themselves be universal and innate in humans. One simple preference that has received great attention in music literature is that for consonance over dissonance. It has widely appreciated since at least the time of the Greeks that some combinations of musical notes are more pleasing than others. Although the fact that consonant and dissonant intervals are perceptually distinct seems to follow from what is known about the peripheral auditory system,[15–17] it remains unclear why consonance is *preferable* to dissonance. This preference is generally acknowledged to be widespread among Westerners, but there is surprisingly little data from other cultures to support a claim of universality.[18,19] Recent work in developmental psychology, however, suggests that the preference for consonance is either innate or acquired very early, as infants as young as two months seem to exhibit the preference.[20–22] There is thus some evidence that the preference is present independent from musical experience, although a larger cross-cultural database would help to augment the existing case.

Given the possibility that this and perhaps other elementary preferences are innate, our research has focused on the question of whether such preferences are unique to humans by testing for them in nonhuman primates. A consonance preference in a nonhuman primate would provide evidence that the preference did not evolve for the purpose of making and/or appreciating music, as nonhuman primates do not naturally make music. Conversely, any feature of music found to be uniquely human becomes a candidate for part of an adaptation for music, particularly if there is evidence that it is specific to music. Nonhuman subjects have the additional advantage of being reared in a laboratory setting, in which their exposure to music can be controlled to an extent not possible in humans for practical and ethical reasons. As a result of this high level of control, many of the concerns often voiced about the role of musical exposure in experimental results from human infants can be decisively addressed. We thus tested for various acoustic preferences, including that for consonance over dissonance, in nonhuman primates.

Our subjects in the experiments to be described are two species of new world monkey—cotton-top tamarins and common marmosets. Both species are native to the South American rain forest; their lineage diverged from that of humans approximately 48 million years ago (FIG. 1). They are generally regarded as the most prim-

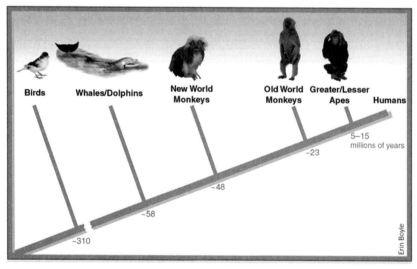

FIGURE 1. Divergence times of some of the relevant taxonomic groups used in studies of the origins and evolution of music. The cotton-top tamarins and common marmosets used in our studies are New World monkeys. (Reproduced with permission from Hauser and McDermott.[10])

itive species of monkey, but are small (weighing on the order of about one pound) and harmless, making them useful experimental subjects. Their hearing characteristics have not been well explored, but the audiograms that have been measured in marmosets are similar to those of humans,[23] and recent auditory physiology work suggests there may be higher-level similarities as well.[24] Recent behavioral work in Japanese monkeys suggests that nonhuman primates can readily discriminate between consonance and dissonance,[25] as one would expect given Helmholtzian theory and the recent physiological results that support it. What is unknown is whether nonhuman primates would also prefer consonance over dissonance as many humans do. All the animals used in our experiments were reared in captivity, and none had ever heard human music prior to the onset of the experiments.

A METHOD TO MEASURE PREFERENCES

To measure preferences in animals, we used a behavioral method in which subjects were placed in a V-shaped maze[26] (FIG. 2); related methods have been developed to test for preferences in birds.[27] Each branch of the maze had a speaker at its end, and a subject's position in the apparatus controlled their auditory environment—one sound was played out of the left speaker when they were in the left branch of the maze, and another out of the right speaker when they were in the right branch. The stimulus for a particular side played continuously as long as the animal was on that side, and switched as soon as they switched sides. If a subject preferred

FIGURE 2. Photo of the apparatus used in nonhuman primate experiments. (Reproduced with permission from McDermott and Hauser.[26])

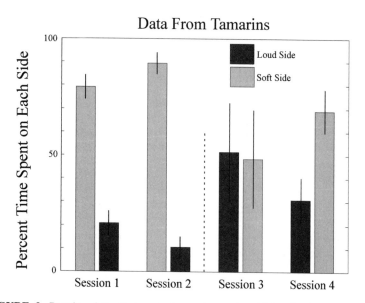

FIGURE 3. Results of the first control experiment, in which animals were presented with a choice between loud and soft white noise. Each *bar* plots the average data from 6 subjects, as a proportion of the total time spent in the apparatus. *Error bars* here and elsewhere denote standard errors. The *dashed line* denotes the reversal of the side assignment that occurred after the second session. (Reproduced with permission from McDermott and Hauser.[26])

one of the two sounds over the other, one might expect them to spend more time in the corresponding side of the apparatus, so as to increase their exposure to the preferred sound. We left animals in the apparatus for five-minute sessions and measured the proportion of time they spent on the left and right.

To verify that the method was appropriate for measuring preferences for sounds, we began by conducting two control experiments. In the first, we presented subjects with a choice between loud (90 dB) and soft (60 dB) white noise. We expected the animals to find the high amplitude noise aversive, and to thus spend more time on the side of the soft noise. The average results from six tamarins over four sessions are shown in FIGURE 3. The animals exhibited a pronounced bias toward the soft side as early as the first session, an effect that increased in the second session. Between the second and third sessions the side–sound pairings were reversed, to rule out effects due to side biases. Following the reversal, the animals spent an average of 50% of the time on each side. Coupled with the increase in the effect from the first session to the second, this indicates that the animals had acquired a side–sound association that took time to be unlearned. By the fourth session (the second after the reversal), however, the effect had reversed, such that they again spent most of the time on the side with the soft noise. The results suggest that the animals learn to associate a side with a sound and modulate their position in the apparatus to reflect their preferences.

In a second control experiment, we presented tamarins with a choice between two classes of species-specific vocalizations: chirps that they emit in the presence of

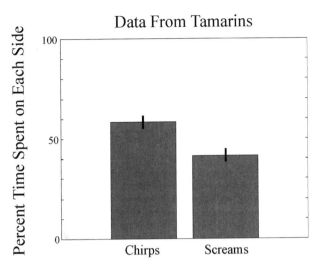

FIGURE 4. Results from the second control experiment, comparing tamarin food chirps with distress screams. Data are averages across sessions. (Reproduced with permission from McDermott and Hauser.[26])

food, and screams that they make when being held by a veterinarian. We reasoned that they would be likely to have negative associations with the screams and positive associations with the chirps, and thus might spend more time on the side with the chirps than that with the screams. Recordings of the two types of vocalizations were equated in amplitude to minimize loudness differences. The same six tamarins were again run in several five-minute sessions. As shown in FIGURE 4, the tamarins spend more time on average with the chirps than with the screams, providing additional evidence that our method provides an appropriate behavioral assay for measuring preferences for sounds.

CONSONANCE AND DISSONANCE

We next proceeded to test for preferences for consonance over dissonance. Before testing our animal subjects with such stimuli, we ran an analogous experiment in humans to confirm that a behavioral method such as ours would demonstrate the consonance preference believed to be widespread in humans. Our human subjects were placed in a room divided in half with a strip of tape (FIG. 5). A concealed speaker was situated on each side of the room, and as in the animal apparatus, each speaker was assigned a particular stimulus. Only one speaker was on at a time, triggered by a subject's position in the room. Our human subjects were given no instructions and were merely told they would be left in the room for five minutes and videotaped. All subjects were naive as to the purpose of the experiment and were involved for a single session. As with the tamarins, we measured the proportion of time spent on each side.

The consonant stimulus in this experiment was a random sequence of two-note chords, the notes of which were separated by either an octave, and fifth, or a fourth.

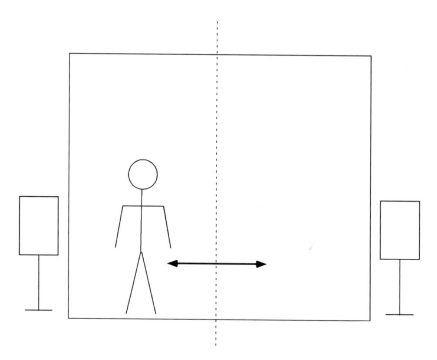

FIGURE 5. Schematic of setup for human control experiments.

The dissonant stimulus was a similar sequence of minor seconds, tritones, and minor ninths. The notes composing the intervals were synthesized complex tones with ten harmonics. The bass note was always middle C. Each interval was 1.5 s in duration.

FIGURE 6 (left) plots the average results for four human subjects, all of whom spent most of their time on the side with the consonant intervals. Typically a human subject would wander around the room until by chance they crossed over the dividing line, thus changing the sound. After moving back and forth across the line a few times, they quickly realized that their position controlled the sound, and thereafter typically spent most of their time on the side of the sound they preferred. These results suggested that our method would be sufficient to demonstrate a consonance preference in nonhuman primates were they to share this with humans,

FIGURE 6 (right) plots the average results for five tamarins. In contrast to the humans, they showed no effect. Note that the animals used in these experiments were the same ones used in the two control experiments, both of which yielded significant effects. Moreover, all 5 animals again showed a preference for loud over soft noise when tested at the conclusion of the consonance experiment, confirming that they had not somehow habituated to the apparatus or method. Rather, it seems that tamarins do not exhibit the preference for consonance over dissonance found in humans, even when tested with analogous methods.

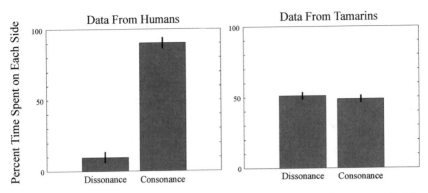

FIGURE 6. Results from experiment comparing consonant and dissonant musical intervals. (*Left*) Results for human subjects. (*Right*) Results for tamarin subjects. (Reproduced with permission from McDermott and Hauser.[26])

SCREECHING

For a second test of whether nonhuman primates might share timbral preferences with humans, we turned to a sound that many humans find highly aversive—the sound of fingernails on a blackboard. We made recordings of a very similar sound produced by scraping a metal garden tool down a glass window; many listeners informally reported the sounds to be very unpleasant. Spectrograms of the sounds we recorded revealed harmonic structure superimposed on broadband noise, similar to what has been previously described.[28] Little is known about why such sounds are so unpleasant, or about the relationship between the perceptual effect they have and that of musical stimuli, but given the strength of the reaction evoked in humans, they seemed a promising stimulus with which to test for timbral preferences in nonhuman primates.

We used a concatenation of several screech recordings as an experimental stimulus. For a control stimulus, we generated white noise with the same amplitude envelope as the screech stimulus. This control stimulus was as loud as the screech stimulus, but otherwise sounded quite different, and we intended it to be much less annoying to human listeners. We again began by running an experiment with human subjects, using the same method as for the consonance experiment. As expected, our method revealed a pronounced preference in humans for the white noise over the screech, shown in FIGURE 7a. In contrast, the tamarins showed no evidence of a preference one way or the other, even when run over many sessions (FIG. 7b). Evidently the screeching sounds that are so annoying to most humans are not particularly aversive for tamarins, at least no more so than our amplitude-matched control stimulus.

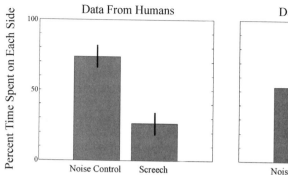
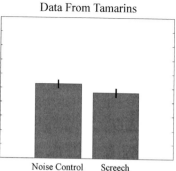

FIGURE 7. Results from experiment comparing a screeching sound to amplitude-matched white noise. (*Left*) Results for human subjects. (*Right*) Results for tamarin subjects. (Reproduced with permission from McDermott and Hauser.[26])

ARE HUMAN ACOUSTIC PREFERENCES UNIQUELY HUMAN?

Two timbral preferences that are pronounced in humans thus appear to be absent in cotton-top tamarins. We have recently replicated the consonance result in common marmosets (McDermott and Hauser, unpublished data), and although it would be ideal to test other species of primates as well, our results raise the possibility that nonhuman primates may lack the timbral preferences that appear to at least partly underlie human appreciation of music.

One key difference between our primate subjects and our human subjects, however, is that the humans all had a lifetime of exposure to music, as do virtually all humans. The consonance preferences apparently present in young infants suggest that a lifetime of exposure is not necessary to develop the preference, but whatever exposure the infants inevitably had may nonetheless be important. Given this, our results suggest three main possibilities: (1) Simple acoustic preferences for consonance and other stimuli could be innate in humans, and unique to them, given the absence of such preferences in the nonhuman primates we have tested. (2) Such preferences might not be unique to humans and could primarily be the result of exposure to musical stimuli, which our nonhuman primate subjects lacked. (3) Such preferences could require exposure to music but might also involve specialized learning mechanisms that could be unique to humans, and perhaps specific to music.

A key issue, therefore, involves determining the role of exposure to music. One important avenue for future research will be to explore the effects of extended musical exposure on nonhuman animals. If nonhuman animals can develop preferences given enough exposure to human music, domain-general learning mechanisms might then also be responsible for human preferences. Conversely, if animals tested after musical exposure still do not exhibit any of the preferences found in humans, the case for uniqueness would be bolstered, for even with similar auditory experience, humans and nonhumans would exhibit different behavior. Further explorations of the effects of musical exposure on humans could help to determine whether exposure coupled with uniquely human learning mechanisms is involved, or whether the preferences in question are, in fact, innate.

MUSIC UNIVERSALS STUDY

In an attempt to assess the effect of the varying musical exposure that occurs in different cultures, one of us (J.M.) has set up an experiment on the internet to measure aspects of music perception in people all over the world. Anyone can participate in the Music Universals Study by visiting <http://music.media.mit.edu>. Our goal is to collect large amounts of data from people with vastly different musical cultures, to examine whether any aspects of music perception are invariant across culture. Differences across cultures would suggest an important role for learning. Web-based experiments are not a replacement for conventional cross-cultural studies, as the subject pool is limited to those with internet access, but they are potentially a useful additional tool with which to ask many questions of interest in music perception.

CONCLUSIONS

The role of musical exposure could also be clarified with a richer cross-cultural database. We propose that evolutionary theories of music's origins will be facilitated by investigating whether aspects of music perception are innate in humans, and, of those, whether any are unique to humans and specific to music. Our studies of preferences in nonhuman primates suggest that many simple acoustic preferences that are pronounced in humans are not shared by our primate relatives. Additional research is needed to investigate the role of musical exposure, but such preferences may thus be innate and unique to humans. Given that some of them appear to be specific to music, they are candidates for part of an adaptation for music. We believe that future research investigating the innateness, uniqueness, and specificity of other aspects of music perception will place strong constraints on the evolutionary origins of music.

ACKNOWLEDGMENTS

We are grateful to Matt Kamen, Altay Guvench, Fernando Vera, Adam Pearson, Tory Wobber, Matthew Sussman, and Alex Rosati for their assistance in running the experiments.

[Competing interests: The authors declare that they have no competing financial interests.]

REFERENCES

1. DARWIN, C. 1871. The Descent of Man and Selection in Relation to Sex. London. John Murray.

2. MERKER, B. 2000. Sychronous chorusing and human origins. *In* The Origins of Music. B. Merker & N. L. Wallin, Eds.: 315–327. The MIT Press. Cambridge, MA.

3. MILLER, G.F. 2001. The Mating Mind: How Sexual Choice Shaped the Evolution of Human Nature, 1st ed. Anchor Books. New York.

4. CROSS, I. 2001. Music, cognition, culture, and evolution. Ann. N. Y. Acad. Sci. **930:** 28–42.

5. HURON, D. 2001. Is music an evolutionary adaptation? Ann. N. Y. Acad. Sci. **930:** 43–61.

6. HAGEN, E.H. & G.A. BRYANT. 2003. Music and dance as a coalition signaling system. Hum. Nat. **14:** 21–51.

7. TREHUB, S.E. 2003. The developmental origins of musicality. Nat. Neurosci. **6:** 669–673.

8. PINKER, S. 1997. How the Mind Works 1st ed. Norton. New York.

9. MCDERMOTT, J. & M.D. HAUSER. 2005. The origins of music: innateness, uniqueness, and evolution. Mus. Percept. In press.

10. HAUSER, M.D. & J. MCDERMOTT. 2003. The evolution of the music faculty: a comparative perspective. Nat. Neurosci. **6:** 663–668.

11. HULSE, S.H. & J. CYNX. 1985. Relative pitch perception is constrained by absolute pitch in songbirds (*Mimus, Molothrus,* and *Sturnus*). J. Comp. Psychol. **99:** 176–196.

12. D'AMATO, M.R. 1988. A search for tonal pattern perception in cebus monkeys: why monkeys can't hum a tune. Mus. Percept. **5:** 453–480.

13. WRIGHT, A.A., J.J. RIVERA, S.H. HULSE, *et al.* 2000. Music perception and octave generalization in rhesus monkeys. J. Exp. Psychol. Gen. **129:** 291–307.

14. BROSCH, M., E. SELEZNEVA, C. BUCKS & H. SCHEICH. 2004. Macaque monkeys discriminate pitch relationships. Cognition **91:** 259–272.

15. HELMHOLTZ, H.V. & A.J. ELLIS. 1954. On the Sensations of Tone as a Physiological Basis for the Theory of Music. 2nd English ed. Dover Publications. New York.

16. FISHMAN, Y.I., I.O. VOLKOV, M.D. NOH, *et al.* 2001. Consonance and dissonance of musical chords: neural correlates in auditory cortex of monkeys and humans. J. Neurophysiol. **86:** 2761–2788.

17. TRAMO, M.J., P.A. CARIANI, B. DELGUTTE & L.D. BRAIDA. 2001. Neurobiological foundations for the theory of harmony in Western tonal music. Ann. N. Y. Acad. Sci. **930:** 92–116.

18. BUTLER, J.W. & P.G. DASTON. 1968. Musical consonance as musical preference: a cross-cultural study. J. Gen. Psychol. **79:** 129–142.

19. MAHER, T.F. 1976. "Need for resolution" ratings for harmonic musical intervals: a comparison between Indians and Canadians. J. Cross Cultural Psychol. **7:** 259–276.

20. ZENTNER, M.R. & J. KAGAN. 1996. Perception of music by infants. Nature **383:** 29.

21. TRAINOR, L.J. & B.M. HEINMILLER. 1998. The development of evaluative responses to music: infants prefer to listen to consonance over dissonance. Infant Behav. Dev. **21:** 77–88.

22. TRAINOR, L.J., C.D. TSANG & V.H.W. CHEUNG. 2002. Preference for sensory consonance in two- and four-month-old infants. Mus. Percept. **20:** 187–194.

23. SEIDEN, H.R. 1958. Auditory acuity of the marmoset monkey (*Hapale jacchus*). Unpublished doctoral dissertation, Princeton University.

24. BENDOR, D. & X. WANG. 2005. The neuronal representation of pitch in primate auditory cortex. Nature **436:** 1161–1165.

25. IZUMI, A. 2000. Japanese monkeys perceive sensory consonance of chords. J. Acoust. Soc. Am. **108:** 3073–3078.

26. MCDERMOTT, J. & M.D. HAUSER. 2004. Are consonant intervals music to their ears? Spontaneous acoustic preferences in a nonhuman primate. Cognition **94:** B11–21.

27. WATANABE, S. & M. NEMOTO. 1998. Reinforcing property of music in Java sparrows (*Padda oryzivora*). Behav. Processes **43:** 211–218.

28. HALPERN, D.L., R. BLAKE & J. HILLENBRAND. 1986. Psychoacoustics of a chilling sound. Percept. Psychophys. **39:** 77–80.

29. FITCH, W.T. 2005. The evolution of music in comparative perspective. Ann. N. Y. Acad. Sci. **1060:** 29–49.

The Conformal Motive in Birdsong, Music, and Language: An Introduction

BJORN MERKER

Department of Psychology, Uppsala University, SE-75142 Uppsala, Sweden

ABSTRACT: Human song and music are part of what may be called a ritual stratum of human culture. As such, they involve special behaviors and neural mechanisms with an uneven but interesting distribution in nature. This may be turned to account not only for shedding light on the biological background of human song and music, but also for elucidating the nature of the cultural traditions within which these arts are practiced and elaborated. The purpose of this chapter is accordingly threefold: (1) to highlight the uniqueness of human culture compared to that of other primates by defining and contrasting two very different types of traditions that coexist within it: those of instrumental culture and those of ritual culture; (2) to outline the crucial role of the mechanism of vocal learning as an enabling device for ritual culture, specifically, by way of the well-studied behavioral biology of birdsong; and (3) to introduce the *conformal motive* as a so-far neglected, but essential, aspect of the neural mechanisms required for vocal learning. Taken together, these considerations will then be used to sketch some of the roles this postulated conformal motive appears to play in human culture, stratified into instrumental, ritual, and language traditions. Some of the implications of this perspective for our understanding of a number of issues related to the human facility for imitation, expressive mimesis, and deliberate teaching are also briefly touched upon.

KEYWORDS: animal song; vocal learning; music; ritual; language; imitation; teaching; conformal motive

INTRODUCTION

The concept of a *conformal motive* to be introduced in this chapter lies implicit in the protracted process through which a songbird acquires the complex patterns of the song culture[1] into which it is born. The stage-wise practice through which the bird perfects its adult song is not externally reinforced or rewarded and requires a motivational mechanism to sustain it. Not every cultural tradition requires such a mechanism for its perpetuation. The contrast between those that do depend upon it and those that do not will therefore help define the behavioral domain within which the conformal motive operates. That domain consists of cultural traditions belonging to what may be called ritual culture. The multiple contrasts between ritual culture and what may be called instrumental culture define two kinds of culture whose dif-

Address for correspondence: Bjorn Merker, Gamla Kyrkv. 44, SE-14171 Segeltorp, Sweden. Voice: +46-86460860; fax: +46-86460860.
gyr694c@tninet.se

Ann. N.Y. Acad. Sci. 1060: 17–28 (2005). © 2005 New York Academy of Sciences.
doi: 10.1196/annals.1360.003

18 ANNALS NEW YORK ACADEMY OF SCIENCES

ferences have pervasive consequences for our understanding of the biological foundations of human music and language.[2]

RITUAL CULTURE AND INSTRUMENTAL CULTURE: CHARACTERISTICS AND CONTRASTS

The distinction between instrumental culture and ritual culture turns on the extent to which invariant aspects of the performance or patterning of the cultural behavior in question are dependent upon—in the sense of being causally constrained or shaped by, derivable from, or obligatory for—the function or purpose of the behavior or not.

Behaviors whose content is largely dictated by or derivable in outline from the goal or purpose of the behavior—such as food washing,[3,4] termite extraction,[5] or nettle stripping,[6] to give examples confined to primates—are instrumental behaviors, and when they are transmitted culturally on a subpopulation-specific basis they define a tradition of instrumental culture. Instrumental behavior is defined by its invariant outcome or end result, and the major part of its difficulty (in initial discovery as well as later learning) is to know that such an outcome is available through behavioral means. This knowledge can be acquired through observational learning once the tradition is a fact. The only motive the learner needs is the desire to partake of the outcome (feasting on termites, say), while the causal coherence of the behaviors that lead up to that outcome provides the backbone of the observational lesson. Success is a matter of having the cognitive capacity to follow its unfolding and the manipulative capacity to give it a try oneself. Efficient execution is promoted by adapting behavior to the causal structure of the task defined by its outcome in accordance with utility, and since only the outcome is fixed, but the initial conditions from which the task is undertaken vary by individual, occasion, and context, widely different concrete behavioral sequences can all be perfectly good modes of execution. Needless to say, considerations of efficiency may favor some modes of execution over others, and these will accordingly dominate performance statistically. Context-attuned performance flexibility is the hallmark of instrumental behavior, while literal duplication of the actual details of execution as performed by an observational model is likely to be counter-productive by impeding rather than promoting instrumental success.

Literal duplication, however, lies at the heart of ritual behavior. When invariant aspects of a behavior are not derivable from its purpose or function (as just outlined), yet are obligatory performance features nevertheless, we are in the presence of a ritual behavior. Its pattern specifics are arbitrary with regard to the ritual's purpose, in the sense that a hypothetical pattern substitution (such as a sequence reversal or inversion) does not on those grounds alone allow one to predict that such a sequence would in principle be any less fit to serve the ritual purpose than the canonical version, yet performing such an alternate version would constitute a violation of the ritual. Ritual behavior has and adheres to a *correct form*, while the achievement of an invariant outcome counts for little if ritual form is violated. In a ritual culture such as our own one can satisfy one's hunger perfectly well by eating without etiquette, yet in so doing one has violated ritual form, and runs the risk of being censured for rudeness. Just as observational learning is the chosen mode for acquiring instrumental behavior, so imitation, in the ordinary dictionary sense, is a mode of acquisition specific to ritual behavior.

The learned, intergenerational transmission of ritual behavior, instantiated by the elaborate song traditions of some bird and whale species, defines a ritual culture. A capacity for high-fidelity duplication of the exact form of a behavioral sequence produced by a model has essentially no uses in nature outside the cultural traditions of ritual carriers. Moreover, the ritual pattern may be performed innumerable times without leading to any outcome at all: an entire mating season may be spent singing without bringing about the mating the singing presumably promotes. An individual, who on the basis of observing such outcome-less efforts on the part of conspecifics concludes that one can spare oneself the trouble by not singing at all, is hardly likely to mate in such a species. In instrumental behavior the individual performing is by and large the judge of the adequacy of performance, while in ritual behavior the audience is the ultimate judge of adequacy. In this sense, ritual is an actual *performance* (display) in a sense not applicable to instrumental behavior.

VOCAL LEARNING FOR SONG AS PARADIGMATIC KEY TO THE NATURE AND ORIGIN OF RITUAL CULTURE

Humans are vocal learners. This means that we readily learn to shape our vocal output to match the pattern of auditory models received through the sense of hearing. We do so with every song we know how to sing; we do so with every word we know how to pronounce, and also when we imitate animate and inanimate sound sources of whatever kind. If this capacity seems unremarkable to us, it is only because we have it: in phylogenetic perspective it is exceedingly rare. Mammals excel in their capacity to learn, yet *vocal* learning is a rarity among them,[7,8] and no ape besides humans possesses it. Beyond a few mammals such as humans, whales, and seals, it is birds that supply the striking examples of the ability to *learn to produce* the patterns of songs or other sounds from auditory models: parrots,[9,10] hummingbirds,[11] and the large group of oscine (but not suboscine) songbird species do so.[12,13]

Vocal learning, which allows a young bird to faithfully reproduce the structural details of song it heard perhaps months earlier (see next section), is an instance of true and spectacularly deferred imitation[14] in the ordinary, dictionary, sense of imitate. The resulting song tradition is subpopulation specific[15] and has a form whose pattern details are not dictated by the purpose served by the behavior, yet for a given individual in a given population, those particulars are obligatory. The behavior has, in other words, all the hallmarks of cultural ritual identified in the foregoing, and this applies to whalesong as well.[16] Across species, song characteristics vary widely; the repertoires of some species feature hundreds and even thousands of individual song types.[17]

Vocal learning for song thus supplies us with paradigmatic examples of ritual culture in animals, namely *forms of imitative learning that faithfully duplicate its model to result in the lasting acquisition of a formally patterned behavior as an integral part of a cultural tradition with ritual characteristics.* By comparison, ape culture has little to offer in this respect, though it abounds in examples of instrumental culture.[5] Human culture, on the other hand, has taken both forms of culture to extreme levels (see final section). The difference between us and our closest ape relatives is most parsimoniously accounted for by assuming that a crucial enabling device was selected for at or after the divergence of the lines that led to humans and chimpanzees, respectively. Vocal learning for song—clearly present in us but not in chimpanzees[8]—supplies that enabling adaptation, I suggest. In this I am only updating, with a focus on ritual culture, a proposal long since made in relation to the origins of language by Darwin,[18] and emphasized by students of birdsong since the 1970s.[19–24]

Neither human song nor speech could exist without vocal learning, and it is presumably not simply a coincidence that the two also share a deep generative similarity.[25] The fact that song, but not language, has arisen again and again in nature entitles us to look to ancestral song as preceding speech, as already proposed by Darwin.[18] That is, long before we spoke, we may have been singing apes in exactly the sense in which whales and songbids are singing animals today. As Darwin, again, surmised, that places the matter of the origin of this capacity squarely in that vast theater of animal communicative signals and manipulative displays that houses all the aesthetic extravaganzas of nature—the peacock's tail no less than the 1800 melodies making up the repertoire of a brown thrasher.[17,26,27] The basic purpose of display is to impress: to show that one is worthy of being taken seriously: as a mate, a fighter, or provider. Doing so by display means demonstrating that one commands resources over and above what is required for merely "getting by." This can be done by engaging in activities requiring unusual amounts of energy, degrees of persistence, levels of complexity, or learning capacity, that is, by unusually loud, persistent, or complex calling in the vocal sphere.[28] Long bouts of complex learned song combine all of these, and we may even guess the fundamental reason why the duplication of detailed pattern characteristics occupies such a prominent place in traditions of ritual culture, as follows.

An individual's mastery of one particular complex song sequence on a learned basis is not just proof of possession of that particular song, but of the perceptual, discriminative, attentional, motor, and learning capacity required to pick out and to master that particular pattern in all its individuality. Its faithful reproduction demonstrates that this was in fact the level of resolution at which one's cognitive and learning capacity operated at the time the song was acquired. The likelihood that an individual should have been in possession of a special-purpose learning capacity capable of acquiring this and no other song variant in a species perpetuating a historically changing song culture is of course vanishingly small. Its faithful duplication is therefore living proof of possession of the general learning capacity sufficient for mastering any member of the pattern space from which the sample is drawn (see Ref. 29, published after these remarks were written). Since learning capacity has a volumetric basis in birds no less than in mammals,[30,31] the demonstrated ability to duplicate the model in all its intricacies is therefore an indirect measure of brain compartment volume. Taken together with the full complement of the behavioral competences required for exercising that capacity—from attention and the rest of the neural prerequisites to a food-gathering ability sufficient to sustain untold hours of practice needed to perfect the performance—makes the duplicative capacity a highly refined and compact measure of fitness in every sense of the word.

The measure of fitness supplied by the capacity for duplication of a complex pattern of ritual culture is thus concerned with the faithfulness with which the duplication takes place, with its fidelity. It allows us to understand the otherwise puzzling question of why biological resources should be invested in copying pattern details that in themselves have no biological utility. At the same time, the capacity for high-fidelity duplication is a necessary prerequisite for the very existence of the cultural tradition supplying the ritual to be mastered, because without a core of fidelity in intergenerational transmission, it would soon degenerate into a set of idiosyncratic performances. Given this central role of the duplicative capacity in ritual culture, we are ready to consider a so-far neglected aspect of its mechanism that will provide a bridge as well to the ubiquitous ritual traditions of humans.

THE CONFORMAL MOTIVE IN SONG LEARNING AND BEYOND

Beginning with the pioneering studies of Thorpe,[32] half a century of intensive study of vocal learning for song in birds has uncovered a wealth of detail regarding the ontogeny of bird song, its behavioral stages and neural mechanisms.[33] The developing bird invests considerable behavioral and neural resources in the duplicative logic of song learning outlined in the foregoing. There is no standard ontogenetic trajectory for doing so that is independent of species, sex, or individual,[34,35] or of environmental and social context.[14,36] The following is therefore only an indication of what is involved, based on some well-studied examples: song learning on the part of a hatchling will be initiated by exposure to the learned pattern of song produced by a (typically) conspecific model, which is stored in memory as a standard ("template") informing later practice. Production learning starts with highly variable jumbles of song elements in a practice phase called subsong, which has been analogized to the babbling phase of human infants.[19,22,37] Its variability not only serves to practice vocal control, but also appears to be integral to the process of pattern learning itself.[38] This is followed by a practice phase called plastic song, which may serve as an arena for practicing pattern production. The animal produces increasingly competent song patterns and variations that can be quite elaborate, but without adhering closely to the patterns of the model. The model template stored in memory is nevertheless in the background throughout this stage of plastic song, because it eventually ends; it settles, as it were, into what is called "crystallized song" of full adult competence, in which the animal finally achieves a high-fidelity replication of the model pattern in its own vocal output. The finished song is less variable than what the bird produced in its "plastic" song phase,[29,39] and high alertness levels or testosterone titers also make production less, rather than more, variable,[35,40] indicating that fidelity is indeed the performance ideal.

The various pattern excursions making up the practice phases accordingly appear to be tacitly governed by a rather specific and detailed performance ideal, revealed only at the end of a protracted process stretching over weeks and months to cover a good part of the bird's first year of life before it is complete. Something must sustain practice through all the preliminary stages and variations by which the bird arrives at its adult competence. The invariable memory record of the song model is one prerequisite for this,[41] and an error signal telling the animal that its own production has not yet matched the model is another.[42] Yet why is the animal concerned with gauging the extent of performance error in the first place? Its plastic song phase contains many examples of song that in formal terms of complexity or esthetics might be considered adequate performances. Yet the animal pushes practice beyond these to the more rigid phase of song crystallization. Moreover, all of this practice is undertaken without being externally reinforced or rewarded in any tangible way whatsoever. In fact, the practice phases of birdsong represent an extreme case of the need for a mechanism to sustain copying efforts when no immediate reward ensues[43] (see also Ref. 40, p. 241).

Both desiderata, that of a mechanism to sustain unreinforced practice effort, and a mechanism that ensures that such practice is sustained until duplication has been accomplished, would seem to lie in the domain of motivational mechanisms, since they aim at an invariant outcome. Moreover, a single motivational mechanism might achieve both ends, one namely that leaves the bird emotionally dissatisfied until duplication has been achieved. Its desired end state is conformity between the animal's own output and the model stored in memory. A fitting name for such a motivational factor might accordingly be *conformal motive,* hereby proposed as a technical term for such a hypothetical motivational mechanism driving practice in vocal learning specifically, and ritual behavior more generally. It would be this conformal motive, in other words, that induces the animal to take the heard pattern "seriously" as a model to be duplicated through practice and then does not let the animal rest until it has achieved with its own voice that which it once heard through its ears and stored in memory. Thus the bird will not settle for any of the variants it produces during the plastic song phase, but instead perseveres until song crystallization has been achieved.

It is only in the domain of ritual behavior, and what is more, of learned ritual, that such a motive is needed. There it is needed in the dual sense of being necessary for maintaining the core of fidelity on which perpetuation of a ritual tradition depends, and in the sense of driving the practice by which the individual carrier of such a tradition demonstrates competence by mastering its pattern standard. Regarding possible evolutionary origins of this conformal motive, it would presumably be derived as a specialization from more widely distributed motivational mechanisms, such as those involved in the incipiently duplicative or imitative behaviors that occur in animal play,[43] or perhaps those of social facilitation,[44] working by way of the memory for song rather than its perception. Be that as it may, the need for a conformal motive to sustain the practice phase of vocal learning generates interesting consequences when examined in light of the human capacity for vocal learning and the nature of human cultural traditions, to which we turn next.

THREE LEVELS OF HUMAN CULTURE IN LIGHT OF THE DUPLICATIVE LOGIC OF VOCAL LEARNING

Innumerable instances of instrumental cultural traditions inform diverse domains of human material culture, subsistence technique, and other practically oriented areas of life. This layer of human culture maintains a continuity in kind (though hardly in degree: compare termite fishing with high technology!) with much of what has been documented as cultural traditions among apes.[5] In addition, however, human culture has a rich layer of ritual culture. There is hardly an area of human life and behavior devoid of some aspect of ritual performance.[45–49]

In the sphere of religion, rituals abound in the form of spells, incantations, mantras, magic rites and divination, liturgy, prayer rites, sacrificial rites, consecration rites, cleansing and exorcism, and healing rites, to name some salient examples. With less of an obligatory connection to religion, there are also rites of passage (e.g., marriage and naming), mortuary and funerary rites, blessing rites, ritual cleanliness, ritual body adornment, festival rites, invocations, oaths, vows and curses, rites of greeting and parting, "ruling rites" (e.g., dynastic or national), and so called rites of rebellion (involving sanctioned violation of customary norms). In the secular sphere, though not necessarily untouched by a religious aspect, we find institutional rites (of, e.g., guilds, associations and corporations), childrens' play rituals, courting rituals, food ceremonials, etiquette, mottos and slogans, rites of the hunt, rites of negotiation, rites of war and peace, and sports and games ceremonials. Again, these are only salient samples of the richness of ritual form exhibited by human cultures, not in scattered instances here and there, but massively and ubiquitously so.

From a biological point of view, this combination of extreme elaboration of instrumental culture with extreme development of ritual culture is of course unprecedented, even without considering the third layer of human culture represented by the singular novelty of human language.[2] It is presumably this unique three-tiered structure that gives human culture its cumulative,[50] genuinely historical character. The joint development of instrumental and ritual culture is, however, not inherently dependent upon human language, though the extent of their elaboration presumably is. Nature attests to the existence of less-advanced versions of a parallel development of both instrumental and ritual culture in traditions of innovative feeding (instrumental culture) and complex learned song (ritual culture) among songbird species.[51]

The paucity of convincing examples of accurate duplication of the detailed *form* of behaviors among great apes in the wild[14,43,50,52] has already been mentioned, though their altered behavior in this regard after intensive contact with humans is not without interest.[53–56] Ultimately our difference from the other apes presumably turns on the mechanism of vocal learning, and in our case too, a conformal motive appears to operate to inform and sustain practice.

In mastering the speaking of our mother tongue during infancy and childhood, the shaping of articulatory output is guided by the duplicative logic of the conformal motive. The massive imprint of dialectal detail on the child's developing speech attests to its acute operation. At a more abstract level, however, the child's willingness, even eagerness, to take *arbitrary* designations (such as "table" for the object which its parents, were they Swedish, would insist on calling "bord") to be obligatory, also would seem to reflect the operation of the conformal motive. At a later stage, our adoption of the terminology, jargon, and modes of expression of whatever subculture or subcultures with which we happen to be identified points in the same direction.

Music, no less than language, consists of culturally transmitted patterns whose specifics are biologically arbitrary yet obligatory in a given tradition. The carriers of such a tradition acquire its patterns through an extended process of exposure (starting in infancy) and practice whose extent varies with the use for which the acquired competence is intended. A musician will spend many hours and days of arduous practice on mastering the performance of a particularly challenging musical passage. All kinds of versions might "do," but nothing will in fact do until the musician has mastered it completely, or "has it" (cf. song crystallization, above), by the duplicative logic of the conformal motive. The approach is the same whether the performer is a singer or an instrumentalist, or a dancer for that matter. This alerts us to the fact that

once the conformal motive is in place its influence need not be confined to the audi-
tory-vocal channel, but may extend to other areas of behavior as well (cf. sign language). In the human case, at least, we have ample reason to think that this is the
case. The human infant has been called an "imitative generalist,"[57] and our facility
for "expressive mimesis" has been emphasized as a specifically human trait by contrast with chimpanzees.[58] The human facility for imitation, virtually absent in chimpanzees, as well as its complement, deliberate teaching—nowhere more necessary
than in the acquisition of complex *arbitrary* patterns featured in advanced ritual—

FIGURE 1. Illustration of conformal motive. (Reproduced with permission from
Keystone-France/IBL Bildbyrå.)

presumably finds its natural origin and explanation in the human conformal motive underpinning our ritual culture (FIG. 1).

This is not the place to analyze the varied manifestations of the conformal motive in human culture, though FIGURE 1 is intended to give a hint of its pervasive presence and influence. Considerable resources are drawn to the ritual sphere of human culture, not only in obvious instances, such as the twelve-day Agnicayana sacrifice of Vedic tradition[59] with its elaborate chanted ceremonies involving 17 priests and requiring months of preparation, but more subtly so. It is easy to overlook the attention and effort that goes into acquiring the ritual patterns of one's natal culture in the course of growing up within it. Its forms become "second nature" to us. They are far from free gifts of nature, however, as we notice when being at a loss for the proper forms of simple things like everyday greetings on first exposure to a foreign culture. What is to be noted here, rather, is the duplicative logic of the conformal motive that lies behind the multifarious culture-specific forms of all these rituals like a hidden agency casting a spell of formal requirements (which inherently could be otherwise but are not) over vast domains of human behavior. Innumerable instrumental ends of human life might be accomplished without the intricacies of ritual form, but not to adhere to the forms of the culture would amount not only to a lack of style and manners, but to falling short of fully human status: we broke with the apes by adding ritual culture to instrumental culture. That in turn might have been a crucial step toward conquering our ultimate biological novelty, language.[2] In all of this, the conformal motive would seem to play a profound yet so-far unrecognized role. For now, this remains no more than a conjectural suggestion, albeit one eminently compatible with much of what we know about our kind, and inviting further exploration in animals as well as in ourselves.

ACKNOWLEDGMENTS

The research upon which this chapter is based was supported by the Bank of Sweden Tercentenary Foundation. Louise Kennedy provided insightful comments that helped me improve an earlier version of this chapter.

[Competing interests: The authors declare that they have no competing financial interests.]

REFERENCES

1. FRAGASZY, D.M. & S. PERRY, Eds. 2003. The Biology of Traditions. Models and Evidence. Cambridge University Press. Cambridge, UK.
2. MERKER, B. 2006. From ape culture to ritual culture and language via vocal learning for song: The singular path to human uniqueness. *In* Communicative Musicality. S. Malloch & C. Trevarthen, Eds. Oxford University Press. Oxford.
3. IMANISHI, K. 1957. Identification: a process of enculturation in the subhuman society of *Macaca fuscata*. Primates **1:** 1–29
4. ITANI, J. 1958. On the acquisition and propagation of a new food habit in the troop of Japanese monkeys at Takasakiyama. *In* Japanese Monkeys: A Collection of Translations. K. Imanishi & S. Altmann, Eds.: 52–65. University of Alberta Press. Edmonton.

5. WHITEN, A., J. GOODALL, W.C. MCGREW, *et al.* 1999. Cultures in chimpanzees. Nature **399:** 682–685.
6. BYRNE, R.W. & J.M.E. BYRNE. 1993. Complex leaf-gathering skills of mountain gorillas (*Gorilla g. beringei*): variability and standardization. Am. J. Primatol. **31:** 241–261.
7. SNOWDON, C.T. & A.M. ELOWSON. 1992. Ontogeny of primate vocal communication. *In* Topics in Primatology, Vol. 1, Human Origins. T. Nishida, F.B.M. de Waal, W. McGrew, P. Marler & M. Pickford, Eds.: 279–290. Tokyo University Press. Tokyo.
8. JANIK, V.M. & P.J.B. SLATER. 1997. Vocal learning in mammals. Adv. Stud. Behav. **26:** 59–99.
9. TODT, D. 1975. Social learning of vocal patterns and modes of their application in grey parrots (*Psittacus erithacus*). Z. Tierpsychol. Suppl. **4:** 1–100.
10. PEPPERBERG, I.M. 1981. Functional vocalizations by an African Grey parrot (*Psittacus erithacus*). Z. Tierpsychol. **55:** 139–160.
11. BAPTISTA, L.F. & K. SCHUCHMANN. 1990. Song learning in the Anna hummingbird (*Calypte anna*). Ethology **84:** 15–26.
12. KROODSMA, D.E. & J.R. BAYLIS. 1982. A world survey of evidence for vocal learning in birds. *In* Acoustic Communication in Birds. D.E. Kroodsma & E.H. Miller, Eds.: 311–337. Academic Press. New York.
13. KROODSMA, D.E. 1988. Song types and their use: developmental flexibility of the male blue-winged warbler. Ethology **79:** 235–247.
14. FREEBERG, T.M. 2000. Culture and courtship in vertebrates: a review of social learning and transmission of courtship systems and mating patterns. Behav. Processes **51:** 177–192.
15. NOTTEBOHM, F. 1972. The origins of vocal learning. Am. Nat. **106:** 116–140.
16. PAYNE, K. 2000. The progressively changing songs of humpback whales: a window on the creative process in a wild animal. *In* The Origins of Music. N.L. Wallin, B. Merker & S. Brown, Eds.: 135–150. The MIT Press. Cambridge, MA.
17. KROODSMA, D.E. & L.D. PARKER. 1977. Vocal virtuosity in the brown thrasher. Auk **94:** 783–785.
18. DARWIN, C. 1871. The Descent of Man and Selection in Relation to Sex. D. Appleton & Company. New York.
19. MARLER, P. 1970. Bird song and speech development: could there be parallels? Am. Sci. **58:** 669–673.
20. NOTTEBOHM, F. 1975. A zoologist's view of some language phenomena, with particular emphasis on vocal learning. *In* Foundations of Language Development. E.H. Lenneberg & E. Lenneberg, Eds.: 61–103. Academic Press. New York.
21. NOTTEBOHM, F. 1976. Discussion paper. Vocal tract and brain: a search for evolutionary bottlenecks. *In* Origins and Evolution of Language and Speech. S.R. Harnad, H.D. Steklis & J. Lancaster, Eds.: Vol. 280: 643–649. Annals of the New York Academy of Sciences. New York, NY.
22. DOUPE, A.J. & P.K. KUHL. 1999. Birdsong and human speech: common themes and mechanisms. Annu. Rev. Neurosci. **22:** 567–631.
23. WILBRECHT, L. & F. NOTTEBOHM. 2003. Vocal learning in birds and humans. Ment. Retard. Dev. Disabil. Res. Rev. **9:** 135–148.
24. JARVIS, E.D. 2004. Learned birdsong and the neurobiology of human language. *In* The Behavioral Neurobiology of Birdsong. H.P. Ziegler & P. Marler, Eds.: Vol. 1016: 749–777. Annals of the New York Academy of Sciences. New York, NY.
25. MERKER, B. 2002. Music: The missing Humboldt system. Mus. Scient. **6:** 3–21.
26. ZAHAVI, A. & A. ZAHAVI. 1997. The Handicap Principle: A Missing Piece of Darwin's Puzzle. Oxford University Press. Oxford.
27. ENQUIST, M., A. ARAK, S. GHIRLANDA & C.-A. WACHTMEISTER. 2002. Spectacular phenomena and limits to rationality in genetic and cultural evolution. Trans. R. Soc. Lond. B Biol. Sci. **357:** 1585–1594.
28. CATCHPOLE, C.K. & P.J.B. SLATER. 1995. Bird Song: Biological Themes and Variations. Cambridge University Press. Cambridge, UK.
29. GARDNER, T.J., F. NAEF & F. NOTTEBOHM. 2005. Freedom and rules: the acquisition and reprogramming of a bird's learned song. Science **308:** 1046–1049.

30. DEVOOGD, T.J., J.R. KREBS, S.D. HEALY & A. PURVIS. 1993. Relations between song repertoire size and the volume of brain nuclei related to song: comparative evolutionary analysis amongst oscine birds. Proc. Biol Sci. **254:** 75–82.
31. MERKER, B. 2004. Cortex, countercurrent context, and dimensional integration of lifetime memory. Cortex **40:** 559–576.
32. THORPE, W.H. 1961. Bird song. Cambridge University Press. Cambridge.
33. ZIEGLER, H.P. & P. MARLER, EDS. 2004. Behavioral Neurobiology of Birdsong. Vol. 1016. Annals of the New York Academy of Sciences. New York, NY.
34. DEVOOGD, T.J. 2004. Where is the bird?. *In* The Behavioral Neurobiology of Birdsong. H. P. Ziegler & P. Marler, Eds.: Vol. 1016: 778–786. Annals of the New York Academy of Sciences. New York, NY.
35. WILLIAMS, H. 2004. Birdsong and singing behavior. *In* Behavioral Neurobiology of Birdsong. H.P. Ziegler & P. Marler, Eds.: Vol. 1016: 1–30. Annals of the New York Academy of Sciences. New York, NY.
36. KING, A.P. & M.J. WEST. 2002. The ontogeny of competence. *In* Conceptions of Development. D.J. Lewkcowicz & R. Lickliter, Eds.: 77–104. Psychology Press. Philadelphia.
37. GOLDSTEIN, M.H., A.P. KING & M.J. WEST. 2003. Social interaction shapes babbling: testing parallels between birdsong and speech. Proc. Nat. Acad. Sci. USA **100:** 8050–8055.
38. ÖLVECZKY, B.P., A.S. ANDALMAN & M.S. FEE. 2005. Vocal experimentation in the juvenile songbird requires a basal ganglia circuit. PLoS Biol. **3:** e153.
39. TCHERNICHOVSKI, O., T.J. LINTS, S. DERÉGNAUCOURT, *et al.* 2004. Studying the song development process: rationale and methods. *In* Behavioral Neurobiology of Birdsong. H.P. Ziegler & P. Marler, Eds.: Vol. 1016: 348–363. Annals of the New York Academy of Sciences. New York, NY.
40. THEUNISSEN, F.E., N. AMIN, S.S. SHAEVITZ, *et al.* 2004. Song selectivity in the song system and in the auditory forebrain. *In* Behavioral Neurobiology of Birdsong. H.P. Ziegler & P. Marler, Eds.: Vol. 1016: 222–245. Annals of the New York Academy of Sciences. New York, NY.
41. ADRET, P. 2004. In search of the song template. *In* Behavioral Neurobiology of Birdsong. H.P. Ziegler & P. Marler, Eds.: Vol. 1016: 303–324. Annals of the New York Academy of Sciences. New York, NY.
42. DERÉGNAUCOURT, S., P.P. MITRA, O. FEHÉR, *et al.* 2004. Song development: In search of the error-signal. *In* Behavioral Neurobiology of Birdsong. H.P. Ziegler & P. Marler, Eds.: Vol. 1016: 364–376. Annals of the New York Academy of Sciences. New York, NY.
43. MIKLÓSI, A. 1999. The ethological analysis of imitation. Biol. Rev. **74:** 347–374.
44. PALESTIS, B. 2004. Social facilitation. *In* Encyclopedia of Animal Behavior. M. Bekoff, Ed.: 981–983. Greenwood Publishing Group. Westport, CT.
45. VAN GENNEP, A. 1908/1960. The Rites of Passage. University of Chicago Press. Chicago.
46. DURKHEIM, É. 1912/1995. The Elementary Forms of Religious Life. Free Press. New York.
47. RADCLIFFE-BROWN, A.R. 1961. Structure and Function in Primitive Society. Free Press. Glencoe, Ill.
48. GLUKMAN, M. 1965. Politics, Law and Ritual in Tribal Society. Blackwell Publishers. Oxford.
49. TURNER, V. 1969. The Ritual Process. Aldine. Hawthorne, NY.
50. TOMASELLO, M., A.C. KRUGER & H.H. RATNER. 1993. Cultural learning. Beh. Brain Sci. **16:** 495–552.
51. LEFEBVRE, L. & J. BOUCHARD. 2003. Social learning about food in birds. *In* The Biology of Traditions. D.M. Fragaszy & S. Perry, Eds.: 94–126. Cambridge University Press. Cambridge.
52. WHITEN, A. & R. HAM. 1992. On the nature and evolution of imitation in the animal kingdom: reappraisal of a century of research. Adv. Stud. Beh. **21:** 239–283.
53. RUSSON, A.E. & B.M.F. GALDIKAS. 1993. Imitation in freeranging rehabilitant orangutans. J. Comp. Psychol. **107:** 147–161.

54. RUSSON, A.E. & B.M.F. GALDIKAS. 1995. Constraints on great apes' imitation: model and action selectivity in rehabilitant orangutan (*Pongo pygmaeus*) imitation. J. Comp. Psychol. **109:** 5–17.

55. TOMASELLO, M., S. SAVAGE-RUMBAUGH & A.C. KRUGER. 1993. Imitative learning of actions on objects by children, chimpanzees and exculturated chimpanzees. Child Dev. **64:** 1688–1705.

56. SEGERDAHL, P., W. FIELDS & S. SAVAGE-RUMBAUGH. 2005. Kanzi's Primal Language: The Cultural Initiation of Primates into Language. Palgrave Macmillan. Houndmills, Basingstoke.

57. MELTZOFF, A.N. 1996. The human infant as imitative generalist: a 20-year progress report on infant imitation with implications for comparative psychology. *In* Social Learning in Animals: The Roots of Culture. C.M. Heyes & B.G. Galef, Eds.: 347–370. Academic Press. San Diego, CA.

58. DONALD, M. 1991. Origins of the Modern Mind. Harvard University Press. Cambridge, MA.

59. STAAL, F. 1989. Rules without Meaning. Ritual, Mantras and the Human Sciences. Peter Lang. New York.

The Evolution of Music in Comparative Perspective

W. TECUMSEH FITCH

University of St. Andrews, School of Psychology, Fife KY16 9JP, Scotland

ABSTRACT: In this paper, I briefly review some comparative data that provide an empirical basis for research on the evolution of music making in humans. First, a brief comparison of music and language leads to discussion of design features of music, suggesting a deep connection between the biology of music and language. I then selectively review data on animal "music." Examining sound production in animals, we find examples of repeated convergent evolution or analogy (the evolution of vocal learning of complex songs in birds, whales, and seals). A fascinating but overlooked potential homology to instrumental music is provided by manual percussion in African apes. Such comparative behavioral data, combined with neuroscientific and developmental data, provide an important starting point for any hypothesis about how or why human music evolved. Regarding these functional and phylogenetic questions, I discuss some previously proposed functions of music, including Pinker's "cheesecake" hypothesis; Darwin's and others' sexual selection model; Dunbar's group "grooming" hypothesis; and Trehub's caregiving model. I conclude that only the last hypothesis receives strong support from currently available data. I end with a brief synopsis of Darwin's model of a songlike musical "protolanguage," concluding that Darwin's model is consistent with much of the available evidence concerning the evolution of both music and language. There is a rich future for empirical investigations of the evolution of music, both in investigations of individual differences among humans, and in interspecific investigations of musical abilities in other animals, especially those of our ape cousins, about which we know little.

KEYWORDS: biomusicology; evolution of music; design features of music; comparative data; birdsong; whalesong; ape drumming; linguistics; communication

INTRODUCTION

After a long hiatus, there has recently been a surge of interest in the biology and evolution of music.[1–4] From a biologist's perspective, a logical starting place to begin a discussion of the biology and evolution of human music is provided by the study of the evolution of communication in animals. In particular, a comparative approach, using empirical data from living organisms, rather than speculations based on fossils, provides a rich source of empirical data to ground hypotheses and test predictions. The purpose of this review is to introduce the reader to some core questions

Address for correspondence: W. Tecumseh Fitch, University of St. Andrews, School of Psychology, Fife KY16 9JP, Scotland. Voice: +44-1334-462054; fax: +44-1334-463042.
wtsf@st-andrews.ac.uk

Ann. N.Y. Acad. Sci. 1060: 29–49 (2005). © 2005 New York Academy of Sciences.
doi: 10.1196/annals.1360.004

in evolutionary musicology (biomusicology), to discuss some comparative ethological data relevant to these questions, and to highlight some key topics for future empirical research. I argue for an empirical, multicomponent perspective, breaking music down into several different subcomponents for further study. Two obvious components, song and instrumental music, provide a framework for the current review. Each of these phenomena, and the biological capacities underlying them (e.g., vocal learning and entrainment, respectively), might have a different evolutionary function and history. Indeed, I will argue that the comparative data available necessitate rather different approaches to the evolution of song versus the evolution of instrumental music.

The comparative approach to music has at least three different components: a cross-cultural component (comparative musicology and ethnomusicology); the intraspecific comparison with other human cognitive functions, especially language; and finally the interspecific comparison with the music of other species. I will discuss the first two here, before turning to my main focus on the comparisons with animals. Because I have recently reviewed the biology and evolution of music from a multicomponent comparative perspective in more detail elsewhere,[5] I will remain brief here and focus citations on reviews, when available. Several recent multi-authored volumes provide detailed consideration of some of the relevant ethological data,[1,6] so I highlight some topics that previous commentators appear to have overlooked (e.g., female birdsong or ape drumming). Because music perception has been well reviewed recently by others,[7,8] my focus in this review is mainly on musical production, rather than perception: music making results in behavior that can be recorded and analyzed and is the focus of most ethological investigation. Thus, our comparative data are richest for this topic.

MUSIC AND LANGUAGE: SIMILARITIES AND DIFFERENCES

There are a number of deep similarities between human music and language, and the comparison between these two faculties provides a persistent leitmotif in musicology. I agree with the opinion of Lerdahl and Jackendoff[9,10] that music should be studied empirically as an independent cognitive domain from language (not assumed to be parasitic upon, or derivative of, language, e.g.,as presented in Ref. 11). However, it seems likely that the results of this empirical endeavor will have important implications for language,[12] just as the study of language has already had a significant impact on the study of music cognition.[9,13,14] Both the core similarities, and a few crucial differences, between music and language are relevant to much of the work on the biology of music, particularly when they are viewed from the perspective of animal communication systems.

Although linguists and musicologists often focus attention on the differences between music and language, from a biologist's perspective the similarities appear more striking. At a superficial level, both music and language use the auditory/vocal domain preferentially. Each faculty has, in addition, closely related nonvocal domains of expression (signed languages, instrumental music and dance), and both can be written or notated successfully. At a deeper cognitive level, there are significant formal similarities between musical and linguistic cognition, both in phonology (e.g., metrical phonology and rhythm[15]) and at higher organizational levels (e.g., the existence of hierarchical phrase structure in both domains[10]). Most importantly, however, both music and language represent human universals, found in all human cultures, and both seem to rely on innate biological proclivities that are quite unusual in the animal kingdom. This biological basis is reflected both in very early biases and predispositions in human infants (e.g., the preference for processing relative rather than absolute pitch, already observable at six months of age[8]) and in a considerable overlap of the brain regions involved in processing musical and linguistic stimuli.[16] These similarities seem unlikely to result from chance, suggesting that the study of the biology and evolution of language will have interesting implications for the evolution of music, and vice versa.

However, within this context of shared properties, there are several striking differences between music and language as well. Perhaps the most obvious is the significant disparity in the type of meaning attached to lingustic and musical stimuli. Although language can be used musically (in the sense of poetry or song, where the form of the utterances themselves become a focus of attention), the prototypical linguistic utterance is propositional: it expresses a specific meaning that can be either true or false. Linguistic semantics is largely focused upon meaning as intentional, propositional messages. Such meaning is largely absent in human music (with some relatively peripheral exceptions, e.g., flute trills signifying birdsong). Certainly, a musical phrase is not easily judged as true or false. Although it would be incorrect to say that music lacks meaning entirely, *musical meaning* is notoriously hard to pinpoint, suggesting that its very openness and ambiguity are perhaps among its virtues. Despite many attempts to define musical meaning, and a widespread conviction that music is well suited to expressing mood and emotion, there is no widely accepted characterization of musical meaning. Ian Cross, focusing on the social functions of musical meaning, has suggested the term *floating intentionality*.[17] Cross points out that music seems to "absorb" *aboutness* from whatever context it appears in (evoking strong associations to that context on later hearings), while simultaneously providing aboutness to that context, in the sense of enriching our sense of the meaningfulness of a given context (hence the close association of music with ritual in all of the world's cultures). A rather different formulation is due to Manfred Bierwisch,[18] who coined the apt term *gestural form* to describe the mapping between musical structure and musical interpretation, connoting our ability to make an analogical mapping between the acoustic form of music and either the affective trajectory of emotional experience, or the motor trajectory of dance. In short, however one attempts to characterize musical meaning, the meaning of music and that of language are clearly distinct, and this difference is crucial to understanding these two related faculties.

DESIGN FEATURES OF MUSIC: A MULTICOMPONENT APPROACH

As already suggested, a profitable comparison of music and language or of human and animal musics requires that we break music down into its various components rather than treating music as an indivisible whole. Achieving an apt analysis is obviously far from a trivial task, and "carving nature at the joints" may require many cycles of theorization and test. However, certain very basic distinctions are already quite clear. The first is between production (music making) and perception: while the former is quite variable among individuals, the latter seems to be much

TABLE 1. Hockett's[22] design features of language

Language	Music		
Design feature	Instrumental	Vocal	Innate calls
1. Vocal auditory channel	No	Yes	Yes
2. Broadcast transmission	Yes	Yes	Yes
3. Rapid fading	Yes	Yes	Yes
4. Interchangeability	No	Yes	Yes
5. Total feedback	Yes	Yes	Yes
6. Specialization	Yes	Yes	Yes
7. Semanticity	No	No	No
8. Arbitrariness	No	No	No
9. Displacement	No	No	No
10. Duality of patterning	No	No	No
11. Productivity	Yes	Yes	No
12. Discreteness	Yes	Yes	No
13. Cultural transmission	Yes	Yes	No

NOTE: Innate calls refers to vocalizations of nonhuman primates along with laughter or screams in humans. Brief explanations of nonobvious terms (see Hockett[22] for detailed description, and Fitch[5] for discussion): 4. interchangeability (anyone can say anything they can understand) vs. males-alone singing; 5. total feedback (you hear what you're saying); 6. specialization (speech "triggers" desired results with negligible direct energy expenditure, unlike forcing someone manually); 7. semanticity (words associated with things); 9. displacement (capacity to refer to nonpresent objects or events); 10. duality of patterning (meaningless elements combine for large number of meaningful elements); 11. productivity (novelty, counterfactuality); 12. discreteness (digital vs. analogue).

more universally shared, even among self-professed "unmusical" individuals.[19] We might also profitably subdivide musical mechanisms along more traditional lines of rhythm, melody, and harmony, as is implicit in much of the experimental literature on music psychology.[20] A third prominent subdivision is that between song and instrumental music. Scholars have long hypothesized different origins for these two subcomponents,[21] and this will form a core distinction in the current review.

My starting point[5] for this analysis of the music faculty and comparison with language will be a characterization of design features of human language offered by the linguist Charles Hockett. Hockett provided a list of relevant features of language that have been quite influential in the subsequent study of animal communication,[22] providing a widely discussed and well-known decomposition of linguistic abilities that also allows a specific comparison with human music. Hockett himself discussed instrumental music in this context but did not discuss song (which is arguably more biologically basic, and certainly shares more with speech). I present Hockett's features for language, as compared to music, in TABLE 1. This table shows that most of the features Hockett singled out as particularly relevant for human spoken language are shared by song. The exceptions all tie in to the basic distinction between musical and linguistic meaning discussed above. Thus, from a linguistic perspective, we see that the two domains share physical and physiological features (dependence on the

TABLE 2. Design features of music

Design feature	Spoken language	Innate calls
1. Complexity	Yes	No
2. Generative	Yes	No
3. Culturally transmitted	Yes	No
4. Discrete pitches	No	No
5. Isochronic	No	No
6. Transposability	Yes	?
7. Performative context	No	No
8. Repeatable	No	No
9. A-referentially expressive	No	Yes

NOTE: See Fitch[5] for detailed explanation and discussion.

auditory channel, and vocalization) as well as deeper formal similarities at the phonological and syntactic levels, but that music lacks propositional meaning. From this perspective one might say, as a first approximation, that music is halfway to language.

Of course, from a musicologist's perspective, characterizing music as language minus meaning seems both limiting and somewhat derogatory. Turning the analysis the other way around, we may ask what special characteristics music possesses and then compare language with these. As a step in this direction I have proposed some design features of music, asking whether these are shared by language.[5] A summary of this analysis is presented in TABLE 2. I make no claims that this is the best breakdown, but the proposed features are often cited by scholars as potential universal characteristics of the musics of the world.[23, 24] Here we find, again, a mixed pattern of shared and disparate elements. Despite the first three shared features, music has a number of specific features that are lacking in language, three of them formal or perceptual and three of them social and intentional. Taking these features as provisionally "special" to music, we may thus turn to comparative data to see which of the shared ensemble of musical and linguistic design features might be shared between humans and other species. See Ref. 5 for a more detailed discussion.

THE COMPARATIVE APPROACH TO MUSIC EVOLUTION

Turning from the comparison between music and language to interspecific comparisons, various behaviors in nonhuman animals can be considered analogues of human music making. The most obvious of these is birdsong, already termed *song* by Aristotle.[25] More recently, biologists have learned that several other animal groups, including two independent orders of marine mammals, whales and seals, have evolved learned, complex vocal displays termed song (for reviews, see Ref. 1). Besides song, however, there are other animal displays that may deserve comparisons with human music that are less well known. In particular, I suggest below that African ape percussive behavior or "drumming" may represent a homologue and thus indicate a potential precursor of human instrumental music.

The comparative method, which involves using data from living organisms to draw inferences about extinct ancestors and/or past selective forces, is one of the most powerful tools available to the biologist. Experimentation in evolutionary biology is limited by the great time spans required for evolution, so (in contrast to physics and chemistry, but in common with meteorology, cosmology, or geology) biology relies heavily on the careful observation, collation, and comparison of "experiments of nature." Each living species can be considered a separate data point, each having its own independent evolutionary history since its divergence from other related species. By gathering data from living organisms and organizing them phylogenetically, we gain access to a huge body of information relevant to understanding evolution. There are two basic types of inferences one can draw from comparative data, which derive from two fundamentally different ways in which organisms can come to resemble one another.

One common form of similarity results from two different lineages independently evolving a similar solution to the same problem. Following a long tradition in comparative biology, starting with Richard Owen, such convergently evolved traits can be termed *analogues* (in contrast to *homologues,* which result from common ancestry, see below). Examples include the independent evolution of flapping wings for flight in insects, birds, and bats; of streamlined form in fast swimming vertebrates such as dolphins, ichthyosaurs, and sharks; or of short, powerful arms adapted for digging in placental and marsupial moles.[26,27] In many cases, striking resemblances in form have arisen in organisms whose ancestors looked very different from one another (e.g., birds arose from a bipedal dinosaur, while bats evolved from a quadrupedal mammal). In such cases, similar lineages clearly represent independent data points, with quite separate evolutionary histories. Here, convergent evolution provides one of the most powerful clues to adaptation, because we can exclude potential confounds such as phylogenetic inertia (the tendency for closely related species to remain similar after their divergence for reasons of history rather than adaptation). Thus, when we observe convergent evolution in nature, we are immediately led to postulate an adaptation to the task at hand[28] (see Ref. 29, however).

The other pervasive form of similarity was termed *homology* by Owen. In current usage, this term indicates similarities in two or more species that are derived from their common ancestor. Thus a homologous trait is one that was present, though perhaps in a somewhat different form or with different function, in the common ancestor of the lineages possessing the homologue. Homology plays a key role in evolutionary biology, because homologous traits allow us to build and test phylogenetic hypotheses.[30] More importantly in the present context, homologous traits allow us to reconstruct the traits of extinct ancestors, even in the absence of fossil evidence, by examining shared characteristics of their living descendents. Summarizing, both homologous and analogous traits provide useful insights into the evolutionary past,[31] and in this review I will offer potential examples of each.

Convergent Evolution: Vocal Learning and Song in Nonhuman Animals

The analogy between birdsong and human music has been recognized since ancient times, and birdsong receives ever-increasing scientific attention as the most sophisticated vocal communication system in the animal kingdom besides human song and language.[32–34] More recently, with the invention of the hydrophone, songlike vocalizations have been discovered in two independent lineages of marine mammal, the cetaceans (whales and dolphins) and the pinnipeds (seals and sea lions).[35,36] Darwin knew that songbirds required exposure to conspecific song in order to sing "properly." This ability to imitate novel sounds, termed *vocal learning*, has since been demonstrated in marine mammals and thus appears to represent a core component of "animal song," as this term is currently used. Because vocal learning is critical in language, many authors have noted the importance of vocal learning in animals as a critical analogue of human language acquisition;[37–39] it is less often noted that it is just as relevant to cultural transmission of musical song.[40] Indeed, I join Marler[41] in suggesting, contra Hauser and McDermott,[7] that comparison of the deep similarities between human and animal song is highly relevant to the evolution of music, possibly more so than the traditional comparison between animal song and human language. From this perspective, a long tradition of research on vocal learning in nonhuman animals has interesting implications for the evolution of both song and speech in humans.

By considering only animal vocal displays that are both learned and complex enough to deserve the appellation *song*, I exclude such phenomena as frog or cricket "song" that, although pleasant to listen to, are neither complex nor learned (for more discussion, see Ref. 5). More controversially, this restriction excludes vocalizations classically termed song in gibbons or suboscine birds, which is complex but not learned.[34,42] Such vocalizations may serve a similar adaptive function to songbird or human song but seem to rely on different neural mechanisms. Vocal learning of novel sounds is not possible for our nearest cousins, the apes, or apparently other nonhuman primates.[35] Despite some limited vocal learning,[44] nothing like the human (or songbird) ability to learn complex, novel vocalizations from the environment has been demonstrated in any nonhuman primate, despite repeated strenuous efforts.[45] Despite some similarities in form and function (for an excellent review, see Ref. 42), and clear homology at the level of the vocal production system, the lack of extensibility of primate calls renders them categorically different from human music and speech. This difference between humans and other primates appears to be underlain by fundamental differences in the neural control of vocalization.[46] I thus concur with Marler[41] that primate calls are a poor choice for comparison to human song (for a different viewpoint, see Ref. 42).

Vocal learning represents a clear case of analogy: neither the ancestral mammal, nor the ancestral bird, nor the common ancestor of birds and mammals, possessed vocal learning or song. Complex vocal learning has arisen independently in at least three clades of birds (oscine passerines or "songbirds," hummingbirds, and parrots) and three clades of mammals (cetaceans, pinnipeds, and hominids), and this list is probably not exhaustive.[35,47] Thus, any similarities in these convergently evolved vocal learning systems may represent adaptations, evolved to solve particular functional problems. A suggestive example is provided by babbling. Both young songbirds and human children go through a stage of private vocal experimentation termed *babbling* in humans and *subsong* in birds, widely believed to allow the youngster to calibrate its vocal apparatus and auditory system, and providing the arena in which it can learn to successively approximate the vocalizations of adult models.[48] A preadult babbling stage that closes the auditory/vocal loop seems to be a prerequisite for vocal learning and is something that both young birds and humans do avidly, without external reinforcement. Babbling/subsong may thus be a key mechanistic component of the innate capacity for complex vocal learning that underlies both human and animal song, as well as human language. This is a testable hypothesis: To my knowledge, no one has yet documented a babbling stage in nonhuman mammalian vocal learners, and if more detailed studies of singing whales or seals revealed no such stage, we could reject this "vocal learning entails babbling" hypothesis.

In terms of function (the adaptive problem "solved" by song), the repeated convergent evolution of learned song in the animal kingdom provides a rich source of data. Darwin recognized that, with some exceptions, it is mostly male birds who sing and that they do so most commonly in the mating season. He thus saw sexual selection for mate choice as a critical factor in the evolution of bird song.[49] Among marine mammals, it appears that only baleen male whales produce complex song, and song is mostly restricted to male seals (although vocal learning is present in bottlenosed dolphins of both sexes); again, song is common only during the mating season. Partly by analogy with birdsong, these displays are thus widely believed to function as sexual displays.[50,51]

However, it is important to note that the preponderance of male song is only a first approximation and that two deviations from this rule are commonly observed in birds. In the first, both males and females contribute to "duets:" complex songs with different, often tightly interlocking, male and female parts.[52,53] These are particularly common in nonmigratory tropical birds, and for historical reasons duetting is less well known than the male song typical of birds from temperate climates. Second, with the rise of detailed studies on identified, sexed birds, it has become increasingly clear that female birds, even in temperate regions, may sing independently of males.[54,55] For example, in robins (*Erithacus rubecula*) the male alone sings during the spring mating season, but in the autumn females establish separate winter territories that they sing to protect[56]—an example of natural, not sexual, selection. There has been a recent surge of interest in female bird song, but our understanding of it is still limited at present.[55] Given the lack of any clear sex differences in singing ability in our own species, where both women and men have highly developed singing abilities, song in female birds is clearly relevant to the evolution of human song. Thus, caution is warranted regarding extrapolation from male-specific bird or whalesong to the sexually egalitarian song of human beings, at least regarding the adaptive function of these vocalizations.

Ape Drumming as a Potential Homologue of Instrumental Music

Turning from analogy to homology, we find a number of plausible perceptual homologies between auditory perception in humans and other vertebrates.[7] At the level of basic pitch and timbre perception, important components of music perception appear to be widely shared. For instance, goldfish and pigeons can both learn to identify musical styles (e.g., blues vs. baroque music) and extrapolate to new pieces outside the training set.[57,58] However, regarding music making, we must turn to our own mammalian order to find behaviors that represent plausible homologues. With regard to the anatomical basis for song production, there are no fundamental differences between humans and most other mammals.[59] To go beyond this very basic, and very deep, set of homologies, we must thus look to the way this system is used, particularly at the level of its neural control. Here it is the human ability to imitate novel sounds (vocal learning) that seems the most critical aspect of our vocal behavior, and as already discussed, this ability is not shared by other primates. If gibbon "song" or chimpanzee long calls are not homologous to human singing, are there any aspects of primate behavior that are potential homologues to human music making? I believe that the answer is yes and that the percussive behavior commonly observed in great apes, often called "drumming," represents a plausible but heretofore overlooked homologue to human instrumental music making.

Bimanual percussion on resonant objects (drumming) is a common, easily observed behavior in African great apes (chimpanzees, bonobos, and gorillas). Behaviors analogous to drumming or other instrumental music are quite rare in animals, the most prominent other examples being palm cockatoos (which use sticks to drum on hollow trees[60]), many species of woodpecker (who seek out resonant trees for display drumming),[61,62] or various desert rodent species who "drum" with their hind feet on the ground.[63] Great ape drumming has been the topic of surprisingly little research and has been largely overlooked in recent discussions of the evolution of music. In gorillas, drumming behavior is prototypically seen in the agonistic displays of adult male silverbacks, where a vocal "hoo" display is commonly combined with bimanual beating on the chest (and the inflated vocal air sacs, increasing the resonance of drumming).[64] However, gorilla drumming is also observed in females and young, often in a contagious, playful, and unstereotyped fashion, and young gorillas beat on many surfaces (including their bellies, the bodies of other gorillas, the floor, and on hollow objects; Fitch & Gomez, unpublished video data). This more playful and creative context is much more suggestively similar to much of human music. In chimpanzees and bonobos, drumming is typically observed on resonant structures in the environment (rather than their own bodies), often as part of the climax of the male dominance display.[65–67] Wild chimpanzees seek out particular tree buttresses and in zoo contexts sometimes discover and drum upon resonant structures (e.g., hollow walls; J. Call, personal communication). Finally, bonobos have a variety of clapping and drumming displays,[68] and enculturated bonobos, such as the language-trained Kanzi, produce and apparently enjoy drumming on resonant objects bimanually in a highly coordinated fashion (S. Savage-Rumbaugh, personal communication).

Although the discovery of tool use by wild chimpanzees[65] generated an uproar, dethroning humans from their sole status of "toolmakers," the existence of drumming in apes has remained largely unnoticed by musicologists. It has long been speculated that vocal and instrumental music have independent origins and evolutionary histories,[21] and the sharp difference between ape drumming (with its similarities to human instrumental music) and ape vocal capabilities (which show no evidence of the vocal learning and control required for song) provide strong support for this hypothesis. In my opinion, ape drumming represents a striking parallel to human percussive behavior, and its appearance in our closest living relatives (but not, apparently, among orangutans or other primates) strongly suggests the possibility of an overlooked and important homology for human instrumental music making. Unfortunately, there has been little empirical research on great ape drumming, and we are currently unable to answer even very basic questions that would help evaluate this hypothesis. In particular, despite very questionable statements in the popular literature,[69] it remains unclear whether apes can entrain their drumming to a regular, externally given beat. Vocally, gibbons may be able to entrain their calling to a metronome (see Ref. 42), and bonobos may entrain their group calling,[68] but here too the data are inadequate at present. Since both entrainment and isochrony have been proposed as design features of music and are thought by some researchers to be uniquely human capabilities, this is an open question that critically needs empirical resolution. Thus, the hypothesis that ape drumming is homologous to human instrumental music remains a tantalizing possibility that cries out for controlled experimental study.

Concluding, even this rather cursory exploration of the comparative database demands that we move beyond simplistic models of music, considered as an undifferentiated whole unique to our species. Once we begin breaking the human music faculty into its component parts, such as song and drumming, we find abundant parallels in the animal world, including very deep and basic homologies (e.g., at the level of vocal production or basic auditory perception) and fascinating analogies (e.g., vocal learning in birds or seals). Further empirical research may well reveal a residue of musical traits that are unusual or possibly unique to our own species (e.g., isochronous entrainment).[40] Of course, the only way to discover such uniquely human characters is to first carefully investigate all potentially relevant animal behaviors. Both the psychology and anthropology literatures are replete with cautionary tales in the form of claims of uniquely human characters that were later falsified by comparative data.[70,71] Thus, the study of animal music is a fascinating and rewarding field in its own right, but it is also a logical prerequisite to any claims about uniquely human musical capabilities.

THE ADAPTIVE FUNCTIONS OF MUSIC:
BEYOND SEXUAL SELECTION

The adaptive function(s) of music represents an area of considerable discussion in the contemporary literature.[1,11,72,73] In my opinion, this issue has been overemphasized, because a specification of the adaptive function(s) of music is neither necessary nor sufficient for a rich understanding of the biology and evolution of music. Several classes of question asked by evolutionary and behavioral biologists are independent of one another and represent complements to one another (rather than alternatives).[74] These include mechanistic questions about how the behavior is implemented genetically, physiologically, and neurally (e.g., the wealth of current research on the mechanisms of birdsong), ontogenetic questions about how it develops (e.g., the parallels between babbling and subsong), phylogenetic questions about the origin and subsequent evolutionary pathway of the behavior, and finally functional questions about the adaptive value (or values) that the behavior served or serves. Answers to such functional questions must ultimately be framed in terms of increased survival and reproductive success. Although functional questions are undeniably interesting, they are notoriously difficult to answer, and represent neither the only, nor the primary, questions that biologists ask about evolved traits. Thus, I do not see the question, Is music an adaptation? to be central to biomusicology. I suggest that debate on this topic has shed more heat than light and should not continue to occupy center stage in this young field.

A key difficulty for functional hypotheses is the ever-present possibility of change of function. Although careful observation and experiments can provide empirical information about the current adaptive function of a trait, we must be quite cautious about extrapolating from such findings to the past ancestral function(s) of the trait. For example, regarding morphological data (where we do have clear fossil evidence), evolutionary history is full of systems that have changed their function. To choose three examples directly relevant to music, the mammalian middle ear bones started as jaw supports (for feeding) but now function as an impedance-matching system for audition. The vertebrate larynx started out as gill supports for underwater respiration in our aquatic ancestors but now is adapted to sound production in terrestrial vertebrates. Finally, our lungs are used in aerial respiration, and drive vocalization, but are homologous to the swim bladder in fish, which serves in underwater posture maintenance and floatation. Clearly, over evolutionary time, we cannot expect current function to flawlessly predict past function or assume that the function of music in today's world is identical with that of our long-extinct ancestors. Indeed, such functional lability has led some theorists to advocate ignoring past function entirely in discussions of adaptation,[75] although this suggestion has not been widely adopted. We should keep this possibility in mind as we evaluate current hypotheses concerning musical function.

Music as Cheesecake

The logical starting point for any discussion of the adaptive value of a trait is that it is not an adaptation at all. This is the null hypothesis to be rejected first in any empirical investigation of adaptation. There are many reasons that any particular trait might be nonadaptive or even maladaptive.[76] The trait might be an automatic by-product of some bona fide adaptation, generated by developmental or architectural constraints (such traits are often termed *spandrels*, following the discussion of Gould and Lewontin[77]). It has been suggested that music is an automatic by-product of language in this way,[11] much as cheesecake is nonadaptive but fulfills a desire for sugars and fat that is (or was) adaptive. Alternatively, a trait may be a "hangover" like the appendix, left over from a past time when it did serve an adaptive function (phylogenetic inertia). This notion of music as a phylogenetic hangover from the communication system of earlier hominids is implicit in Darwin's hypothesis of a music-like "protolanguage" (see below).

Finally, a supposed trait might actually be an artifact of our way of carving up an organism, but not itself under genetic control or subject to past selection. To take an uncontroversial example, a chess grand master may spend all of his time playing chess, relying on his chess skills for both survival and increased mating success, and scientists can even study the genetics and neurology of chess.[78] Such demonstrable utility obviously does not make "chess" an adaptation. Although various components of chess-playing ability might be properly considered adaptations (e.g., the ability to form complex perceptual and symbolic representations, encode rules, and plan ahead strategically), it would be silly to reify the "chess faculty" as an adaptation in its own right. Of course, music is historically much older than chess, is found in all human cultures, and appears to entail perceptual specializations that are either innate or very early developing, and for all these reasons the human music capacity seems a much more promising candidate as a Darwinian adaptation than chess. However, we should treat this reification of music as very provisional: the term *music* is a recent one in English and is absent in many languages.[40] Cross-culturally, perhaps the relevant adaptive complex is music + dance. In any case, the first goal of a scholar interested in the adaptive value of music should be to reject the null hypothesis—that music lacks any biologically adaptive value—by demonstrating some systematic, widespread, and plausibly ancient current function of music. Even with regard to the current function of human music, potentially a topic of empirical research, hard data is surprisingly scarce and vastly outweighed by the theoretical speculation on this topic. This lack of solid data represents a clear and easily filled lacuna in our current understanding of music evolution, as I will stress below.

Music and Sexual Selection

Perhaps the most widespread hypothesis for the adaptive function of music dates back to Darwin, who suggested, by analogy with birdsong, that "musical notes and rhythm were first acquired by the male or female progenitors of mankind for the sake of charming the opposite sex."[49] Combined with the contemporary prominence of love songs, the idea that music (either song or instrumental music) functions mainly in mate choice, and sexual selection more generally, seems intuitive and is often repeated.[79,80] This makes it surprising that there is, to my knowledge, no clear evidence supporting this supposition for humans, and its wide appeal seems to rest mainly upon intuition. We might expect, for example, demonstrations that skilled musicians, across cultures, have greater reproductive success than nonmusicians (or less-skilled musicians). Experiments involving attractiveness ratings of videos, using musical skill as a controlled variable, could easily be performed. Results from Web-based surveys querying large numbers of musicians about their sex habits, although demanding circumspection, would also provide a valuable initial source of data. Even statistical data showing that famous Western musicians have significantly more offspring (or better offspring survival) than others would be worthwhile. For every Bach who spawned a large family, however, there may be a Beethoven who died childless. Despite there being no obvious hurdles in the way, data supporting the sexual selection hypothesis are, to my knowledge, currently unavailable, and we must look to nonhuman animals for relevant empirical data.

The comparative database on animal song, at first glance, offers little support for a primary role of sexual selection in the evolution of human singing or vocal learning. In the vast majority of well-studied species, song is the exclusive province of males and occurs mainly during the mating season after the attainment of sexual maturity. Human singing, by contrast, is done equally well by males and females, in many different behavioral contexts. However, as discussed above, there is a growing literature on female bird song that may weaken the impact of this particular criticism. Among primates, duetting species tend to be monogamous,[81] and monogamy is an extremely rare mating system among mammals (less than 5% of species). The human tendency (admittedly imperfect) toward long-term mating partnerships and male parental care makes this analogy particularly interesting. A much more telling difficulty for the sexual selection hypothesis, in my opinion, is the very early maturation of music perception and singing behavior in human infants, and the universal use of song between mothers and their infants.[82] Such early maturation is quite atypical of sexually selected traits in the animal kingdom, which typically appear in their mature form only upon sexual maturity, when they are needed. Thus, despite an intuitive appeal, there is currently little data supporting the role of sexual selection and mate choice as the sole or primary selective force underlying the specific human capacity for song and vocal learning, or other subcomponents of the music faculty, and skepticism seems warranted until such data have been gathered.

Music and Social Groups

A second possibility for the function of song derives from its potential role in increasing group solidarity, consistent with the apparent role of music in defusing tension and cementing individual and group relationships in today's world. This argument has been presented in detailed form by Robin Dunbar,[83,84] who suggested that an increase in group size over the course of human evolution necessitated ever-more sophisticated mechanisms for maintaining group harmony. In most primates, the social bonds underlying group solidarity are maintained by grooming (via its physiological concomitants, such as endorphin release). By this hypothesis, the intensive, one-on-one nature of the grooming relationship became inadequate once group size increased beyond a certain limit and was replaced by "vocal grooming" that allowed a single vocalizer to simultaneously service multiple listeners, and thus cement multiple relationships simultaneously. Given the centrality of individual relationships and group membership in chimpanzees, and the apparently increasing importance of group solidarity later in hominid evolution, the notion that some novel mechanism helped maintain group cohesion during human evolution is plausible, and Dunbar's hypothesis is consistent with a considerable mass of comparative, neural, and archaeological data. Although Dunbar situates his hypothesis in the context of the evolution of language rather than music, he makes clear that in the early stages of his hypothetical phylogeny, vocal grooming was devoid of propositional meaning and thus more akin to song than to speech. Thus, this group cohesion function is consistent with Darwin's notion of a music-like protolanguage (see below). Again, however, I know of no empirical data clearly demonstrating a specific role for music in enhancing group harmony or coordination in the scientific literature, providing another promising topic for empirical research.

Music and Caregiving

A third possible function of music, specifically song, concerns its role in parent–offspring communication and, in particular, the mothers' song to regulate infant arousal.[85–87] The mother–infant relationship represents a specific, intense social bond, rendered particularly important in humans (and apes) by our long childhood and lengthy interbirth interval, making each child an unusually valuable investment. The use of lullabies to quite effectively soothe infants to sleep is apparently a human universal, practiced in all cultures.[82] Given the potential detrimental effects of crying or upset infants throughout our evolutionary history, the success of parents in achieving the goal of keeping their children quiet could be of considerable adaptive relevance to the evolution of song. Similarly, the use of play songs to arouse infants, focus their attention, and strengthen the mother–infant bond is both widespread and potentially adaptively relevant. Thus the considerable data supporting the effects of music on mood and arousal are compatible with this hypothesis.[88,89] The extremely early development of music perceptual abilities, while incompatible with sexual selection, is obviously nicely explained by the caregiving hypothesis. A documented infant preference for song over speech provides an argument against the null hypothesis that song is simply a nonadaptive by-product of speech.[43] Finally, the childcare hypothesis is compatible with Dunbar's vocal grooming hypothesis, as well as related arguments for the evolutionary significance of motherese.[90,91] The efficacy of song in general social bonding might derive evolutionarily from its prior development in mother–infant communication (much as many courtship displays are evolutionarily derived from parent–offspring displays, e.g., begging displays). Thus, many of the arguments for vocal grooming and group cohesion can be applied, *mutatis mutandis*, to the caregiving hypothesis. I conclude that the childcare hypothesis represents the account of the adaptive function of music currently most firmly grounded in data.

Summarizing this section, debates about the adaptive function(s) of music will remain difficult to resolve. Certainty about the past adaptive function of music will be hard to achieve, because it is likely that the function(s) of music may have changed over evolutionary time, or that music has served (or continues to serve) multiple adaptive functions sequentially or simultaneously. Despite its interest, the question, Is music an adaptation? is thus not necessarily the most productive focus for discussions of the biology and evolution of music. In striking contrast to the amount of published speculation, empirical data on function are scarce, but a wide variety of data could potentially inform the issue, and might readily be collected. Thus, this topic provides clear opportunities for future empirical study, and the adaptive hypotheses already on offer will hopefully spur a new crop of experimental and comparative studies rather than further speculation.

THE PHYLOGENY OF MUSICAL ABILITIES

For behavioral traits, questions concerning phylogenetic history are often the most difficult to answer, and the phylogeny of human musical abilities is no exception. The lack of relevant fossils for most behavioral traits puts strict limits on what we can know with certainty about past history of a trait. There are few fossil clues that are relevant to the evolution of singing. The most convincing possibility so far is due to MacLarnon and Hewitt,[92] who examined the size of the thoracic intervertebral canal in extant primates and various fossil hominids. Because some of the motor neurons involved in the fine control of breathing are housed in the thoracic canal, these authors proposed that the enlargement of this bony space during the evolution of the genus *Homo* indicates an increase in the fine respiratory motor control involved in vocalization. Based on the fossil data, this occurred at some point after *Homo ergaster*, an early member of our own genus. Although the authors situated their arguments in the context of speech evolution, they are equally (if not more) relevant to singing.[59] Singing requires both greater respiratory capacity (both in terms of air volume and subglottal pressures) and finer control than that necessary for normal conversational speech.[93] Thus, these data can be reinterpreted, perhaps more convincingly, as pushing the onset of singing back to before the split between Neanderthals and modern humans.

Regarding instrumental music, we are on firmer archaeological ground.[94] A wide variety of fossil musical instruments have been discovered (incidentally making our material evidence for instrumental music far stronger, and far older, than fossil evidence for language). The oldest indubitable musical instrument is a fine bone flute found in Geissenklösterle in Germany, associated with modern human remains, reliably dated to 36,000 years ago.[95] More tantalizing, but far more controversial, is a multiply pierced cave bear bone found associated with Neanderthal remains in Divje Baba, Slovenia, and considered by its discoverers to be a flute. Although not much older than the Geissenklösterle flute (the Divje Baba artifact is dated to 40,000 years ago), its association with Neanderthals would push the evolution of instrumental music to before the split between Neanderthals and modern *Homo sapiens* (at least 200,000, and more likely 500,000 years ago).[96] According to some scholars, this split predates the evolution of spoken language, and a Neanderthal flute would then provide quite strong evidence that music preceded language in our evolutionary history. Unfortunately, both pieces of evidence are controversial. First, the fossil evidence for the evolution of human speech is very tenuous,[37] and the timing of the origin of spoken language remains highly controversial. Furthermore, critics of the Divje Baba find[97] claim that it represents not a hominid artifact, but a bone pierced by carnivore teeth (for discussion and further references, see Ref. 98). Thus, the currently available paleontological data do not strongly constrain hypotheses about the origin and phylogenetic history of human music.

Given this, it is unsurprising that speculative hypotheses fill the literature on this topic. This older literature is well reviewed by Révész,[99] with a brief synopsis in English by Kunst.[100] Rather than review these many hypotheses here, I will end by discussing a phylogenetic hypothesis that seems, to me, quite plausible, and which more importantly illustrates the value of incorporating comparative principles—a fitting end for the current review. This is Darwin's hypothesis of a music-like protolanguage.[49] Although Darwin's comments on this topic were brief, they concisely lay out the idea (often rediscovered or rehearsed without attribution by modern scholars[101]) that an intermediate stage of human evolution, before the evolution of language, was characterized by a vocal communication system more similar to modern song than to modern speech.

Recognizing that music is a human universal "present...in men of all races, even the most savage," Darwin clearly felt that human musical ability warranted an evolutionary explanation. However, he concluded that "as neither the enjoyment nor the capacity of producing musical notes are faculties of the least use to man in reference to his daily habits of life, they must be ranked amongst the most mysterious with which he is endowed." After briefly discussing the comparative evidence regarding learned birdsong and gibbon vocalizations, Darwin concluded that "primeval man, or rather some early progenitor of man, probably first used his voice in producing true musical cadences, that is in singing." Regarding function, and again drawing on the comparative data, Darwin suggested, based on "widely-spread analogy, that this power would have been especially exerted during the courtship of the sexes, would have expressed various emotions, such as love, jealousy, triumph, and would have served as a challenge to rivals." Finally, Darwin contrasts his theory to that of Spencer and Diderot that music is derived from speech. Thus, Darwin's musical protolanguage hypothesis holds that a music-like communication system, based crucially on vocal learning and lacking propositional meaning, predates true language.

Darwin's hypothesis has much to recommend it,[12] and many biologically oriented researchers find it plausible.[5,40,41] First, the frequent convergent evolution of song-like communication systems in many vertebrate lineages suggests that, given the proper conditions, complex, learned song evolves relatively easily. This is in sharp contrast to spoken language per se that, as a vehicle for coveying discrete propositional meaning, appears to be uniquely human. The many similarities between music and language mean that, as an evolutionary intermediate, music really would be halfway to language, and would provide a suitable intermediate scaffold for the evolution of intentionally meaningful speech. Darwin's hypothesis is compatible with the paleontological data reviewed above and with the observation of considerable, but not total, overlap in the brain mechanisms underlying music and language.[2-4]

The hypothesis also makes some testable predictions: Darwin's hypothesis predicts a correlation of individuals' skills in components shared by music and language (e.g., phonetic and phonological skills), but deviation in those that are not shared (e.g., semantic interpretative abilities). This can easily be tested, for instance by examining individual variability among modern humans in particular subcomponents of the musical versus linguistic faculties. As more genetic data becomes available, and our understanding of the genetic bases for music and language improves, we can further predict a partial overlap of genetic determinants of the two traits. Furthermore, to the extent that different genes are involved in the two domains, the stamp of selection, as estimated via techniques like those applied to language,[102] should be much older in genes underlying music than those involved in speech, and we would expect to see indicators of relaxed selection pressures on musical capabilities during recent postlinguistic human evolution. Regardless of the ultimate fate of Darwin's hypothesis, such data could lead to a useful integration between biolinguistics and biomusicology, as well as valuable steps forward in our mechanistic understanding of the biology and evolution of music. Indeed, phylogenetic hypotheses are perhaps most valuable as sources of clear empirical predictions that spur the collection of relevant data.

CONCLUSION

In summary, I have argued that the comparative approach has great promise to enrich our understanding of the biology of music, and that data from animal behavior provide both inspiration for, and constraints on, theories of the evolution of music. More importantly, a more detailed exploration of the neural mechanisms and ontogenetic development of musical behaviors in other species potentially offers important insights into the analogous human capabilities. Data on birdsong have already provided an extremely fruitful source of insights into the genetics, endocrinology, physiology, and neural control of the most complex vocalization system known outside of human language and music. In sharp contrast, the potentially fascinating homology between great ape drumming and human instrumental music has been largely overlooked, but may offer equally useful insights into another major component of human musical behavior. The intraspecific comparative approach between music and other cognitive faculties seems equally promising, especially with language (though by no means excluding other arts, especially dance). Theories about the adaptive value and phylogenetic history of music have much in common with those for language, and there appears to be considerable room for mutual constraint and cross-pollination in discussing the evolution of music and language together. In this vein, Darwin's hypothesis of a music-like protolanguage holds considerable appeal and appears to be consistent with much of the available comparative and neural data. The study of the biology and evolution of music (biomusicology) and language (biolinguistics) are cognate fields, with substantial potential for cross-fertilization and integration. Thus, the budding field of biological and evolutionary musicology seems to hold considerable promise, both for understanding music itself and for a deeper understanding of other aspects of complex, biologically based, but culturally contingent human cognition.

ACKNOWLEDGMENTS

I thank Ian Cross, Robin Dunbar, Ann MacLarnon, Bjorn Merker, and Sandra Trehub for discussions and/or comments on an earlier version of this manuscript.

[Competing interests: The author declares that he has no competing financial interests.]

REFERENCES

1. WALLIN, N.L., B. MERKER & S. BROWN. 2000. The Origins of Music. The MIT Press. Cambridge, MA.
2. AVANZINI, G. et al., EDS. 2003. The Neurosciences and Music. Vol. 999. Annals of the New York Academy of Sciences. New York.
3. PERETZ, I. & R.J. ZATORRE, EDS. 2003. The Cognitive Neuroscience of Music. Oxford University Press. Oxford.
4. ZATORRE, R.J. & I. PERETZ, EDS. 2001. The Biological Foundations of Music. Vol. 930. Annals of the New York Academy of Sciences. New York.
5. FITCH, W.T. The biology and evolution of music: a comparative perspective. Cognition. In press. e-pub 10 jan; PMID: 16412411.
6. HAUSER, M.D. & M. KONISHI. 1999. The Design of Animal Communication. Bradford/MIT Press. Cambridge, MA.
7. HAUSER, M.D. & J. MCDERMOTT. 2003. The evolution of the music faculty: a comparative perspective. Nature Neuroscience. 6: 663–668.
8. TRAINOR, L.J. 2005. Are there critical periods for musical development? Dev. Psychobiol. 46: 262–278.
9. JACKENDOFF, R. & F. LERDAHL. The capacity for music: what is it, and what's special about it. Cognition. In press.
10. LERDAHL, F. & R. JACKENDOFF. 1983. A Generative Theory of Tonal Music. MIT Press. Cambridge.
11. PINKER, S. 1997. How the Mind Works. Norton. New York.
12. FITCH, W.T. The Evolution of Language: A Comparative Review. Biology and Philosophy. In press.
13. SLOBODA, J.A. 1985. The musical mind: the cognitive psychology of music. Clarendon. Oxford.
14. TEMPERLEY, D. 2001. The Cognition of Basic Musical Structures. MIT Press. Cambridge, MA.
15. SELKIRK, E. 1984. Phonology and Syntax: The Relation between Sound and Structure. MIT Press. Cambridge, MA.
16. KOELSCH, S. et al. 2004. Music, language, and meaning: brain signatures of semantic processing. Nat. Neurosci. 7: 511–514.
17. CROSS, I. 2003. Music, cognition, culture, and evolution. In The Cognitive Neuroscience of Music. I. Peretz & R.J. Zatorre, Eds.: 42–56. Oxford University Press. Oxford.

18. BIERWISCH, M. 1978. Gestische form als bedeutung musikalischer zeichen. Jahrbuch Peters: 161–178.
19. KOELSCH, S. *et al.* 2000. Brain indices of music processing: "nonmusicians" are musical. J. Cognit. Neurosci. **12:** 520–541.
20. DEUTSCH, D. 1999. The Psychology of Music. Academic Press. San Diego.
21. SACHS, C. 1926. Anfänge der Musik. Bull. Union Musicol. **6:** 136–236.
22. HOCKETT, C.F. 1960. Logical considerations in the study of animal communication. *In* Animal Sounds and Communication. W.E. Lanyon & W.N. Tavolga, Eds. American Institute of Biological Sciences. Washington, D.C.
23. AROM, S. 2000. Prologomena to a biomusicology. *In* The Origins of Music. N.L. Wallin, B. Merker & S. Brown, Eds.: 27–29. The MIT Press. Cambridge, MA.
24. NETTL, B. 2000. An ethnomusicologist contemplates universals in musical sound and musical culture. *In* The Origins of Music. N.L. Wallin, B. Merker & S. Brown, Eds.: 463–472. The MIT Press. Cambridge, MA.
25. ARISTOTLE. c. 350 B.C. The History of Animals.
26. POUGH, F.H., J.B. HEISER & W.N. MCFARLAND. 1996. Vertebrate Life. Prentice-Hall. Upper Saddle River, NJ.
27. HILDEBRAND, M. 1974. Analysis of Vertebrate Structure. John Wiley & Sons. New York.
28. GOULD, S.J. 1976. In defense of the analog: a commentary to N. Hotton. *In* Evolution, Brain, and Behavior: Persistent Problems. R.B. Masterton, W. Hodos & H. Jerison, Eds.: 175–179. John Wiley & Sons. New York.
29. WAKE, D.B. 1991. Homoplasy: the result of natural selection, or evidence of design limitations? Am. Naturalist **138:** 543–567.
30. HALL, B.K. 1994. Homology: The Hierarchical Basis of Comparative Biology. Academic Press. San Diego, CA.
31. SANDERSON, M.J. & L. HUFFORD, EDS. 1996. Homoplasy: the recurrence of similarity in evolution. Academic Press. San Diego.
32. MARLER, P. & H. SLABBEKOORN. 2004. Nature's Music: The Science of Birdsong. Elsevier Academic Press. San Diego, CA.
33. KROODSMA, D.E. & E.H. MILLER. 1996. Ecology and Evolution of Acoustic Communication in Birds. Academic Press. New York.
34. CATCHPOLE, C.K. & P.L.B. SLATER. 1995. Bird song: themes and variations. Cambridge University Press. New York.
35. JANIK, V.M. & P.B. SLATER. 1997. Vocal learning in mammals. Adv. Study Behav. **26:** 59–99.
36. THOMSON, D.H. & W.J. RICHARDSON. 1995. Marine mammal sounds. *In* Marine Mammals and Noise. W.J. Richardson *et al.*, Eds.: 159–204. Academic Press. San Diego.
37. FITCH, W.T. 2000. The evolution of speech: a comparative review. Trends Cognit. Sci. **4:** 258–267.
38. MARLER, P. 1976. An ethological theory of the origin of vocal learning. Ann. N. Y. Acad. Sci. **280:** 386–395.
39. NOTTEBOHM, F. 1976. Vocal tract and brain: a search for evolutionary bottlenecks. Ann. N. Y. Acad. Sci. **280:** 643–649.
40. MERKER, B. 2000. Synchronous chorusing and human origins. *In* The Origins of Music. N.L. Wallin, B. Merker & S. Brown, Eds.: 315–327. The MIT Press. Cambridge, MA.
41. MARLER, P. 2000. Origins of music and speech: insights from animals. *In* The Origins of Music. N.L. Wallin, B. Merker & S. Brown, Eds.: 31–48. The MIT Press. Cambridge, MA.
42. GEISSMANN, T. 2000. Gibbon song and human music from an evolutionary perspective. *In* The Origins of Music. N.L. Wallin, B. Merker & S. Brown, Eds.: 103–123. The MIT Press. Cambridge, MA.
43. TREHUB, S.E. 2003. Musical predispositions in infancy: an update. *In* The Cognitive Neuroscience of Music. I. Peretz & R.J. Zatorre, Eds.: 3–20. Oxford University Press. Oxford.
44. CROCKFORD, C. *et al.* 2004. Wild chimpanzees produce group-specific calls: a case for vocal learning? Ethology **110:** 221–243.
45. HAUSER, M.D. 1996. The Evolution of Communication. MIT Press. Cambridge, MA.

46. JÜRGENS, U. 1998. Neuronal control of mammalian vocalization, with special reference to the squirrel monkey. Naturwissenschaften **85:** 376–388.
47. POOLE, J.H. *et al.* 2005. Elephants are capable of vocal learning. Nature **434:** 455–456.
48. DOUPE, A.J. & P.K. KUHL. 1999. Birdsong and human speech: common themes and mechanisms. Annu. Rev. Neurosci. **22:** 567–631.
49. DARWIN, C. 1871. The Descent of Man and Selection in Relation to Sex. John Murray. London.
50. PAYNE, R. & S. MCVAY. 1971. Songs of humpback whales. Science **173:** 583–597.
51. PAYNE, K. 2000. The progressively changing songs of humpback whales: a window on the creative process in a wild animal. *In* The Origins of Music. N.L. Wallin, B. Merker & S. Brown, Eds.: 135–150. The MIT Press. Cambridge, MA.
52. WICKLER, W. 1980. Vocal duetting and the pair bond: 1. Coyness and partner commitment: a hypothesis. Z. Tierpsychol. **52:** 201–209.
53. LANGMORE, N.E. 1998. Functions of duet and solo songs of female birds. Trends Ecol. Evol. **13**.
54. LANGMORE, N.E. 2000. Why female birds sing. *In* Signalling and Signal Design in Animal Communication. Y. Espmark, T. Amundsen & G. Rosenqvist, Eds.: 317–327. Tapir Academic Press. Trondheim, Norway.
55. RIEBEL, K. 2003. The "mute" sex revisited: vocal production and perception learning in female songbirds. Adv. Study Behav. **33–86**.
56. KRINER, E. & H. SCHWABL. 1991. Control of winter song and territorial aggression of female robins (*Erithacus rubecula*). Ethology **87:** 37–44.
57. PORTER, D. & A. NEURINGER. 1984. Music discriminations by pigeons. J. Exp. Psychol. Anim. Behav. Processes **10:** 138–148.
58. CHASE, A.R. 2001. Music discriminations by carp (*Cyprinus carpio*). Anim. Learn. Behav. **29:** 336–353.
59. FITCH, W.T. Mammalian vocal production. *In* Encyclopedia of Language and Linguistics. N. Kennedy, Ed. Elsevier. Oxford. In press.
60. WOOD, G.A. 1984. Tool use by the palm cockatoo *Probosciger aterrimus* during display. Corella **8:** 94–95.
61. DODENHOFF, D.J., R.D. STARK & E.V. JOHNSON. 2001. Do woodpecker drums encode information for species recognition? Condor **103:** 143–150.
62. STARK, R.D., D.J. DODENHOFF & E.V. JOHNSON. 1998. A quantitative analysis of woodpecker drumming. Condor **100:** 350–356.
63. RANDALL, J.A. 1997. Species-specific footdrumming in kangaroo rats: *Dipodomys ingens, D. deserti, D. spectabilis*. Anim. Behav. **54:** 1167–1175.
64. SCHALLER, G.B. 1963. The Mountain Gorilla. University of Chicago Press. Chicago.
65. GOODALL, J. 1986. The Chimpanzees of Gombe: Patterns of Behavior. Harvard University Press. Cambridge, MA.
66. ARCADI, C., D. ROBERT & C. BOESCH. 1998. Buttress drumming by wild chimpanzees: temporal patterning, phrase integration into loud calls, and preliminary evidence for individual distinctiveness. Primates **39:** 505–518.
67. ARCADI, A.C., D. ROBERT & F. MUGURUSI. 2004. A comparison of buttress drumming by male chimpanzees from two populations. Primates **45:** 135–139.
68. DE WAAL, F.B.M. 1988. The communicative repertoire of captive bonobos (*Pan paniscus*), compared to that of chimpanzees. Behaviour **106:** 183–251.
69. WILLIAMS, L. 1967. The Dancing Chimpanzee: A Study of the Origins of Primitive Music. Norton. New York.
70. HAUSER, M., N. CHOMSKY & W. T. FITCH. 2002. The language faculty: what is it, who has it, and how did it evolve? Science **298:** 1569–1579.
71. FITCH, W.T., M.D. HAUSER & N. CHOMSKY. The evolution of the language faculty: clarifications and implications. Cognition. In press.
72. HURON, D. 2001. Is music an evolutionary adaptation? Ann. N. Y. Acad. Sci. **930:** 43–61.
73. BALTER, M. 2004. Seeking the Key to Music. Science **306:** 1120–1122.
74. TINBERGEN, N. 1963. On aims and methods of ethology. Z. Tierpsychol. **20:** 410–433.
75. REEVE, H.K. & P. SHERMAN. 1993. Adaptation and the goals of evolutionary research. Quart. Rev. Biol. **68:** 1–32.

76. MAYNARD SMITH, J. *et al.* 1985. Developmental constraints and evolution. Quart. Rev. Biol. **60:** 265–287.
77. GOULD, S.J. & R.C. LEWONTIN. 1979. The spandrels of San Marco and the panglossian paradigm: a critique of the adaptationist programme. **205:** 581–598.
78. CRANBERG, L.D. & M.L. ALBERT. 1988. The chess mind. *In* The Exceptional Brain: Neuropsychology of Talent and Special Abilities. L.K. Obler & D. Fein, Eds.: 156–190. Guilford Press. New York.
79. MILLER, G.F. 2001. The Mating Mind: How Sexual Choice Shaped the Evolution of Human Nature. Doubleday. New York.
80. MILLER, G.F. 2000. Evolution of music through sexual selection. *In* The Origins of Music. Wallin, N.L., B. Merker & S. Brown, Eds.: 329–360. The MIT Press. Cambridge, MA.
81. HAIMOFF, E.H. 1986. Convergence in the duetting of monogamous Old World primates. J. Hum. Evol. **15:** 51–59.
82. TREHUB, S.E. 2000. Human processing predispositions and musical universals. *In* The Origins of Music. N. L. Wallin, B. Merker & S. Brown, Eds.: 427–448. The MIT Press. Cambridge, MA.
83. DUNBAR, R. 1996. Grooming, Gossip, and the Evolution of Language. Harvard University Press. Cambridge, MA.
84. DUNBAR, R. 1993. Coevolution of neocortical size, group size, and language in humans. Behav. Brain Sci. **16:** 681–735.
85. DISSANAYAKE, E. 2000. Antecedents of the temporal arts in early mother–infant interaction. *In* The Origins of Music. N.L. Wallin, B. Merker & S. Brown, Eds.: 389–410. The MIT Press. Cambridge, MA.
86. TREHUB, S.E. & L.J. TRAINOR. 1998. Singing to infants: lullabies and play songs. Adv. Infant Res. **12:** 43–77.
87. TREHUB, S.E. 2003. The developmental origins of musicality. Nat. Neurosci. **6:** 669–673.
88. HUSAIN, G., W.F. THOMPSON & E.G. SCHELLENBERG. 2002. Effects of musical tempo and mode on arousal, mood, and spatial abilities. Mus. Percept. **20:** 151–171.
89. JUSLIN, P. & J.A. SLOBODA. 2001. Music and emotion: theory and research. Oxford University Press. Oxford.
90. FALK, D. 2004. Prelinguistic evolution in early hominids: whence motherese? Behav. Brain Sci. **27:** 491–450.
91. FERNALD, A. 1992. Human maternal vocalizations to infants as biologically relevant signals: an evolutionary perspective. *In* The Adapted Mind. J. Barkow, L. Cosmides & J. Tooby, Eds.: 391–428. Oxford University Press. New York.
92. MACLARNON, A. & G. HEWITT. 1999. The evolution of human speech: the role of enhanced breathing control. Am. J. Phys. Anthropol. **109:** 341–363.
93. SUNDBERG, J. 1987. The Science of the Singing Voice. Northern Illinois University Press. DeKalb, IL.
94. D'ERRICO, F. *et al.* 2003. Archaeological evidence for the emergence of language, symbolism, and music: an alternative multidisciplinary perspective. J. World Prehistory **17:** 1–70.
95. HAHN, J. & S. MÜNZEL. 1995. Knochenflöten aus den Aurignacien des Geissenklösterle bei Blaubeuren, Alb-Donau-Kreis. Fundberichte aus Baden-Württemberg **20:** 1–12.
96. KRINGS, M. *et al.* 1997. Neandertal DNA sequences and the origin of modern humans. Cell **90:** 19–30.
97. D'ERRICO, F. *et al.* 1998. A Middle Palaeolithic origin of music? Using cave-bear bone accumulations to assess the Divje Babe I bone "flute." Antiquity **72:** 65–76.
98. KUNEJ, D. & I. TURK. 2000. New perspectives on the beginnings of music: archaeological and musicological analysis of a Middle Paleolithic bone "flute." *In* The Origins of Music. N.L. Wallin, B. Merker & S. Brown, Eds.: 235–268. The MIT Press. Cambridge, MA.
99. RÉVÉSZ, G. 1941. Der Ursprung der Musik. International Arch. Ethnogr. **50:** 65.

100. KUNST, J. 1959. Ethnomusicology: A Study of Its Nature, Its Problems, Methods, and Representative Personalities to Which is Added a Bibliography. Martinus Nijhoff. The Hague.
101. BROWN, S. 2000. The "musilanguage" model of music evolution. *In* The Origins of Music. N.L. Wallin, B. Merker & S. Brown, Eds.: 271–300. The MIT Press. Cambridge, MA.
102. ENARD, W. *et al.* 2002. Molecular evolution of FOXP2, a gene involved in speech and language. Nature **418:** 869–872.

Neuronal Mechanisms Underlying the Perception of Pitch and Harmony

GERALD LANGNER

Neuroacoustics, TU-Darmstadt, 64287 Darmstadt, Germany

ABSTRACT: Temporal processing of periodic acoustic signals in the auditory brain stem provides an explanation for pitch perception and the natural preference of our hearing system for harmonic relationships in music. Experimental evidence is reviewed for a corresponding neuronal model of correlation analysis and the spatial representation of pitch information along the second neural axis of the auditory system.

KEYWORDS: pitch perception; hearing; neuronal processing; auditory system

INTRODUCTION

Resonances, characterized by harmonic frequency relationships, are found throughout the natural world, and harmonic sounds resulting from such resonances are essential elements of animal communication, of speech and, of course, of music. Harmonic sounds are characterized by periodic envelope or amplitude modulations that reflect the vibration of musical instruments as well as of vocal cords. While the timbre of a periodically modulated sound depends strongly on its often complex spectral composition and therefore may be quite different from the smooth timbre of a pure tone, the perceived pitch of a periodic sound is the same, provided both have the same period.[1] On the basis of von Helmholtz's resonance and place theories about the function of the ear and pitch perception, it was theorized that pitch is a secondary attribute of sounds, the outcome of cochlear filtering and of hypothetical neuronal processing of the frequency spectrum of acoustic signals.[2] Nowadays it is widely accepted that the neuronal mechanisms underlying the perception of periodicity pitch rely on temporal encoding of the periodicity of acoustic signals, which is available in the auditory nerve up to about 5 kHz, the upper limit of musical pitch, as defined by our ability to recognize musical intervals. Accordingly, in light of neurophysiological results,[3] pitch reflects a primary, temporal aspect of acoustic signals: the periodicity of harmonic sounds. Since all frequency components (harmonics) of a harmonic sound are multiples of its fundamental frequency, the period of the fundamental is also encoded in the cochlea in amplitude modulations

Address for correspondence: Gerald Langner, Neuroacoustics, TU-Darmstadt, Schnittspahnstr. 3, 64287 Darmstadt, Germany. Voice: +49-6151-16-3605; fax: +49-6151-16-3803.
gl@bio.tu-darmstadt.de

Ann. N.Y. Acad. Sci. 1060: 50–52 (2005). © 2005 New York Academy of Sciences.
doi: 10.1196/annals.1360.043

resulting from superpositions of frequency components above the third harmonic.[1] As a consequence the period of the fundamental is coded temporally in spike intervals in the auditory nerve and analysed by neurons in the auditory brain stem (cochlear nucleus: CN) and midbrain (inferior colliculus: IC).[4]

REVIEW OF THEORETICAL AND EXPERIMENTAL RESULTS

In order to explain how our brain extracts periodicity pitch, and how neurons in the CN and the IC respond to amplitude-modulated (periodic) signals, a model was proposed that performs a correlation between signal fine-structure and envelope.[4] Its processing elements are a trigger, an oscillator, a reducer, and a coincidence neuron, which are supposed to have their counterparts in well-described on-type, chopper neurons, and pauser neurons in the CN and disc cells in the IC. The trigger unit synchronizes the responses of oscillator and reducer cycle to the modulation. While the oscillator responds with short bursts of regular intrinsic oscillations to each modulation period, the reducer generates intervals precisely related to the signal fine structure. By integrating synchronized activity of many nerve fibers the reducer is able to code frequencies to the upper limit of phase coupling. The coincidence unit is activated by simultaneous inputs from oscillator and reducer and responds best when signal fine-structure and envelope are correlated and the envelope period matches the reducer delay. Thereby it responds best to a periodically modulated sound (BMF) and is simultaneously representing a certain frequency and a certain pitch. Such a combined frequency and periodicity analysis is in line with modern experiments and discussions of this topic.[5]

However, as a mathematical result of the model, coincidence neurons also respond to integer multiples of BMF. This harmonicity effect is also observed in neuronal responses of IC neurons, although the resulting comb filters are mostly restricted to the onset. The suppression of responses to harmonics of BMF requires a well-timed input to IC neurons that is synchronized to the envelope period, This is provided by the ventrolateral lemniscus (VNLL), which is activated by on-type neurons, the physiological correlates of the trigger neurons in the model. The VNLL seems to suppress harmonic responses in the IC, which otherwise are a necessary side effect of the correlation process.

Anatomical evidence suggests that the VNLL has a helical organization.[6] This is a unique anatomical structure, which, together with its function in harmonic processing, suggests that it is organized as a pitch helix. This hypothesis was verified in gerbils using 2-deoxy-glucose (2-DG) functional labelling. The results indicate that the VNLL is helically organized with neurons involved in the processing of 8 octaves, one octave per turn, reminiscent of the pitch helix well-known in music psychology.

In the IC, periodicity information is represented in a topographic map along a second neural axis, orthogonal to the well-known tonotopic axis.[4] The tonotopic axis represents the cochlear frequency analysis and codes the frequency composition of a sound, thus representing major aspects of its timbre, while the periodicity axis represents pitch. Such maps have been verified with optical recording in cat cortex, with 2-DG mapping in the IC and the cortex of gerbils, and with magnetoencephalography in the auditory cortex of humans.[7]

CONCLUSIONS

Our investigations imply that fundamental aspects of harmonic perception are by no means just culturally inherited. Instead the recognition of musical harmony, or consonance, is an intrinsic property of our brains. Although realized by neuronal components, the processing of periodic sounds may be directly compared to resonances in musical instruments. Just as temporal aspects of acoustic resonances underlie the harmonicity of musical sounds, temporal aspects of neuronal correlation mechanisms may explain our preference for harmonic relationships in music. Pythagoras observed a correspondence between the mathematical laws underlying our perception of consonance and those governing universal harmony of the physical world. He suggested a universally valid principle: consonance, or harmony, whether auditory or cosmic, depends on integers and simple integer ratios. These ancient philosophical views may be supported by the neurophysiology of pitch perception.

[Competing interests: The author declares that he has no competing financial interests.]

REFERENCES

1. SCHOUTEN, J.F. *et al.* 1962. Pitch of the residue. J. Acoust. Soc. Am. **34:** 1418–1424.
2. GOLDSTEIN, J.L. 1973. An optimum processor theory for the central formation of the pitch of complex tones. J. Acoust. Soc. Am. **54:** 1496–1516.
3. LANGNER, G. 1983. Evidence for neuronal periodicity detection in the auditory system of the guinea fowl: implications for pitch analysis in the time domain. Exp. Brain Res. **52:** 333–355.
4. LANGNER, G. 1992. Periodicity coding in the auditory system. Hearing Res. **60:** 115–142.
5. HARTMANN, W.M. & S.L. DOTY. 1996. On the pitches of the components of a complex tone. J. Acoust. Soc. Am. **99:** 567–578.
6. MERCHÁN, M.A. & P. BERBEL. 1996. Anatomy of the ventral nucleus of the lateral lemniscus in rats A nucleus with a concentric laminar organization. J. Comp. Neurol. **372:** 245–263.
7. LANGNER, G. *et al.* 1997. Frequency and periodicity are represented in orthogonal maps in the human auditory cortex: evidence from magnetoencephalography. J. Comp. Physiol. **181:** 665–676.

Tonality and Nonlinear Resonance

EDWARD W. LARGE AND AMY ELIZABETH TRETAKIS

Center for Complex Systems and Brain Sciences, and Department of Psychology, Florida Atlantic University, Boca Raton, Florida 33431, USA

ABSTRACT: We outline a theory of tonality that predicts tonal stability, attraction, and categorization based on the principles of nonlinear resonance. Perception of tonality is the natural consequence of neural resonance, arising from central auditory nonlinearities.

KEYWORDS: auditory processing; tonality; hearing; cochlea

INTRODUCTION

Tonality is the organized relationship of tones in music.[1] A tonal system includes a central tone, or tonic, and a collection of related tones, each with a prespecified fundamental frequency. Within a tonal context, certain tones are perceived as more stable than others, such that less stable tones provide points of stress, and more stable tones provide points of repose. Less stable tones are heard relative to more stable ones, such that more stable tones are said to attract less stable tones.[2] What nervous system processes give rise to such perceptions in music?

Recent evidence suggests that the cochlea performs active frequency transformation of sounds by nonlinear resonance, using a network of locally coupled outer–hair cell oscillators.[3,4] In species that lack cochleae, auditory neurons respond selectively to temporal and spectral features of communication sounds,[5] and nonlinear resonance based on excitation and inhibition has been implicated in such responses.[6] In the mammalian auditory system, neural activity in several areas, including the cochlear nucleus, the inferior colliculus, and A1, is phase locked to the stimulus waveform. Inhibition plays a significant role in these responses.[7] Phase locking deteriorates as the auditory pathway is ascended.[8]

HYPOTHESIS

We hypothesize that nonlinear frequency transformation takes place in the mammalian central auditory nervous system (CANS). Nonlinear frequency transformation occurs when a network of coupled nonlinear resonators, each tuned to a distinct eigenfrequency, is driven by an external stimulus. A nonlinear neural resonator can

Address for correspondence: Edward Large, Center for Complex Systems and Brain Sciences, Florida Atlantic University, 777 Glades Rd., Boca Raton, FL 33431. Voice: 561-297-0106; fax: 561-297-3634.

large@ccs.fau.edu

Ann. N.Y. Acad. Sci. 1060: 53–56 (2005). © 2005 New York Academy of Sciences.
doi: 10.1196/annals.1360.046

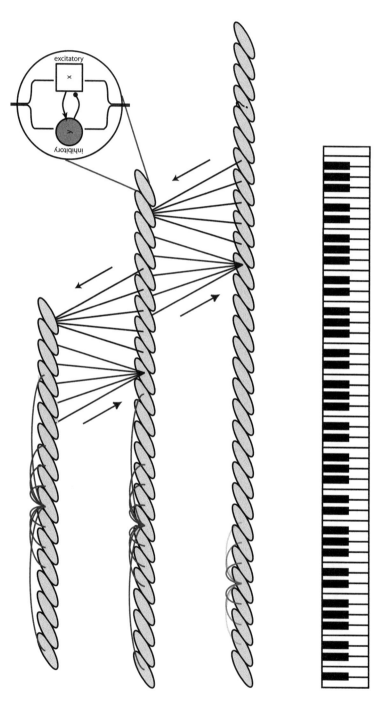

FIGURE 1. Cascaded nonlinear frequency transformation. Initial analysis is performed by the cochlea, and further transformations are performed in the CANS by neural oscillators. Phase locking deteriorates as the pathway is ascended. **Inset:** A neural oscillator, consisting of an excitatory-inhibitory neuron pair.

be modeled as a pair of interacting neurons (or populations), one excitatory and one inhibitory.[9,10] Mathematical analysis of resonator networks reveals generic properties of nonlinear frequency transformation, including extreme sensitivity to weak stimuli, sharp frequency tuning, amplitude compression, frequency detuning, natural phase differences, and nonlinear distortions.[11] These properties are consistent with psychoacoustic phenomena, such as hearing thresholds, frequency discrimination, loudness scaling, Stevens' rule, harmonic distortion, and combination tones. Nonlinear resonance is also a plausible neural mechanism for pitch perception in humans.[12]

A simple nonlinear CANS model is shown in FIGURE 1. After initial nonlinear frequency analysis by the cochlea, networks of neural resonators further transform the stimulus. Such a cascaded analysis is not redundant, because nonlinear transformations cannot be collapsed into a single stage, as is the case with linear transformations. A recent analysis suggests that as the series of nonlinear transformations is ascended, a complex web of resonances arises, including objective stimulus frequencies and a large number of nonlinear "distortions."[13] A dynamic field emerges, with active resonances in various regions embodying a musical scale.

PREDICTIONS

The Arnol'd tongues bifurcation diagram (FIG. 2) allows the study of these resonances, by postulating a specific connectivity.[13] The horizontal axis gives the ratio of each oscillator's frequency to the tonic, while the vertical gives the strength of afferent input at the tonic frequency. The resonance regions, or tongues, show the neural areas that will resonate for various input strengths. As stimulus strength is increased, larger patches of neurons resonate, indicated by the increasing width of the tongues. Moreover, each area resonates at a specific frequency ratio with the tonic.

The widest tongues reflect the most stable resonances. A stable resonance is resistant to change in frequency due to interactions with other parts of the dynamic field. The relative stability of tones in the key of C major is predicted by the relative widths of the resonance regions in the bifurcation diagram. The theory also makes predictions about tonal attraction. In regions where the tongues overlap, more stable resonances tend to overpower less stable ones, such that the neural population in the overlap region will tend to oscillate at the frequency of the more stable resonance.

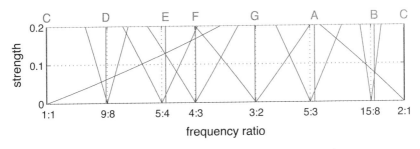

FIGURE 2. Arnol'd tongues diagram showing predicted resonances for the major scale. The width of the resonance regions predicts stability of the corresponding tones. Overlapping of regions predicts tonal attraction, with more stable regions dominating.

Historical notions of consonance and dissonance hold that musical consonance is determined by the ratios of small whole numbers.[14] However, the principle of small integer ratios does not explain the acceptability of equal temperament, because equal-tempered (ET) intervals are irrational. Also it does not explain the acceptability of large variations in intonation commonly observed in musical performance.[15] Perceptual categorization of musical intervals[15] explains both observations, but contradicts the principle of small integer ratios.

The gray vertical lines in FIGURE 2 denote ET frequencies. Resonance center frequencies do not precisely match ET tone frequencies. However, as stimulus strength increases, larger neural populations resonate, eventually encompassing the equal tempered ratios. Our theory predicts both perceptual categorization and the importance of small integer ratios. Resonance predicts how much variation in intonation is acceptable before a pitch is heard as mistuned.

ACKNOWLEDGMENTS

This research was supported by NSF CAREER Award BCS-0094229.

[Competing interests: The authors declare that they have no competing financial interests.]

REFERENCES

1. PISTON, W. 1978. Harmony. W. W. Norton & Company. New York.
2. LERDAHL, F. 2001. Tonal Pitch Space. Oxford University Press. New York.
3. RUGGERO, M.A. 1992. Responses to sound of the basilar membrane of the mammalian cochlea. Curr. Opin. Neurobiol. 2: 449–456.
4. CHOE, Y., M.O. MAGNASCO & A.J. HUDSPETH. 1998. A model for amplification of hair-bundle motion by cyclical binding of Ca2+ to mechanoelectrical-transduction channels. Proc. Natl. Acad. Sci. 95: 15321–15336.
5. CRAWFORD, J.D. 1997. Feature detection by auditory neurons in the brain of a sound-producing fish. J. Comp. Physiol. A 180: 439–450.
6. LARGE, E.W. & J.D. CRAWFORD. 2002. Auditory temporal computation: interval selectivity based on postinhibitory rebound. J. Comp. Neurosci. 13: 125–142.
7. GROTHE, B. & G. M. KLUMP. 2000. Temporal processing in sensory systems. Curr. Opin. Neurobiol. 10: 467–473.
8. LANGNER, G. 1992. Periodicity coding in the auditory system. Hear. Res. 60: 115–142.
9. WILSON, H.R. & J.D. COWAN. 1973. A mathematical theory of the functional dynamics of cortical and thalamic nervous tissue. Kybernetik 13: 55–80.
10. HOPPENSTEADT, F.C. & E.M. IZHIKEVICH. 1996. Synaptic organizations and dynamical properties of weakly connected neural oscillators I: Analysis of a canonical model. Biol. Cybern. 75: 117–127.
11. LARGE, E.W. 2006. A generic nonlinear model for auditory perception. In Auditory Mechanisms: Processes and Models. A.L. Nutall, Ed.: 506–508. World Scientific. Singapore.
12. CARTWRIGHT, J.H.E., D.L. GONZÀLEZ & O. PIRO. 1999. Nonlinear dynamics of the perceived pitch of complex sounds. Phys. Rev. Lett. 82: 5389–5392.
13. LARGE, E.W. & A.E. TRETAKIS. 2006. A dynamic field theory of tonality. In Brain, Mind and Culture. A. Ionnides, Ed. In press.
14. HUNT, F V. 1992. Origins in Acoustics. Acoustical Society of America. Woodbury, NY.
15. BURNS, E.M. 1999. Intervals, scales, and tuning. In The Psychology of Music. D. Deutsch, Ed.: 215–264. Academic Press. San Diego.

Part II: Language and Music—A Comparison

Introduction

MIREILLE BESSON[a] AND ANGELA FRIEDERICI[b]

[a]Institut de Neurosciences Cognitives de la Méditerranée,
Centre National de la Recherche Scientifique, Marseille, France

[b]Max Planck Institute for Human Cognitive and Brain Sciences,
Leipzig, Germany

The comparison between language and music is an old issue that has been of strong interest to biologists, philosophers, and writers for a long time. Darwin,[1] for instance, considered that music derived from a protolanguage aimed at sexual selection: "Musical notes and rhythm were first acquired by the male or female progenitors of mankind for the sake of charming the opposite sex." Rousseau,[2] by contrast, considered that music came first and that primitive languages were sung rather than spoken. Spencer[3] and Descartes[4] both considered that there was a common origin for language and music, with a strong connection to motion; the strongest feelings were expressed through music or speech, with larger and more rapid movements to produce them. More recently, Hauser, Chomsky, and Fitch[5] have argued that the human capacity to process language and music may be based on the species-specific ability to process recursive structures.

Interestingly, the language–music comparison has received renewed interest over the last 15 years, with the recent development of brain-imaging methods. Several papers and chapters have been devoted to this issue (see, for instance, the books by Wallin, Merker, and Brown,[6] and Peretz and Zatorre.[7]) Fascinating similarities and differences have been highlighted, such as the parallel evolution of language and music, their importance for our understanding of human cultures, and their universality, on one side, and on the other, the fact that in contrast to language, music is a self-referential system that ignores the signifier–signified contrast. From a neuroscience perspective, the most challenging findings from brain imaging studies may be that common networks are activated in tasks that were first thought to involve specialized brain areas and mechanisms.[8] The neural networks for language and music processing do show a large overlap.[9]

Taking upon the issues mentioned above, Patel and Daniele[10] were able to demonstrate cultural influences of linguistic rhythm on musical rhythm, thereby emphasizing the similarities between linguistic and musical rhythmic production. In the follow-up studies described here, they provide evidence that the language-specific intonation of English and French is reflected in the turn-of-the century classical

Address for correspondence: Mireille Besson, Centre de Recherche en Neurosciences Cognitives, CNRS-CRNS, 31-Chemin Joseph Aiguier, 13402-Marseille Cedex 20, France. Voice: +33-491-164-305; fax: +33-491-774-969.

Mireille.Besson@inom.cnrs-mrs.fr

Ann. N.Y. Acad. Sci. 1060: 57–58 (2005). © 2005 New York Academy of Sciences.
doi: 10.1196/annals.1360.061

instrumental music of England and France, respectively. Moreover, in further testing for the relationship between language and music, they investigated the processing of musical syntax in Broca's aphasia. These patients who suffer from syntactic processing difficulties in language also showed difficulties in processing musical syntax.

A different approach to test further the relationship between language and music is to investigate the neurocognition of song processing. Indeed, songs provide a unique and intimate combination of speech and music. Since the time of early opera (e.g., Monteverdi's *L'Orfeo*, first performed in Venice in 1607), the issue of whether the words or the music are more important has been the subject of a long-lasting controversy.[11–13] However, while vocal music is considered the oldest and still the most popular form of music, little is known about the biological basis of song perception and cognition. The chapter by Daniele Schön and collaborators describes several brain-imaging experiments, using both fMRI and ERPs, that were specifically designed to investigate the relationship between the linguistic and musical dimensions in song perception. Spoken language, sung language, and vocalises were compared in the first experiment: results showed a strong overlap of the regions involved in language, music, and song processing. Moreover, results of the second and third experiments, revealing an effect of the irrelevant on the relevant dimension (i.e., music when paying attention to language or vice-versa), seem to indicate that the linguistic and musical dimensions of songs are processed in interaction.

REFERENCES

1. DARWIN, C. 1871. The Descent of Man. Princeton University Press. Princeton, NJ.
2. ROUSSEAU, J.-J. 1781/1993. Essai sur l'Origine des Langues. Flammarion. Paris.
3. SPENCER, H. 1857. The origin and function of music. Fraser's Magazine **56**: 396–408.
4. DESCARTES, R. 1618. Abrégé de musique. Compendium Musicae. Épiméthée. P.U.F. Paris.
5. HAUSER, M.D., N. CHOMSKY & W.T. FITCH. 2002. The faculty of language: what is it, who has it, and how did it evolve? Science **298**: 1569–1579.
6. WALLIN, N.L., B. MERKER & S. BROWN, EDS. 2000. The Origins of Music. MIT Press. Cambridge, MA.
7. ZATORRE, R. & I. PERETZ, EDS. 2003. The Biological Foundations of Music. Oxford University Press. Oxford.
8. POSNER, M.I. & M.K. ROTHBART. 2005. Influencing brain networks: implications for education. Trends Cogn. Sci. **9**: 99–103.
9. KOELSCH, S., T.C. GUNTER, D.Y. VON CRAMON, *et al.* 2002. Bach speaks: a cortical "language-network" serves the processing of music. NeuroImage **17**: 956–966.
10. PATEL, A.D. & J.R. DANIELE. 2003. An empirical comparison of rhythm in language and music. Cognition **87**: B35–45.
11. BOULEZ, P. 1966. Son et verbe. *In* Relevés d'Apprenti. Le Seuil. Paris.
12. LEVMAN, B.G. 1992. The genesis of music and language. Ethnomusicology **36**: 147–170.
13. STENDHAL, L. 1824. La Vie de Rossini. Flammarion. Paris.

The Relationship of Music to the Melody of Speech and to Syntactic Processing Disorders in Aphasia

ANIRUDDH D. PATEL

The Neurosciences Institute, San Diego, California 92121, USA

ABSTRACT: Two new empirical studies address the relationship between music and language. The first focuses on melody and uses research in phonetics to investigate the long-held notion that instrumental music reflects speech patterns in a composer's native language. The second focuses on syntax and addresses the relationship between musical and linguistic syntactic processing via the study of aphasia, an approach that has been explored very little. The results of these two studies add to a growing body of evidence linking music and language with regard to structural patterns and brain processing.

KEYWORDS: melody; prosody; aphasia; syntax; language; music

INTRODUCTION

Human cultures make use of two organized sound systems: those of music and language. While these systems have many obvious differences, both employ rhythmic and melodic patterns and rule-governed sequences. Scholars have thus long been interested in possible links between the domains. Indeed, the issue has engaged a wide range of thinkers over the centuries, including music theorists, linguists, poets, and philosophers. Although the topic is an old one, recent years have marked a watershed in the history of music–language studies, with the rise of empirical studies rooted in cognitive science and neuroscience. Here I illustrate the empirical approach with two new studies addressing questions of current interest in the relationship between music and language.

The first question is whether a composer's instrumental music reflects the prosody of his or her native language. A new quantitative model of intonation perception is used to study this issue. The second question is whether music and language overlap with regard to the neural bases of syntactic processing. This is addressed via the study of music perception in aphasia, an approach that has been virtually unexplored. Full scientific details of these studies are available in papers that have been submitted for publication by Patel, Iversen and Rosenberg,[1] and Patel, Iversen, and Hagoort.[2] Here I describe these studies in a larger historical context and provide an overview of the methods and findings.

Address for correspondence: Aniruddh D. Patel, The Neurosciences Institute, 10640 John Jay Hopkins Drive, San Diego, CA 92121. Voice: 858-626-0285; fax: 858-626-2099.
apatel@nsi.edu

Ann. N.Y. Acad. Sci. 1060: 59–70 (2005). © 2005 New York Academy of Sciences.
doi: 10.1196/annals.1360.005

MUSICAL STRUCTURE AND LINGUISTIC PROSODY

Background

A number of musicologists and linguists have claimed that the rhythms and melodies of a culture's native language are reflected in its instrumental music.[3,4] For example, the noted English musicologist Gerald Abraham voiced this idea,[3] citing as one example Ralph Kirkpatrick's comment on French keyboard music:

"Both Couperin and Rameau, like Fauré and Debussy, are thoroughly conditioned by the nuances and inflections of spoken French. On no Western music has the influence of language been stronger."

Kirkpatrick (a harpsichordist and music scholar) was effectively saying that something about French keyboard music sounds like the French language. Similar claims have been made about the instrumental music of other cultures, including England.[5] Surprisingly, these provocative claims have gone largely untested from a scientific standpoint. Why is this the case?

The issue seems to be a practical problem, namely a lack of good tools for quantifying rhythmic and melodic patterns in language in a way that can be directly compared to music. Fortunately, new tools from phonetics are helping overcome this obstacle. For example, a measure of temporal patterning in sentences, the normalized pairwise variability index (nPVI), has recently been developed to explore rhythmic differences between "stress-timed" and "syllable-timed" languages.[6] The nPVI computes the degree of durational contrast between neighboring events in a sequence, and has been used to examine temporal patterns of vowel duration in sentences from a variety of languages.[7,8] The salient finding is that the nPVI is generally higher for languages classified as stress-timed (e.g. British English, Dutch, and Thai) than for languages classified as syllable-timed (e.g. French, Spanish, and Singapore English). This likely reflects a greater degree of vowel reduction in stress-timed languages.[9]

In previous research, we applied the nPVI to note durations in instrumental classical themes from England and France, and found that English music had a higher nPVI than French music.[10,11] This earlier study illustrates our general approach to comparing prosody to musical structure. Specifically, we choose two cultures, A and B, and compare their speech prosody using a quantitative method. If a difference is found, then instrumental music from A and B is compared using precisely the same method. The question of interest is whether the musics of A and B show also show a significant difference in the same direction as the linguistic difference (e.g., A > B for both speech and music). If so, then this suggests that musical structure reflects speech prosody.

Examining Melody in Speech and Music

Having conducted cross-domain research on rhythm, we have now turned our attention to melody. Using the same databases of speech and music from our previous work,[10,12] we ask if there is a difference between English and French speech intonation that is reflected in the music of the two cultures. When studying rhythm we had the benefit of an existing quantitative measure that differentiated between English and French speech and that could be applied to music in a straightforward way (the nPVI). In the case of intonation, no such measure was available. Furthermore, there has been very little work examining empirical differences between intonation in British English and French.

In searching for methods that could help us overcome these problems, our interest was captured by a recent computational model of speech intonation perception.[13,14] The central notion behind the "prosogram" model is that the raw fundamental frequency (Fo) contour of a sentence, although an accurate physical description of the speech signal, is not the most accurate representation of intonation as it is perceived by human listeners. In particular, empirical research suggests that pitch perception in speech is subject to two perceptual transformations. The first is perceptual segregation of the Fo contour into syllable-sized units due to the rapid spectral and amplitude fluctuations in the speech signal.[15] The second is temporal integration of Fo within a syllable, meaning that the perceived pitch of a syllable is actually a time-weighted average of the intrasyllabic Fo movement.[16] The prosogram instantiates this second transformation via an automatic algorithm (the first stage—phonetic segmentation—is provided by the user). As a result of these transformations, the original Fo contour of a sentence is converted to a sequence of discrete tonal segments. An example of the model's output is given in FIGURE 1.

This figure reveals why the prosogram is useful to those interested in comparing speech and music. The perceptual representation of intonation produced by the prosogram is quite music like, consisting mostly of level pitches. (Syllables are assigned pitch glides if the amount of intrasyllabic Fo change is large enough to exceed a perceptual threshold. In our corpus, only 3% of syllables were assigned glides). On a cognitive level, this is interesting because it implies that the auditory image of speech intonation in a listener's brain has more in common with music than has

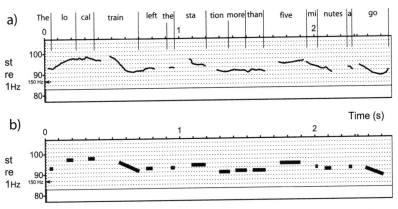

FIGURE 1. Illustration of prosogram analysis of speech intonation. Part **a** shows the original Fo contour of the British English sentence "The local train left the station more than five minutes ago." Syllables and their temporal boundaries are marked at the top of the figure. Part **b** shows a prosogram analysis of this sentence, with the tonal elements assigned by the prosogram, that is, level pitches or glides (1 per syllable). Units on the y axis are pitches in semitones (st) re 1 Hz; thin *horizontal dashed lines* mark 2 st intervals. The st value corresponding to 150 Hz is shown for reference. The prosogram runs under Praat, a freely available program for speech analysis: <http://bach.arts.kuleuven.ac.be/pmertens/prosogram/>.

generally been believed. This has implications for cross-domain transfer of statistical learning, as discussed in the next section. On a practical level, the dominance of level pitches means that intonation patterns in different languages can be compared using tools that can also be applied to music, for example, statistical measurements of pitch height or interval patterns. We adopt this approach in our studies as described below. Before turning to the details of our measurements, however, it is worth discussing the concept of statistical learning and how it guided our choice of what to measure.

Statistical Learning of Tonal Patterns in Speech and Music

Statistical learning (SL) refers to tracking patterns in the environment and acquiring implicit knowledge of their statistical properties, without any direct feedback. Research in music cognition has demonstrated statistical learning of tonal patterns in novel musical sequences. For example, listeners show sensitivity to the distribution of different pitches and to interval patterns.[17-19] These studies have shown that SL in music can occur with atonal or culturally unfamiliar materials, meaning that it is not confined to tonal patterns that follow familiar musical conventions.

Synthesizing these findings with the music-like representation of intonation provided by the prosogram, we hypothesized that SL of tonal patterns occurs for speech intonation in one's native language, especially since one has extensive exposure to such patterns from an early age. SL of speech patterns has been demonstrated for phonetic/syllabic patterns,[20] so it seems plausible that it could apply to speech tonal patterns as well. If this is the case, then composers (like other members of their culture) will have implicit knowledge of the statistics of their native language's intonation patterns, which could influence their creation of pitch patterns in another domain, namely music.

The question of interest then is What aspects of intonation patterns are learned and reflected in music? Since the speech intonation does not conform to any musical scale, SL of speech tone sequences is unlikely to involve categorical patterns (e.g., transitions between certain stable intervals). Thus we decided to investigate a simple statistical aspect of spoken pitch patterns, namely their variability.

Quantifying Variability in Spoken and Musical Pitch Patterns

Using the prosogram representation of intonation, we quantified the pitch variability of each sentence in two ways. First, we measured the variation of individual pitches about their mean. Second, we measured the variability of pitch intervals, where intervals were defined as the frequency distance between successive tonal elements. Semitone units were used in both measures to reflect the perceptual scaling of intonation.[21] In performing these measurements we only measured level pitches: glides were ignored (for interval measures, only intervals between immediately adjacent level tones were computed).

We then applied the same measurements of pitch variability to music, examining the ~300 instrumental classical themes that we had examined previously for rhythm.[10] These were taken from *A Dictionary of Musical Themes*[22] and represented the work of 6 English and 10 French composers whose lives spanned the turn of the 20th century, an era of musical nationalism (composers included, e.g., Elgar and

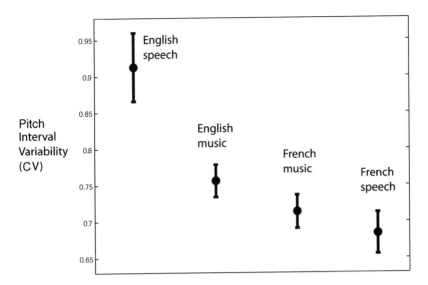

FIGURE 2. Pitch interval variability for language and music. Pitch interval variability is defined as the coefficient of variation (CV, SD/mean) of absolute interval size for a sequence, whether a sentence or a musical theme. Variability in language was measured from intervals between level tones in prosograms of British English and French sentences. Variability in music was measured from music notation of English and French classical instrumental themes. Semitones were used in both measurements. British English speech has significantly greater variability than French speech, and English music has significantly greater variability than French music. *Error bars* show standard errors.

Debussy). Musical measurements were made in semitone units directly from music notation.

Results

The primary result of interest was that music reflected a specific aspect of speech intonation, namely the variability of pitch interval size (which we have termed *melodic interval variability* or MIV). This variability was significantly higher in English than in French sentences, and in English than in French musical themes, as shown in FIGURE 2. By contrast, pitch variability about the mean did not differ between the languages or the musics.

Discussion

Our new work provides evidence that speech intonation is reflected in turn-of-the-century classical instrumental music in England and France. This provides further support for those scholars who have intuited a link between a culture's music and its linguistic prosody. Interestingly, the aspect of intonation that is reflected in music is pitch interval variability, which is lower for French than for English. That is, as the voice moves from one syllable to the next in a sentence, the size of each pitch change is more uniform in French than in English speech. Similarly, as a melody moves from one note to the next in musical themes, the size of each pitch change is more uniform in French than in English music.

Although we have focused on just two cultures in our studies, a principal goal of this work has been to develop methods that can be applied much more broadly in the study of music's relation to linguistic prosody. In the case of the nPVI, studies of speech and music in other cultures have already begun and are yielding promising results. The pitch interval variability can also be examined in other cultures, using the freely available prosogram as a tool (see legend for FIG. 1).

We have focused on a very simple aspect of the statistical patterning of pitch in this study (variability). Prosogram representations of intonation, with their sequences of level tones, clearly call for more sophisticated analyses of pitch patterns in speech, to see if other aspects of speech intonation that differentiate languages are reflected in music.

MUSICAL SYNTACTIC PROCESSING IN APHASIA

Background

Music and language employ sequences of perceptually discrete elements organized in principled ways. That is, both are syntactic systems. A growing body of research from neuroimaging points to overlap in the syntactic processing of language and music.[23–25] Yet this work must be reconciled with evidence for neuropsychological dissociations between musical and linguistic syntactic abilities.[26] In an attempt to unify these observations, Patel[27] suggested that language and music involve distinct and domain-specific syntactic representations stored in long-term memory (such as nouns and verbs vs. chords and their harmonic relations), but that neural resources involved in activating these representations during online processing are shared. One implication of this shared syntactic integration resource hypothesis (SSIRH) is that representations can be selectively damaged (leading to dissociations), but that damage to activation-related neural resources should result in deficits in both domains. Patel pointed out that a crucial test of this idea involved aphasics with linguistic disorders of syntactic comprehension. If, as has been hypothesized by some neurolinguists,[28,29] such disorders reflect a problem with activation of structural information (vs. a loss of stored syntactic representations), then according to the SSIRH, these aphasics should exhibit a musical syntactic deficit as well.

Music Perception in Aphasia: A Largely Unexplored Area

Remarkably, there has been virtually no work on musical syntactic (e.g., harmonic) processing in aphasia. This is particularly striking since an early study by Francès *et al.*[30] suggested that aphasic individuals with linguistic comprehension disorders also have a deficit in the perception of musical tonality. The researchers studied a large group of persons with aphasia and had them judge whether two short, isochronous melodies were same or different. The melodies were either tonal or atonal. Under these circumstances, normal participants (even those with no musical training) show superior performance on the tonal stimuli. Aphasic individuals failed to show this tonal superiority effect, leading the authors to suggest that the perception of tonality "seems to engage preestablished circuits existing in the language area."

This idea has lain fallow for decades, with no further studies of tonality perception in aphasia. Why might this be? Good tools for testing linguistic comprehension in aphasia and for probing the perception of tonal relationships have long been available, yet no one has attempted to replicate or extend these results. This is made even more puzzling by the fact that the findings of Francès et al. were somewhat clouded by methodological issues, and thus naturally called for further work.

It is likely that the absence of research on this topic reflects historical and conceptual forces. In particular, within music cognition there has been an emphasis on cases of dissociation between aphasia and amusia. The most oft-cited example is that of Shebalin,[31] a Russian composer who continued to write music after becoming severely aphasic. Citing this and other reports of aphasia without amusia, Marin and Perry[32] concluded that "these cases of total dissociation are of particular interest because they decisively contradict the hypothesis that language and music share common neural substrates" (p. 655).

If this conclusion were true, then there would clearly be no reason to pursue the issue of music perception in aphasia. However, there is, in fact, a serious problem with this conclusion. As pointed out by Tzortzis et al.,[33] virtually all cases of aphasia without amusia represent composers or conductors, that is, individuals with an extraordinarily high degree of musical training. Modern research on neural plasticity has revealed that the brains of professional musicians differ from those of nonmusicians in a variety of ways, including increased gray matter density in specific regions of the frontal cortex and increased corpus callosum size.[34,35] Thus generalizations about language–music relationships in aphasia cannot be drawn on the basis of case studies of professional musicians.

The purpose of the current study was to apply well-established tests from cognitive psychology to explore the relationship between linguistic and musical syntactic processing in nonmusician aphasics. As with the other study described in this article (on prosody), full details are available in a paper submitted for publication.[2] Here I give an overview of the methods and results.

Participants and Tasks

Nine Dutch-speaking aphasics and twelve age- and education-matched controls were tested on linguistic and musical tasks. Some participants had played musical instruments as a hobby, but none had been a professional musician. The aphasics were classified as Broca's type based on the Aachen Aphasia Test and on clinical interviews, and had a mean age of 60.1 years. All were at least nine months poststroke and had left hemisphere lesions with variable locations, including frontal and temporal regions, not necessarily including Broca's area. Such variability is well known from studies of Broca's aphasia[36,37] and prevented us from addressing issues of localization. We focused instead on cognitive relationships between music and language based on performance of syntactic tasks in both domains.

For language, syntactic comprehension abilities were assessed with a sentence-picture matching task, which has been much used in studies of aphasia.[38] In this task the participant hears one sentence at a time and must point to the corresponding picture on a sheet with four different pictures. Sentences varied across five levels of syntactic complexity. For example, a sentence with an intermediate level of complexity (level 3) was the passive structure: "The girl on the chair is greeted by the man" (FIG. 3). Deter-

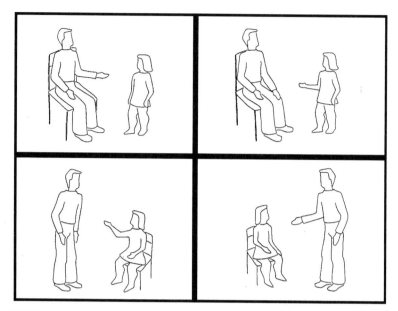

FIGURE 3. Example panel from the language syntax test (the sentence-picture matching task). In this case the participant heard the sentence: "The girl on the chair is greeted by the man." The task is to point to the corresponding picture.

mining who did what to whom in such sentences relies on syntactic information (e.g., simple word-order heuristics such as "first noun = agent" do not work).

For music, we used a harmonic priming task. Harmonic priming is a well-studied paradigm in music cognition and tests the influence of a preceding harmonic context on the processing of a target chord. Much research has shown that a target chord is processed more rapidly and accurately if it is close to (vs. distant from) the tonal center created by the prime. This indicates implicit knowledge of the harmonic conventions of tonal music and has been repeatedly demonstrated in nonmusician listeners in Western cultures.[39] We used the original two-chord version of the harmonic priming task,[40] with a single chord serving as the prime. (Prime and target were 1 s long each, separated by 50 ms). This places minimal demands on attention and memory and is thus suitable for use with aphasics. The harmonic distance between prime and target was regulated by the circle of fifths for musical keys. Harmonically close versus distant targets were 2 versus 4 steps clockwise steps away from the prime on the circle. This directly pits conventional harmonic distance against psychoacoustic similarity, since the distant target shared a common tone with the prime[41] (FIG. 4). The participants' task was to judge whether the second chord was tuned or mistuned (on 50% of the trials, it is mistuned by flattening one note in the chord). The main focus of interest, however, is in reaction times (RTs) to well-tuned targets as a function of their harmonic distance from the prime. A faster RT to close versus distant chords is evidence of harmonic priming.

Participants also completed two control experiments that tested for the ability to discriminate tuned from mistuned chords and tested auditory short-term memory.

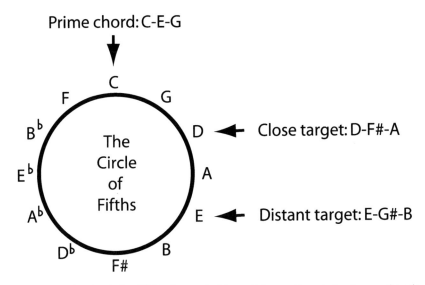

FIGURE 4. The circle of fifths for musical keys. Prime and target chords were the principal chord of each key. An example of a prime chord is shown (the C major triad), along with its close versus distant harmonic targets (the D major triad and E major triad, respectively). Note that the prime chord shares a tone with the distant chord (the note E in this case) but not with the close chord. This pits psychoacoustic similarity against conventional harmonic syntax with regard to distance along the circle of fifths.

Results

The language experiment showed that the aphasics had a linguistic syntactic comprehension deficit and thus were suitable for studying cross-domain syntactic processing. Turning to music, the control experiments revealed that aphasics and controls did not differ on basic auditory discrimination and short-term memory abilities. In contrast, the priming task revealed a significant difference between aphasics and controls. Controls showed normal harmonic priming, with faster reaction times to harmonically close versus distant well-tuned targets. Aphasics, however, failed to show a priming effect and even showed a nonsignificant trend to be faster on distant targets, suggestive of responses driven by psychoacoustic similarity rather than by harmonic knowledge. To check if aphasics would show priming if given enough time between prime and target (e.g., due to slow activation of structural information), we conducted an additional experiment in which the silent interval between prime and target chords was increased to 1 second. Aphasics still failed to show priming.

Discussion

Aphasics with syntactic comprehension problems in language also have a musical syntactic deficit; that is, they seem not to activate the implicit knowledge of harmonic relationships that Western nonmusicians normally exhibit. Importantly, this is not due to low-level acoustic deficits (e.g., difficulty discriminating tuned from mistuned chords), nor is it a generalized consequence of brain damage, since there are cases of individuals with bilateral cortical lesions who show normal harmonic priming.[42] This supports the SSIRH[27] for language and music. It also favors a "processing view" of syntactic disorders in aphasia, that is, a general problem activating stored syntactic representations (e.g., verbs together with their lexical category and thematic role information) rather than a language-specific disruption of these representations.[29] One direction for future work is to make tasks as comparable as possible across domains, for example, comparing performance on syntactic priming tasks in language[43] to harmonic priming tasks in music. When comparable tasks are used it will be particularly interesting to examine individual variation among aphasics to determine if there is a relationship between the severity of deficits in the two domains. More generally, the time is ripe for systematic work on musical syntactic processing in aphasia, especially since the results are of theoretical relevance to both music cognition and neurolinguistics.

CONCLUSION

The studies presented here provide new evidence for the relationship between linguistic prosody and musical structure, and between syntactic processing in music and language. In each case, a new approach is introduced that can be used for further work. Specifically, a good deal more can be done to compare melody in speech and music based on analyses of tonal patterns provided by prosogram representations of intonation. Similarly, much remains to be learned about the relationship between linguistic and musical syntactic processing via the study of aphasia, since many well-studied linguistic and musical tasks remain to be probed. Future work in these areas can help advance our understanding of basic principles in the human brain's processing of structured sound.

ACKNOWLEDGMENTS

I thank Meagan Curtis for comments on this paper, and Claus Heeschen for discussions of music and aphasia. This work was supported by the Neurosciences Research Foundation as part of its program on music and the brain at The Neurosciences Institute, where ADP is the Esther J. Burnham Fellow. The aphasia research was also supported by Grant 400-56-384 from the Netherlands Organization for Scientific Research (NWO) to Peter Hagoort. We are grateful to Anke Kooijman for testing the participants in the aphasia study.

[Competing interests: The author declares that he has no competing financial interests.]

REFERENCES

1. PATEL, A.D., J.R. IVERSEN & J.C. ROSENBERG. Comparing the rhythm and melody of speech and music: the case of British English and French. Submitted for publication.

2. PATEL, A.D., J.R. IVERSEN & P. HAGOORT. Impaired syntactic processing of language and music in Broca's aphasia. Submitted for publication.
3. ABRAHAM, G. 1974. The Tradition of Western Music. University of California Press. Berkeley. p. 83.
4. WENK, B.J. 1987. Just in time: on speech rhythms in music. Linguistics 25: 969–981.
5. HALL, R.A. JR. 1953/1972. Elgar and the intonation of British English. The Gramophone, June 1953: 6–7. Reprinted in Intonation: Selected Readings. D. Bolinger, Ed.: 282–285. Penguin. Harmondsworth.
6. LOW, E.L., E. GRABE & F. NOLAN. 2000. Quantitative characterizations of speech rhythm: syllable-timing in Singapore English. Lang. Speech 43: 377–401.
7. GRABE, E. & E.L. LOW. 2002. Durational variability in speech and the rhythm class hypothesis. In Laboratory Phonology 7. C. Gussenhoven & N. Warner, Eds.: 515–546. Mouton de Gruyter. Berlin.
8. RAMUS, F. 2002. Acoustic correlates of linguistic rhythm: Perspectives. In Proceedings of Speech Prosody, Aix-en-Provence. B. Bell & I. Marlien, Eds.: 115–120. Laboratoire Parole et Langage. Aix-en-Provence.
9. DAUER, R.M. 1983. Stress-timing and syllable-timing reanalyzed. J. Phonetics 11: 51–62.
10. PATEL, A.D. & J.R. DANIELE. 2003. An empirical comparison of rhythm in language and music. Cognition 87: B35–B45.
11. HURON, D. & J. OLLEN. 2003. Agogic contrast in French and English themes: further support for Patel and Daniele. Music Percept. 21: 267–271.
12. NAZZI, T., J. BERTONCINI & J. MEHLER. 1998. Language discrimination in newborns: toward an understanding of the role of rhythm. J. Exp. Psychol. Hum. Percept. Perform. 24: 756–777.
13. D'ALESSANDRO, C. & P. MERTENS. 1995. Automatic pitch contour stylization using a model of tonal perception. Computer Speech and Language 9: 257–288.
14. MERTENS, P. 2004. The prosogram: semi-automatic transcription of prosody based on a tonal perception model. In Proceedings of Speech Prosody 2004, Nara (Japan), March 23–26. B. Bel & I. Marlien, Eds.
15. HOUSE, D. 1990. Tonal Perception in Speech. Lund University Press. Lund.
16. D'ALESSANDRO, C. & M. CASTELLENGO. 1994. The pitch of short-duration vibrato tones. J. Acoust. Soc. Am. 95: 1617–1630.
17. ORAM, N. & L.L. CUDDY. 1995. Responsiveness of Western adults to pitch-distributional information in melodic sequences. Psychol. Res. 57: 103–118.
18. KRUMHANSL, C.L. 2000. Tonality induction: a statistical approach applied cross-culturally. Music Percept. 17: 461–479.
19. SAFFRAN, J.R., E.K. JOHNSON, R.N. ASLIN & E.L. NEWPORT. 1999. Statistical learning of tone sequences by human infants and adults. Cognition 70: 27–52.
20. SAFFRAN, J.R., R.N. ASLIN & E.L. NEWPORT. 1996. Statistical learning by 8-month-old infants. Science 274: 1926–1928.
21. NOLAN, F. 2003. Intonational equivalence: an experimental evaluation of pitch scales. In Proceedings of the 15th International Congress of Phonetic Sciences: 771–774. Barcelona, Spain.
22. BARLOW, H. & S. MORGENSTERN. 1983. A Dictionary of Musical Themes, revised edition. Faber and Faber. London.
23. PATEL, A.D., E. GIBSON, J. RATNER, et al. 1998. Processing syntactic relations in language and music: an event-related potential study. J. Cognit. Neurosci. 10: 717–733.
24. KOELSCH S., T.C. GUNTER, D.Y VON CRAMON, et al. 2002. Bach speaks: a cortical "language-network" serves the processing of music. Neuroimage 17: 956–966.
25. MAESS, B., S. KOELSCH, T. GUNTER, et al. 2001. Musical syntax is processed in Broca's area: an MEG study. Nat. Neurosci. 4: 540–545.
26. PERETZ, I. 1993. Auditory atonalia for melodies. Cognit. Neuropsychol. 10: 21–56.
27. PATEL, A.D. 2003. Language, music, syntax, and the brain. Nat. Neurosci. 6: 674–681.
28. KOLK, H.H. & A.D. FRIEDERICI. 1985. Strategy and impairment in sentence understanding by Broca's and Wernicke's aphasics. Cortex 21: 47–67.
29. KOLK, H.H. 1998. Disorders of syntax in aphasia: linguistic-descriptive and processing approaches. In Handbook of Neurolinguistics. B. Stemmer & H.A. Whitaker, Eds.: 249–260. Academic Press. San Diego.

30. FRANCÈS, R., F. LHERMITTE & M. VERDY. 1973. Le déficit musical des aphasiques. Rev. Int. Psychol. Appl. **22:** 117–135.
31. LURIA, A., L. TSVETKOVA & J. FUTER. 1965. Aphasia in a composer. J. Neurol. Sci. **2:** 288–292.
32. MARIN, O.S.M. & D.W. PERRY. 1999. Neurological aspects of music perception and performance. *In* The Psychology of Music, 2nd edition. D. Deutsch Ed.: 653–724. Academic Press. San Diego.
33. TZORTZIS, C., M-C. GOLDBLUM, M. DANG, *et al.* 2000. Absence of amusia and preserved naming of musical instruments in an aphasic composer. Cortex **36:** 227–242.
34. GASER, C. & G. SCHLAUG. 2003. Brain structures differ between musicians and non-musicians. J. Neurosci. **23:** 9240–9245.
35. SCHLAUG G., L. JANCKE, Y. HUANG, *et al.* 1995. Increased corpus callosum size in musicians. Neuropsychologia **33:** 1047–1055.
36. WILLMES, K. & K. POECK. 1993. To what extent can aphasic syndromes be localized? Brain **116:** 1527–1540.
37. CAPLAN, D., N. HILDEBRANDT & N. MAKRIS. 1996. Location of lesions in stroke patients with deficits in syntactic processing in sentence comprehension. Brain **119:** 933–949.
38. HAGOORT, P., M. WASSENAAR & C. BROWN. 2003. Real-time semantic compensation in patients with agrammatic comprehension: electrophysiological evidence for multiple-route plasticity. Proc. Natl. Acad. Sci. USA **100:** 4340–4345.
39. BIGAND, E., B. POULIN, B. TILLMAN, *et al.* 2003. Sensory versus cognitive components in harmonic priming. J. Exp. Psychol. Hum. Percept. Perform. **29:** 159–171.
40. BHARUCHA, J.J. & K. STOECKIG. 1986. Reaction time and musical expectancy. J. Exp. Psychol. Hum. Percept. Perform. **12:** 403–410.
41. TEKMAN, H.G. & J.J. BHARUCHA. 1998. Implicit knowledge versus psychoacoustic similarity in priming of chords. J. Exp. Psychol. Hum. Percept. Perform. **24:** 252–260.
42. TRAMO, M.J., J.J. BHARUCHA & F.E. MUSIEK. 1990. Music perception and cognition following bilateral lesions of auditory cortex. J. Cognit. Neurosci. **2:** 195–212.
43. FRIEDERICI, A.D. & K. KILBORN. 1989. Temporal constraints on language processing: syntactic priming in Broca's aphasia. J. Cogn. Neurosci. **1:** 262–272.

Musical and Linguistic Processing in Song Perception

DANIELE SCHÖN,[a] REYNA LEIGH GORDON,[a,b] AND MIREILLE BESSON[a]

[a]Institute of Cognitive Neurosciences of the Mediterranean, CNRS, Marseille, France

[b]Center for Complex Systems and Brain Sciences, Florida Atlantic University, Boca Raton, Florida 33431, USA

ABSTRACT: One approach to comparing the neural bases of language and music is through the use of song, which is a unique and ecological combination of these two cognitive domains. In song, language and music are merged into one acoustic signal with two salient dimensions. By manipulating either the linguistic or musical dimensions (or both) of song and studying their relationships, it is possible to gain important information about the neural networks underlying language and music cognition. We will present a brief review followed by recent behavioral, electrophysiological, and neuroimaging studies concerned with the functional and structural relationships of music and language. These results, together with the previous studies in the field, help understanding whether the different levels of music and language processing are independent or interactive.

KEYWORDS: singing; linguistics; music; cognition; neuroanatomy; brain; neuroimaging

INTRODUCTION

Singing, like music, language, and other human activities is a universal human behavior. Nonetheless, it has been too scarcely exploited as a subject of neurocognition research until now.[1] This is somewhat surprising since singing is particularly well suited to the study of the relationship between language and music. However, most of the studies investigating the relationship between music and language have used language and music separately.[2–5] This is somewhat problematic insofar as it is difficult to compare results issued from different tasks, different subjects, different types of analyses, and different statistical thresholds to define what is considered significant. Other experiments have been carried out using musical and linguistic tasks with the same participants,[6] or a similar task with different materials.[7–10] The comparison is easier since participants are the same and the tasks are as similar as possible. However, this approach also has some drawbacks. First, the models that are used to describe the two tasks as "similar" are often too simple and reductive. For

Address for correspondence: Daniele Schön, INCM-CNRS, 31, Chemin J. Aiguier, 13402 Marseille, France.
schon@incm.cnrs-mrs.fr

Ann. N.Y. Acad. Sci. 1060: 71–81 (2005). © 2005 New York Academy of Sciences.
doi: 10.1196/annals.1360.006

instance, the comparison of musical and linguistic syntax rarely goes beyond a general and loosely defined concept of syntax (see Patel[11] for an interesting discussion). Second, the materials sometimes sacrifice some of their ecological validity because of the need to make tasks/materials as comparable as possible.

The advantage of singing is that both linguistic and musical information are merged into one acoustic signal with two salient dimensions, allowing for a direct comparison within the same experimental material. In other words, it is possible to ask subjects to perform a linguistic task while manipulating the musical dimension, or vice-versa, to study potential interferences from one dimension on the other. The flip side of the coin is that in song, the phonological and metrical structures of language are strongly influenced by the type of melody that is used to sing. For instance, vowels sung in a very high register have the tendency to be deformed and less recognizable.[12] Moreover, the duration and accentuation of syllables will depend upon the musical structure.[13] These complications are further increased when songs are in tone languages, in which pitch concurrently contains lexical and musical information.[14,15]

In the following section, we will review several studies that have investigated song cognition and discuss the findings relevant to the comparison of language and music processing. We seek to show how recent research has dealt with the interaction or independence of language and music processing, modulation of attention, and functional mapping of brain areas involved in singing. Then, we will briefly present three experiments that we recently carried out in our laboratory on linguistic and musical processing in song perception.

RELATIONSHIP BETWEEN THE LINGUISTIC
AND MUSICAL DIMENSIONS OF SONG

Results showing independent processing of language and music in song would provide support for the idea that separate neural substrates are responsible for processing these two different types of information. By contrast, results showing interactive processing would argue in favor of common neural resources to process language and music. The early neuropsychological and imaging studies of song perception examined the linguistic and musical aspects of song separately, by comparing normal speech and vocalises (nonlinguistic, melodic singing).[16–18] More recently, researchers have compared sung and spoken language.[19–22] Behavioral and electrophysiological methods have granted the possibility of using a more fine-grained approach, by studying the musical and linguistic dimensions of song as they occur in sung language.[23–28] Just as attentional processes generally exert a powerful influence over the functional organization of virtually all cognitive behaviors, attention has played a key role in the latter group of studies.

A series of experiments has used song to study how harmony in music interacts with phonology and semantics in language.[25,28] In both experiments, musician and nonmusicians listened to sung chord progressions that ended on the expected (tonic) chord, or the less expected but still congruous (subdominant) chord. However, participants were asked to ignore the harmonic dimension and focus their attention on the linguistic one (phonology or semantics). In the first study, the behavioral data showed that the less-expected harmonic stimulus did interfere with performance on the phoneme-monitoring task, in which subjects focused on nonword syllables sung by four voices.[25] The follow-up study's results also showed that the less-expected musical cadence interfered with semantic priming, measured by performance in a lexical decision task.[28] Measured implicitly, these interactions between language and music in song not only provide evidence for shared cognitive resources but also demonstrate how musical and linguistic priming occur in song.

Interestingly, another series of experiments on harmony and semantics in song perception yielded findings that point to independence of language and music processing.[24,26] Listeners showed distinct event-related brain potential (ERP) components for semantic (N400) and harmonic (P300) violations when simultaneously attending to the linguistic and musical dimensions of very ecological opera excerpts.[24] Indeed, by successfully fitting the data to an additive model of the two components, the authors concluded that the perception of lyrics and melodies was independent. Further support for the independence of semantic and harmonic processing comes from another experiment using the same stimuli, in which participants showed similar levels of performance when they simultaneously attended the lyrics and melodies, compared to when attending only one dimension.[26]

Although no direct comparison can be made, it seems likely that crucial differences in stimuli and experimental designs would account for these differing findings. While a subtle musical change, which was acceptable but not highly expected in the harmonic context, was introduced in the former set of studies, the latter used harmonically and semantically incongruous endings that might have captured the listener's attention more dramatically to the point of causing independent effects.

The above-mentioned research seems to substantiate the idea that the dynamic relationship between the multiple levels of structural and functional organization of language and music in song hinges on the allocation of attentional resources to different dimensions of song. In addition, findings obtained with the dichotic listening method show task- and attention-related effects when the musician's attention was modulated between phonemes and melodies in sung syllables/digits.[23,29] Finally, results revealed typical lateralization patterns of left-ear (right hemisphere) advantage for the melodic task and right-ear (left hemisphere) advantage for the phonological task.

Another approach to investigating music and language specificity through song is to consider the role of human voice specificity. Following up on some recent research using brain imaging to reveal a voice-selective area in the supratemporal sulcus (STS; see Belin[30] for a review), Levy, Granot, and Bentin[31,32] used sung tones to distinguish ERPs occurring in response to musical pitches produced by voices and various instruments. In an oddball paradigm, listeners were first instructed to respond when they heard a *piano* tone presented among a series of sustained pitches played by instruments or sung by different voices on a neutral vowel. Results showed that a positive component, peaking at 320 ms (P320), called a voice-specific response (VSR), was elicited only by sung tones, even when subjects were attending a piano tone.[31] However, in follow-up experiments, the VSR occurred only when participants were attending to timbre but not when they were attending to duration.[32] Again, the use of singing builds a nice bridge between language and music, in addition to highlighting how task and attention to different acoustic parameters serve to allocate neural resources to different types of auditory stimuli.

CEREBRAL STRUCTURES INVOLVED IN SONG PROCESSING

One of the main goals of studying the nature of the relationship between the linguistic and musical dimensions of song is to determine whether music and language are processed by separate or integrated cerebral structures. Recent functional magnetic resonance imaging (fMRI) and positron emission tomography (PET) studies of singing production have sought to establish the neural bases of language and music in song. In particular, the spatial locations correlating with the linguistic and musical dimensions of song do seem to follow the classic patterns of left dominance for language and right dominance for music, heretofore described in the literature on language and music processing.[33,34]

Recent methodological advances with transcranial magnetic stimulation (TMS), which allows experimenters to collect pseudoneuropsychological data by provoking temporary, reversible lesions of localized cortical areas in healthy subjects, have brought forth evidence for right hemisphere dominance in singing.[20,35,36] Speech is inhibited by applying TMS to certain areas of the left hemisphere; conversely, when TMS is performed on the right cortex, some participants are unable to sing melodies. Speech and nonlinguistic melodic singing were also compared by Riecker and colleagues,[18] using fMRI. They showed opposite hemispheric activations with divergent networks for speaking (left lateralization) and singing (right lateralization). Thus, it seems that compared to speech, the production of sung vocalises uses more right hemispheric resources. However, their design did not include a more ecological condition of sung language, and it is therefore difficult to assess whether the right hemispheric dominance was due to the vocal motor aspects or to the musical aspects of singing.

Sung language was also studied by Jeffries and colleagues[21] using PET. Nonmusicians performed two different tasks: singing an "overlearned" song and speaking the text of the same song. As predicted, results also showed hemispheric lateralization (right dominance for singing, left for speaking). Thus, the authors were able to speculate that a right-lateralized network of brain area, typically involved in music perception and cognition, is also crucial for producing language in song. However, since no "vocalises" were included in the protocol, it is impossible to determine whether the right-hemispheric activations reflect music processing or linguistic production specific to song, or a combination of the two. Again, a direct comparison would be necessary to clarify the status of sung language.

Brown and colleagues[37] took an alternative approach to studying nonlinguistic singing, by focusing on the different aspects of the structural organization of music. Indeed, they were able to use PET to differentiate brain activity (in the planum polare, BA38), associated with complex musical tasks such as harmonization and melody repetition, from activity associated with a simpler musical task (e.g., monotonic vocalization). While this elegant study did show the brain's sensitivity to various musical elements of singing, the phonological and semantic conditions of singing also have to be investigated to determine if that same musical vocalization network would also be mobilized to sing songs with words.

In summary, song perception and production do indeed provide a powerful model for comparing the neurocognition of language and music in the brain. While previous studies of song demonstrate some lateralization evidence for left-dominant language processing and right-dominant musical processing, on the whole it remains to be determined whether these two dimensions of song are interactive or independent. It is essential to keep in mind that task demands and attention modulations exert an influence on this relationship. Furthermore, results also differ as a function of which levels of processing are being examined. Finally, song perception has not been studied outside of the context of a singing production task with either the fMRI or PET methods, so the spatial localization of the perception of singing also remains to be explored. Most importantly, no one study has yet directly compared spoken language, sung language, and vocalises within the same design and with the same participants.

EXPERIMENTS ON THE NEUROCOGNITION OF SINGING

In this last part, we will present recent data of three experiments that we recently carried out to investigate the relationship of the linguistic and musical dimensions in song perception. In the first study, we directly compared spoken language, sung language, and vocalises. The main question was whether we would confirm a previous finding of different and lateralized networks for language and music. Nonmusician participants listened to pairs of spoken words (trisyllabic French words, spoken), sung words (trisyllabic words sung on three-note melodies), and vocalises (the syllable "vi" sung on three-note melodies). They had to judge whether the two stimuli of a pair were the same or different. Note that, in order to make the task more demanding, the first syllable or the first note of the two stimuli in a given pair was always the same. We also used pairs of noises as a control condition. fMRI results for each experimental condition contrasted with the control condition (noise) showed a very similar bilateral pattern of activation, mostly in temporal regions (see FIG. 1). Thus, results did not reveal a clearly lateralized pattern of activation. Interestingly, careful examination of the activity of a single voxel in the right middle temporal gyrus shows that its level of activity is modulated by the experimental condition (FIG. 2). Indeed, this is highest when participants are listening to sung words (top curve), and decreases for vocalises (middle curve) and spoken words (low curve). Note that this does not mean that this voxel is "specific" to song perception, but rather that it is involved in all three conditions, with maximal amplitude of the blood oxygen level–dependent (BOLD) response for sung words. When we directly compared sung and spoken words, for instance, many voxels are significantly more involved in sung than spoken word processing. While only performing such a type of comparison may have led us to believe that these voxels are specifc to song perception, the use of a control condition clearly allowed us to show that this is not the case. Nonetheless, if we do not overinterpret differences, it is interesting to make direct comparisons of the experimental conditions. The comparison of sung words to vocalises gives a pattern of activation lateralized to the left, mostly in regions well known to be involved in lexical semantic processing. The comparison of sung words to spoken words gives a pattern of activation more lateralized to the right, mostly in regions well known to be involved in pitch processing, pitch memory, and voice processing. Interestingly, the largest differences between sung and spoken words are found in the anterior portion of the STS, a region that seems to be dedicated to voice processing and related areas.[38] What may be singing, both from a perceptual and emotional point of view, is a sort of supervoice!

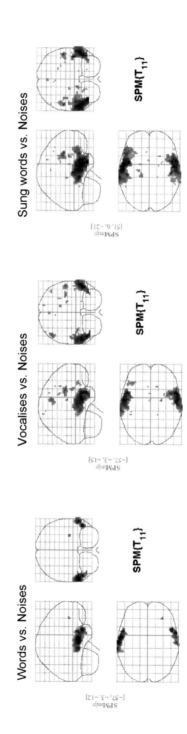

FIGURE 1. Each experimental condition is contrasted with the control condition. Activations are projected onto a transparent brain in the three different planes.

Sung words vs noise

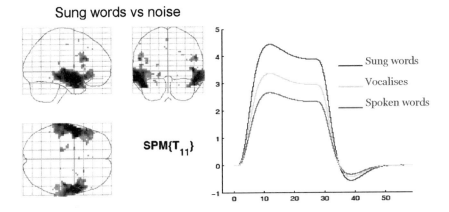

FIGURE 2. Adjusted amplitude of the BOLD response of a single voxel in the right middle temporal gyrus.

In summary, we found a strong overlap of the regions involved in language, music, and song processing. Moreover, the lateralization that was predicted on the basis of previous results was rather small. It only appears as a difference in the degree of activation when sung words are compared to vocalises (especially in linguistic areas) and to spoken words (especially in musical areas).

While these first results are interesting, they did not shed light on the relationship between musical and linguistic dimensions in song perception. To determine whether these two dimensions of songs are treated independently or interactively, we conducted an ERP study.

Nonmusician participants listened to pairs of sung words and, in different blocks of trials, had to focus their attention on the words to decide whether they were same or different, or on the melody to decide whether the notes were same or different. Therefore, depending upon the instructions given to the subjects, the linguistic or the musical dimension was relevant to the task at hand. For instance, if subjects have to pay attention to language, language is the relevant dimension and music the irrelevant one. We used four experimental conditions (both dimensions same; music same/language different; music different/language same; both dimensions different).

Behavioral data showed small and similar error rates in both attentional conditions. By contrast, reaction times were longer in the musical task. Moreover, reaction times were longer when the irrelevant dimension was incongruent with the relevant dimension, and this effect was stronger when participants paid attention to music. In other words, we found some interference effects of the irrelevant dimension on the relevant one, and language produced more interference on music than vice-versa. ERP results also point to an effect of the irrelevant dimension on the relevant one, in both attentional situations. As expected from the literature and colleagues,[39,40] when participants focused attention on linguistic information, the N400 to the second word of the pair was larger for other than the same words (FIG. 3, dotted line). Most interestingly, a change in the irrelevant dimension (different melody/same word) elicits

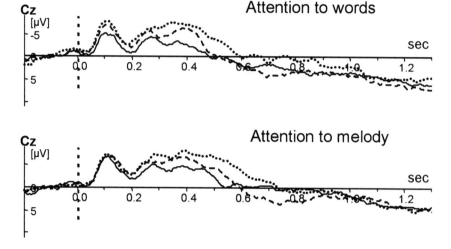

FIGURE 3. *Top*: ERPs when participants are focusing their attention on linguistic information. *Bottom*: ERPs when participants are focusing their attention on musical information. *Solid line* = both dimensions same; *dotted line* = music same/language different; *dashed line* = music different/language same.

an early negative effect, followed by a positive-going component (FIG. 3, dashed line). Thus, nonmusician participants seem to process the music dimension even when irrelevant for the task at hand.

When participants paid attention to the melody, a change in the relevant dimension (different melody) is, as expected from the literature and colleagues[41] associated with a late positive-going component (FIG. 3, dashed line). Interestingly again, a change in the irrelevant dimension (different word/same melody) also elicits a more negative-going component than in the control condition (all same). Thus, linguistic information seems to be processed (as reflected by what we interpret as an N400-like component) even when irrelevant for the task at hand (FIG. 3, dotted line).

To summarize, results show that, in both the linguistic and musical tasks, processing of the relevant dimension seems influenced by the irrelevant dimension. In other words phonological/lexical processing and pitch processing cannot be processed independently.

In order to determine which cerebral structures are involved in these interactions, we conducted the same study using fMRI. Specifically, the aim was to test the following three anatomical hypotheses. First, would lateralization differ depending upon the orientation of attention on the linguistic or melodic dimensions? Second, when language is the irrelevant dimension, would an effect be found in typical linguistic areas? Conversely, when music is the irrelevant dimension, would an effect be found in musical areas? Finally, and perhaps most interestingly, would language, when irrelevant, influence the degree of activation in musical areas, and would music, when irrelevant, influence activation in linguistic areas?

Regarding the first hypothesis, results show that when paying attention to language, linguistic processing is rather bilateral, except for a left frontal opercular activation. By contrast, when paying attention to music, melodic processing is rather right lateralized to middle and superior temporal regions. Thus, linguistic processing in song seems less lateralized than typically reported for spoken words. Regarding the second hypothesis, language as an irrelevant dimension induces strong and bilateral activation, not restricted to the typical linguistic areas. Music as an irrelevant dimension induces more activation in the right hemisphere. Finally, regarding the third hypothesis, when language is irrelevant and music relevant, overlaps of activation are found over bilateral temporal and prefrontal opercular regions, and they are larger in the right hemisphere. When music is irrelevant and language relevant, there is an overlap of activations over bilateral temporal regions. Therefore, it seems that both language and music, when irrelevant, influence the degree of activation in areas that are typically activated by the relevant dimensions.

CONCLUSION

Using different and complementary brain imaging methods provided interesting insights into the neurocognition of singing. Overall, the results obtained with fMRI and ERPs were compatible, in that they showed that the linguistic and musical dimensions of songs are processed by similar, overlapping brain areas. Moreover, both the time course of linguistic and musical processing, and the degree of activation in areas typically involved in the processing of the relevant dimension, are influenced by the irrelevant dimension. Thus, taken together, these findings support the hypothesis that the linguistic and melodic components of songs are processed in interaction. While they support and extend the results, recently reported by Bigand and colleagues[25] and Poulin-Charronnat and colleagues,[28] they stand in contrast to those previously reported by Besson *et al.*[24] and Bonnel *et al.*[26] However, as mentioned above, both the task at hand and the stimuli play a crucial role in the ongoing perceptual and cognitive processes involved in human information processing. The materials used by Besson *et al.*[24] were ecological and complex (excerpts from opera sung a cappella), and participants were only asked to listen to these excerpts. Because of the methodological constraints linked with fMRI data acquisition, and since it was very important to use the same task for the fMRI and ERPs experiments, here we used very simple materials (pairs of stimuli), which may only reflect part of the high-level processes involved in song perception. Using words as stimuli may, for instance, have limited linguistic processing for the phonological and lexical aspects. Similarly, by using three notes, we limited musical processing to very simple melodic processing. Therefore, methodological advances are strongly needed to be able to use more ecological materials with fMRI. Finally, musical expertise has been shown to exert a strong influence on brain structural and functional organization.[42] Differences between musicians and nonmusicians are even more likely to emerge when participants are asked to focus attention on one dimension of songs. In future experiments, it will be important to test musicians to further track the intriguing nature of the relationship between two of the most human cognitive abilities, music and language, when they are naturally combined in songs.

ACKNOWLEDGMENTS

The series of experiments reported in this chapter were conducted thanks to the support of the Human Frontier Science Program (HSFP 2/15/06RGP0053) and by a grant from the Ministère de la Recherche (Programme Cognitique, ACI "Imagerie cérébrale") to Mireille Besson. We would like to thank Monique Chiambretto and Vanina Luigi for methodological assistance; the two singers, Véronique Kirkyacharian and Serge Charron, for their precious voices; as well as all the participants in the experiments.

REFERENCES

1. PERETZ, I. *et al.* 2004. Singing in the brain: insights from cognitive neuropsychology. Music Percept. **21:** 373–390.
2. MAESS, B. *et al.* 2001. Musical syntax is processed in Broca's area: an MEG study. Nat. Neurosci. **4:** 540–545.
3. KOELSCH, S. *et al.* 2002. Bach speaks: a cortical "language-network" serves the processing of music. Neuroimage **17:** 956–966.
4. LEVITIN, D.J. & V. MENON. 2003. Musical structure is processed in "language" areas of the brain: a possible role for Brodmann Area 47 in temporal coherence. Neuroimage **20:** 2142–2152.
5. KNOSCHE, T.R. *et al.* 2005. Perception of phrase structure in music. Hum. Brain Mapping **24:** 259–273.
6. BESSON, M. & F. MACAR. 1987. An event-related potential analysis of incongruity in music and other non-linguistic contexts. Psychophysiology **24:** 14–25.
7. SCHÖN, D., C. MAGNE & M. BESSON. 2004. The music of speech: music training facilitates pitch processing in both music and language. Psychophysiology **41:** 341–349.
8. MAGNE, C., D. SCHÖN & M. BESSON. Musician children detect pitch violations in both music and language better than non-musician children: behavioral and electrophysiological approaches. J. Cognit. Neurosci. In press.
9. PATEL, A.D. *et al.* 1998. Processing syntactic relations in language and music: an event-related potential study. J. Cognit. Neurosci. **10:** 717–733.
10. PATEL, A.D. *et al.* 1998. Processing prosodic and musical patterns: a neuropsychological investigation. Brain Lang. **61:** 123–144.
11. PATEL, A.D. 2003. Language, music, syntax and the brain. Nat. Neurosci. **6:** 674–681.
12. SCOTTO DI CARLO, N. 1994. L'Intelligibilité de la voix chantée. Médecine des Arts **10:** 2–15.
13. PALMER, C. & M. KELLY. 1992. Linguistic prosody and musical meter in song. J. Mem. Language **31:** 525–542.
14. CHEN-HAFTECK, L. 1999. Discussing text-melody relationship in children's song-learning and singing: a Cantonese-speaking perspective. Psychol. Music **27:** 55–70.
15. WONG, P.C. M. & R.L. DIEHL. 2002. How can the lyrics of a song in a tone language be understood? Psychol. Music **30:** 202–209.
16. GORDON, H.W. & J.E. BOGEN. 1974. Hemispheric lateralization of singing after intracarotid sodium amylobarbitone. J. Neurol. Neurosurg. Psychiatry **37:** 727–738.
17. RYDING, E., B. BRADVIK & D.H. INGVAR. 1987. Changes of regional cerebral blood flow measured simultaneously in the right and left hemisphere during automatic speech and humming. Brain. **110** (Pt 5): 1345–1358.
18. RIECKER, A. *et al.* 2000. Opposite hemispheric lateralization effects during speaking and singing at motor cortex, insula and cerebellum. Neuroreport **11:** 1997–2000.
19. CADALBERT, A. *et al.* 1994. Singing with and without words: hemispheric asymmetries in motor control. J. Clin. Exp. Neuropsychol. **16:** 664–670.
20. EPSTEIN, C.M. *et al.* 1999. Localization and characterization of speech arrest during transcranial magnetic stimulation. Clin. Neurophysiol. **110:** 1073–1079.
21. JEFFRIES, K.J., J.B. FRITZ & A.R. BRAUN. 2003. Words in melody: an H(2)15O PET study of brain activation during singing and speaking. Neuroreport **14:** 749–754.

22. STAGER, S.V., K.J. JEFFRIES & A.R. BRAUN. 2003. Common features of fluency-evoking conditions studied in stuttering subjects and controls: an H(2)15O PET study. J. Fluency Disord. **28:** 319–336.
23. BARTHOLOMEUS, B. 1974. Effects of task requirements on ear superiority for sung speech. Cortex **10:** 215–223.
24. BESSON, M. *et al.* 1998. Singing in the brain: independence of lyrics and tunes. Psychol. Sci. **9:** 494–498.
25. BIGAND, E. *et al.* 2001. The effect of harmonic context on phoneme monitoring in vocal music. Cognition **81:** 11–20.
26. BONNEL, A.M. *et al.* 2001. Divided attention between lyrics and tunes of operatic songs: evidence for independent processing. Percept. Psychophys. **63:** 1201–1213.
27. KOLINSKY, R. *et al.* Perceptual interactions between lyrics and tunes: vowels sing but consonants speak. In preparation.
28. POULIN-CHARRONNAT, B. *et al.* 2005. Musical structure modulates semantic priming in vocal music. Cognition **94:** 67–78.
29. GOODGLASS, H. & M. CALDERON. 1977. Parallel processing of verbal and musical stimuli in right and left hemispheres. Neuropsychologia **15:** 397–407.
30. BELIN, P., S. FECTEAU & C. BEDARD. 2004. Thinking the voice: neural correlates of voice perception. Trends Cognit. Sci. **8:** 129–135.
31. LEVY, D.A., R. GRANOT & S. BENTIN. 2001. Processing specificity for human voice stimuli: electrophysiological evidence. Neuroreport **12:** 2653–2657.
32. LEVY, D.A., R. GRANOT & S. BENTIN. 2003. Neural sensitivity to human voices: ERP evidence of task and attentional influences. Psychophysiology **40:** 291–305.
33. ZATORRE, R.J., P. BELIN & V.B. PENHUNE. 2002. Structure and function of auditory cortex: music and speech. Trends Cogn. Sci. **6:** 37–46.
34. TERVANIEMI, M. *et al.* 2000. Lateralized automatic auditory processing of phonetic versus musical information: a PET study. Hum. Brain Mapp. **10:** 74–79.
35. STEWART, L. *et al.* 2001. Transcranial magnetic stimulation produces speech arrest but not song arrest. Ann. N. Y. Acad. Sci. **930:** 433–435.
36. LO, Y.L. & S. FOOK-CHONG. 2004. Ipsilateral and contralateral motor inhibitory control in musical and vocalization tasks. Exp. Brain Res. **159:** 258–262.
37. BROWN, S. *et al.* 2004. The song system of the human brain. Brain Res. Cogn. Brain Res. **20:** 363–375.
38. BELIN, P. *et al.* 2000. Voice-selective areas in human auditory cortex. Nature **403:** 309–312.
39. BENTIN, S., G. MCCARTHY & C.C. WOOD. 1985. Event-related potentials, lexical decision and semantic priming. Electroencephalogr. Clin. Neurophysiol. **60:** 343–355.
40. HOLCOMB, P.J. & H.J. NEVILLE. 1990. Auditory and visual semantic priming in lexical decision: a comparison using event-related brain potentials. Lang. Cogn. Processes **5:** 281–312.
41. BESSON, M. & F. FAITA. 1995. An event-related potential study of musical expectancy: comparison of musicians with non-musicians. J. Exp. Psychol. Hum. Percept. Perform. **21:** 1278–1296.
42. PANTEV C. *et al.* 2003. Music learning-induced cortical plasticity. Ann. N. Y. Acad. Sci. **999:** 438–450.

Neural Correlates of Rapid Spectrotemporal Processing in Musicians and Nonmusicians

N. GAAB,[a,b] P. TALLAL,[c] H. KIM,[a] K. LAKSHMINARAYANAN,[c]
J. J. ARCHIE,[a] G. H. GLOVER,[b] AND J. D. E. GABRIELI[a]

[a]Department of Psychology, Stanford University, Stanford, California, USA

[b]Department of Radiology, Stanford University, Stanford, California, USA

[c]Center for Molecular and Behavioral Neuroscience,
Rutgers University, Newark, New Jersey, USA

ABSTRACT: Our results suggest that musical training alters the functional anatomy of rapid spectrotemporal processing, resulting in improved behavioral performance along with a more efficient functional network primarily involving traditional language regions. This finding may have important implications for improving language/reading skills, especially in children struggling with dyslexia.

KEYWORDS: music; language; fMRI; rapid temporal processing

INTRODUCTION

Language and reading impairments have often been characterized by phonological deficits, particularly in the ability to process rapidly presented sounds or rapid acoustic changes within sounds.[1] These findings have been accompanied by fMRI studies that show involvement mainly of left-hemispheric language and auditory areas in discrimination of speech and nonspeech stimuli characterized by rapid temporal acoustic cues.[2–4] Furthermore, musical training has been shown to improve verbal memory in adults[5] and in children,[6] as well as reading ability[7–9] and phonological segmentation.[10] Nevertheless, it remains unclear why musical training leads to improved language and reading skills.

Musical training has been shown to improve various behavioral aspects of auditory processing such as rhythm, pitch, or melody processing, and to lead to alterations of the functional brain anatomy used while performing various auditory tasks.[11] Given the link between spectrotemporal processing and language abilities, as well as between musical training and language/reading skills, we hypothesized that musical training may specifically enhance the ability to process rapid spectrotemporal acoustic cues and furthermore to alter its underlying functional anatomy. This in turn may have an influence on the acoustic/phonetic analysis skills essential to language and reading.

Address for correspondence: Nadine Gaab, Ph.D., Department of Brain and Cognitive Sciences, Massachusetts Institute of Technology, 77 Massachusetts Avenue, Room NE46-4037C, Cambridge, MA 02139. Voice: 617-253-3415.
gaab@mit.edu

Ann. N.Y. Acad. Sci. 1060: 82–88 (2005). © 2005 New York Academy of Sciences.
doi: 10.1196/annals.1360.040

METHODS

Participants

Twenty healthy right-handed nonmusicians (NM) and 20 musicians (M) (20 women and 20 men, age range: 18–33) were recruited for this study.

Stimuli and Experimental Task

As depicted in FIGURE 1, subjects listened to three-tone sequences comprising two complex tones and were asked to reproduce the order of the tones manually. Each tone was 75 ms long and had a fundamental frequency of either 100 Hz or 300 Hz. The interstimulus interval (ISI) between the three tones varied in four steps (5 ms, 20 ms, 50 ms, and 300 ms). The sequences required either sequencing (S+: e.g., 100-300-100 Hz) or nonsequencing (S–: e.g., 100-100-100 Hz). Performance scores (% correct) and reaction times (RT) were obtained for each condition and ISI.

Imaging Procedure

The fMRI data (see FIG. 1) was collected with a 3.0T GE Signa scanner using a spiral in/out T2* pulse sequence with 30 slices and a sparse temporal sampling design.[12]

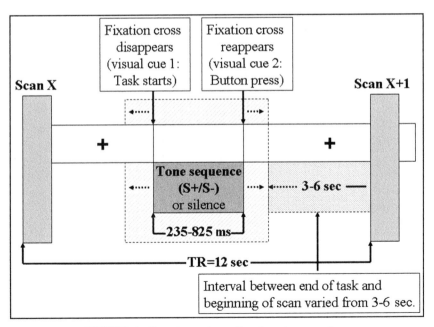

FIGURE 1. Experimental paradigm/imaging procedure.

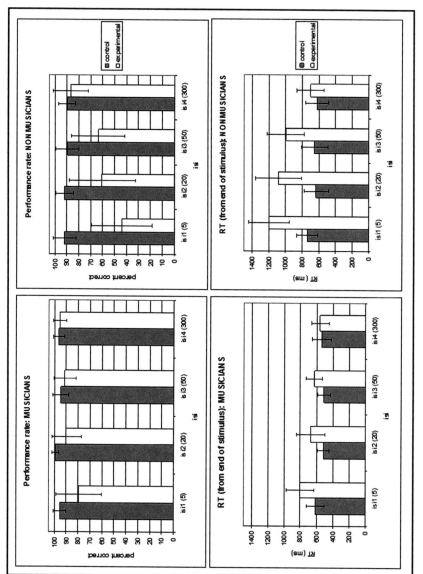

FIGURE 2. Behavioral results.

TABLE 1. Correlations in musicians' group ($P < .01**$)

	Years played	Hours played last 5 years
Overall % correct (S+)	0.50**	0.45**
Overall % correct (S−)	0.41**	0.27
ISI: 5 ms (S+), % correct	0.52**	0.48**
ISI: 5 ms (S−), % correct	0.18	0.19
ISI: 300 ms (S+), % correct	0.24	0.24

fMRI Data Analysis

Following preprocessing, statistical analyses were performed using a general linear model (SPM2). A finite-impulse-response model was specified for each subject and each ISI was modeled as a separate condition. One overall (S+ versus S−) and one parametric contrast was specified for each subject. For the parametric contrast, the natural logarithm of each ISI was scaled such that four values with a mean of zero were obtained. These values were then used to test a parametric relationship (signal increase with respect to decreased ISI), and random-effect models (one- and two-sample *t* tests) were specified.

RESULTS

Behavioral Results

M showed a significantly better performance for all four ISI in S+ and for the ISIs of 20 ms and 300 ms in S−, as well as faster RTs for S+ and S− (FIG. 2). Both groups showed enhanced performance for longer versus shorter ISIs in S+. Within M, the number of years an instrument was played significantly correlated with overall performance in S+ and S− and performance in ISI 5 ms in S+. Hours played in the last 5 years significantly correlated with overall and ISI 5 ms performance in S+. No significant correlations were found for performance in ISI 5 ms in S− or ISI 300 ms in S+ (TABLE 1).

IMAGING RESULTS

The Sequencing Effect

Within the NM group, significant activations for S+ versus S− were observed in bilateral superior temporal gyrus, inferior/middle frontal gyrus, inferior parietal regions, pre/postcentral and central gyrus, the cuneus, right lingual gyrus and the anterior cingulate (see FIG. 3). Within the M group, this contrast revealed bilateral activation of the superior temporal gyrus and the left postcentral gyrus (all $P < .05$, corrected).

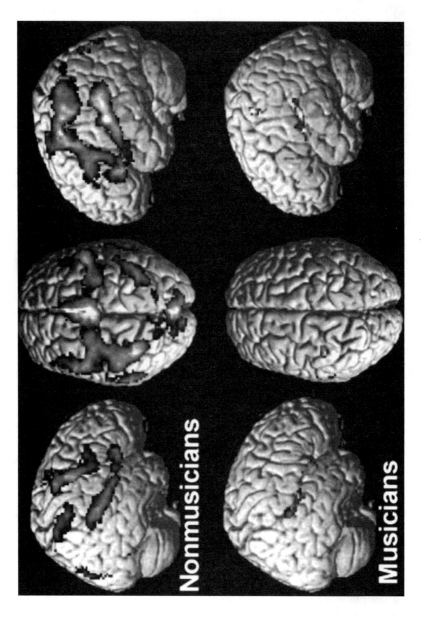

FIGURE 3. Imaging results: sequencing effect. (*See online version for color figure.*)

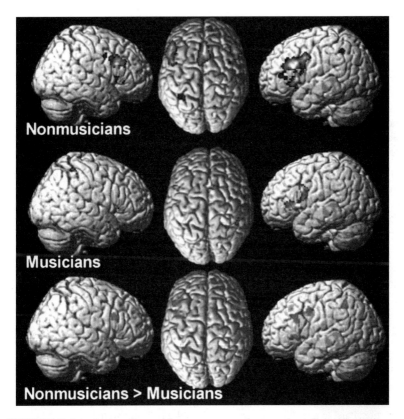

FIGURE 4. Imaging results: rapid spectrotemporal processing effect. (*See online version for color figure.*)

The Rapid Temporal Sequencing Effect

As shown in FIGURE 4, among NM, this analysis revealed activation of bilateral inferior/middle/medial frontal gyrus, the anterior cingulate, and a left inferior parietal region ($P < .001$). Among M, this analysis revealed activation of bilateral inferior frontal regions, the anterior cingulate, and a right inferior parietal region (all $P < .001$). The contrast NM > M revealed increased activation of left inferior frontal gyrus, bilateral middle frontal and left medial frontal gyrus, the anterior cingulate and a left inferior parietal region ($P < .005$).

DISCUSSION

Overall, this study suggests that musical training not only improves nonverbal rapid spectrotemporal processing, but also changes the neural network involved in rapid spectrotemporal processing so that it overlaps primarily with brain areas traditionally associated with language processing (e.g., Broca's region).

Results from previous studies indicate that acoustical training aimed to improve rapid auditory processing ability is beneficial to children struggling with language and reading.[1] Our findings further show that musical experience can improve the processing of auditory stimuli that require rapid spectrotemporal processing and therefore might enhance the acoustic/phonetic skills essential to language/reading as has been suggested by behavioral studies.[5-10]

Further studies should focus on the exact neural mechanisms underlying the relationship between nonverbal and verbal rapid spectrotemporal processing, its brain correlates (especially the role of the inferior frontal gyrus), and the potential role musical training may play in improving language and literacy skills.

[Competing interest: We have potential conflicts of interest. Paula Tallal is a cofounder and director of Scientific Learning Corporation. She is also a consultant for Posit Science Corporation and Mind Streams Corporation, and serves on their scientific advisory boards.]

REFERENCES

1. TALLAL, P. 2004. Improving language and literacy skills is a matter of time. Nat. Rev. Neurosci. **5:** 721–728.
2. ZATORRE, R.J. & P. BELIN. 2001. Spectral and temporal processing in human auditory cortex. Cereb. Cortex **11:** 946–953.
3. JOANISSE, M.F. & J.S. GATI. 2003. Overlapping neural regions for processing rapid temporal cues in speech and nonspeech signals. Neuroimage **19:** 64–79.
4. ZAEHLE, T. *et al.* 2004. Evidence for rapid auditory perception as the foundation of speech processing: a sparse temporal sampling study. Eur. J. Neurosci. **20:** 2447–2456.
5. CHAN, A.S. *et al.* 1998. Music training improves verbal memory. Nature **396:** 128.
6. HO, Y.C. *et al.* 2003. Music training improves verbal but not visual memory: cross-sectional and longitudinal explorations in children. Neuropsychology **17:** 439–450.
7. BARWICK, J. *et al.* 1989. Relations between reading and musical abilities. Br. J. Educ. Psychol. **59:** 253–257.
8. LAMB, S.J. & A.H. GREGORY. 1993. The relationship between music and reading in beginning readers. Educ. Psychol. **13:** 19–27.
9. DOUGLAS, S. & P. WILLATTS. 1994. The relationship between musical ability and literacy skills. J. Res. Read. **17:** 99–107.
10. OVERY, K. 2003. Dyslexia and music. From timing deficits to musical intervention. Ann. N. Y. Acad. Sci. **999:** 497–505.
11. PERETZ, I. & R.J. ZATORRE. 2005. Brain organization for music processing. Annu. Rev. Psychol. **56:** 89–114.
12. GAAB, N. *et al.* 2003. Functional anatomy of pitch memory-an fMRI study with sparse temporal sampling. Neuroimage **19:** 1417–1426.

Spectral and Temporal Processing in the Human Auditory Cortex—Revisited

MARC SCHÖNWIESNER,[a,b] RUDOLF RÜBSAMEN,[a] AND
D. YVES VON CRAMON[b]

[a]Institute of Zoology, Faculty of Biology II, University of Leipzig, Germany

[b]Max Planck Institute of Human Cognitive and Brain Sciences, Leipzig, Germany

ABSTRACT: We use novel noise-like sound stimuli to identify cortical areas in which the functional magnetic resonance signal covaries with spectral and temporal acoustic complexity. The results support a model of hemispheric functional asymmetry for fine-grained spectral and fast temporal processing.

KEYWORDS: hemispheric dominance; lateralization; fMRI

INTRODUCTION

Here we investigate the acoustic basis of the hemispheric asymmetry for the processing of speech and music. Experiments on this question ideally involve stimuli that are perceptually unrelated to speech and music, but contain acoustic characteristics of both. We introduce a new class of noise-like sound stimuli with no resemblance to speech or music that permits independent parametric variation of spectral and temporal acoustic complexity. We seek cortical areas in which the functional magnetic resonance (fMRI) signal covaries with the number of simultaneous spectral components (spectral complexity) or the temporal modulation rate (temporal complexity) of the stimuli. Responses from the left and right Heschl gyrus covaried with the spectral parameter, and responses from the left superior temporal gyrus (STG) covaried with the temporal parameter. The equivalent region on the right STG responded exclusively to the spectral parameter. These findings indicate that hemispheric specialization depends on acoustic rather than on semantic stimulus characteristics.

MATERIAL AND METHODS

The experiment comprised 10 stimulus conditions in a parametric design with five levels of spectral and temporal complexity. The stimuli differed in temporal

Address for correspondence: Marc Schönwiesner, Neuroscience Unit, Institute of Zoology, Faculty of Biology II, University of Leipzig, Talstraße 33, 04103 Leipzig, Germany. Voice: +49-341-9736-745; fax: +49-341-9736-848.

marcs@uni-leipzig.de

Ann. N.Y. Acad. Sci. 1060: 89–92 (2005). © 2005 New York Academy of Sciences.
doi: 10.1196/annals.1360.051

modulation rate (temporal complexity) and number of independently modulated spectral components (spectral complexity), but not in bandwidth and energy.

The five levels of parametric variation of spectral complexity were 4, 6, 8, 12, and 16 independently modulated components. The five levels of temporal modulation rate were 5, 8, 14, 20, and 30 Hz. All stimuli in the spectral variation condition had a fixed temporal modulation rate of 3 Hz and all stimuli in the temporal variation condition had a fixed number of three spectral components. The stimulus parameters were chosen to yield an approximately linear increase of perceived acoustic complexity.

We acquired 284 functional volumes from each of the 19 normal subjects using sparse imaging, while the stimulus conditions and a silent condition were presented in random order. Functional images were preprocessed and subjected to two independent fixed-effects group analyses for the spectral and the temporal factors using a general linear model. Second-level random effects analyses were implemented by extracting signal changes from regions of interest on the superior temporal plane.

RESULTS

The analysis of covariation for the spectral parameter revealed two regions of significant response covariation, one on the left and one on the right superior temporal plane (FIG. 1A). The activated cortical volume was mainly confined to Heschl's gyrus (HG), but also included anterolateral portions of the planum temporale and, on the right side, a portion of the superior temporal gyrus just lateral from HG. Analysis of covariance for the temporal parameter revealed three regions, the most prominent one located on the left superior temporal gyrus, lateral from HG, and lateral and slightly anterior from the spectral focus. In rhesus monkeys, this region would correspond to nonprimary anterolateral belt cortex.[1]

To visualize the precise relationship between parameter value and BOLD response in the ROIs, we extracted the mean percentage signal change evoked by the different stimuli in the primary auditory cortices and superior temporal gyri (FIG. 1B). The left and right HG showed essentially the same responses, a roughly linear increase in BOLD response with increasing spectral complexity, and no discernible effect of temporal modulation rate. The right superior temporal gyrus also responded with linearly increasing activation with increasing spectral complexity, but not with temporal complexity. The activation in left anterolateral belt showed a significant positive correlation with the temporal modulation rate. Spectrally complex stimuli appeared to activate the left anterolateral belt slightly as well, but only in a constant manner. The signal change did not follow the increasing spectral modulation beyond parameter values of two.

To check the consistency of activations across subjects, average effect sizes were extracted from the left and right ROIs on HG (FIG. 1C, lower panel) and STG (upper panel) of all individual subjects. Each black bar shows the correlation strength between brain activation and stimulus complexity of one subject. Temporal and spectral covariation was significantly lateralized in the STG ROI, where the temporal covariation effects were greater in the left side than in the right side ROI ($P < .05$), and the spectral covariation effects greater in the right than in the left ROI ($P < .01$). The small filled circles in the upper panel give the lateralization index (difference

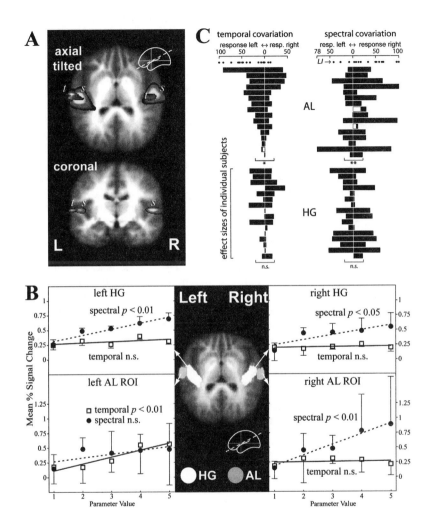

FIGURE 1. (**A**) Areas of significant BOLD response covariation with the spectral and the temporal parameter shown on the average structural image of the group. The height threshold for activation is $P < .001$, uncorrected. Areas sensitive to spectral modulation are labeled with a small *s*, and those sensitive to temporal modulation with a *t*. (**B**) Slopes of BOLD signal changes in four regions of interest (ROI) as a function of spectral and temporal input parameters. Symbols indicate average signal changes and standard deviation, with lines indicating a least-squares linear regression. The *middle panel* shows location and extent of the ROIs (HG, Heschl's gyrus; AL, anterolateral ROI) on the average structural image. (**C**) Effect sizes were extracted from the left and right ROIs on HG and from the anterolateral belt area on the superior temporal gyrus (AL, see **B** for ROI location) of all individual subjects. Each *black bar* shows the correlation strength between brain activation and stimulus complexity of one subject. Effect sizes in left-side auditory structures are plotted to the left and in right-side structures to the right, making the center of mass of each horizontal bar an estimate of the hemispheric lateralization of the individual responses.

between left- and right-side responses in percent of their sum) for the STG ROIs in all subjects. Lateralization was high and consistent in direction (temporal: left, spectral: right) in the majority of subjects.

DISCUSSION

The parametric analysis of the BOLD responses to the five levels of spectral and temporal modulation revealed cortical areas in which brain activation covaried with the perceived stimulus complexity: the auditory cortex on Heschl's gyrus and a region at the border of the anterolateral planum temporale with the superior temporal gyrus. A subsequent region-of-interest (ROI) analysis in these areas revealed a roughly linear, positive relationship between the brain responses and the perceived level of spectral and temporal complexity. An asymmetry of the response to spectral and temporal modulation was found in nonprimary auditory cortex. Activity in an area of the left STG covaried preferentially with temporal modulation, while the equivalent region on the right side responded primarily to spectral modulation. The data of individual subjects confirmed the consistency of the results in the group. Because of the advantages of random spectrogram sounds over stimuli derived from speech or tonal samples (noiselike, broadband, spectral and temporal acoustic features can be manipulated independently without altering the noiselike perceptional quality), the present study provides clear evidence for an acoustic basis of the hemispheric lateralization of speech/music perception. The findings agree with a model of hemispheric functional asymmetry for fine-grained spectral and fast temporal processing,[2] and permit a generalization of the model to include the processing of complex spectral profiles in secondary auditory areas.

[Competing interests: The authors declare that they have no competing financial interests.]

REFERENCES

1. TIAN, B. *et al.* 2001. Functional specialization in rhesus monkey auditory cortex. Science **292:** 290–293.
2. ZATORRE, R.J. & P. BELIN. 2001. Spectral and temporal processing in human auditory cortex. Cereb. Cortex **11:** 946–953.

Influence of Musical Training on Pitch Processing

Event-Related Brain Potential Studies of Adults and Children

SYLVAIN MORENO AND MIREILLE BESSON

Institut de Neurosciences Cognitives de la Méditerranée (INCM)–CNRS, Marseille, France

ABSTRACT: The aim of this experiment was to determine whether eight weeks of musical training based on pitch processing could help eight-year-old children detect pitch changes in language. Results show that a relatively short exposure (eight weeks) to pitch processing in music exerts some influence on pitch processing in language. Therefore, these results are in line with the hypothesis that common processes may underlie pitch processing in language and in music.

KEYWORDS: learning; brain plasticity; language; music; children; ERPs

The influence of extended musical practice on the level of performance in domains that are not directly linked to music has been described in both adults and children.[1–3] However, these results should be considered with caution insofar as other experiments have failed to demonstrate beneficial effects of music training in nonmusical domains.[4] Also, as pointed out by Schellenberg,[5] most studies are problematic because of uncontrolled differences between the music and control groups, such as differences in mood, arousal, and motivation;[6] number of years of education;[7] extra-scholarly activities; pedagogy; and personal contact with the instructor. Finally, as noted by Thompson et al.,[8] although most studies were successful in showing positive correlations between music and other cognitive domains, very few studies have aimed at testing specific hypotheses regarding the causal links underlying these effects.

In a recent series of experiments with both adults[9] and eight-year-old children,[10] we tested the hypothesis that common processes underlie pitch perception in music and in language. Consequently, musicians (adult and children) should perceive pitch deviations better than nonmusicians not only in music, but also in language. Although detailed analysis of the event-related brain potentials (ERPs) revealed some differences between adults and children, and between music and language (see Magne et al.[10]), results overall were in line with this hypothesis.

Address for correspondence: Sylvain Moreno, INCM-CNRS, 31-Chemin Joseph Aiguier, 13402 Marseille, Cedex 20, France. Voice: +33-4-91-16-41-13; fax: +33-4-91-77-49-69.
moreno@incm.cnrs-mrs.fr

Ann. N.Y. Acad. Sci. 1060: 93–97 (2005). © 2005 New York Academy of Sciences.
doi: 10.1196/annals.1360.054

The specific aim of this experiment was to determine whether changes in the pattern of brain electrical activity can be found after only eight weeks of musical training. We hypothesized that no differences between groups should be found before training, whereas after training detection of pitch violations in language should be better performed by children who received music rather than painting training. If we were able to demonstrate that significant changes are found in the pattern of brain electrical activity after only eight weeks of musical training, such results would support the importance of musical training for performance in other cognitive domains.

PARTICIPANTS

We tested 20 nonmusician eight-year-old children (i.e., without any specific musical training) from the same elementary school, matched by age, years of education, and socioeconomic background,[11] and who were all involved in extra-scholarly activities other than music. All children performed the same task (see below). Then, one group of 10 children took eight weeks of music training and the other group of 10 children eight weeks of painting training (see heading NOTE 1 below for details). Finally, children were tested again using the same task.

MATERIAL AND PROCEDURE

Children were presented with short sentences from children's books, and the fundamental frequency (F0) of the sentence's final word was manipulated to create a weak (35% increase) or a strong (120% increase) pitch violation. They were asked to determine if the final words sounded normal or strange. EEG was recorded from 28 scalp electrodes, and EEG acquisition was synchronized with final word onset.

RESULTS AND DISCUSSION

The main results can be summarized as follows. Although main effects of pitch violation and training were found on both reaction times (RTs) and error rate measures, these effects were not significantly different between the music and painting training groups (FIG. 1).

By contrast, analysis of the ERPs revealed significant differences between music and painting groups. Clearly, whereas the strong incongruities elicited very similar effects in both groups before training, a decrease in the amplitude of a late positivity after training was found only in the music group (see FIG. 2). The difference was largest over parietal regions. One possible interpretation is that because children in the music group were specifically trained on pitch, pitch processing became more automatic. The decreased amplitude of the late positivity would consequently reflect the automation of underlying processes that thereby require fewer and fewer neurons (see Ref. 12 for a similar interpretation).

In conclusion, the present results show some evidence for an influence of music training on pitch processing in language after only eight weeks of musical training.

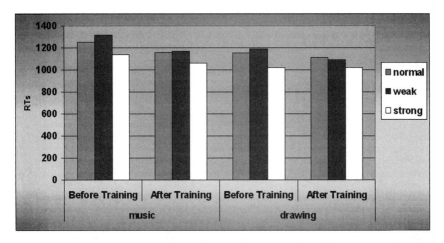

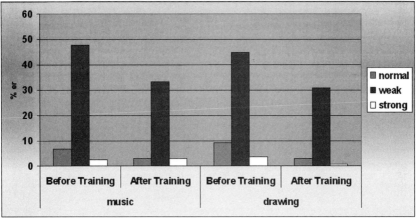

FIGURE 1. Reaction times (RTs; *top panel*) and error rate (*bottom panel*) to congruous final words and to weak and strong incongruities are presented for the music and painting training groups.

It will be interesting in further experiments to train children over longer periods to further track these transfer effects.

NOTE 1

Training

Music training was based on a new technology: "The Musical Garden" (FIG. 3), developed by C. Napoleoni and the Machin'Art association. The Musical Garden is a 4 × 4 meter carpet that comprises pads of different colors. Each pad is connected

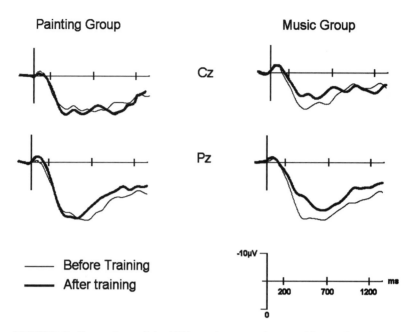

FIGURE 2. Comparison of the ERPs to the strong incongruities in the painting and music groups before (*gray curve*) and after (*black curve*) training. Data are presented for central (Cz) and parietal (Pz) electrodes.

FIGURE 3. Schematic representation of the "Musical Garden" (Designer: C. Napoleoni).

to a computer, so that when the child walks on the different pads, different sounds are played depending on the software in use.

The aim of the musical training was to train children to perform finer and finer pitch discrimination. The painting group served as a control group, to be able to isolate any specific effects due to musical training.

ACKNOWLEDGMENTS

This research was supported by a grant from the Human Frontier Science Program (HSFP #RGP0053) and by a grant from the Ministère de la Jeunesse, de l'Éducation Nationale et de la Recherche (03 2 443) to Mireille Besson. Sylvain Moreno benefits from a doctoral fellowship from the same ministry. We would like to thank Mitsuko Aramaki, Monique Chiambretto, Cyrille Magne, Vanina Luigi, Marjorie Geva, and Daniele Schön for their precious help, and the Machin'Art association, in particular, Christian Napoléoni, for their collaboration with the "Musical Garden," as well as all the children who participated in the study.

[Competing interests: The authors declare that they have no competing financial interests.]

REFERENCES

1. ANVARI, S.H., L.J. TRAINOR, J. WOODSIDE & B.A. LEVY. 2002. Relations among musical skills, phonological processing, and early reading ability in preschool children. J. Exp. Child Psychol. **83:** 111–130.
2. COSTA-GIOMI, E. 2004. Effects of three years of piano instruction on children's academic achievement, school performance, and self-esteem. Psychol. Mus. **32:** 139–152.
3. SCHELLENBERG, E.G. 2004. Music lessons enhance IQ. Psychol. Sci. **15:** 511–514.
4. HASSLER, M., N. BIRBAUMER & A. FEIL. 1987. Musical talent and visual-spatial ability: onset of puberty. Psychol. Mus. **15:** 141–151.
5. SCHELLENBERG, E.G. 2001. Music and nonmusical abilities. Ann. N. Y. Acad. Sci. **930:** 355–371.
6. STANDLEY, J.M. & J.E. HUGHES. 1997. Evaluation of an early intervention music curriculum for enhancing prereading/writing skills. Mus. Ther. Perspect. **15:** 79–85.
7. CHAN, A.S, Y.C. HO & M.C. CHEUNG. 1998. Music training improves verbal memory. Nature **396:** 128.
8. THOMPSON, W.F., E.G. SCHELLENBERG & G. HUSAIN. 2004. Decoding speech prosody: do music lessons help? Emotion **4:** 46–64.
9. SCHÖN, D., C. MAGNE & M. BESSON. 2004. The music of speech: music facilitates pitch processing in language. Psychophysiology **41:** 341–349.
10. MAGNE, C., D. SCHON & M. BESSON. Musician children detect pitch violations in both music and language better than nonmusician children: behavioral and electrophysiological approaches. J. Cogn. Neurosci. In press.
11. LAROQUE, P. 1968. Les Classes Sociales. Presses Universitaires de France. Paris.
12. JANCKE, L., N.J. SHAH & M. PETERS. 2000. Cortical activations in primary and secondary motor areas for complex bimanual movements in professional pianists. Brain Res. Cogn. Brain Res. **10:** 177–183.

Part III: Mental Representations of Music— Combining Behavioral and Neuroscience Tools

Introduction

ERICH SCHRÖGER

Institut für Psychologie I, Universität Leipzig, Seeburgstrasse Leipzig, Germany

The session entitled Mental Representations, chaired by Giuliano Avanzini, comprised six contributions, each of them concerned with the investigation of how the human brain represents musical information. Mental representations may be defined as hypothetical entities that guide our processing of music. In other words, how we perceive, understand, and appreciate music is determined not only (even mainly) by the musical input per se, but also by what we do with this musical input, in particular, and by what we know about music in general. The establishment of a current mental representation of musical information cannot be understood in isolation apart from the listener's neurocognitive architecture of musical knowledge. The following contributions apply this basic approach to different realms, using different methodologies. However, an important communality of these seminal papers is that they include measures of the organ that is responsible for the fascinating ability to understand music, namely, the human brain.

Barbara Tillman combines behavioral priming paradigms with functional magnetic resonance imaging (fMRI) to investigate tonal knowledge of nonmusicians. For example, she found evidence for the hypothesis that musical expectations on musical structure processing are not dependent on explicit musical knowledge. Petr Janata uses fMRI to identify the cortical networks that process musical structure. He found, for example, that the medial prefrontal cortex reveals activation in the perception of music's movement through tonal space. He argues, however, that depending on task and previous expertise, the neural correlates of music processing can vary considerably. Josef Rauschecker is concerned with both bottom-up and top-down processes resulting in the perceptual organization of sound sequences and complex sounds in humans and monkeys. Interestingly, this "streaming" seems to happen outside the primary auditory cortex (BA10) where also the coding and retrieval of musical sequences takes place. Hervé Platel applies the classic distinction between episodic and semantic memory to the realm of music. He presents neuroimaging results supporting the existence of a musical semantic and episodic memory. Mark Jude Tramo and Peter A. Cariani investigate the functional neuroanatomy of music perception, especially the neural representation of pitch in the primary auditory cortex. Importantly, they convincingly dispel the widely distributed

Address for correspondence: Erich Schröger, Institut für Psychologie I, Universität Leipzig, Seeburgstr. 14-20, D-04103 Leipzig, Germany. Voice: +49-341-97-359-88; fax: +49-341-97-359-69.
schroger@psychologie.uni-leipzig.de

Ann. N.Y. Acad. Sci. 1060: 98–99 (2005). © 2005 New York Academy of Sciences.
doi: 10.1196/annals.1360.062

misconception that A1 is not essential for humans and other animals in the discrimination of pure tone frequency. Donald A. Hodges and Jonathan H. Burdette show that music conductors are more accurate in pitch discrimination, temporal-order judgments, locating targets in space, and in making use of auditory information when localizing visual targets than nonconductors. Tentative fMRI data suggest that BA37 and BA39/40 are concerned with improvements of multisensory processing in conductors.

In general, the combination of behavioral measures from psychoacoustics and experimental psychology with modern brain imaging tools (e.g., fMRI and PET), neuropsychological approaches (e.g., lesion studies), and recordings at the neuronal level (e.g., spike activity) opens a new field for the investigation of how music is represented not only in our brains but also in our minds. Not only does it produce new data, but also the combination of these rather different approaches triggers important synergetic effects. For example, it inspired new experimental questions that could hardly have been expected to be developed if the domains mentioned above had remained isolated from each other. Although this field is relatively new, it has already generated interesting results, as demonstrated in the following contributions. In the future this field will enrich our understanding of music representation and our scientific approach to the mental representation of music.

Implicit Investigations of Tonal Knowledge in Nonmusician Listeners

BARBARA TILLMANN

CNRS-UMR 5020 & IFR 19, Lyon, France

ABSTRACT: By mere exposure to musical pieces in everyday life, Western listeners acquire sensitivity to the regularities of the tonal system and to the context dependency of musical sounds. This implicitly acquired tonal knowledge allows nonmusician listeners to perceive relationships among musical events and to develop expectations for future events that then influence the processing of these events. The musical priming paradigm is one method of the indirect investigation of listeners' tonal knowledge. It investigates the influence of a preceding context (with its musical structures and relationships) on the processing of a musical target event, without asking participants for direct evaluations. Behavioral priming data have provided evidence for facilitated processing of musically related events in comparison to unrelated and less-related events. The sensitivity of implicit investigations is further shown by I.R., a patient with severe amusia, showing spared implicit knowledge of music. Finally, the priming paradigm allows us to investigate the neural correlates of musical structure processing. Two fMRI studies reported the implication of inferior frontal regions in musical priming, contrasting related and unrelated events, as well as finer structural manipulations contrasting in-key events.

KEYWORDS: implicit learning; musical priming; musical expectations; musical expertise; musically untrained listeners; cognition; neuroimaging

IMPLICIT LEARNING: A COGNITIVE CAPACITY AT THE ORIGIN OF TONAL ACCULTURATION

Implicit learning processes enable the acquisition of highly complex information without complete verbalizable knowledge of what has been learned.[1] This capacity of the cognitive system is studied in the laboratory with artificial material containing statistical structures, such as finite state grammars or artificial languages.[2–5] Two examples of highly structured systems in our natural environment are language and music. Listeners become sensitive to the underlying regularities just by mere exposure to linguistic and musical material in everyday life. The implicitly acquired knowledge then influences perception and interaction with the environment.

Tonal acculturation reflects this cognitive capacity to become sensitive to regularities in the environment.[6] The overall pattern of results in music cognition

Address for correspondence: Université Claude Bernard-Lyon I, CNRS UMR 5020, Neurosciences et Systèmes Sensoriels, 50 Av. Tony Garnier, F-69366 Lyon Cedex 07, France. Voice: +33-4-3728-7493; fax: +33-4-3728-7601.

barbara.tillmann@olfac.univ-lyon1.fr

Ann. N.Y. Acad. Sci. 1060: 100–110 (2005). © 2005 New York Academy of Sciences.
doi: 10.1196/annals.1360.007

research suggests that mere exposure to Western musical pieces suffices to develop implicit, but nevertheless sophisticated, knowledge of the tonal system.[6-8] Just by listening to music in everyday life, nonmusician listeners[a] become sensitive to the regularities and structures of the tonal system without being necessarily able to verbalize them. The implicit knowledge embodies the functions of tones and chords in a key,[9-11] the relationships among different keys,[12-15] and the change in tonal functions depending on the context.[16-18] On the basis of this implicit knowledge, listeners process musical structures and develop musical expectations for future events. Because of the temporal nature of sound, expectations take a central position in music perception: which sounds are most probable to come next and at what moment should they occur?[19] Expectations are part of cognitive functions in general, they shape our interaction with the environment, and—depending on their nature—facilitate or delay the processing of events. Musical expectations not only influence efficiency of processing but have a role in musical expressivity. Composers (or improvising performers) fulfill listeners' perceptual expectations more or less early or only partially in a musical piece. Following Meyer,[20] playing with musical expectations is at the origin of musical expressivity that is communicated through musical compositions. The musical priming paradigm, in the center of this article, focuses on the influence of musical expectations on processing (speed and accuracy) and its neural correlates. It also provides further insight into the role of expectations for musical expressivity, notably the resistance of schematic expectations to repeated listening.[21,22]

INVESTIGATING MUSICAL EXPECTATIONS: THE MUSICAL PRIMING PARADIGM

While listening to a musical piece, listeners develop musical expectations for future events, and these expectations influence perception. The priming paradigm is an implicit investigation method that studies the influence of perceivers' expectations on the efficiency of perception (i.e., accuracy or processing speed). It has been extensively used in psycholinguistics[23] and has been introduced into the music perception domain by Bharucha and Stoeckig[24] and considerably developed since then. This implicit investigation method is an adequate approach to probe nonmusicians' musical knowledge, which might be more sophisticated than can be revealed by direct investigation methods using explicit judgments or memory tasks.

The basic setup consists of a prime context (i.e., a chord or a chord sequence) and a target event (i.e., a chord). The relationships between the prime and the target are systematically manipulated: this manipulation concerns the musical relatedness as defined by music theory. The underlying hypothesis is that the prime context allows the perceiver to develop expectations for future events, with more strongly related events being more expected. These expectations influence event processing, with facilitated processing of expected events over unexpected or less-expected ones. A sensory priming approach predicts facilitated processing for target events having occurred in the context (repetition priming) or sharing perceptual features with the

[a]In the present article, nonmusician listeners are defined as musically untrained listeners who had never received explicit instruction in music theory or on playing an instrument.

context: the stronger the acoustical overlap between prime and target, the more the target's processing is facilitated. A cognitive priming approach postulates expectations based on listeners' knowledge about possible relationships between events independently of repetition: the higher the musical event is situated in the tonal hierarchy, the more it is expected and the more its processing is facilitated.

The central feature of the priming paradigm is the indirect investigation of the context's influence on event processing: participants are not required to make direct judgments on the relationship between prime context and target, but their task focuses on a perceptual feature of the target chord. In parallel to the lexical-decision task in psycholinguistics, a frequently used task is based on sensory consonance/dissonance judgments:[17,24] half of the targets are consonant (i.e., well-tuned, correctly constructed chords), half of the targets are rendered acoustically dissonant (i.e., either by mistuning or by adding an out-of-key tone), and participants make speeded consonance/dissonance judgments on the target chord.[b] Alternative priming tasks require judgments of temporal asynchrony,[25] phoneme discrimination,[26] timbre discrimination,[27] and loudness discrimination,[28] as well as lexical decisions of sung words/nonwords.[29] The various tasks provide evidence for the influence of musical relationships on the speed of processing musical events, notably by showing facilitated processing for musically related targets.

First, musical priming studies investigated the influence of short, one-chord prime contexts, mostly in association with strong tonal expectancy violations opposing in-key and out-of-key events.[24,30–32] Bigand and Pineau[17] extended musical priming effects to longer musical contexts and, additionally, to finer tonal manipulations. In eight-chord sequences, the last chord defined the target. Its relationship to the global context (chords 1 to 6) was manipulated while holding constant the local context (chord 7) to control for local sensory influences on target processing. The target acted either as an important chord of the key, and was supposed to be highly expected (the tonic chord), or as a less important chord, supposed to be less expected (the subdominant chord). The experimental task requested participants to make speeded sensory judgments on the targets, which were either consonant or rendered acoustically dissonant. The critical finding was to show that consonant/dissonant judgments were more accurate and faster when targets acted as tonic rather than as subdominant chords. This outcome, observed for musician and nonmusician participants, suggests that the processing of harmonic spectra is facilitated for events that are the most predictable in the current context. Together with Bigand et al.,[33] the priming data provide evidence that musical expectations do not only occur from chord to chord, but also involve the processing of higher levels of musical relationships and of temporal structures. Recent research started to use the musical priming paradigm for the investigation of both tonal and temporal expectations.[34]

Nonmusician listeners understand not only the difference between tonic and subdominant, but also more detailed differences reflecting the harmonic hierarchy. In a recent musical priming study,[35] the tonal prime contexts ended on one of three targets placed at the top of the harmonic hierarchy: the tonic, the dominant, or the

[b]When the experimental manipulations contrast related and unrelated target chords, the experimental trials consist of musical sequences, with 25% of the trials ending on related consonant chords, 25% on related dissonant chords, 25% on unrelated consonant chords, and 25% on unrelated dissonant chords.

subdominant. The response time patterns imitated the chords' positions in the hierarchy: the tonic was processed the fastest, followed by the dominant, and then the subdominant. This study (together with Tillmann et al.[36]) additionally investigated whether the processing of the tonal center of the prime context represents a cost or a benefit for target chord processing. For this purpose, a procedure from psycholinguistic research[37] was adapted: target processing in tonal contexts was compared to target processing in "neutral" global prime contexts (baseline sequences), which did not orient participants' musical expectations. These baseline contexts consisted of chord combinations that did not install a tonal center. The comparison of baseline contexts to tonal contexts provided evidence that the processing of the prime's tonal center (and the resulting tonal expectations) represented a benefit only for related tonic targets, but a cost for unrelated, out-of-key targets, and even a cost for less-related, subdominant targets, which still belong to the activated tonal center. The dominant targets were situated at the level of the baseline, representing neither cost nor benefit of the processed tonal center. This data pattern suggests a privileged role of the tonic chord, in agreement with its previously suggested role as a cognitive reference point.[6,8,38] It further shows nonmusicians' sensitivities to fine differences in the harmonic functions of three in-key chords at the top of the harmonic hierarchy.

A set of musical priming studies provided evidence that cognitive priming prevails over sensory priming.[31–33,39,40] To investigate the strength of the cognitive priming component in music, Bigand et al.[39] pitted sensory influences against cognitive influences in chord sequences—a rationale comparable to Tekman and Bharucha[32] for single-chord primes. The target (i.e., the last chord of eight-chord sequences) was either strongly related or less-related (i.e., tonic versus subdominant chord) to the prime context. The critical point was to simultaneously manipulate the targets' frequencies of occurrence in the prime context. In the "no-target-in-context condition," neither related nor less-related targets occurred in the prime context, equalizing the contribution of sensory priming in the two conditions. In the "subdominant-target-condition," the strength of sensory priming was boosted by adding less-related targets into the prime context. In this condition, sensory priming was expected to be stronger and thus predicting facilitated processing for less-related, subdominant targets. The results, however, showed facilitation for related tonic targets in both conditions (i.e., no effect of the added subdominant chords). The strength of cognitive priming in contrast to sensory priming was confirmed for relatively fast tempi (i.e., 300 ms and 150 ms). Converging evidence has been provided by another study showing that response time patterns are determined by musical relatedness and not by immediate repetition (the strongest form of sensory priming).[40] This finding raises considerable difficulties for sensory models of music perception because the processing of a musical event is facilitated when it is preceded by a different, but musically related chord in comparison to when it is preceded by an identical (repeated) chord. The data with musical material are different from language or visual stimuli, for which the influence of repetition is stronger and more persistent than the influence of semantic relationships.[41]

Overall findings demonstrate the influence of knowledge-driven expectations on musical structure processing independently of the listener's musical expertise. Factors governing musical expectations are based on cognitive processes that do not require explicit musical knowledge. The strength of cognitive components in musical priming can be further investigated with melodic material.[42] The musical priming paradigm is an implicit investigation method that provides evidence for sophisticated tonal knowledge, and not only for nonmusician listeners, but also for children and patients. The musical priming method showed that six-year-old children differentiate between tonic and subdominant chords just as do adult listeners.[43] This outcome stands in contrast to explicit investigations suggesting that not until the age of 10 can children differentiate among harmonic functions of chords (e.g., tonic and dominant chords).[44]

SPARED IMPLICIT MUSICAL KNOWLEDGE IN AN AMUSIC PATIENT

In various domains, priming paradigms provided evidence for some spared implicit processing or knowledge in brain-damaged patients (i.e., suffering from alexia or prosopagnosia) who manifest dramatic deficits when explicit judgments are requested. For example, a prosopagnosic patient may be unable to explicitly recognize a face while the face's familiarity influences the processing of a word (i.e., the name of an associatively related person) presented thereafter.[45] Our goal was to investigate an amusic patient, I.R., for eventually spared implicit knowledge on tonal relationships and its influence on music perception.[46] I.R. represents a severe case of amusia that has been extensively studied.[47–49] This patient sustained bilateral cerebral damage that severely impaired music perception and memory. I.R. does not recognize familiar excerpts, does not detect dissonance, and fails in melodic and temporal tests asking for direct judgments or comparisons. However, I.R. exhibits normal performance in language tasks, in the recognition of environmental sounds, and has no impairment of basic auditory functions.

Our study tested I.R. with the musical priming paradigm to investigate an eventual implicit access to remaining musical knowledge as has been reported for neuropsychological cases in other domains. Experiments 1 and 2 used priming tasks based on phoneme and timbre discrimination, respectively. Experiments 3 and 4 used explicit judgments of completion on the same musical material as in Experiment 2 and on linguistic stimuli as a control condition (i.e., short sentences that varied in the congruency of their endings). For the two musical priming tasks, I.R. (and her matched controls) showed facilitated processing for related targets over less-related targets and thus replicated musical priming effects observed for healthy college students.[26,27] However, I.R.'s explicit judgments of completion did not differentiate between the conditions: less-related musical sequences were judged as complete as related sequences. This outcome is in agreement with I.R.'s failure in previous explicit testing methods of music perception.[49] The control experiment with linguistic stimuli showed that I.R.'s deficit was specific to musical sequences; she performed like normal controls for completion judgments on sentences.

Our study revealed I.R.'s sensitivity to musical relationships with the implicit priming paradigm, but not with explicit judgments. This outcome suggests some spared musical knowledge in this previously declared amusic patient. I.R.'s data recommend the use of implicit, indirect methods for the testing of listeners with musical disorders. The implicit access might reveal knowledge that would not have been shown by explicit, direct testing methods. In an ongoing research project together with I. Peretz, we are using the musical priming paradigm to test people with congenital amusia (or tone deafness), who present a musical impairment that cannot be explained by hearing loss, brain lesions, or cognitive and social deficits.[50–53]

NEUROPHYSIOLOGICAL CORRELATES OF CONTEXT EFFECTS AND MUSICAL EXPECTATIONS

Neural correlates of musical structure processing have been investigated by focusing on violations of musical expectations developed by the listener. This approach is similar to the investigation of neural correlates in language processing, notably using semantic and syntactic violations.[54,55] Over the last few years, a growing number of studies have investigated structure processing and expectancy violation with musical stimuli.[56–59] Electrophysiological correlates of musical context effects have been reported by Regnault et al.[59] with the priming material of Bigand and Pineau.[17] Less-related targets evoked a P3 component (200–300 ms latency range) with larger amplitude than after strongly related targets. This data pattern was reported for musically trained and untrained listeners: both groups were sensitive to changes in musical function of the target due to the prime context.

Nonmusicians' sensitivities to violations of musical expectations in chord sequences have been further shown with magneto-encephalography MEG[60] and functional magnetic resonance imaging fMRI,[61–64] suggesting the involvement of inferior frontal regions in musical structure processing. An unexpected chord structure (i.e., Neapolitan chords) evoked a bilateral mERAN (magnetic early right anterior negativity), with its source being localized in Broca's area and its right hemisphere homologue.[60] Using fMRI, Koelsch et al.[61] confirmed inferior frontal activation for several types of expectancy violations: strongly expected in-key chords were compared to either dissonant tone clusters (tone combinations without regular chord interval structure), modulating chords installing a new key (and thus containing out-of-key tones), and chords played by different instruments. Using the musical priming paradigm, Tillmann et al.[63] reported stronger activation of bilateral inferior frontal regions for unrelated, out-of-key targets than for related targets. In eight-chord sequences, the last chord defined the target that was either strongly related (a tonic chord) or unrelated (a chord belonging to a distant key). As in previous musical priming studies, half of the targets were rendered acoustically dissonant for the experimental task. Participants were scanned with fMRI while performing speeded intonation judgments (consonant versus dissonant) on the target chords. Behavioral results acquired in the scanner replicated the facilitation effect of related over unrelated consonant targets. The blood oxygen level–dependent (BOLD) signal linked to target processing revealed activation of bilateral inferior frontal regions (i.e., inferior frontal gyrus, frontal operculum, insula) that was stronger for unrelated than for related targets.

Most of the studies investigating neural correlates of musical expectations have used rather strong tonal manipulations, notably with unexpected events being out-of-key chords[55,63] or containing out-of-key tones.[60–62] This strong manipulation allows for two alternative explanations of neural correlates, which do not consider the listener's knowledge-based expectations, but which refer to sensory features. Since only component tones of expected events occurred in the context (and not those of the unexpected events), the phenomenon of reduced neural activity for repeated items (see Refs. 65–67 for words and pictures) might explain the decreased activation for expected events. Furthermore, since out-of-key tones define sensory deviants, increased inferior frontal activation for unrelated events might reflect neural correlates of target detection and of processing novel events.[68,69] With the goal to isolate neural correlates of musical structure violation without sensory repetition or deviance, our recent fMRI study[64] used the musical priming paradigm with relatively subtle manipulations as in previous behavioral studies. Related and less-related targets belonged to the context key, did not create any contextual dissonance, and did not occur in the prime context.[26,27,39] The musical material was either played with musical timbres[27] (i.e., created here by combinations of musical instruments) or sung with artificial syllables.[26] Behavioral data acquired in the scanning session confirmed facilitation for related over less-related targets for both materials in timbre discrimination and phoneme discrimination tasks. Independent of the material carrying the musical information (i.e., timbres versus sung phonemes), activation in the right (with reduced statistical threshold also in the left) inferior frontal regions (in the vicinity of inferior frontal gyrus, frontal operculum, with adjacent anterior insula) was increased for less-related targets in comparison to related targets. This outcome discards the evoked sensory explanations based on repetition suppression or deviance processing. It suggests that inferior frontal regions are sensitive to musical expectancy violations and are involved in the processing of music-syntactic relationships. Increased activation in the anterior insula and orbitofrontal cortex (see also Ref. 62 after stronger musical expectancy violations) can be integrated into the theoretical framework on musical emotion,[20] notably that violations of expectations or delays of resolutions give raise to meaningful and expressive moments in music.

This finding can be integrated into previous results showing that inferior frontal regions participate in nonlinguistic processes beyond phonological, semantic, and syntactic processes. The similarities in neural correlates of musical, linguistic, and acoustical processing might refer to neural correlates necessary for structural integration, notably over time.[70–72] Future research will need to investigate the specificity of inferior frontal regions: their functional relevance might depend on the activation status of other connected brain regions (i.e., the integration in an overall network) and on specialized regions in these relatively large areas (i.e., with some being specialized for different domains and others being more domain general).

For the processing of musical structures, the overlap in activation patterns between our study[64] and Koelsch et al.[62] suggests an overall network that might include inferior frontal areas (frontal operculum and adjacent anterior insula) and orbital gyrus, temporal regions (aSTG, STS/STG, posterior MTG), and the right inferior parietal lobe (SMG). Future research needs to investigate more directly the interaction and connectivity between these regions for the processing of musical structures. This specification of the neural network pattern in music processing becomes crucial in light of the data pattern of the amusic patient I.R.. Despite lesions in the right inferior frontal cortex, musical priming effects have been observed in the behavioral approach, suggesting that I.R. has the ability to develop musical, knowledge-based expectations. Some of the areas of activation observed in our fMRI study are situated in healthy tissue, notably left inferior frontal, posterior STS and MTG, as well as supramarginal activation. This observation illustrates for music perception how the combination of neuropsychological data and normal functional imagining data reveals the need for further understanding of distributed brain systems underlying event processing.[73]

ACKNOWLEDGMENTS

I would like to thank all those who collaborated in this research—in particular, Emmanuel Bigand, Jamshed J. Bharucha, Petr Janata, Isabelle Peretz, Bénédicte Poulin-Charronnat, Philippe Lalitte, Nicolas Escoffier, Géraldine Lebrun-Guillaud, Fréderic Marmel, and Stefan Koelsch.

[Competing interests: The author declares that she has no competing financial interests.]

REFERENCES

1. SEGER, C.A. 1994. Implicit learning. Psychol. Bull. **115**: 163–169.
2. REBER, A.S. 1967. Implicit learning of artificial grammars. J. Verb. Learn. Verb. Behav. **6**: 855–863.
3. REBER, A.S. 1989. Implicit learning and tacit knowledge. J. Exp. Psychol. Gen. **118**: 219–235.
4. SAFFRAN, J.R., E.L. NEWPORT & R.N. ASLIN. 1996. Word segmentation: the role of distributional cues. J. Mem. Lang. **35**: 606–621.
5. ALTMANN, G.T.M., Z. DIENES & A. GOODE. 1995. Modality independence of implicitly learned grammatical knowledge. J. Exp. Psychol. Learn. Mem. Cogn. **21**: 899–912.
6. FRANCÈS, R. 1958. La perception de la musique. Vrin: Paris.
7. DOWLING, W.J. & D.L. HARWOOD. 1986. Music Cognition. Academic Press, San Diego CA.
8. KRUMHANSL, C.L. 1990. Cognitive foundations of musical pitch. Oxford University Press. New York, NY.
9. CUDDY, L.L. & B. BADERTSCHER. 1987. Recovery of the tonal hierarchy: some comparisons across age and levels of musical experience. Percept. Psychophys. **41**: 609–620.
10. HÉBERT, S., I. PERETZ & L. GAGNON. 1995. Perceiving the tonal ending of tune excerpts: the roles of pre-existing representation and musical expertise. Can. J. Exp. Psychol. **49**: 193–209.
11. TILLMANN, B., E. BIGAND & F. MADURELL. 1998. Local versus global processing of harmonic cadences in the solution of musical puzzles. Psychological Research/Psychologische Forschung **61**: 157–174.
12. BARTLETT, J.C. & W.J. DOWLING. 1980. Recognition of transposed melodies: a key-distance effect in developmental perspective. J. Exp. Psychol. Hum. Percept. Perform. **6**: 501–515.
13. CUDDY, L.L. & W.F. THOMPSON. 1992. Asymmetry of perceived key movement in chorale sequences: converging evidence from a probe-tone analysis. Psychol. Res. **54**: 51–59.
14. CUDDY, L.L. & W.F. THOMPSON. 1992. Perceived key movement in four-voice harmony and single voices. Music Percept. **9**: 427–438.
15. THOMPSON, W.F. & L.L. CUDDY. 1989. Sensitivity to key change in chorale sequences: a comparison of single voices and four-voice harmony. Music Percept. **7**: 151–168.
16. BIGAND, E. 1997. Perceiving musical stability: the effect of tonal structure, rhythm, and musical expertise. J. Exp. Psychol. Hum. Percept. Perform. **23**: 808–822.
17. BIGAND, E. & M. PINEAU. 1997. Global context effects on musical expectancy. Percept. Psychophys. **59**: 1098–1107.
18. JANATA, P., J.L. BIRK, B. TILLMANN, et al. 2003. Online detection of tonal pop-out in modulating contexts. Music Percept. **20**: 283–305.

19. JONES, M.R. & M. BOLTZ. 1989. Dynamic attending and responses to time. Psychol. Rev. **96:** 459–491.
20. MEYER, L.B. 1956. Emotion and Meaning in Music. University of Chicago Press. Chicago, IL.
21. JUSTUS, T.C. & J.J. BHARUCHA. 2001. Modularity in musical processing: the automaticity of harmonic priming. J. Exp. Psychol. Hum. Percept. Perform. **27:** 1000–1011.
22. TILLMANN, B. & E. BIGAND. 2004. Musical priming: schematic expectations resist repetition priming. Proceedings of the 8th International Conference of Music Perception and Cognition. Evanston, IL.
23. NEELY, J.H. 1991. Semantic priming effects in visual word recognition: a selective review of current findings and theories. *In* Basic Processes in Reading: Visual Word Recognition. D. Besner & G.W. Humphreys, Eds.: 264–336. Lawrence Erlbaum. Mahwah, NJ.
24. BHARUCHA, J.J. & K. STOECKIG. 1986. Reaction time and musical expectancy: priming of chords. J. Exp. Psychol. Hum. Percept. Perform. **12:** 403–410.
25. TILLMANN, B. & J.J. BHARUCHA. 2002. Effect of harmonic relatedness on the detection of temporal asynchronies. Percept. Psychophys. **64:** 640–649.
26. BIGAND, E., B. TILLMANN, B. POULIN, *et al.* 2001. The effect of harmonic context on phoneme monitoring in vocal music. Cognition **81:** B11–B20.
27. TILLMANN, B., E. BIGAND, N. ESCOFFIER, *et al.* Influence of harmonic context on musical timbre processing. Eur. J. Exp. Psychol. In press.
28. POULIN-CHARRONNAT, B. & E. BIGAND. Context effects of loudness judgments of musical chords. Submitted for publication.
29. POULIN-CHARRONNAT, B., E. BIGAND, F. MADURELL, *et al.* 2005. Musical structure modulates semantic priming in vocal music. Cognition **94:** B67–78.
30. BHARUCHA, J.J. & K. STOECKIG. 1987. Priming of chords: spreading activation or overlapping frequency spectra? Percept. Psychophys. **41:** 519–524.
31. TEKMAN, H.G. & J.J. BHARUCHA. 1992. Time course of chord priming. Percept. Psychophys. **51:** 33–39.
32. TEKMAN, H.G. & J.J. BHARUCHA. 1998. Implicit knowledge versus psychoacoustic similarity in priming of chords. J. Exp. Psychol. Hum. Percept. Perform. **24:** 252–260.
33. BIGAND, E., F. MADURELL, B. TILLMANN, *et al.* 1999. Effect of global structure and temporal organization on chord processing. J. Exp. Psychol. Hum. Percept. Perform. **25:** 184–197.
34. TILLMANN, B. & G. LEBRUN-GUILLAUD. Influence of tonal and temporal expectations on chord processing and on completion judgments of chord sequences. Psychol. Res. In press.
35. TILLMANN, B., P. JANATA, J. BIRK, *et al.* Tonal centers and expectancy: facilitation or inhibition of chords at the top of the harmonic hierarchy? In preparation.
36. TILLMANN, B., P. JANATA, J. BIRK, *et al.* 2003. The costs and benefits of tonal centers for chord processing. J. Exp. Psychol. Hum. Percept. Perform. **29:** 470–482.
37. STANOVICH, K.E. & R.F. WEST. 1983. On priming by a sentence context. J. Exp. Psychol. Gen. **112:** 1–36.
38. BHARUCHA, J.J. 1984. Anchoring effects in music: the resolution of dissonance. Cognit. Psychol. **16:** 485–518.
39. BIGAND, E., B. POULIN, B. TILLMANN, *et al.* 2003. Cognitive versus sensory components in harmonic priming effects. J. Exp. Psychol. Hum. Percept. Perform. **29:** 159–171.
40. BIGAND, E., B. TILLMANN, D. MANDERLIER, *et al.* Repetition priming: is music special? Q. J. Exp. Psychol. In press.
41. DANNENBRING, G.L. & K. BRIAND. 1982. Semantic priming and the word repetition effect in a lexical decision task. Can. J. Psychol **36:** 435–444.
42. MARMEL, F. & B. TILLMANN. Cognitive priming in melody perception. In preparation.
43. SCHELLENBERG, E.G., E. BIGAND, B. POULIN, *et al.* Children's implicit knowledge of harmony in Western music. Dev. Sci. In press.
44. IMBERTY, M. 1969. L'Acquisition des Structures Tonales chez l'Enfant. Klincksieck. Paris, France.

45. YOUNG, A.W., D. HELLAWELL & E.H.F. DEHAAN. 1988. Cross-domain semantic priming in normal subjects and a prosopagnosic patient. Q. J. Exp. Psychol. **40A:** 561–580.
46. TILLMANN, B., I. PERETZ & E. BIGAND. Musical priming in an amusic patient: the power of implicit tasks. In preparation.
47. PATEL, A.D., I. PERETZ, M. TRAMO, et al. 1998. Processing prosodic and musical patterns: a neuropsychological investigation. Brain Lang. **61:** 123–144.
48. PERETZ, I., A.J. BLOOD, V. PENHUNE, et al. 2001. Cortical deafness to dissonance. Brain **124:** 928–940.
49. PERETZ, I., S. BELLEVILLE & S. FONTAINE. 1997. Dissociations entre musique et langage après atteinte cérébrale. Rev. Can. Psychol. Exp. **51:** 354–367.
50. PERETZ, I. & K.L. HYDE. 2003. What is specific to music processing? Insights from congenital amusia. Trends Cognit. Sci. **7:** 362–367.
51. PERETZ, I., J. AYOTTE, R.J. ZATORRE, et al. 2002. Congenital amusia: a disorder of fine-grained pitch discrimination. Neuron **33:** 185–191.
52. FOXTON, J.M., J.L. DEAN, R. GEE, et al. 2004. Characterization of deficits in pitch perception underlying "tone deafness." Brain **127:** 801–810.
53. AYOTTE, J., I. PERETZ & K.L. HYDE. 2002. Congenital amusia: a group study of adults afflicted with a music-specific disorder. Brain **125:** 238–251.
54. KUTAS, M. & S.A. HILLYARD. 1980. Event-related brain potentials to semantically inappropriate and surprisingly large words. Biol. Psychol. **11:** 99–116.
55. PATEL, A.D., E. GIBSON, J. RATNER, et al. 1998. Processing syntactic relations in language and music: an event-related potential study. J. Cogn. Neurosci. **10:** 717–733.
56. BESSON, M. & F. FAÏTA. 1995. An event-related potential (ERP) study of musical expectancy: comparison of musicians with nonmusicians. J. Exp. Psychol. Hum. Percept. Perform. **21:** 1278–1296.
57. JANATA, P. 1995. ERP measures assay the degree of expectancy violation of harmonic contexts in music. J. Cogn. Neurosci. **7:** 153–164.
58. KOELSCH, S., T. GUNTER, A.D. FRIEDERICI, et al. 2000. Brain indices of music processing: "nonmusicians" are musical. J. Cogn. Neurosci. **12:** 520–541.
59. REGNAULT, P., E. BIGAND & M. BESSON. 2001. Different brain mechanisms mediate sensitivity to sensory consonance and harmonic context: evidence from auditory event-related brain potentials. J. Cogn. Neurosci. **13:** 241–255.
60. MAESS, B., S. KOELSCH, T.C. GUNTER, et al. 2001. Musical syntax is processed in Broca's area: an MEG study. Nat. Neurosci. **4:** 540–545.
61. KOELSCH, S., T.C. GUNTER, D.Y. VON CRAMON, et al. 2002. Bach speaks: a cortical "language-network" serves the processing of music. NeuroImage **17:** 956–966.
62. KOELSCH, S., T. FRITZ, K. SCHULZE, et al. 2005. Adults and children processing music: an fMRI study. NeuroImage **25:** 1068–1076.
63. TILLMANN, B., P. JANATA & J.J. BHARUCHA. 2003. Activation of the inferior frontal cortex in musical priming. Brain Res. Cogn. Brain Res. **16:** 145–161.
64. TILLMANN, B., S. KOELSCH, N. ESCOFFIER, et al. Cognitive priming in sung and instrumental music: activation of inferior frontal cortex. Submitted for publication.
65. BUCKNER, R.L., J. GOODMAN, M. BUROCK, et al. 1998. Functional-anatomic correlates of object priming in humans revealed by rapid presentation event-related fMRI. Neuron **20:** 285–296.
66. MACCOTTA, L. & R.L. BUCKNER. 2004. Evidence for neural effects of repetition that directly correlate with behavioral priming. J. Cogn. Neurosci. **16:** 1625–1632.
67. WAGNER, A.D., W. KOUSTAAL, A. MARIL, et al. 2000. Task-specific repetition priming in left inferior prefrontal cortex. Cereb. Cortex **10:** 1176–1184.
68. KIEHL, K.A., K.R. LAURENS, T.L. DUTY, et al. 2001. Neural sources involved in auditory target detection and novelty processing: an event-related fMRI study. Psychophysiology **38:** 133–142.
69. LINDEN, D.E.J., D. PRVULOVIC, E. FORMISANO, et al. 1999. The functional neuroanatomy of target detection: an fMRI study of visual and auditory oddball tasks. Cereb. Cortex **9:** 815–823.
70. COULL, J.T. 2004. fMRI studies of temporal attention: allocating attention within, or towards, time. Brain Res. Cogn. Brain Res. **21:** 216–226.

71. FUSTER, J.M. 2.001. The prefrontal cortex—An update: time is of the essence. Neuron
    30: 319–333.
72. JANATA, P. & S.T. GRAFTON. 2003. Swinging in the brain: shared neural substrates for
    behaviors related to sequencing and music. Nat. Neurosci. **6:** 682–687.
73. PRICE, C.J. & K.J. FRISTON. 2001. Functional neuroimaging of neuropsychologically
    impaired patients. *In* Handbook of Functional Neuroimaging of Cognition.
    R. Cabeza & A. Kingstone, Eds.: 379–399. MIT Press. Cambridge, MA.

Brain Networks That Track Musical Structure

PETR JANATA

Center for Mind and Brain, University of California, Davis,
Davis, California 95616, USA

ABSTRACT: As the functional neuroimaging literature grows, it becomes increasingly apparent that music and musical activities engage diverse regions of the brain. In this paper I discuss two studies to illustrate that exactly which brain areas are observed to be responsive to musical stimuli and tasks depends on the tasks and the methods used to describe the tasks and the stimuli. In one study, subjects listened to polyphonic music and were asked to either orient their attention selectively to individual instruments or in a divided or holistic manner across multiple instruments. The network of brain areas that was recruited changed subtly with changes in the task instructions. The focus of the second study was to identify brain regions that follow the pattern of movement of a continuous melody through the tonal space defined by the major and minor keys of Western tonal music. Such an area was identified in the rostral medial prefrontal cortex. This observation is discussed in the context of other neuroimaging studies that implicate this region in inwardly directed mental states involving decisions about the self, autobiographical memory, the cognitive regulation of emotion, affective responses to musical stimuli, and familiarity judgments about musical stimuli. Together with observations that these regions are among the last to atrophy in Alzheimer disease, and that these patients appear to remain responsive to autobiographically salient musical stimuli, very early evidence is emerging from the literature for the hypothesis that the rostral medial prefrontal cortex is a node that is important for binding music with memories within a broader music-responsive network.

KEYWORDS: attention; tonality; torus; expectancy; emotion; Alzheimer disease; autobiographical memory; event-related potential; rostral medial prefrontal cortex

INTRODUCTION

As a phenomenon at the core of humanity, music engages the brain in manifold ways. Music-oriented behaviors range from inattentive listening and peripheral awareness of music playing in the background, for instance, as we converse with a partner at a dinner table, to full engagement, as in entranced performing in a drum circle for hours on end. It can be argued that in order to support this broad continuum of musical interactions, music necessarily engages the full spectrum of psychological processes that are studied today, ranging from sensation to action, via attention, memory, and emotion. Although such an unsupported argument might seem self-

Address for correspondence: Petr Janata, Center for Mind and Brain, University of California, Davis, One Shields Avenue, Davis, CA 95616. Voice: 530-297-4471; fax: 530-297-4400.
pjanata@ucdavis.edu

Ann. N.Y. Acad. Sci. 1060: 111–124 (2005). © 2005 New York Academy of Sciences.
doi: 10.1196/annals.1360.008

evident, it is nonetheless unclear as to how music selects among and integrates these functions and why it does it in such a potent way as to be a stimulus and activity that most humans actively seek out. In addition, the statement that music engages the whole brain is a very fuzzy starting point at best, and additional constraints are required in order to match up music-specific functions with domain-general brain functions. In the most restricted sense, music is an auditory phenomenon and is therefore the domain of the auditory cortex. To the extent that it is structured in time, as is the other remarkably human phenomenon, language, music also becomes the domain of the right-hemisphere homologue of Broca's area. The perceptual and syntactic aspects of music perception might be thought of as served by two functional *über*-modules that comprise one or more brain regions.[1,2]

The concept of functional modularity is a very useful construct because it imposes constraints on the mapping between components of musical behaviors, music's artful movement along underlying physical dimensions, and the functional neuroanatomy of the brain. Proposals based on diverse neuroimaging and neuropsychological data have been forwarded about the relationships among postulated modules and their neuroanatomical and neurophysiological substrates.[1–3] Though there is a temptation of those more distant from the field or unappreciative of music's alluring complexities to view music as some sort of unitary phenomenon or module that resides in a very circumscribed set of brain areas, as one delves further into understanding how the different components of music interact one encounters the need to recruit more brain areas in order to provide all of the component functions for describing/explaining musical behaviors. Thus, as much of a challenge as trying to pinpoint the neural locus of one specific musical function is the challenge of characterizing how different properties of music and its associated behaviors are distributed in a network of coupled and dynamically interacting brain areas.

In this paper, I focus on studies that have examined how attention and expectancy act in tonal contexts. Tonality can be thought of as one of the three major perceptual dimensions of music on which music is organized, the other two being rhythm and timbre. The music of different cultures differentially weights these dimensions. By far the most extensively studied musical tradition in cognitive psychology is that of Western tonal music, and that is the tradition from which the studies I will discuss in this paper arise.

DOMAIN-GENERAL NETWORKS FOR ATTENTIVE LISTENING TO MUSIC

Although we can perceive music peripherally as we direct most of our attention elsewhere, we ultimately want to understand the brain states that are correlated with motivated, attentive, and direct engagement with music, be it as listeners or performers. Music provides a complex soundscape for attention to roam on. Perhaps one of the most common things we do as listeners when listening to an ensemble, be it a latin salsa band or a piano trio, is that we shift our attention among the instruments, following for a while the rhythmic harmonies of the piano before switching our attention to the lilting melody of the violin, and so on. How does such attention-orienting and switching behavior operate in music when we try to maintain focus on a single instrument's part, or try to detect when any one of the players makes a mistake? Does it recruit the parietal cortex, which is known to underlie the orienting of other forms of attention, such as visuospatial attention? Will orienting of attention to different regions of timbral space enhance activity of timbre-selective regions of the cortex?

In order to begin answering these questions, we performed a neuroimaging experiment with fMRI in which listeners were cued to orient their attention to one of multiple instruments or to distribute their attention across the different instruments. In one task, included in order to root the study in the dominant (target detection) paradigm for studying selective auditory attention, listeners were requested to perform timbral violation detection while listening to a 15-s excerpt from a Schubert piano trio and focusing their attention on either a single instrument or dividing it

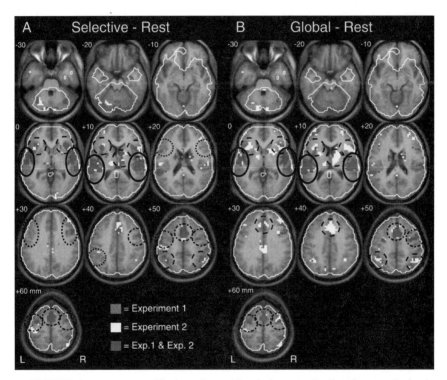

FIGURE 1. A summary of brain regions active during attentive listening to polyphonic music. During selective attention (**A**), subjects were cued to orient their attention to one of the instruments, either to track the part the instrument was playing as though they were trying to memorize its melody (Exp. 1), or to listen for timbral deviances (Exp. 2). During divided attention conditions (**B**), subjects were instructed to listen holistically to the overall sound (Exp. 1) or to detect deviants that could occur in any of the instrumental parts (Exp. 2). Regions that comprise domain-general networks are circled with various *dashed* and *dotted lines*. Activations of sensory (auditory) cortex are circled with a *solid line*. *Dashed circles:* regions that are generally active in attention-demanding and target-detection tasks. *Dotted circles:* regions involved in working memory tasks. *Dash-dotted circles:* premotor areas known also to be activated during expectancy-generating perceptual tasks. Adapted from Janata *et al.*[4] Please see the color figure in the online version.

across instruments in order to detect slight changes in the timbral quality of the instrument. In recognition of the fact that we do not always listen to music from a target detection perspective, even though we might be listening attentively, we asked another group of listeners to focus their attention on individual melody lines of baroque flute duets rendered with two very different timbres, as though they were trying to memorize the part the instrument was playing, and in the separate nonselective attention trial to listen holistically and appreciate the music as a whole. The details of these experiments are given elsewhere,[4] and a summary of the results is shown in FIGURE 1.

Regardless of whether the attentional task was structured as a target-detection task or as a voluntary orienting task without the requirement of detecting a particular target, the music and tasks recruited a network of brain areas extending bilaterally well beyond the auditory cortex. A domain-general attention network was recruited, involving parietal and ventrolateral prefrontal cortex, together with the cerebellum, basal ganglia, and thalamus. In addition, premotor areas were recruited, most notably and consistently the presupplementary motor area (preSMA), and to a lesser extent the anterodorsal aspects of the precentral gyrus. These results extend observations from other ethological tasks, such as action observation,[5] and more abstract tasks using multimodal object sequences,[6,7] that premotor cortex is active in primarily perceptual tasks that have strong and directed anticipatory components to them. When viewed as internal, goal-directed actions, it is no surprise that expectations activate those same brain areas that are involved in action planning of sequences that will ultimately be realized in overt motor output. Mental imagery studies provide prime examples that this form of expectation drives premotor regions.[8–10] Furthermore, these results help reduce the conceptual independence of perception and action, and instead encourage the viewing of music in a perception/action cycle framework that integrates multiple domain-general functions. The observation that the inferior frontal sulcus region was more active under task conditions emphasizing memorization is consistent with the critical role of this region in memory encoding and working memory, that is, the need to integrate past information with current information and expectations of how the future should behave,[11] also with respect to musical material.[12,13] More importantly, the differences in the activation patterns in the two attentional experiments described above (FIG. 1) indicate the flexibility with which the brain is recruited for processing musical stimuli. In other words, the task demands shape the activity that we label as the brain's processing of music.

DETECTING TONAL EXPECTANCY VIOLATIONS

One of the most common ways of examining how the brain responds to music has been to record what happens when expectations based on an established musical context are violated. Generally the contexts are tonal contexts comprising short chord progressions or melodies with target events usually, though not always, situated at the terminal position. Recordings of event-related brain potentials (ERPs) document differences in the processing of tonally related (expected) and unrelated (unexpected) targets. These effects manifest themselves in the P300, late-positive component (LPC), and P600 components of ERPs, which appear as positive peaks in voltage above parietal regions of the brain.[14–20] Although localization of the gen-

erators of the P300 has proven challenging, intracortical and fMRI recordings indicate that activity in this time window following salient events permeates attentional
networks and target detection networks, including the ventrolateral prefrontal cortex
(VLPFC) and parietal cortex.[21,22] Activity in the ventrolateral prefrontal areas in response to tonal violations is bilateral, with a right-hemisphere bias, and manifests
itself as a negative deflection in the ERP that has been termed a right anterotemporal
negativity (RATN[23]) or early right anterior negativity (ERAN).[24,25] The role of ventrolateral prefrontal regions in the processing of tonal violations is further supported
by fMRI data that show stronger blood oxygen level–dependent (BOLD) responses
to deviant targets.[26,27] It is important to note that, in the domain of music, these areas
do not serve solely to detect violations per se, but might serve in a more integrative
capacity that helps maintain a sense of coherence across a sequence of events. Thus
the VLPFC shows stronger BOLD responses to passages of music played normally
compared with passages of the same music that have been scrambled by reversing
the acoustical signal in short time windows.[28]

TRACKING MOVEMENT THROUGH TONAL SPACE

Much of the research on tonal expectancy violations in music has been motivated
by the hypothesis that the ongoing manipulation of expectations in tonal space
shapes our emotional experience of music.[29] However, when considering responses
to rather stark expectancy violations in short contexts, we need to ask ourselves how
well these short contexts reflect or model our affective responses to extended passages of music. What brain areas are responsive to more subtle movements through
tonal space that create the ebb and flow of our emotional responses?

To address this question, we capitalized on the fact that tonal space in Western
tonal music is represented parsimoniously on the surface of a torus.[30–32] Using the
distributional information about the relative activation among pitch classes in short
windows of time, that is, what notes are being played how often, one can view the
time-varying activation in tonal space on the surface of a torus.[33,34] In order to identify brain areas that follow the movement pattern of a musical stimulus through tonal
space, we composed a melody that systematically moves through the 24 major and
minor keys,[35] and asked subjects to perform either a tonal-deviance judgment task[35]
or timbral-deviance detection task in order to maintain their attention on the melody
while we recorded the BOLD signal in an fMRI experiment (see more information,
see Ref. 34). Each subject participated in at least three fMRI sessions, so that we
could identify those brain areas that consistently showed correlates of tonality tracking in the BOLD signal.

In addition to including subject responses to deviants and general task block
effects in our regression model, we included respiration and heart rate recordings, as
well as a parametric description of the melody's movement through tonal space, obtained using the IPEM Toolbox.[36] Because activation on the torus was modeled with
a 16×12 grid consisting of 192 locations, we wanted to find a way of reducing the
number of regressors that was necessary for describing the time-varying activity on
the toroidal surface while maintaining the fidelity of the activation patterns that
could arise. Decomposing the activation patterns using spherical harmonics proved
very useful in this regard. Equation 1 shows the formula for the decomposition,[37]

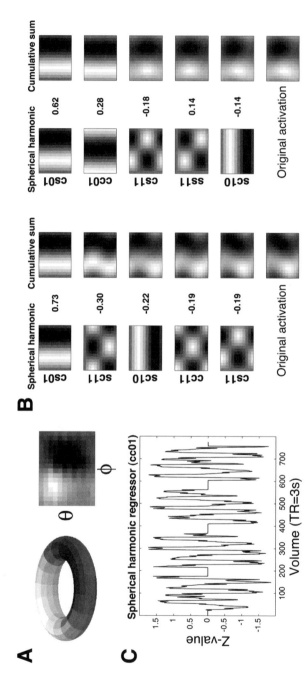

FIGURE 2. Examples of how an activation pattern on a toroidal surface is decomposed into its constituent spatial frequencies about each of the axes (θ and ϕ) using the spherical harmonics equation (Eq 1). (**A**) The activation pattern on the torus at one moment in time as it appears in a folded and unfolded (rectangular) version. (**B**) The original activation patterns at two different moments in time are composed of the weighted sum of relatively few spherical harmonics components. The components are shown on the *left* and labeled with the corresponding term in the equation. For example, "sc10" corresponds to the $\sin(1\theta)\cos(0\phi)$. Because $\cos(0) = 1$, the values are constant across the phi dimension. The amplitude for each term is shown between each pair of columns of images, and the cumulative sum images on the *right* show the convergence on the original activation pattern. *White* and *black* represent areas of strong (positive) and weak (negative) activation on the torus, respectively. (**C**) An example of the time-varying amplitude values of one spherical harmonic component that are entered as a regressor in the multiple regression model with which the fMRI BOLD activation data are analyzed. This regressor shows the values of the "cc01" component during 4 runs through the melody, where the runs are separated by a section of flat line. Because the melody started in different keys during the different runs, the pattern of positive and negative values in the waveform is phase-shifted from run to run.

$$f(\theta, \phi) = \sum_{m, n}^{\infty} a_{mn}^{cc} \cos(m\theta)\cos(n\phi) + \sum_{m, n}^{\infty} a_{mn}^{cs} \cos(m\theta)\sin(n\phi) +$$

$$\sum_{m, n}^{\infty} a_{mn}^{sc} \sin(m\theta)\cos(n\phi) + \sum_{m, n}^{\infty} a_{mn}^{ss} \sin(m\theta)\sin(n\phi)$$

(1)

where theta (θ) and phi (ϕ) are the phase angles along each of the circular axes of the torus, and m and n are the integers, starting at 0, that correspond to the spatial frequency along the theta and phi dimensions, respectively. The amplitude (contribution) of any given spherical harmonic term is given by a. The superscripts and subscripts on a simply track which term (coscos, cossin, etc.) and pair of spatial frequencies is being designated by a. FIGURE 2B illustrates that the activation pattern

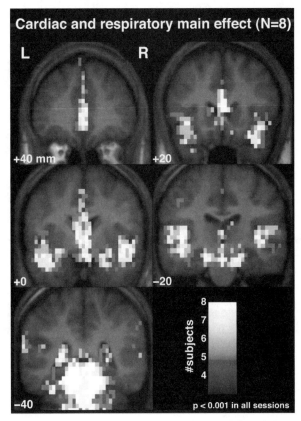

FIGURE 3. Group summary maps of regions where BOLD activity was significantly correlated with heart rate and respiration in all sessions that a subject participated in. *Color scale* intensity reflects the number of subjects for whom the particular voxel was consistently activated. See text for more details, and the online version for the color figure.

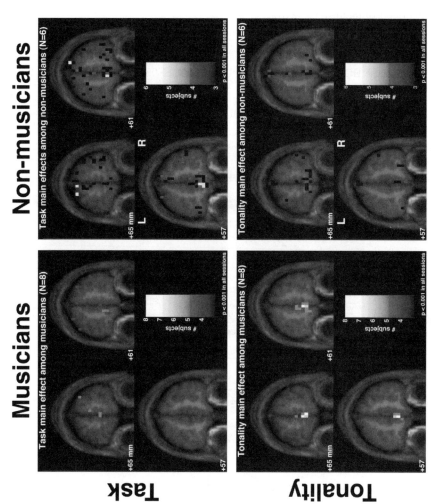

FIGURE 4. *See following page for legend.*

on the surface of the torus at any given moment in time is the weighted sum of relatively few spherical harmonics terms. FIGURE 2C shows the time-varying amplitudes of one spherical harmonic term. In the analysis of the fMRI data, 35 of the spherical harmonics terms (starting with the lowest spatial frequencies) were used, and the proportion of variance they explain was tested for them as a group.

Although heart rate and respiration were recorded in order to remove the variance associated with them from the data, thereby increasing sensitivity to detect the tonality effects of interest, the activation maps for these "nuisance" parameters potentially contain important information regarding interactions of the musical stimulus to which the subject is listening and physiological parameters that are known to vary with the emotional state of a person.[38–40] FIGURE 3 shows the distribution of voxels in which the BOLD response was significantly ($P < .001$) correlated with the heart rate and respiration regressors in every session in which the subject participated. The statistical maps likely reflect a number of influences, including pulsation effects along the brain stem and ventral midline. Additional extensive activation was observed bilaterally throughout the insula, a region known to regulate heart rate.[41,42] Interestingly, significant activations were also observed repeatedly in the auditory cortex along the superior temporal gyrus (STG). Though preliminary, these analyses delineate a network of interest for future experiments on autonomic responses to musical stimuli.

Summary activation maps for the task and tonality parameters are shown in FIGURE 4 in sections through rostral aspects of the medial prefrontal cortex. These maps show the number of subjects showing significant activation ($P < .001$) in all sessions at each voxel. Both main task effects and tonality effects were observed in musicians and nonmusicians in BA10/11. We were led to focus on this region because BOLD responses within it were consistently correlated with the tonality regressors in our cohort of musicians. FIGURE 5C shows activation patterns within this region showing more detail for two subjects from the original group of musically trained subjects. For each voxel that was significantly active in all scanning sessions for the subject, information is given about its average responsiveness to the melody as it moved through tonal space. A "tonality sensitivity surface" (TSS) was calculated for each voxel by entering the parameter estimates (beta values) from the regression model for each of the 35 spherical harmonic regressors into the spherical harmonic equation (Eq. 1). Thus, these TSSs represent the average activation pattern on the toroidal surface that the voxel was sensitive to. The TSSs were correlated with the average pattern on the torus that the stimulus created when it dwelled in each key (FIG. 5A), and the voxel was then color coded as belonging to one of three key regions (FIG. 5B). Example TSSs are shown in FIGURE 5C for several voxels.

FIGURE 4. Group summary maps of regions where BOLD activity was significantly correlated with task and tonality regressors in all sessions that a subject participated in. *Color scale* intensity reflects the number of subjects for whom the particular voxel was consistently activated. Please see the online version for the color figure.

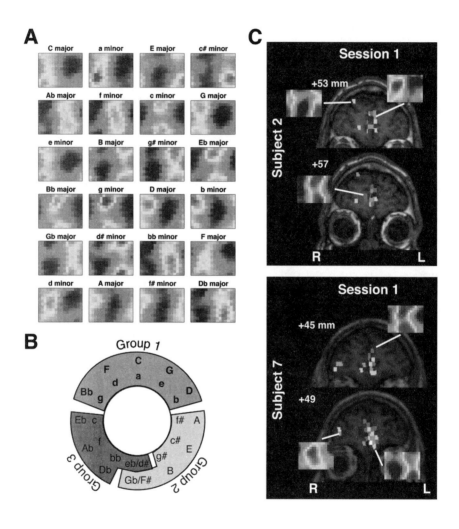

FIGURE 5. A more detailed view of the responses of rostral medial prefrontal cortex to the melody's movement through tonal space. (**A**) The average activation pattern created on the torus by the melody while it dwelled in each key. (**B**) The major and minor keys were divided into three regions that were used to code which region a voxel was, on average, most sensitive to. (**C**) Voxels in which the BOLD activity was significantly correlated in every session for the subject are coded with the key region that they responded most strongly to, on average, within one session. Examples of the tonality sensitivity surfaces for that voxel are shown in the insets. See text for more details, and the color figure in the online version. Adapted from Janata *et al.*[34]

FUNCTIONS OF THE ROSTRAL MEDIAL PREFRONTAL CORTEX

The observation of tonality tracking by the rostral medial prefrontal cortex (RMPFC) was initially puzzling. However, since the time of these initial observations, a considerable functional neuroimaging literature has emerged that pertains to these areas and provides a context for our findings that is consistent with the view that music's movement through tonal space is what shapes our personal affective experience of the music.

The RMPFC appears to be particularly active when subjects are asked to make self-referential judgments[43,44] or evaluative judgments about how they feel about a particular stimulus or concept suggested by a stimulus.[45,46] Metanalyses indicate that this region is generally involved in the cognitive control and evaluation of emotion.[47,48] The observations that the RMPFC is more active during introspective self-directed mental states is also in accord with findings of higher activity in medial areas during the resting or "default state."[49,50] Overall, the dichotomy between internally directed and externally directed mental states maps on quite well to the medial/lateral activation dichotomy.[51] Yet other studies have described a functional interaction between medial and lateral prefrontal areas in the context of attentional switching between monitoring of external stimuli and monitoring of internal goal states or stimulus-independent tasks.[52–56]

Three other sets of observations start to fill out a picture of how music might interact with the RMPFC. First, parts of the RMPFC and primarily regions ventral to it are sensitive to the relative amounts of consonance/dissonance in chords accompanying melodies.[57] Second, familiarity judgments of sets of stimuli in which 50% are, on average, familiar to listeners lead to increased activity in RMPFC.[58] Third, there is a pair of observations that tentatively suggests a combined role of RMPFC in music and autobiographical memory. Largely anecdotal evidence (that is slowly gaining empirical support) suggests that Alzheimer disease (AD) patients respond very positively to familiar music from their childhood, often singing along and readily detecting deviances implanted in the musical stimuli.[59] Interestingly, the rostral and ventral aspects of the medial prefrontal cortex are among the last in which significant cortical atrophy is observed in AD patients.[60] These observations together with those discussed earlier regarding the role of RMPFC in self and emotion judgments generate a hypothesis that the RMPFC is a locus at which music and autobiographical memories are bound together. We look forward to addressing this hypothesis as well as the mechanisms of interaction between the expectancy violation–sensing regions of the VLPFC and tonality-tracking regions of the RMPFC in our future research.

ACKNOWLEDGMENTS

The work described in this paper benefitted from the assistance and helpful comments of my colleagues Barbara Tillmann, Jeffrey Birk, Jack Van Horn, Marc Leman, and Jamshed Bharucha, to whom I am grateful. The work was also supported by National Institutes of Health Grants P50 NS17778-18 and R03 DC05146-01.

[Competing interests: The author declares that he has no competing financial interests.]

REFERENCES

1. KOELSCH, S. 2005. Neural substrates of processing syntax and semantics in music. Curr. Opin. Neurobiol. **15:** 207–212.
2. PATEL, A.D. 2003. Language, music, syntax, and the brain. Nat. Neurosci. **6:** 674–681.
3. PERETZ, I. & M. COLTHEART. 2003. Modularity of music processing. Nat. Neurosci. **6:** 688–691.
4. JANATA, P., B. TILLMANN & J.J. BHARUCHA. 2002. Listening to polyphonic music recruits domain-general attention and working memory circuits. Cogn. Affect. Behav. Neurosci. **2:** 121–140.
5. GREZES, J. & J. DECETY. 2001. Functional anatomy of execution, mental simulation, observation, and verb generation of actions: a meta-analysis. Hum. Brain Mapp. **12:** 1–19.
6. SCHUBOTZ, R.I. & D.Y. VON CRAMON. 2002. Dynamic patterns make the premotor cortex interested in objects: influence of stimulus and task revealed by fMRI. Brain Res. Cogn. Brain Res. **14:** 357–369.
7. SCHUBOTZ, R.I. & D.Y. VON CRAMON. 2001. Functional organization of the lateral premotor cortex: fMRI reveals different regions activated by anticipation of object properties, location, and speed. Cogn. Brain Res. **11:** 97–112.
8. HALPERN, A.R. & R.J. ZATORRE. 1999. When that tune runs through your head: a PET investigation of auditory imagery for familiar melodies. Cerebr. Cortex **9:** 697–704.
9. LANGHEIM, F.J.P., J.H. CALLICOTT, V.S. MATTAY, et al. 2002. Cortical systems associated with covert music rehearsal. Neuroimage **16:** 901–908.
10. MEISTER, I.G., T. KRINGS, H. FOLTYS, et al. 2004. Playing piano in the mind: an fMRI study on music imagery and performance in pianists. Cogn. Brain Res. **19:** 219–228.
11. FUSTER, J.M. 2000. Executive frontal functions. Exp. Brain Res. **133:** 66–70.
12. GAAB, N., C. GASER, T. ZAEHLE, et al. 2003. Functional anatomy of pitch memory: an fMRI study with sparse temporal sampling. Neuroimage **19:** 1417–1426.
13. ZATORRE, R. J., A. C. EVANS & E. MEYER. 1994. Neural mechanisms underlying melodic perception and memory for pitch. J. Neurosci. **14:** 1908–1919.
14. BESSON, M. & F. MACAR. 1987. An event-related potential analysis of incongruity in music and other nonlinguistic contexts. Psychophysiology **24:** 14–25.
15. PALLER, K.A., G. MCCARTHY & C.C. WOOD. 1992. Event-related potentials elicited by deviant endings to melodies. Psychophysiology **29:** 202–206.
16. JANATA, P. 1995. ERP measures assay the degree of expectancy violation of harmonic contexts in music. J. Cogn. Neurosci. **7:** 153–164.
17. BESSON, M. & F. FAÏTA. 1995. An event-related potential (ERP) study of musical expectancy: comparison of musicians with nonmusicians. J. Exp. Psychol. Hum. Percept. Perform. **21:** 1278–1296.
18. PATEL, A.D., E. GIBSON, J. RATNER, et al. 1998. Processing syntactic relations in language and music: an event-related potential study. J. Cogn. Neurosci. **10:** 717–733.
19. BEISTEINER, R., M. ERDLER, D. MAYER, et al. 1999. A marker for differentiation of capabilities for processing of musical harmonies as detected by magnetoencephalography in musicians. Neurosci. Lett. **277:** 37–40.
20. REGNAULT, P., E. BIGAND & M. BESSON. 2001. Different brain mechanisms mediate sensitivity to sensory consonance and harmonic context: evidence from auditory event-related brain potentials. J. Cogn. Neurosci. **13:** 241–255.
21. HALGREN, E., K. MARINKOVIC & P. CHAUVEL. 1998. Generators of the late cognitive potentials in auditory and visual oddball tasks. Electroencephalogr. Clin. Neurophysiol. **106:** 156–164.
22. KIEHL, K.A., K.R. LAURENS, T.L. DUTY, et al. 2001. Neural sources involved in auditory target detection and novelty processing: an event-related fMRI study. Psychophysiology **38:** 133–142.
23. PATEL, A.D. 1998. Syntactic processing in language and music: different cognitive operations, similar neural resources? Mus. Percept. **16:** 27–42.
24. KOELSCH, S., T. GUNTER, A.D. FRIEDERICI, et al. 2000. Brain indices of music processing: "nonmusicians" are musical. J. Cogn. Neurosci. **12:** 520–541.
25. MAESS, B., S. KOELSCH, T. C. GUNTER, et al. 2001. Musical syntax is processed in Broca's area: an MEG study. Nat. Neurosci. **4:** 540–545.

26. KOELSCH, S., T.C. GUNTER, D.Y. VON CRAMON, et al. 2002. Bach speaks: a cortical "language-network" serves the processing of music. Neuroimage 17: 956–966.
27. TILLMANN, B., P. JANATA & J.J. BHARUCHA. 2003. Activation of the inferior frontal cortex in musical priming. Cogn. Brain Res. 16: 145–161.
28. LEVITIN, D.J. & V. MENON. 2003. Musical structure is processed in "language" areas of the brain: a possible role for Brodmann Area 47 in temporal coherence. Neuroimage 20: 2142–2152.
29. MEYER, L.B. 1956. Emotion and Meaning in Music. University of Chicago Press. Chicago.
30. KRUMHANSL, C.L. & E.J. KESSLER. 1982. Tracing the dynamic changes in perceived tonal organization in a spatial representation of musical keys. Psychol. Rev. 89: 334–368.
31. KRUMHANSL, C.L. 1990. Cognitive Foundations of Musical Pitch. Oxford University Press. New York.
32. LEMAN, M. 1995. Music and Schema Theory: Cognitive Foundations of Systematic Musicology. Springer-Verlag. Berlin.
33. TOIVIAINEN, P. & C.L. KRUMHANSL. 2003. Measuring and modeling real-time responses to music: the dynamics of tonality induction. Perception 32: 741–766.
34. JANATA, P., J.L. BIRK, J.D. VAN HORN, et al. 2002. The cortical topography of tonal structures underlying Western music. Science 298: 2167–2170.
35. JANATA, P., J. L. BIRK, B. TILLMANN, et al. 2003. Online detection of tonal pop-out in modulating contexts. Mus. Percept. 20: 283–305.
36. LEMAN, M. 1995. A model of retroactive tone-center perception. Mus. Percept. 12: 439–471.
37. BOYD, J.P. 2001. Chebyshev and Fourier Spectral Methods. Dover. New York.
38. CRITCHLEY, H.D., P. ROTSHTEIN, Y. NAGAI, et al. 2005. Activity in the human brain predicting differential heart rate responses to emotional facial expressions. Neuroimage 24: 751–762.
39. KUNIECKI, M., A. URBANIK, B. SOBIECKA, et al. 2003. Central control of heart rate changes during visual affective processing as revealed by fMRI. Acta Neurobiol. Exp. 63: 39–48.
40. KRUMHANSL, C.L. 1997. An exploratory study of musical emotions and psychophysiology. Can. J. Exp. Psychol. 51: 336–353.
41. VOLKOW, N.D., G.J. WANG, J.S. FOWLER, et al. 2000. Increased activity of the temporal insula in subjects with bradycardia. Life Sci. 67: 2213–2220.
42. CRITCHLEY, H.D., D.R. CORFIELD, M.P. CHANDLER, et al. 2000. Cerebral correlates of autonomic cardiovascular arousal: a functional neuroimaging investigation in humans. J. Physiol. (Lond.) 523: 259–270.
43. GUSNARD, D. A., E. AKBUDAK, G. L. SHULMAN, et al. 2001. Medial prefrontal cortex and self-referential mental activity: relation to a default mode of brain function. Proc. Natl. Acad. Sci. USA 98: 4259–4264.
44. KELLEY, W.M., C.N. MACRAE, C.L. WYLAND, et al. 2002. Finding the self? An event-related fMRI study. J. Cogn. Neurosci. 14: 785–794.
45. ZYSSET, S., O. HUBER, E. FERSTL, et al. 2002. The anterior frontomedian cortex and evaluative judgment: an fMRI study. Neuroimage 15: 983–991.
46. ZYSSET, S., O. HUBER, A. SAMSON, et al. 2003. Functional specialization within the anterior medial prefrontal cortex: a functional magnetic resonance imaging study with human subjects. Neurosci. Lett. 335: 183–186.
47. OCHSNER, K.N., S.A. BUNGE, J.J. GROSS, et al. 2002. Rethinking feelings: an fMRI study of the cognitive regulation of emotion. J. Cogn. Neurosci. 14: 1215–1229.
48. OCHSNER, K.N. & J.J. GROSS. 2005. The cognitive control of emotion. Trends Cogn. Sci. 9: 242–249.
49. RAICHLE, M.E., A.M. MACLEOD, A.Z. SNYDER, et al. 2001. A default mode of brain function. Proc. Natl. Acad. Sci. USA 98: 676–682.
50. GREICIUS, M.D., B. KRASNOW, A.L. REISS, et al. 2003. Functional connectivity in the resting brain: a network analysis of the default mode hypothesis. Proc. Natl. Acad. Sci. USA 100: 253–258.
51. FOX, M.D., A.Z. SNYDER, J.L. VINCENT, et al. 2005. The human brain is intrinsically organized into dynamic, anticorrelated functional networks. Proc. Natl. Acad. Sci. USA 102: 9673–9678.

52. KOECHLIN, E., G. CORRADO, P. PIETRINI, *et al.* 2000. Dissociating the role of the medial and lateral anterior prefrontal cortex in human planning. Proc. Natl. Acad. Sci. USA **97:** 7651–7656.
53. KOECHLIN, E., C. ODY & F. KOUNEIHER. 2003. The architecture of cognitive control in the human prefrontal cortex. Science **302:** 1181–1185.
54. GILBERT, S.J., C.D. FRITH & P.W. BURGESS. 2005. Involvement of rostral prefrontal cortex in selection between stimulus-oriented and stimulus-independent thought. Eur. J. Neurosci. **21:** 1423–1431.
55. BURGESS, P. W., J. S. SIMONS, I. DUMONTHEIL, *et al.* 2005. The gateway hypothesis of rostral prefrontal cortex (area 10) function. *In* Measuring the Mind: Speed, Control, and Age. J. Duncan, P. McLeod & L. Phillips, Eds.: 251–246. Oxford University Press. Oxford.
56. RAMNANI, N. & A.M. OWEN. 2004. Anterior prefrontal cortex: insights into function from anatomy and neuroimaging. Nat. Rev. Neurosci. **5:** 184–194.
57. BLOOD, A.J., R.J. ZATORRE, P. BERMUDEZ, *et al.* 1999. Emotional responses to pleasant and unpleasant music correlate with activity in paralimbic brain regions. Nat. Neurosci. **2:** 382–387.
58. PLATEL, H., J.C. BARON, B. DESGRANGES, *et al.* 2003. Semantic and episodic memory of music are subserved by distinct neural networks. Neuroimage **20:** 244–256.
59. CUDDY, L.L. & J. DUFFIN. 2005. Music, memory, and Alzheimer's disease: is music recognition spared in dementia, and how can it be assessed? Med. Hypotheses **64:** 229–235.
60. THOMPSON, P.M., K.M. HAYASHI, G. DE ZUBICARAY, *et al.* 2003. Dynamics of gray matter loss in Alzheimer's disease. J. Neurosci. **23:** 994–1005.

Neural Encoding and Retrieval of Sound Sequences

JOSEF P. RAUSCHECKER

Laboratory of Integrative Neuroscience and Cognition, Department of Physiology and Biophysics, Georgetown University Medical Center, Washington, DC 20007, USA

ABSTRACT: Although considerable progress has been made recently in our understanding of the coding of complex sounds in the cerebral cortex, the processing and storage of tone sequences is still poorly understood. We have used functional magnetic resonance imaging to identify brain mechanisms involved in the encoding and retrieval of melodies by studying the anticipation of familiar music. The results suggest a specific role for each of the following brain structures: the anterior part of the right superior temporal cortex, the right inferior frontal cortex and anterior insula, the left anterior prefrontal cortex, the lateral cerebellum, and the anterior cingulate. In a separate study, we investigated single-neuron responses in the auditory cortex of awake behaving monkeys to alternating tone sequences that in humans evoke the perception of "streaming." Depending on the frequency separation between the tones, an initial single stream may segregate into two streams after a build-up period of several seconds. The neural responses in the monkeys' primary auditory cortex (A1) mirror the psychophysical time course extremely well, suggesting that habituation within A1 may be one reason for stream segregation. However, the higher auditory and prefrontal areas found to be activated by musical melodies are expected to interact with primary areas in both bottom-up and top-down fashion to bring about the perceptual organization of sound sequences.

KEYWORDS: complex sounds; tone sequences; melodies; music perception; auditory cortex; functional magnetic resonance imaging (fMRI); single-neuron recording; nonhuman primates; memory; neuroanatomy

INTRODUCTION

The last decade has brought considerable progress in our understanding of how complex sounds are processed by the brain, both in humans and in appropriate animal models.[1,2] This increased interest in the processing of complex sounds is motivated in part by the recognition that simple pure tones, as they have traditionally been used in auditory physiology, do not excite many neurons in advanced stations of the auditory pathway. Instead, these neurons respond preferentially to more complex sounds. Higher areas of auditory cortex are, presumably, responsible also for

Address for correspondence: J.P. Rauschecker, Laboratory of Integrative Neuroscience and Cognition, Georgetown University Medical Center, New Research Building, Room WP15, Box 571460, Washington, DC 20057-1460. Voice: 202-687-1580; fax: 202-687-0617.

rauschej@georgetown.edu

Ann. N.Y. Acad. Sci. 1060: 125–135 (2005). © 2005 New York Academy of Sciences.
doi: 10.1196/annals.1360.009

processing the sounds that make up speech or music. The motivation to understand neural representations of speech and music, therefore, requires the use of sound stimuli that are of comparable complexity. Both speech and music typically consist of sound sequences, so studies on how such sequences are processed and stored by the brain are gaining particular importance.

Noninvasive studies with functional magnetic resonance imaging (fMRI) can be used to investigate the effects of speech and music directly in the human brain. This approach can highlight areas involved in certain aspects of speech and music processing. However, the actual neural mechanisms underlying these extraordinary abilities can only be understood by studying the neural networks of the brain at the level of single neurons. This can be accomplished by microelectrode studies of awake behaving animals, most notably nonhuman primates, whose brains are most similar to that of humans. The present chapter will present two examples of studies relevant for the neural basis of music processing: a neuroimaging study on anticipation, storage, and recall of musical sequences (melodies); and a study in awake behaving monkeys on the neural correlates of the perceptual segregation between two parallel tone sequences.

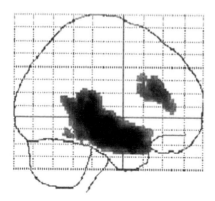
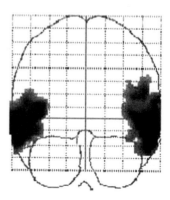

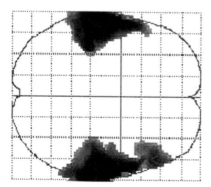

FIGURE 1. Activation of the cerebral cortex while listening to music that the subjects had never heard before (unfamiliar music). Auditory areas of the superior temporal cortex in both hemispheres are activated. Right midfrontal cortex (BA 46) is activated only by unfamiliar (but not familiar) music, suggesting that it participates in the encoding process of new melodies. Statistical parametric maps of fMRI group data are shown as sagittal, coronal, and axial projections. All voxels significant at the $P < .01$ level, with correction for multiple comparisons, are shown.

PROCESSING OF MUSICAL SEQUENCES

It is self-evident that music of any kind will activate the auditory pathways throughout all stages. Auditory cortex is activated by various kinds of music, although many studies (e.g., Ref. 3) have shown that cortex in the right hemisphere is generally activated more by it than cortex in the left hemisphere, which is more specialized for language. This may have to do with a preference for different time constants in the two hemispheres.[4,5] Different aspects of music (melody, rhythm) are processed differentially by the two hemispheres.[6–8] In summary, it is clear that both cerebral hemispheres participate in the processing of music, although perhaps to a different extent and in different roles. FIGURE 1 clearly demonstrates vast activation by music of auditory cortical areas in the superior temporal cortex on both sides.

It is also intuitively clear that extended tone sequences, as they are typically contained in music, require a different kind of coding than, for instance, single isolated sounds (e.g., chords). Despite their being drawn out over seconds or even minutes, melodies are usually perceived as a single entity or "auditory object." That means that the tone sequences that make up a melody or tune must be stored together and can be recalled as one entity. Indeed, studies with musical imagery have shown that playing just a few tones of a well-known melody will enable people to complete the tune quite easily.[9] To find out what this means for the way melodies (or sound sequences in general, as this is equally relevant for speech) are encoded and stored in the brain is one of the goals of our current research.

ANTICIPATION OF MUSICAL SEQUENCES

The following observation is well known to every music lover: When two melodies are concatenated with a silent interval of several seconds in-between and are frequently heard in that sequence (for instance, consecutive movements of a classical symphony, or sound tracks on a compact disc), the beginning of the second melody can be anticipated quite vividly at the end of the first, that is, during the silent interval preceding the second melody. In fact, this anticipation can be so vivid that the second melody pops up in one's "inner ear" almost automatically and is virtually impossible to suppress. This common observation demonstrates that tone sequences are stored and integrated over long durations. It also shows that perception of the preceding, highly familiar sequence triggers the recall of the subsequent sequence, almost as in a domino effect. Such "anticipatory readout" requires a neural circuit in which perceptual coding and retrieval mechanisms are intimately intertwined. Memory storage of musical melodies necessitates the preservation of time as well as frequency intervals in the correct order, and their retrieval has to reproduce that information. Neurons in nonprimary auditory cortex of macaques show the capacity for temporal integration over several 100 milliseconds.[1,10] However, such a time window is not long enough to store "templates" of melodic sequences with durations of several seconds or minutes. It has been suggested that the only way this can be done is by means of chained associations.[11]

Using fMRI, we have studied brain activation during silent anticipation of music and have contrasted it with activation while actually listening to music. Subjects were first screened for the reliability of their anticipatory imagery using sound tracks from their favorite compact disc. Each subject listened to the final 32 s of each track in random order and was asked to produce the opening bars of the following track, which did not contain any lyrics. Only subjects who scored >75% correct were used for further study. While subjects listened to music, whether familiar or unfamiliar, extensive regions of superior temporal cortex in both hemispheres were reliably activated (FIG. 1). Unfamiliar (but not familiar) music activated the midfrontal region (BA 46) exclusively on the right. This may be indicative of a semantic encoding process, which leads to the storage of the novel stimuli in memory. Although in most cases encoding into long-term memory has been associated with left prefrontal activation,[12] encoding of nonverbal material has been shown to activate right dorsolateral prefrontal cortex.[13–15] According to our data, this seems to include the encoding of musical melodies.

Activation during silent anticipation of the next track following familiar music ("familiar silence") was compared with activation during silence following unfamiliar music, which the subjects had not heard before ("unfamiliar silence"). During familiar but not during unfamiliar silence, subjects demonstrated significant activation in the following brain regions (FIG. 2): (1) in the left hemisphere: anterior prefrontal cortex (BA 10); (2) in the right hemisphere: inferior frontal cortex (BA 47)

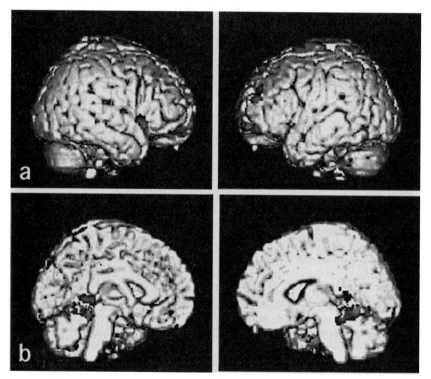

FIGURE 2. Brain activation during silent anticipation of a familiar musical sequence. Statistical parametric maps have been rendered onto the lateral (**a**) and medial (**b**) surfaces of a representative 3-D brain volume. All voxels significant at the $P < .01$ level, with correction for multiple comparisons, are shown. For further description see text.

near the frontal operculum extending into the anterior insular cortex and, in a separate focus of activation, anterior superior temporal corresponding to rostral auditory association cortex (BA 22); (3) bilaterally, but predominantly on the left: posterior lateral cerebellum. Some, but not all, subjects activated the anterior cingulate region (BA 32) bilaterally.

Activation of the left anterior prefrontal cortex (BA 10) is typically associated with the retrieval of information from semantic memory.[12] Activity in the inferior frontal region (BA 47) and the frontal operculum and anterior insula on the left has been implicated in the formulation of an articulatory plan in speech,[16] and the same regions on the right are activated during singing.[17,18] Consistent with previous studies of auditory imagery,[9,19,20] no activation of primary auditory cortex was seen. Finally, the consistent activation in all subjects of the lateral cerebellar hemispheres during anticipatory musical imagery underscores the general importance of the cerebellum in sensory and cognitive processing,[21,22] especially when timing information is involved.[23,24] A more detailed discussion of the role of the cerebellum and its connections with prefrontal cortex[25] will be developed elsewhere (Rauschecker *et al.*, in preparation).

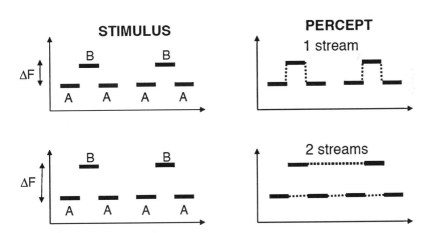

FIGURE 3. Auditory streams: stimuli and percepts. (A) Stimulus sequences used to measure perceptual auditory streaming (*left*) and corresponding auditory percepts (*right*). The stimulus sequences consisted of repeating tone triplets, ABA, where A and B represent tones of different frequencies. The *top panels* illustrate the case of a small frequency separation (ΔF) between the A and B tones; the corresponding percept is that of a single stream of connected tones, with a distinctive galloping rhythm: ABA-ABA-..., where the dash stands for a silent gap. The *lower panel* illustrates the case of a large ΔF between the A and B tones; this typically evokes a percept of two mono-tonic streams with different tempi playing in parallel: A-A-A-A... on the one hand, -B---B on the other. (Reproduced with permission from Micheyl *et al.*[26]).

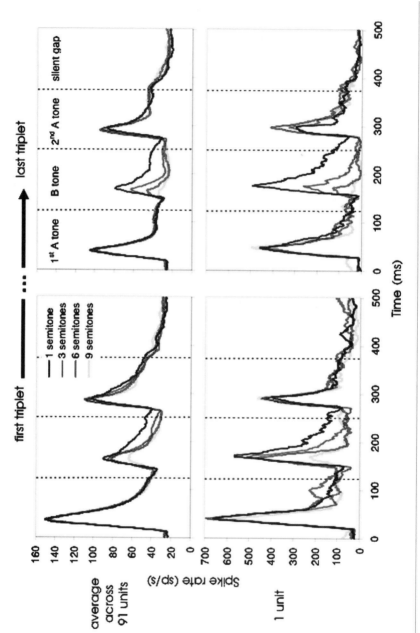

FIGURE 4. *See following page for legend.*

PERCEPTUAL ORGANIZATION OF TONE SEQUENCES
IN AWAKE BEHAVING MONKEYS

In order to investigate the neural coding of tone sequences at the single-cell level, rhesus monkeys with a chronic recording cylinder were trained to sit in a primate chair while they were listening to the tone sequences, and activity of single neurons was recorded from their auditory cortex.[26] In addition, we made use of a well-established perceptual phenomenon that arises when one listens to tones alternating between two frequencies, A and B, as illustrated in FIGURE 3. When the frequency separation (ΔF) between A and B is small, and the tones alternate slowly, listeners hear a coherent sequence of tones whose pitch jumps up and down. By contrast, when ΔF is large and/or the repetition rate (RR) is fast, two parallel but separate streams of constant-pitch tones are heard. Despite long-ranging efforts, the brain mechanisms underlying this perceptual phenomenon remain unclear.

In our recent study,[26] we have taken advantage of the fact that stream segregation generally takes some time to build up:[37–39] Initially, sound sequences tend to be heard as a single, fused stream, and only after several seconds the sounds appear to split into distinct streams that can be individually followed. This systematic switch of the auditory percept makes it possible to compare neural responses associated with dramatically different percepts without any change in the evoking stimulus, simply by recording neural responses at various points during an ongoing sequence of sounds.

Alternating tone sequences (ABA) were offered as stimuli. Single-unit responses in the primary auditory cortex (A1) of two monkeys were recorded. The frequency of the A tone was adjusted to correspond to the best frequency (BF) of the unit being tested, as estimated from preceding tuning-curve measurements. As in the psychophysical experiment, the frequency of the B tone was set 1, 3, 6, or 9 semitones above that of the A tone and remained constant within a sequence.

Two major trends were apparent in the poststimulus time histograms of neuronal responses to the ABA tone patterns (FIG. 4). The first trend was a general decrease of the B-tone responses as ΔF increased. This effect can be explained simply in terms of frequency selectivity: since the neurons, by definition, had their BF at the A-tone frequency, increasing ΔF amounted to moving the B tones further and further away from the most sensitive region of the frequency tuning curve. The second trend was a general decrease in response magnitude between the first and last triplets (compare left and right in FIG. 4). This response decay affected both A- and B-tone responses, at all frequency separations. To quantify this habituation and characterize its time course, we counted the number of spikes evoked in response to each of the succes-

---

FIGURE 4. Peristimulus time histograms of neural responses in primary auditory cortex of alert macaque monkeys to ABA triplets. *Left*: neural responses to the first triplet in the sequence. *Right*: neural responses to the last (20th) triplet in the sequence. *Top*: average response recorded in the two animals. *Bottom*: example response from a single unit. The *vertical dashed lines* mark the onset or offset times of the individual tones, relative to the response. The temporal positions of the A and B tones and of the intertriplet gap are indicated in the *upper right panel*. The different colors (*online version*) indicate responses obtained using different frequency separations (ΔF) between A and B, as shown in the legend in the *upper left panel*. (Reproduced with permission from Micheyl et al.[26]).

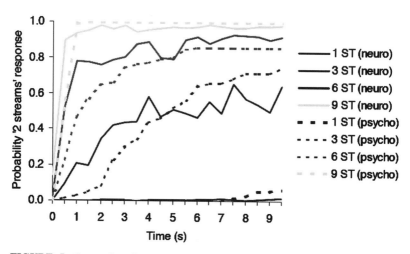

FIGURE 5. Comparison between psychometric and neurometric functions representing probabilities of "two-stream" responses as a function of time and ΔF. Psychometric functions from human psychophysics are plotted as *dashed lines*, to facilitate comparison with the neurometric functions from the monkeys, which are shown as *solid lines*. A good correspondence was found. The color-coding scheme (*online version*) for ΔF is the same as in FIGURE 4. See Ref. 26 for an updated version.

sive tones in the 10-s stimulus sequence and studied how the counts varied as a function of time after sequence onset. Overall, spike counts were found to decrease significantly as a function of triplet position (i.e., from the first to the last triplet in the sequence stimulus) [$F(1, 90) = 41.63$, $P < .0005$].

FROM SPIKES TO PERCEPTS: A SIMPLE MODEL OF AUDITORY STREAMING

In order to relate meaningfully the neural responses to the psychophysical data, we devised a simple model of how spike trains evoked by the ABA tone sequences could lead to the percept of one or two auditory streams. Details on the functioning of the model are given in Ref. 26. The probabilities of "two-stream" responses predicted as a function of both ΔF and time based on the measured neural responses, or "neurometric functions," are shown in FIGURE 5. The human psychometric functions are plotted as dashed lines to facilitate comparison. It can be seen that the neurometric functions replicate all the major features of the psychometric functions measured in human listeners. Specifically, (1) at all ΔFs except the smallest, the predicted probability of a "two-streams" response increased over time; (2) this increase was longer at the two intermediate ΔFs, 3 and 6 semitones; (3) at the largest ΔF (9 semitones), the predicted probability of a "two-streams" response increased abruptly after sequence onset. This good qualitative agreement between the neurometric and the psychometric functions was paralleled by a good quantitative agreement, as demonstrated by the fact that the former usually fell within the 95% confidence intervals around the latter; thus, the predictions based on neural data from the two monkeys fall within the range of the across-subject variability in the psychophysical data.

These results demonstrate that neural responses recorded in A1 while awake primates listen to tone sequences can account both qualitatively and quantitatively for several essential features of an important phenomenon of perceptual auditory organization in humans. The predictions derived from the neural data replicate successfully the increase in the probability of perceiving a tone sequence as two segregated auditory streams when the frequency separation (ΔF) between consecutive tones is increased. More importantly, the neurometric functions faithfully replicate the observed increases in the probability of hearing segregation (i.e., two streams) during the first 10 s following sequence onset at intermediate and large ΔFs. This result is noteworthy, because, unlike those obtained in previous studies,[27–29] our study demonstrates a correspondence between changes in neural responses and changes in perception, which occur without any concomitant change in the physical stimulus.

THE BRAIN LOCUS OF AUDITORY SEQUENCE PERCEPTION

Although the results presented here demonstrate that neural responses in A1 can readily account for some important features of auditory streaming, including its build-up, this does not mean that A1 is necessarily the final processing stage involved in auditory streaming. Judging from our fMRI studies, cortical areas beyond A1 are likely involved in the perceptual organization of sound sequences. In fact, the model proposed above posits that the perceptual decision between two alternative auditory objects (e.g., one or two streams) is based on the activity of neural structures that "read" the neural responses coming from A1. These neurons may thus be located in higher areas of the auditory object-identification ("what") pathway that originates in the anterolateral belt in nonhuman primates[2,10,30] or the equivalent areas in humans.[31–35] The anterolateral belt projects to the same areas of anterior prefrontal (and inferior frontal) cortex (BA 10)[31,36] that have been involved in the coding and retrieval of musical sequences in our first study. The remarkable convergence between findings in humans and nonhuman primates creates hope that at least some of the more basic neural mechanisms of music processing can be unraveled in animal models at the single- and multiunit levels.

ACKNOWLEDGMENTS

The following individuals have helped with the collection of the data mentioned in this chapter or have provided critical input: Jennifer Van Lare and Brandon Zielinski (fMRI studies); Bob Carlyon, Christophe Micheyl, Biao Tian (monkey studies). The work was funded by NIH Grant R01-DC03489 and NSF Grant 0350041.

[Competing interests: The author declares that he has no competing financial interests.]

REFERENCES

1. RAUSCHECKER, J.P., B. TIAN & M. HAUSER. 1995. Processing of complex sounds in the macaque nonprimary auditory cortex. Science 268: 111–114.
2. RAUSCHECKER, J.P. 1998. Cortical processing of complex sounds. Curr. Opin. Neurobiol. 8: 516–521.
3. ZATORRE, R.J., A.C. EVANS & E. MEYER. 1994. Neural mechanisms underlying melodic perception and memory for pitch. J. Neurosci. 14: 1908–1919.
4. BOEMIO, A. et al. 2005. Hierarchical and asymmetric temporal sensitivity in human auditory cortices. Nat. Neurosci. 8: 389–395.
5. ZATORRE, R.J. & P. BELIN. 2001. Spectral and temporal processing in human auditory cortex. Cereb. Cortex 11: 946–953.
6. PERETZ, I. 1990. Processing of local and global musical information in unilateral brain damaged patients. Brain 13: 1185–1205.
7. PLATEL, H. et al. 1997. The structural components of music perception. A functional anatomical study. Brain 120: 229–243.
8. PATEL, A.D. et al. 1998. Processing prosodic and musical patterns: a neuropsychological investigation. Brain Lang. 61: 123–144.
9. HALPERN, A.R. & R.J. ZATORRE. 1999. When that tune runs through your head: a PET investigation of auditory imagery for familiar melodies. Cereb. Cortex 9: 697–704.
10. TIAN, B. et al. 2001. Functional specialization in rhesus monkey auditory cortex. Science 292: 290–293.
11. EBBINGHAUS, H. 1964. Memory: A Contribution to Experimental Psychology. Dover. New York.
12. NYBERG, L., R. CABEZA & E. TULVING. 1996. PET studies of encoding and retrieval: The HERA model. Psychon. Bull. Rev. 3: 135–148.
13. KLINGBERG, T. & P.E. ROLAND. 1998. Right prefrontal activation during encoding, but not during retrieval, in a nonverbal paired-associates task. Cereb. Cortex 8: 73–79.
14. KELLEY, W.M. et al. 1998. Hemispheric specialisation in human dorsal frontal cortex and medial temporal lobe for verbal and nonverbal memory encoding. Neuron 20: 927–936.
15. WAGNER, A.D. et al. 1998. Material-specific lateralization of prefrontal activation during episodic encoding and retrieval. Neuroreport 9: 3711–3717.
16. WISE, R.J. et al. 1999. Brain regions involved in articulation. Lancet 353: 1057–1061.
17. PERRY, D.W. et al. 1999. Localization of cerebral activity during simple singing. Neuroreport 10: 3979–3984.
18. RIECKER, A. et al. 2000. Opposite hemispheric lateralization effects during speaking and singing at motor cortex, insula, and cerebellum. Neuroreport 11: 1997–2000.
19. ZATORRE, R.J. et al. 1996. Hearing in the mind's ear: a PET investigation of musical imagery and perception. J. Cogn. Neurosci. 8: 29–46.
20. GRIFFITHS, T.D. 2000. Musical hallucinosis in acquired deafness. Phenomenology and brain substrate. Brain 123: 2065–2076.
21. FIEZ, J.A. 1996. Cerebellar contributions to cognition. Neuron 16: 13–15.
22. PARSONS, L.M. & P.T. FOX. 1997. Sensory and cognitive functions. In The Cerebellum and Cognition. J.D. Schmahmann, Ed.: 255–271. Academic Press. San Diego.
23. IVRY, R. & S.W. KEELE. 1989. Timing functions of the cerebellum. J. Cogn. Neurosci. 1: 136–152.
24. PENHUNE, V.B., R. ZATORRE & A.C. EVANS. 1998. Cerebellar contributions to motor timing: a PET study of auditory and visual rhythm reproduction. J. Cogn. Neurosci. 10: 752–765.
25. MIDDLETON, F.A. & P.L. STRICK. 2001. Cerebellar projections to the prefrontal cortex of the primate. J. Neurosci. 21: 700–712.
26. MICHEYL, C. et al. 2005. Perceptual organization of tone sequences in the auditory cortex of awake macaques. Neuron 48: 139–148.
27. BEE, M.A. & G.M. KLUMP. 2004. Primitive auditory stream segregation: a neurophysiological study in the songbird forebrain. J. Neurophysiol. 92: 1088–1104.
28. FISHMAN, Y.I. et al. 2001. Neural correlates of auditory stream segregation in the primary auditory cortex of the awake monkey. Hear. Res. 151: 167–187.

29. FISHMAN, Y.I., J.C. AREZZO & M. STEINSCHNEIDER. 2004. Auditory stream segregation in monkey auditory cortex: effects of frequency separation, presentation rate, and tone duration. J. Acoust. Soc. Am. **116:** 1656–1670.
30. KAAS, J.H. & T.A. HACKETT. 2000. Subdivisions of auditory cortex and processing streams in primates. Proc. Natl. Acad. Sci. USA **97:** 11793–11799.
31. RAUSCHECKER, J.P. & B. TIAN. 2000. Mechanisms and streams for processing of "what" and "where" in auditory cortex. Proc. Natl. Acad. Sci. USA **97:** 11800–11806.
32. BINDER, J.R. et al. 2000. Human temporal lobe activation by speech and nonspeech sounds. Cereb.Cortex **10:** 512–528.
33. BINDER, J.R. et al. 2004. Neural correlates of sensory and decision processes in auditory object identification. Nat. Neurosci. **7:** 295–301.
34. WESSINGER, C.M. et al. 2001. Hierarchical organization of human auditory cortex revealed by functional magnetic resonance imaging. J. Cogn. Neurosci. **13:** 1–7.
35. ZATORRE, R.J., M. BOUFFARD & P. BELIN. 2004. Sensitivity to auditory object features in human temporal neocortex. J. Neurosci. **24:** 3637–3642.
36. ROMANSKI, L.M. et al. 1999. Dual streams of auditory afferents target multiple domains in the primate prefrontal cortex. Nat. Neurosci. **2:** 1131–1136.
37. ANSTIS, S. & S. SAIDA. 1985. Adaptation to auditory streaming of frequency-modulated tones. J. Exp. Psychol. **11:** 257–271.
38. BREGMAN, A.S. 1978. Auditory streaming is cumulative. J. Exp. Psychol. Hum. Percept. Perform. **4:** 380–387.
39. CARLYON, R.P., R. CUSACK, J.M. FOXTON & I.H. ROBERTSON. 2001. Effects of attention and unilateral neglect on auditory stream segregation. J. Exp. Psychol. Hum. Percept. Perform. **27:** 115–127.

Functional Neuroimaging of Semantic and Episodic Musical Memory

HERVÉ PLATEL

Inserm E.0218-Université de Caen, Centre Cyceron,
Laboratoire de Neuropsychologie, CHU Côte de Nacre, France

ABSTRACT: The distinction between episodic and semantic memory has become very popular since it was first proposed by Tulving in 1972. So far, very few neuropsychological, psychophysical, and imaging studies have related to the mnemonic aspects of music, notably on the long-term memory features, and practically nothing is known about the functional anatomy of long-term memory for music. Numerous functional imaging studies have shown that retrieval from semantic and episodic memory is subserved by distinct neural networks. For instance, the HERA model (hemispheric encoding/retrieval asymmetry) ascribes to the left prefrontal cortex a preferential role in the encoding process of episodic material and the recall of semantic information, while the right prefrontal cortex would preferentially operate in the recall of episodic information. However, these results were essentially obtained with verbal and visuospatial material. We have done a study to determine the neural substrates underlying the semantic and episodic components of music using familiar and nonfamiliar melodic tunes. Two distinct patterns of activations were found: bilateral activation of the middle and superior frontal areas and precuneus for episodic memory, and activation of the medial and orbital frontal cortex bilaterally, left angular gyrus, and the anterior part of the left middle and superior temporal gyri for semantic memory. We discuss these findings in light of the available neuropsychological data obtained in brain-damaged subjects and functional neuroimaging studies.

KEYWORDS: music; semantic memory; episodic memory; functional neuroimaging; HERA model

The neuropsychological study of music, first from clinical cases then more recently with the functional neuroimaging methods, is a unique and creative way to better understand links between cognitive processes (i.e., perception, attention, memory, and emotion) and the brain.[1] Numerous works of functional neuroimagery on the elementary aspects (i.e., pitch, rhythm, or timbre) of musical perception are now published.[2–4] So far, very few neuropsychological, psychophysical, and imaging studies have related to the mnemonic aspects of music, notably on the long-term memory features, and the concept of *musical memory* is considered very little in the psychological and neuropsychological literature. Experimental psychology and

Address for correspondence: Hervé Platel, Ph.D., EMI-E.0218 Inserm-Université de Caen, U.F.R. de Psychologie, Université de Caen, Esplanade de la Paix, 14032 Caen Cedex, France. Voice: +33-02-31-56-65-91.

herve.platel@unicaen.fr

Ann. N.Y. Acad. Sci. 1060: 136–147 (2005). © 2005 New York Academy of Sciences.
doi: 10.1196/annals.1360.010

psychoacoustic investigations mainly relate to working rather than long-term memory processes.[5-7] Experimental investigations regarding both recall of well-known tunes and new music generally concurred in showing the importance of melodic relative to rhythmic information in both the encoding and retrieval processes for musical material.[8-10] This supports the idea of a melodic advantage in the constitution of a musical lexicon.

Findings in brain-damaged patients suggest that although identification and recognition of a musical piece appear to involve both hemispheres, integrity of the left hemisphere is, however, critical,[11] as illustrated by a left brain–damaged patient who exhibited impaired melodic identification despite intact melodic discrimination.[12] Interestingly, these disturbances in melodic identification were found to be dissociated from the language impairment observed following cerebral lesions involving specifically the verbal semantic processes.[13] In addition, the studies carried out by Zatorre[14] and Samson and Zatorre[15] in patients with right or left temporal lobectomies revealed a major role of the right temporal areas in the recognition of unknown melodies, suggesting that such recognition, which is not based on a semantic strategy, is subtended by a perceptive analysis and a comparison of the melodies. The observation of patient C.N.[16] is particularly relevant, because, several years after having sustained bilateral temporal lesions, this patient was unable to memorize new musical tunes. Furthermore, there was a lack of priming effect for musical material in this patient, which suggested abolished access and encoding deficit for music. For Isabelle Peretz,[16] this clinical case argues for the existence of a long-term memory subsystem specific to musical material.

The distinction between episodic and semantic memory was first proposed by Tulving in 1972. Episodic memory is conceived as the memory of life events linked to their spatial and temporal context of acquisition, whereas semantic memory corresponds to the memory of concepts, transcending a particular context.[17] Earlier studies of brain-damaged patients, and more recently functional neuroimaging in the normal subject, have considerably increased our insight into the cognitive and functional anatomical features of these two memory systems. Clinical neuropsychological studies have strongly contributed to the recognition of the role of the hippocampus in episodic memory.[18-19] Neuroimaging studies have confirmed the importance of the medial temporal and frontal lobes in the encoding and retrieval of episodic information.[20-21] Mayes and Montaldi,[22] however, note that functional neuroimaging has been much less successful at confirming the roles of the midline diencephalon and basal forebrain structures in episodic memory tasks. Explanatory hypotheses stressed both technical and methodological limitations.[23] The group of Tulving[24] suggested an anteroposterior functional gradient in the hippocampus, with encoding sustained by the anterior and retrieval by the posterior hippocampus (the HIPER model). However, other structures, such as the prefrontal cortex, the anterior cingulate, and the precuneus have been found to engage during episodic memory tasks.[20,23] With regard to semantic memory, studies in brain-damaged patients have shown the importance of the left temporoparietal association cortex,[25,26] which was confirmed by numerous PET studies in healthy subjects.[27-29] Neuroimaging studies also stressed the role of the prefrontal cortex and led to the development of a functional model specifying the structures involved in the encoding and retrieval of episodic information and in the recall in semantic memory.[30,31] For instance, the HERA model (hemispheric encoding/retrieval asymmetry) ascribes to the left prefrontal cortex a preferential role in the encoding process of episodic material and the recall of semantic information, while the right prefrontal cortex would preferentially operate in the recall of episodic information. Nyberg[32] further proposed that the right anterior prefrontal cortex would be involved in all memory tasks, and the right posterior prefrontal cortex in the more difficult retrieval conditions. So far, however, experimental data have almost exclusively been concerned with verbal or visual material, and practically nothing is known about the functional anatomy of long-term memory for music.

Very few neuroimaging studies have addressed musical perception thus far. In their review of 275 PET and fMRI studies, Cabeza and Nyberg[20] quote only two PET studies on musical perception.[3,33] The article by Zatorre et al.[33] is also cited in the "working memory" section of this review, and another study from the same group[34] is also quoted in the "mental imagery" section. Contrasting with the large number of functional neuroimaging studies that have assessed working, episodic, or semantic memory of verbal and visual material,[20,23] very few have dealt with musical material. One of them is the PET study of Zatorre et al.,[33] which supported the idea of a preferential involvement of the right hemisphere in the memorization of nonfamiliar melodies by nonmusical subjects. The comparison between condition 4 (pitch judgment, comparing the first to the last note of sequences of eight notes) and condition 2 (passive listening of the same nonfamiliar melodies) disclosed a complex pattern of cortical and subcortical activations, notably involving the frontal and temporal lobes on the right side, and the insular and parietal regions bilaterally. According to the authors, these findings indicate that the simple comparison of pitch involves a neural network that includes the right prefrontal cortex, whereas the active short-term memory of pitch within a melody engages a much more distributed network, including, among other regions, both the frontal and temporal lobes on the right side.

Although it dealt with long-term rather than short-term musical memory, the PET study of Holcomb et al.[35] revealed findings similar to those of Zatorre et al.[33] In the control task, 12 healthy nonmusical subjects passively listened to a series of 60 notes spaced by 2 seconds and pressed a button alternatively on the right and left sides; in the active task, they were asked to detect two precise notes within the same series of 60 notes, one of high pitch with the instruction to press the right button, and one of low pitch with the instruction to press the left button. These nonmusical subjects had to be extensively trained into identifying these two notes over the three days preceding the actual PET session. The comparison between the identification task and the passive listening task revealed significant activation in the right inferior and middle frontal gyri, insula, anterior cingulate gyrus, and supplementary motor area. The left insula and part of the left cerebellum were also activated. Furthermore, the reaction times had a significant positive correlation with activity in the right middle and inferior frontal gyri (i.e., the longer the identification response time, the greater the functional response), indicating that the neuronal activity in these two regions was, in part, modulated by the "difficulty" in performing the task. These results therefore suggest that pitch recognition engages preferentially the prefrontal areas of the right hemisphere.

Although these two studies are consistent with each other and globally agree with the neuropsychological literature about the role of the right hemisphere in musical memory, they did not provide any information as to whether a specific network for musical episodic or semantic memory would exist, as distinct from verbal memory.

In our PET study on the perception of the various components of music,[3] the familiarity of musical melodies was one of the four components investigated. Among the presented sequences, 50% could be judged as familiar based on pre-experimental studies ("target sequences"). Target sequences were musical excerpts difficult to verbalize. We excluded songs and used either excerpts from classical music or other known tunes for which neither the title nor the composer would be familiar to French people, such as the national anthem of the United States. All these targets were selected based on extensive pre-experimental tests on a large sample of healthy subjects matched for sex and lack of musical experience to the PET experimental group. The comparison between this task and the three other perceptual tasks of the protocol (namely, pitch, timbre, and rhythm) revealed specific activations, mainly in the left hemisphere and involving the inferior frontal gyrus (BA 47) and the anterior part of the middle and superior temporal gyrus (BA 21 and 22). These findings highlight the left hemisphere's role in musical familiarity judgment and are consistent with clinical observations showing that left hemisphere damage is associated with impaired music identification of the associative agnosia type.[12,36,37] Interestingly, this left middle, and superior temporal region was also activated in a study by Sergent et al.[4] of musicians playing scores. Importantly, the activated areas only partially overlapped with areas known to activate during verbal semantic memory tasks,[38] which also involve the left superior or middle temporal gyrus, but more posteriorly than with melodies.[20,27] It remains that, due to the particular paradigm used in our study (same stimuli for all tasks, each one serving as reference for the others), we cannot exclude the involvement, in our familiarity task, of additional regions similarly activated in the other conditions, including right hemisphere areas.

The study by Halpern and Zatorre[39] raises important issues regarding the neuronal substrates of musical memory. In this PET study of mental imagery of familiar melodies, the authors assessed the ability of nonmusical subjects to mentally imagine the subsequent notes of both beginnings of very familiar melodies and tone sequences made up from notes from the familiar melodies. Three conditions were carried out in the following order: control condition, where the subjects simply listened to nonfamiliar musical sequences; mental imagery task, in which the subjects listened to the beginning of familiar melodies and were instructed to mentally imagine the subsequent notes (over a few seconds); music memory condition, in which the subjects were exposed again to the same nonfamiliar sequences presented in the first condition, but were asked to mentally replay the entire sequence (mental repetition).

The subjects of this study were selected beforehand so as to have strong feelings of familiarity for the chosen musical extracts, and in addition they were refamiliarized with these melodies on the day of the PET session so that they knew exactly at which time they would have to start mentally imagining the subsequent notes. According to the authors, comparing condition 2 to condition 3 would eliminate common processes regarding early auditory analysis, working memory, and mental imagery, leaving out the process of retrieval from semantic memory only. This comparison disclosed activations in the right inferior and middle frontal gyri (BAs 45, 46, and 47), in the superior and inferior right temporal gyri (BAs 37 and 22), and in the right anterior cingulate and parietal regions (BAs 32 and 40). Though less conspicuous than that in the right hemisphere, some activation also involved the left hemisphere, precisely in the middle and inferior frontal gyri (BAs 45/46/9).

Although these findings concerning the involvement of the right frontal regions in melody perception are entirely consistent with previous data by the same group,[33] they are in apparent conflict not only with our own findings detailed above,[3] but also with the extensive functional imaging literature regarding retrieval from semantic memory of verbal material, which consistently show a preferential involvement of left hemisphere regions.[20] Halpern and Zatorre interpret their findings in the general framework of the right hemisphere being dominant for the treatment of musical stimuli. They, however, raise the issue of the possible involvement of episodic memory processes in their mental imagery task of familiar melodies (which served as a control task), which could produce the participation of right prefrontal regions, as it is classically obtained in neuroimaging studies for episodic tasks.

In order to specifically study semantic memory and episodic memory of music, we carried out a novel PET study.[40] On the basis of our previous PET study, as well as on the neuropsychological literature, we had predicted that musical semantic processes will induce specific activation of the left inferior frontal and left anterior temporal areas, without excluding the possible right-side contribution of the same cortical regions. Concerning musical episodic memory, we had anticipated activation of the classic episodic memory network, namely involving, without clear-cut lateralization, the prefrontal cortex, the anterior cingulate gyrus, the precuneus, and potentially also the hippocampal regions.

In this work, we have defined musical semantic memory as that referring to "well-known" excerpts of music stored in memory without it being possible to retrieve the temporal or spatial circumstances surrounding their encounter. Semantic memory allows us to identify, or to have a strong feeling of knowing, familiar songs or melodies. Musical semantic memory may represent a musical lexicon, separate of a verbal lexicon, even though strong links exist between them. Episodic memory for musical information refers here to the capacity to recognize a musical excerpt (whether familiar or not) for which the spatiotemporal context surrounding its former encounter (i.e., when, where, and how) can be recalled.

The paradigm included four different conditions: one semantic condition, one episodic condition, a perceptual control condition, and a rest measurement. Musical material was specially created for this study and comprised 128 short melodies (5 seconds) played without orchestration, but with the same timbre of instrument (flute). All were real melodies, extracted from the classic and modern repertoires, but excluding songs so as to limit verbal associations. We also excluded extracts that might spontaneously evoke autobiographical memories, such as the "wedding march" or melodies used in popular TV spots. The 128 tunes comprised 64 familiar and 64 unfamiliar tunes. The familiar melodies were those judged very familiar by more than 70% of subjects in a pilot study of 150 subjects matched with the experimental sample, whereas the unfamiliar melodies were those judged as unknown by more than 80% of the subjects from the same population. Both samples were extracted from a larger database of melodies that was used in this pre-experimental study.

In the semantic memory condition, the subjects were asked to judge if the extract was familiar or unfamiliar. Half of the sequences were target melodies selected for their statistically significant familiarity, wheras the remaining 50% were distractors corresponding to unfamiliar melodies based on the same criteria. Two episodic memory tasks were carried out after the semantic task, one with exclusive familiar melodies and the other with unfamiliar melodies. In each, half of the items were targets (previously heard melodies), and half were distractors. The instruction was to recognize the sequences already presented in the semantic task, which was always given before the episodic task and without the subjects knowing that they would be subsequently tested for their memory of the sequences. In the perceptual control condition, the instruction was to judge if the last two notes of the melodies were of a similar or different pitch. Since by necessity the semantic condition contained both familiar and unfamiliar items, two distinct control tasks had to be constructed, one containing only familiar and the other only unfamiliar melodies. None of the melodies used in these two perceptual control tasks were employed in the other conditions.

Nine young healthy men were selected out of a population of university students. They were all fully right-handed and free of any psychiatric or organic pathology, and had normal hearing. To avoid specific cognitive strategies related to musical expertise, these subjects were selected so as to belong to class I of Wertheim and Botez,[41] that is, "musical people without theoretical musical studies and musical knowledge." We therefore studied common listener subjects so that our findings would generalize to the largest population, whereas musicians have specific strategies for listening to music, which depend, in particular, on the played instrument. The subjects who participated in this research were selected on the basis of two principal criteria: first, they were to be "common listeners" (i.e., not music lovers, who tend to listen to a specific type of music only), and second, they were to have normal performances in a test of pitch perception. The subjects selected for the PET study were very similar to the subjects who took part in the selection of the familiar and unfamiliar melodies; thus, the general musical culture of these two groups was equivalent. They were told that the experiment in which they were to take part related to the perception of the music, but never informed that the experiment related to musical memory, so that they did not train themselves to memorize melodies before the experiment.

Comparing the perceptual control tasks with rest revealed extensive bilateral activation, mainly of the lateral temporal areas (FIG. 1). The temporal activations included the primary and secondary auditory cortices and were particularly conspicuous on the right hemisphere. The activation pattern of these perceptive tasks is an awaited result and consolidates the idea of a right temporal preferential contribution in the perception of melody and pitch judgment.[33]

Compared to the perceptive control tasks, the activation patterns observed for the semantic and episodic tasks were clearly independent (FIG. 2). Only a very limited overlap of activation may be considered in the medial frontal regions. These distinct and new results, obtained here with musical material, confirm the significant functional independence of semantic and episodic memory processes, already documented with verbal and visuospatial material.[20,22]

Comparing the episodic task to the perceptive control tasks, bilateral activations of the middle and superior frontal gyri and the precuneus were found, with right-sided predominance. Comparing the episodic versus semantic tasks confirmed, first, the right dominance for the retrieval processes, and, second, suggested that our musical semantic memory task weakly engages the right hemisphere, given that the subtraction of the semantic processes in the episodic tasks does not decrease or remove these activations of the right hemisphere.

Regarding activation of the precuneus, although previous functional imaging studies on perception and processing of musical material have reported activation

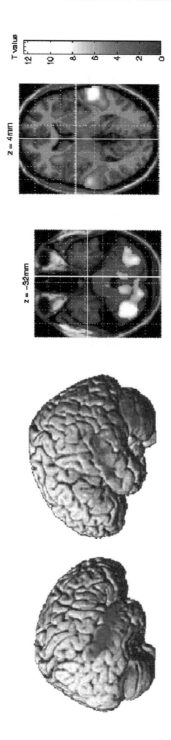

FIGURE 1. Control tasks versus rest. Significantly activated regions at the threshold of $P < .05$ corrected for multiple comparisons displayed with surface rendering (*left*) and illustrative cuts (*right*) of the SPM99 T1-weighted MRI template.

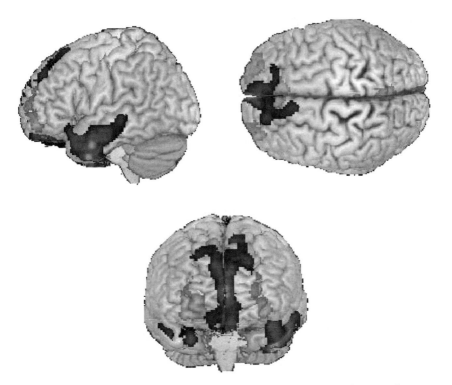

FIGURE 2. Superimposition of contrasts. Black (red, *online version*): semantic versus control tasks; gray (green, *online version*): episodic versus control tasks. Significantly activated regions at the threshold of $P < .05$ are corrected for multiple comparisons.

near this area,[3,4] similar findings have concerned nonmusical material, and thus this finding has been interpreted as reflecting a process of mental imagery triggered by the particular task.[42] However, precuneus activation has also been observed in episodic memory tasks, regardless of whether or not the items presented imageable characteristics.[43] According to Kapur *et al.*,[44] this cerebral region would be particularly involved in episodic retrieval, and more precisely in the success of episodic recall. This interpretation would fit well with our results, as our musical material does not exhibit particularly imageable features. Further arguments in favor of this interpretation would be the lack of activation of the precuneus in our semantic task, despite use of the same musical material as for the episodic task, and the fact that, at debriefing, no subject mentioned having employed a specific mental representation strategy for episodic recall, whether the melodies were familiar or not. Moreover, we found a significant positive correlation between performances and episodic retrieval only for the right precuneus region. This isolated significant correlation reinforces the crucial role of this cortical region in the success in episodic retrieval. However, the features of the cognitive processes sustained by the precuneus remain uncertain; beyond the mental imagery hypothesis, we could propose that, regarding the connectivity between this cortical region and the hippocampus, this region would play a specific role in linking and maintaining events in a particular chronology. To conclude on the episodic results, this activation pattern is entirely consistent with previous findings regarding episodic retrieval of verbal or visual spatial items,[23,45] and therefore does not suggest any specific network for musical episodic memory.

The comparison between the semantic memory tasks and the perceptive control tasks revealed bilateral activations of the medial frontal regions as well as the left angular gyrus and the anterior part of the left middle and superior temporal gyri. The semantic versus episodic contrast, however, revealed an additional right middle temporal gyrus activation (BA 21), suggesting a contribution of these specific right-sided areas in semantic memory processes.

Similar medial frontal activations have been documented in tasks of categorization of semantic association between faces and names,[46] suggesting, again, nonspecificity. Thus, activation of neither the medial frontal areas nor the left angular gyrus appears to be very specific to musical semantic memory. In contrast, could the activation of the left middle temporal and inferior frontal regions characterize the access to a musical semantic memory? First, this latter finding replicates what we obtained previously in a task of judgment of musical familiarity.[3] Furthermore, activation of the left inferior frontal gyrus, although less extensive than in our earlier study, was obtained here with a different musical material and experimental paradigm. Additional elements support the hypothesis that this activation points to cortical regions that underlie musical semantic memory. One metanalysis of activation peaks from PET studies relating to the perception of language and music (data from Petersen *et al.*[27] and Sergent *et al.*[4]) revealed an incomplete overlap of activations.[38] Thus, the perception of simple sounds, melodies, or timbre produces mainly temporal and prefrontal activations that only partially overlap those obtained for the perception of phonemes, logatoms, or real words,[3,4,27,47,48] and activation peaks obtained with musical material appear more anteriorly located than for linguistic material. We suggest, therefore, that the anterior temporal and inferior frontal regions subtend specific nonverbal auditory processing abilities, and that they might underlie a musical lexicon.

Why should there be a left hemisphere specialization for musical semantic memory? The prevalent idea that the right hemisphere preferentially underlies the perception of music arises primarily from experimental studies (in particular with the dichotic listening technique) using unfamiliar melodies. Meanwhile, the few neuropsychological case studies in which a specific impairment of musical identification has been observed (in the absence of perceptual deficit), almost exclusively concerned left hemisphere lesions.[12,49,50] The few functional neuroimaging studies that have focused on musical memory[33,35,39,51] showed mainly bilateral, but preferentially right-sided, frontal and anterior temporal activations. According to Halpern and Zatorre,[39] the observed asymmetry in favor of the right hemisphere would be related to the musical specificity of the cognitive processes. There might, however, be some difficulty in the interpretation of this finding, given a contamination by episodic processes, acknowledged by the authors in their paradigm. However, on the basis of the present work and our earlier study,[3] we believe that the anterior part of the left temporal regions is particularly involved in nonverbal semantic processes and could sustain musical semantic representations. Nonetheless, the possibility that verbalization mechanisms (e.g., trying to find the title of a melody or the name of the composer) have interfered with purely musical semantic memory processes cannot be excluded. For this reason, the direct contrast between musical and verbal semantic memory access presents a considerable interest; for the moment, only event-related brain potentials studies,[52] which show differential ERP effects for semantic processes when subjects focus their attention only on the lyrics or on the music of opera excerpts, gave some arguments in favor of a possible separate semantic memory subsystem between language and music.

Overall, the results of the present investigation would be in agreement with the HERA model proposed by Tulving and collaborators,[31] which is based on functional asymmetry in favor of the left hemisphere for semantic memory research, and right hemisphere dominance for episodic retrieval. Despite the fact that the HERA model remains controversial due to conflicting results from some functional neuroimaging studies, the review of the literature carried out by Cabeza and Nyberg[20] highlighted the fact that the majority of the published data reported such more or less marked asymmetry. To conclude, these findings about semantic and episodic memory of music are altogether consistent with the results established earlier with verbal or visuospatial material, even though we think that some functional specificity for musical memory emerges from our results. On the basis of our data, the anterior temporal cortex (mainly of the left hemisphere) would appear particularly involved in semantic memory for musical material.

[Competing interests: The author declares that he has no competing financial interests.]

REFERENCES

1. PERETZ, I. & R.J. ZATORRE. 2005. Brain organization for music processing. Annu. Rev. Psychol. **56:** 89–114.
2. ZATORRE, R.J. 2001. Neural specializations for tonal processing. Ann. N. Y. Acad. Sci. **930:** 193–210.
3. PLATEL, H., C. PRICE, J.C. BARON, et al. 1997. The structural components of music perception: a functional anatomical study. Brain **120:** 229–243.
4. SERGENT, J., E. ZUCK, S. TERRIAH, et al. 1992. Distributed neural network underlying musical sight reading and keyboard performance. Science **257:** 106–109.
5. DEUTSCH, D. 1970. Tones and numbers: specificity of interference in immediate memory. Science **168:** 1604–1605.
6. SEMAL, C. & L. DEMANY. 1991. Dissociation of pitch from timbre in auditory short-term memory. J. Acoust. Soc. Am. **89:** 2404–2410.
7. SEMAL, C. & L. DEMANY. 1993. Further evidence for an autonomous processing of pitch in auditory short-term memory. J. Acoust. Soc. Am. **94:** 1315–1322.
8. HALPERN, A. 1984. Organization in memory for familiar songs. J. Exp. Psychol. Learn. Mem. Cogn. **3:** 496–512.
9. DOWLING, W., S. KWAK & M. ANDERS. 1995. The time course of novel melodies. Percept. Psychophys. **57:** 136–149.
10. HÉBERT, S. & I. PERETZ. 1997. Recognition of music in long-term memory: are melodic and temporal patterns equal partners? Mem. Cogn. **25:** 518–533.
11. LECHEVALIER, B., H. PLATEL & F. EUSTACHE. 1995. Neuropsychologie de l'identification musicale. Rev. Neurol. **151:** 505–510.
12. EUSTACHE, F., B. LECHEVALIER, F. VIADER & J. LAMBERT. 1990. Identification and discrimination disorders in auditory perception: a report on two cases. Neuropsychologia **28:** 257–270.
13. SIGNORET, J.L., P. VAN EECKOUT, M. PONCET & P. CASTAIGNE. 1987. Aphasie sans amusie chez un organiste aveugle. Rev. Neurol. **143:** 172–181.

14. ZATORRE, R.J. 1985. Discrimination and recognition of melodies after unicerebral excisions. Neuropsychologia **23:** 31–41.
15. SAMSON, S. & R.J. ZATORRE. 1991. Recognition memory for text and melody of songs after unilateral temporal lobe lesion: evidence for dual encoding. J. Exp. Psychol. Learn. Mem. Cogn. **17:** 793–804.
16. PERETZ, I. 1996. Can we lose memory for music? A case of music agnosia in a nonmusician. J. Cogn. Neurosci. **8:** 481–496.
17. TULVING, E. 1985. Memory and consiousness. Can. Psychol. **26:** 1–11.
18. SCOVILLE, W.B. & B. MILNER. 1957. Loss of recent memory after bilateral hippocampal lesion. J. Neurol. Neurosurg. Psychiatry **20:** 11–21.
19. SQUIRE, L.R. 1992. Memory and the hippocampus: a synthesis from findings with rats, monkeys, and humans. Psychol. Rev. **99:** 195–231.
20. CABEZA, R. & L. NYBERG. 2000. Imaging cognition II: empirical review of 275 PET and fMRI studies. J. Cogn. Neurosci. **12:** 1–47.
21. FLETCHER, P.C. & R.N.A. HENSON. 2001 Frontal lobes and human memory. Insights from functional neuroimaging. Brain **124:** 849–881.
22. MAYES, A. & D. MONTALDI. 2001. Exploring the neural bases of episodic and semantic memory: the role of structural and functional neuroimaging. Neurosci. Biobehav. Rev. **25:** 555–573.
23. DESGRANGES, B., J.C. BARON. & F. EUSTACHE. 1998. The functional neuroanatomy of episodic memory: the role of the frontal lobes, the hippocampal formation, and other areas. NeuroImage **8:** 198–213.
24. LEPAGE, M., R. HABIB & E. TULVING. 1998. Hippocampal PET activation of memory encoding and retrieval: the HIPPER model. Hippocampus **8:** 313–322.
25. WARRINGTON, E.G. & T. SHALLICE. 1984. Category specific semantic impairments. Brain **107:** 829–854.
26. HODGES, J.R. & K. PATTERSON. 1995. Is semantic memory consistently impaired early in the course of Alzheimer's disease? Neuroanatomical and diagnostic implications. Neuropsychologia **33:** 441–459.
27. PETERSEN, S.E., P.T. FOX, M.I. POSNER, et al. 1988. Positron emission tomographic studies of the cortical anatomy of single word processing. Nature **331:** 585–589.
28. FRITH, C.D., K. FRISTON, P.F. LIDDLE & R.S.J. FRACKOWIAK. 1991. A PET study of word finding. Neuropsychologia **29:** 1137–1148.
29. CAPPA, SF, D. PERANI, T. SCHNUR, et al. 1998. The effects of semantic category and knowledge type on lexical–semantic access: a PET study. NeuroImage **8:** 350–359.
30. SHALLICE, T., P. FLETCHER, C.D. FRITH, et al. 1994. Brain regions associated with acquisition and retrieval of verbal episodic memory. Nature **368:** 633–635.
31. TULVING, E., S. KAPUR, F.I.M. CRAIK, et al. 1994. Hemispheric encoding/retrieval asymmetry in episodic memory: positron emission tomography findings. Proc. Natl. Acad. Sci. USA **91:** 2016–2020.
32. NYBERG, L. 1998. Mapping episodic memory. Behav. Brain Res. **90:** 107–114.
33. ZATORRE, R.J., A.C. EVANS & E. MEYER. 1994. Neural mechanisms underlying melodic perception and memory for pitch. J. Neurosci. **14:** 1908–1919.
34. ZATORRE, R.J., A. HALPERN, D. PERRY, et al. 1996. Hearing in the mind's ear: a PET investigation of musical imagery and perception. J. Cogn. Neurosci. **8:** 29–46.
35. HOLCOMB, H., D. MEDOFF, P. CAUDILL, et al. 1998. Cerebral blood flow relationships associated with a difficult tone recognition task in trained normal volunteers. Cereb. Cortex **8:** 534–542.
36. AYOTTE, J., I. PERETZ, I. ROUSSEAU, et al. 2000. Patterns of music agnosia associated with middle cerebral artery infarcts. Brain **123:** 1926–1938.
37. LECHEVALIER, B., F. EUSTACHE & Y. ROSSA. 1985. Les Troubles de la Perception de la Musique d'Origine Neurologique. Masson. Paris.
38. DRURY, H.A. & D.C. VAN ESSEN. 1997. Functional specializations in human cerebral cortex analyzed using the visible man surface-based atlas. Hum. Brain Mapp. **5:** 233–237.
39. HALPERN, A.R. & R.J. ZATORRE. 1999. When that tune runs through your head: a PET investigation of auditory imagery for familiar melodies. Cereb. Cortex **9:** 697–704.
40. PLATEL, H., J.C. BARON, B. DESGRANGES, et al. 2003. Semantic and episodic memory of music are subversed by distinct neural networks. NeuroImage **20:** 244–256.

41. WERTHEIM, N. & M.I. BOTEZ. 1959. Plan d'investigation des fonctions musicales. Encéphale **48**: 246–255.
42. KOSSLYN, S.M., W.L. THOMPSON & N.M. ALPERT. 1997. Neural systems shared by visual imagery and visual perception: a positron emission tomography study. NeuroImage **6**: 320–334.
43. KRAUSE, B.J., D. SCHMIDT, F. MOTTAGHY, et al. 1999. Episodic retrieval activates the precuneus irrespective of the imagery content of word pair associates: a PET study. Brain **122**: 225–263.
44. KAPUR, S., F.I.M CRAIK., C. JONES, et al. 1995. Functional role of the prefrontal cortex in retrieval of memories: a PET study. NeuroReport **6**: 1880–1884.
45. BERNARD, F., B. DESGRANGES, H. PLATEL, et al. 2001. Contributions of frontal and medial temporal regions to verbal episodic memory: a PET study. NeuroReport **12**: 1737–1741.
46. TEMPINI, M.L., C. PRICE, O. JOSEPHS, et al. 1998. The neural systems sustaining face and proper-name processing. Brain **121**: 2103–2118.
47. DÉMONET, J.F., C. PRICE, R. WISE & R.S.J. FRACKOWIAK. 1994. Differential activation of right and left posterior sylvian regions by semantic and phonological tasks: a positron emission tomography study in normal human subjects. Neurosci. Lett. **182**: 25–28.
48. ZATORRE, R.J., A.C. EVANS, E. MEYER & A. GJEDDE. 1992. Lateralization of phonetic and pitch discrimination in speech processing. Science **256**: 846–849.
49. DUPRÉ, E. & M. NATHAN 1911. Le langage musical. In Etude Médico-Psychologique. pp. 64–75. Alcan. Paris.
50. SOUQUES, A. & H.BARUK. 1930. Autopsie d'un cas d'amusie (avec aphasie) chez un professeur de piano. Rev. Neurol. **1**: 545–557.
51. MATTEIS, M., M. SILVESTRINI, E. TROISI, et al. 1997. Transcranial doppler assessment of cerebral flow velocity during perception and recognition of melodies. J. Neurol. Sci. **149**: 57–61.
52. BESSON, M. & D. SCHÖN. 2001. Comparison between language and music. Ann. N. Y. Acad. Sci. **930**: 232–258.

Neurophysiology and Neuroanatomy of Pitch Perception: Auditory Cortex

MARK JUDE TRAMO,[a,b,c] PETER A. CARIANI,[a,e] CHRISTINE K. KOH,[a,b] NIKOS MAKRIS,[a,d] AND LOUIS D. BRAIDA[b]

[a]*Department of Neurology, Harvard Medical School, and Massachusetts General Hospital; The Institute for Music and Brain Science, Auditory Neuroscience Program, Boston, Massachusetts 02114, USA*

[b]*Sensory Communication Group, Research Laboratory of Electronics, Massachusetts Institute of Technology, Cambridge, Massachusetts 02139, USA*

[c]*Eaton-Peabody Laboratory of Auditory Physiology, Massachusetts Eye and Ear Infirmary, Boston, Massachusetts 02114, USA*

[d]*Center for Morphometric Analysis, Martinos Center for Biomedical Imaging, Massachusetts General Hospital, Charlestown, Massachusetts 02129, USA*

[e]*Department of Physiology, Tufts Medical School, Boston, Massachusetts 02111 USA*

ABSTRACT: We present original results and review literature from the past fifty years that address the role of primate auditory cortex in the following perceptual capacities: (1) the ability to perceive small differences between the pitches of two successive tones; (2) the ability to perceive the sign (i.e., direction) of the pitch difference [higher (+) vs. lower (−)]; and (3) the ability to abstract pitch constancy across changes in stimulus acoustics. Cortical mechanisms mediating pitch perception are discussed with respect to (1) gross and microanatomical distribution; and (2) candidate neural coding schemes. Observations by us and others suggest that (1) frequency-selective neurons in primary auditory cortex (A1) and surrounding fields play a critical role in fine-grained pitch discrimination at the perceptual level; (2) cortical mechanisms that detect pitch differences are neuroanatomically dissociable from those mediating pitch direction discrimination; (3) cortical mechanisms mediating perception of the "missing fundamental frequency (F0)" are neuroanatomically dissociable from those mediating pitch perception when F0 is present; (4) frequency-selective neurons in both right and left A1 contribute to pitch change detection and pitch direction discrimination; (5) frequency-selective neurons in right A1 are necessary for normal pitch direction discrimination; (6) simple codes for pitch that are based on single- and multiunit firing rates of frequency-selective neurons face both a "hyperacuity problem" and a "pitch constancy problem"—that is, frequency discrimination thresholds for pitch change direction and pitch direction discrimination are much smaller than neural tuning curves predict, and firing rate patterns change dramatically under conditions in which pitch percepts remain invariant; (7) cochleotopic organization of frequency-selective neurons bears little if any relevance to perceptual acuity and pitch constancy;

Address for correspondence: Mark Jude Tramo, M.D., Ph.D., Director, The Institute for Music and Brain Science, 175 Cambridge Street, Suite 340, Boston, MA 02114. Voice: 617-726-5409. mtramo@hms.harvard.edu; http://www.brainmusic.org

Ann. N.Y. Acad. Sci. 1060: 148–174 (2005). © 2005 New York Academy of Sciences. doi: 10.1196/annals.1360.011

and (8) simple temporal codes for pitch capable of accounting for pitches higher than a few hundred hertz have not been found in the auditory cortex. The cortical code for pitch is therefore not likely to be a function of simple rate profiles or synchronous temporal patterns. Studies motivated by interest in the neurophysiology and neuroanatomy of music perception have helped correct longstanding misconceptions about the functional role of auditory cortex in frequency discrimination and pitch perception. Advancing knowledge about the neural coding of pitch is of fundamental importance to the future design of neurobionic therapies for hearing loss.

KEYWORDS: pitch; missing fundamental; psychophysics; periodicity; autocorrelation; auditory cortex; lesion effects; neural coding

INTRODUCTION

Empirical work on the neurophysiology and neuroanatomy of pitch perception in humans and animals has a long and rich history. The topic is of fundamental interest in neuroscience, for it is well-suited to probing the relationships among (1) physical features of sensory stimuli (e.g., the frequency of a sine tone); (2) perceptual attributes of sensory stimuli (e.g., the pitch of the tone); (3) perceptual constancy across changes in stimulus physics (e.g., pitch constancy across changes in tone intensity); (4) neural coding of stimulus features and perceptual attributes (e.g., action potential firing patterns in single neurons and populations of neurons in the auditory nerve, brain stem, and cortex); and (5) gross and microanatomical mapping of sensory, perceptual, and cognitive functions (e.g., lateralization and localization of regional metabolic changes).

Pitch is the auditory percept associated with the frequency (f) or, equivalently, the period ($T = 1/f$) of sound wave vibrations in the audible frequency range (~20–20 kHz). In Western music notation, a note (e.g., A_4) symbolizes a distinctive pitch, independent of its loudness, timbre, or other perceptual attributes. The position of the note on the staff and its clef indicate how high or low the pitch is. By convention, the pitch of A_4 in the Western scale of equal temperament has the same pitch as a 440-Hz sinusoidal tone (a.k.a. pure tone). The oft-cited, half-century-old definition of pitch by the American National Standards Institute[104]—"that attribute of auditory sensation by which sounds may be ordered on a scale extending from low to high"—is incomplete, for it implies a single psychological dimension (pitch height). The well-studied perceptual phenomena of octave similarity (pitch chroma) and within-octave pitch-class hierarchies (e.g., tonic–dominant relationships) establish pitch as a multidimensional percept.

In speech, pitch contrasts convey voiced/unvoiced distinctions, prosodic inflections, and speaker identity. In music, two or more simultaneous pitches comprise harmonic intervals and chords; two or more successive pitches comprise melodic intervals and melodies. Pitch percepts are evoked by a wide range of periodic acoustic signals. Particularly strong pitches are evoked by pure tones and complex tones whose frequencies belong to the same harmonic series (harmonic tones). When frequency components are not harmonically related (inharmonic tones), pitch percepts are weaker, and intervals and chords sound more dissonant.

Stimuli with very different power spectra can produce the same pitch (pitch equivalence or pitch constancy, FIG. 1). The pitch of a harmonic tone corresponds to

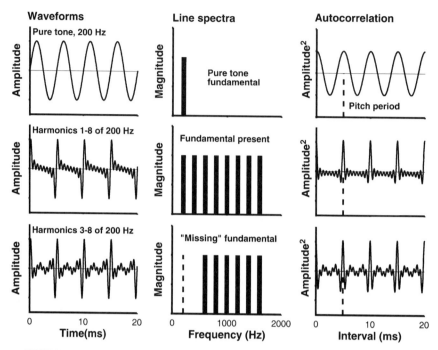

FIGURE 1. Waveforms (**left column**), line magnitude spectra (**middle column**), and auto-correlation functions (**right column**) of a pure tone (**top row**), harmonic tone with spectral energy at F0 (**middle row**), and harmonic tone missing F0 (**bottom row**) that evoke the same pitch. $f = F0 = 200$ Hz. In this example, all frequency components have the same amplitude and phase (sine), but these are not necessary conditions for the three tones' pitch equivalence.

the pitch of a pure tone at the former's F0, even when there is no energy at F0 or low-pass noise masks F0.[1–5] In the present paper, we use the terms *F0 pitch* when the pitch is evoked by a pure tone or harmonic tone with spectral energy at F0, and *missing-F0 pitch* when the pitch is evoked by a harmonic tone without spectral energy at F0. Both F0 pitch and missing-F0 pitch are perceived by a wide range of animals, including monkeys,[8] cats,[11] birds,[12] frogs,[13] and fish.[14] Old World monkeys also perceive octave similarity[9] and, like rats[15] and birds,[16] are sensitive to acoustic features of simultaneous harmonic tones that cue perception of consonance and dissonance in the vertical dimension by humans.[7,10] On the basis of various theoretical assumptions about the underlying neural processing mechanisms, some authors refer to F0 pitch as *spectral pitch*, and missing-F0 pitch as *virtual pitch*, *periodicity pitch*, *residue pitch*, and *synthetic pitch*. However, virtual pitch is virtual only in the sense that F0 is missing in the frequency spectrum of the tone; in the time domain, the fundamental period ($T_{F0} = 1/F0$) is present as the dominant periodicity in the autocorrelation function of the stimulus (FIG. 1). Missing-F0 pitch is related to Rameau's concept of the *basse fondamentale* in his *Treatise on Harmony*.[6]

The capacity to perceive pitch is a basic function of the auditory nervous system that supports melody and harmony perception in music, prosody perception in speech, voice recognition, environmental sound recognition, and language acquisition. In light of evidence from comparative neuroscience, ethnomusicology, and developmental neuroscience, it seems reasonable to propose that the underlying auditory mechanisms are innate and that they were necessary (though not sufficient) for the evolution of music among humans.

GROSS NEUROANATOMY AND NEUROPHYSIOLOGY

Knowledge about the neurophysiology and neuroanatomy of the auditory cortex at the gross, macroscopic level provides insights into the spatial organization of distributed neural systems mediating different aspects of music perception. In addition, it guides the placement of microelectrodes used to study neural coding and neural circuitry at the cellular level.

Most current knowledge about the cortical neuroanatomy of pitch perception is derived from two sources: (1) behavioral experiments with humans and animals who have focal brain lesions; and (2) behavioral experiments and passive stimulation experiments with normal humans that measure changes in blood flow, metabolism, and electrical or magnetic field potentials in normal humans. The strengths of the two methods complement each other.

Lesion effect experiments tell us which gray and white matter structures are necessary for normal performance and allow us to test for functional dissociations. They do not provide information about the full anatomical extent of neurons and axons participating in task performance. In nonhuman primates and other animals, structures of interest can be lesioned selectively by mechanical (e.g., aspiration) or chemical (e.g., ibotenic acid) methods, and the location of the lesion can be pinpointed microanatomically via postmortem inspection of its local histochemical and cytoarchitectonic boundaries as well as its far-reaching effects on the distal axons of damaged neurons (anterograde degeneration) and on the somas of damaged axons (retrograde degeneration). In humans, naturally occurring lesions rarely respect anatomical or physiological boundaries, and *in vivo* spatial resolution is typically coarse. Because the histopathology of ischemic infarcts and excisions is well circumscribed and homogeneous weeks to years after onset, these types of lesions are better suited for structure–function studies than intracerebral hemorrhages, Alzheimer disease, brain tumors, and other diseases associated with heterogeneous-focal, multifocal, or diffuse cortical pathology. Knowledge about the anatomical distribution of gross structures that are necessary for the generation of electrical and magnetic field potentials evoked by passive acoustic stimulation has also been gained through lesion studies. Few studies have examined blood flow and metabolism during task performance in patients with ischemic strokes because measurements in the damaged area can be misleading in the face of luxury perfusion, cellular infiltration, and the uncoupling of blood flow and metabolism. In normal human volunteers, analyses of blood flow, metabolism, and field potentials can detect structure–function correlates with good spatial resolution (≤ 10 mm, depending on the method), provide coarse information about temporal resolution within and across gross structures, and demonstrate the entire distribution of structures that are active during task performance or passive stimulation. However, these methods cannot establish whether any one node in the distributed network of activation plays an essential role in task performance.

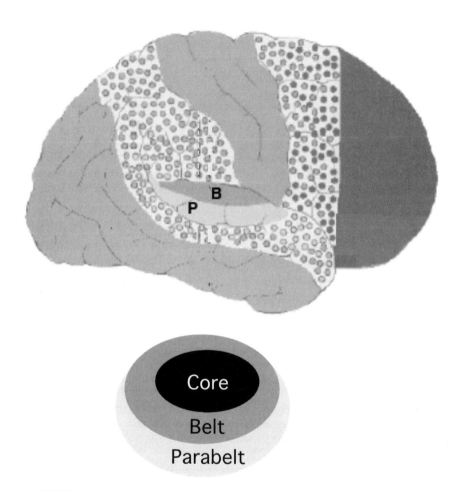

FIGURE 2. (Top) Lateral view, right cerebral hemisphere, human. **(Bottom)** Schematic of the corresponding core area (*black*), belt area (*dark gray*), and parabelt area (*light gray*) on the superior and lateral surfaces of the right temporal lobe. In humans, apes, and Old World monkeys (e.g., macaques), the core area is buried in the sylvian fissure and cannot be seen on a lateral view. Adapted from Tramo.[107]

The term *auditory cortex* does not refer to a specific brain structure(s) that can be identified directly via static magnetic resonance images, pathological specimens, or histological sections. By definition, auditory cortex refers to gray matter of the cerebral cortex whose neurons respond to auditory stimuli but not to visual, tactile, or other sensory stimuli. In humans and our close anthropoid relatives, Old World monkeys (including macaques), pathoanatomical, electrophysiological, and metabolic studies collected for over one hundred years have established that the vast majority of auditory cortex neurons are housed in the superior portion of the temporal lobe, inside and below the lateral fissure (FIG. 2). On the superior surface of the superior temporal gyrus (STG, also known as the first temporal convolution, T1), near the junction of its anterior two-thirds and posterior third, lies the transverse gyrus(i) of Heschl (TG), prominent in humans, less prominent in apes, and rudimentary in macaques.[17–19] Most of TG is populated by small, densely packed neurons (konio-cortex) and myelinated axons, which bring afferent auditory input from the medial geniculate nucleus (MGN) of the thalamus. TG stains heavily throughout much of its radial and longitudinal extent for cytochrome oxidase, acetylcholinesterase, par-valbumin, Nissl substance, and myelin.

Setting aside differences in physiological and microanatomical criteria and nomenclature that have evolved in the hundred years since Campbell's "audito-sensory" and "audito-psychic" dichotomy (c.f. Refs. 17, 20–22), the terms *primary auditory cortex* (A1) and *core area* are commonly used to refer to one or more adja-cent koniocortical fields containing frequency-selective Layer IV neurons whose to-pographic organization mirrors the one-dimensional frequency map of the cochlea and whose afferent input comes almost entirely from the ventral division of MGN, which is also populated by frequency-selective, cochleotopically órganized neurons. The nomenclature can be confusing: "A1" has been used to refer to the set of all such fields by some authors, including us, and to one specific cochleotopic field within the core area by Hackett and others. A1 is "primary" in the sense that it receives the bulk of short-latency, afferent input from the brain stem's major ascending (lemnis-cal) pathway via the ventral division of MGN. Around A1 in TG and extending throughout most of the STG is the auditory association cortex (a.k.a., nonprimary auditory cortex). This stretch of gray matter is "associative" in the sense that its neu-rons synapse with other STG neurons and with neurons in the temporal, frontal, and parietal cortices that respond to stimuli in two or more sensory modalities (multimo-dal cortex) or that fire without sensory stimulation (supramodal cortex). Many A1 neurons send their axons into a ring of surrounding belt fields of auditory association cortex (FIG. 2), which also receive afferent input from the ascending lemniscal-adjunct pathway via the medial MGN, dorsal MGN, and other thalamic nuclei. Like A1 neurons, many belt neurons are frequency selective, that is, they respond over a restricted range of the animal's audible spectrum. Most A1 neurons have spectral bandwidths for on-excitation that are narrower than those of belt neurons.[23] Many belt neurons send axons into the next surrounding ring of parabelt fields, whose neu-rons are hard to excite or inhibit with pure tones, a response property shared by the vast majority of cells elsewhere in STG. Neurons in the auditory association cortex are reciprocally connected with multimodal and supramodal neurons in frontal, pa-rietal, and temporal cortices, the basal ganglia, and the cerebellum to form a widely distributed neural system for music cognition.

Effect of Auditory Cortex Lesions on Pure-Tone Pitch Perception

Despite abundant evidence of neuronal frequency selectivity and cochleotopic organization in A1 and surrounding fields in multiple animal species, the regnant view among twentieth-century neuropsychologists, neurophysiologists, and neuro-anatomists held that the auditory cortex was not necessary for normal performance on pure-tone pitch discrimination tasks. In 1963, on the basis of unpublished selec-tive-ablation experiments in cats, Guttman and Diamond[24] argued, "tonotopic orga-nization at the cortical level is not necessary for the perception of tones." In 1975, in their authoritative review of animal and human lesion effects in the *Handbook of*

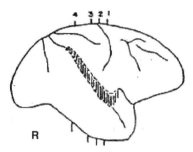 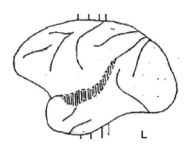

FIGURE 3. Lateral view, right (R) and left (L) cerebral hemispheres of a rhesus monkey (*Macaca mulatta*), whose gross-microanatomical correlates are similar to those of other Old World monkeys but not New World monkeys. The *dark area* in each hemisphere marks the ablation site in the lateral surface of STG; this lateral view does not show the lesions made in the superior surface of STG, which houses A1 in Old World monkeys, apes, and humans. *Curved lines* within each hemisphere are the major fissures and sulci of the cerebral hemispheres. *Vertical lines* above and below each hemisphere indicate the locations of postmortem coronal sections that were inspected macro- and microscopically for hemispheric and thalamic lesions. Adapted from Evarts.[27]

Sensory Physiology,[24] Neff, Diamond, and Casseday concluded, "frequency discrimination may be learned or relearned after bilateral lesions involving all or nearly all of primary auditory cortex in animals such as the cat and monkey and in human patients." The Neff doctrine had a profound influence on current opinion about structure–function and physiological–perceptual correlates throughout the cognitive neuroscience boom of the late twentieth century. For example, after finding a deficit in missing-F0 pitch discrimination in a subpopulation of right temporal lobectomy patients, Zatorre[38] stated, "simple frequency discrimination (i.e., with pure tones, or with complex tones when the fundamental [frequency] is present) is not permanently disrupted even by large bilateral lesions of auditory cortex." Scrutiny of the experiments in Old World monkeys and humans on which these claims were based yields important lessons about methodology that have heuristic value for future work in the field. (For a critical review of selective-ablation studies in cats, see Refs. 25 and 26).

Among primate lesion effect studies cited by Neff *et al.*[24] was Evarts'[27] condition-ablation experiment in macaques[27] (FIG. 3). A go/no-go, one-interval, two-alternative, forced-choice, operant-conditioning task required each monkey to respond differentially to a 350 Hz pure tone and a 3500 Hz pure tone of the same duration and similar intensity. After near-complete bilateral ablations of superior temporal cortex, one monkey (M-3) needed 600 trials to learn the task and reach the response criterion of 80% accuracy over 50 consecutive trials. Two other monkeys (M-19 and M-20) were studied preoperatively as well as postoperatively. Before surgery, one monkey needed 450 trials and the other 700 trials to reach the response criterion. After bilateral ablation of TG and all but a small anterior portion of STG, one monkey reached the response criterion in the first 50 trials, and the other was able to relearn the task with less training than it needed preoperatively. Evarts remarked that the results were "difficult to reconcile with the strict tonotopic organization" of primate A1 and speculated that small remnants of remaining auditory cortex were "of great functional importance." However, interpretation of the results is confounded by a methodological flaw: the task failed to assess pure-tone pitch discrimination anywhere near psychophysical threshold. The frequency difference (Δf) between the tones was 3150 Hz, and the Weber fraction was 164% [Δf/mean Δf) $\times$ 100]. In other words, the behavioral task was too insensitive, so the experiment was biased in favor of supporting the null hypothesis.

In fact, clinical case reports, audiological assessments, and neuropsychological experiments on F0 pitch throughout the nineteenth and twentieth centuries suffered from the same methodological flaw. For example, the oft-cited cases of Jerger and colleagues[28,29] and Zatorre[38] were examined using pure-tone Δf's and harmonic-tone ΔF0's corresponding to Weber fractions of ~40%, about 40 times the normal threshold (Weber fraction ~1%).[30] In addition, the sensitivity, specificity, and spatial resolution of their anatomical methods precluded precise definition of lesion sites and sizes. Neurologists and other physicians never carry tuning forks that are less than an octave apart (Weber fraction = 67%), and when keyboards are used to test patients in the hospital or laboratory, the minimum ΔF0 they can test is constrained by the Western scale of equal temperament, which has a minimum step size of approximately 6%. In general, the test method used by Evarts, Jerger, and many others—the method of constant stimuli—is not well suited to measuring psychophysical thresholds, because the stimuli are "canned:" they are designed and generated before performance is tested. This makes it difficult to place observations near each listener's threshold unless pilot studies can provide useful constraints.

Experiments carried out in recent years at separate laboratories with different neurological patients, many of which were motivated by interest in the neuroanatomy of music perception, have rendered the Neff doctrine untenable. Together, the results have forced a fundamental change in current opinion about auditory cortex function and a reappraisal of the functional relevance of neuronal frequency-selectivity in A1.

In 1989, my colleagues and I began a series of experiments on pitch, harmony, and melody perception with a middle-aged, ambidextrous man, MHS, who has chronic bilateral auditory cortex infarcts. His subjective complaints of impaired music, speech, and environmental sound perception were precipitated by his second stroke, which was very small but unfortunately placed in his left TG and posterior STG (FIG. 4). His previous right-sided infarct involved most of the right middle cerebral artery territory, including TG, STG, and multiple temporal, frontal, and parietal gray and white matter structures. The first clue that he had impaired pitch discrimination came from the Seashore Measures of Musical Talents Pitch Discrimination Test,[31] which measures the accuracy of pitch direction discrimination using a two-interval, two-alternative, forced-choice paradigm and the method of constant stimuli. The test contains five blocks of ten trials in which Δf decreases over successive blocks. With stimuli presented over loudspeakers at a comfortable listening level, MHS's response accuracy fell to chance as Δf decreased.[32] We subsequently used an adaptive procedure and well-calibrated pure tones to measure Δf thresholds for pure-tone pitch perception.[30,33] In one experiment, MHS was asked to judge whether the pitch of the second tone was higher or lower than the pitch of the first tone (pitch direction discrimination); in another, he was asked to judge whether the pitch of the second tone was same as or different from the pitch of the first tone (pitch change detection). Normal and patient controls performed well on both tasks [Weber

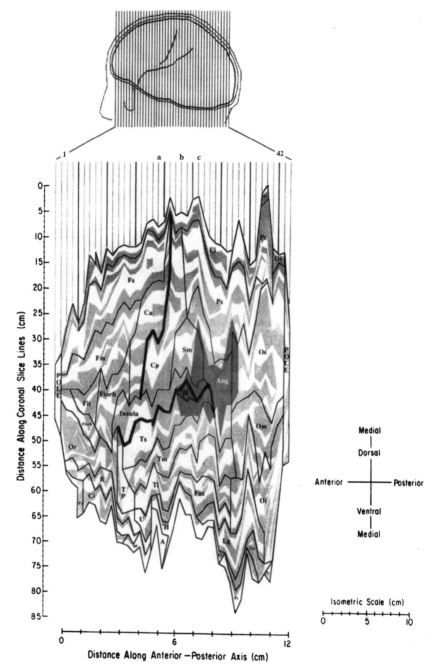

FIGURE 4. Flat map of the unfolded surface of the left cerebral cortex of case MHS (a.k.a., case A1+) reconstructed from MRIs. *Bold lines* indicate the lateral and central fissures. The *dark gray area* near the lateral fissure marks the infarct. *Light gray areas* throughout the map indicate intrasulcal surfaces, *white areas* pial surfaces. Adapted from Tramo *et al.*[105]

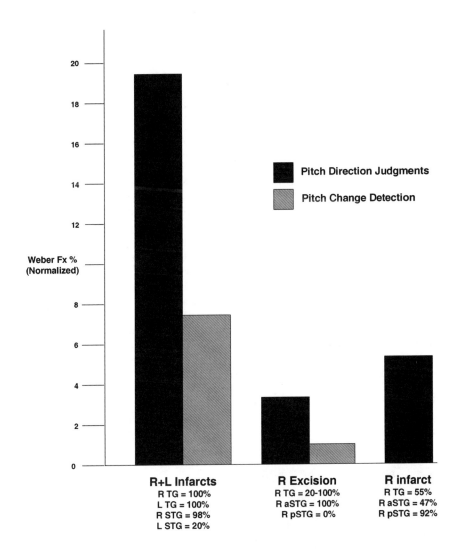

FIGURE 5. Bar graph showing Weber fractions for pitch direction discrimination (higher–lower judgments, *black bars*) and pitch change detection (same–different judgments, *gray bars*) in patients with auditory cortex lesions. (**Left bars**) Case MHS. (**Middle bars**) Mean thresholds from six temporal lobectomy patients with partial right TG and partial STG lesions reported by Johnsrude *et al.*[35] (**Right bars**) Case WKF. Because the tasks used to measure thresholds in MHS, WKF, and the temporal lobectomy patients differed (albeit slightly), normalized Weber fractions were computed by dividing patient Weber fractions by the mean Weber fraction measured in separate groups of normal controls.

fraction ~1% at 1 kHz, intensity = 40 dB, sensation level (SL) at each ear, tone duration = 500 ms, interstimulus interval (ISI) = 200 ms]. MHS's performance differed from that of controls in two ways: (1) his Weber fractions were seven or more times higher than those of controls on both tasks; and (2) his Weber fraction for pitch direction discrimination was twice that for pitch change detection (FIG. 5). Kazui *et al.*[34] also reported elevated thresholds for pitch change detection in a stroke patient with bilateral TG and left STG lesions; pitch direction discrimination was not tested.

The results of recent condition-ablation experiments in Old World monkeys are consistent with the effects of bilateral auditory cortex lesion effect in humans. Harrington *et al.*[41] reported that Δf thresholds for pitch change detection in macaques with bilateral complete TG lesions and near-complete STG lesions were more than twice the Δf thresholds in normal macaques.

Comparisons between the effects of bilateral versus unilateral auditory cortex lesions on pitch perception may shed light on hemispheric specialization and interhemispheric integration. Divenyi and Robertson[36] reported elevated pitch change detection thresholds in four patients with right hemisphere strokes. Pitch direction discrimination was not tested, and lesion localization was not reported. Johnsude *et al.*[35] used an adaptive procedure and a two-interval two-alternative forced choice paradigm to measure Δf thresholds for pure-tone pitch change detection and pitch direction discrimination in epilepsy patients with anterior temporal lobectomies (*f* = 800–1200 Hz; intensity = 75 ± 2 dBA A.U.; ISI = 1 s). On the direction discrimination task, seven of eight patients with lesions involving right TG (as well as right anterior STG and other temporal lobe structures) had Weber fractions that were three or more times greater than normal (FIG. 5); the mean performance of lobectomy patients with left- or right-sided lesions that spared right TG was within normal limits. By contrast, on the change detection task, most right TG patients, like other patients, had normal Weber fraction. We have also observed elevated Weber fractions for pitch direction discrimination following a partial right TG lesion in stroke patient WKF, a right-handed musician whose chronic infarct also involves right posterior STG, and neighboring gyri but spares much of right anterior STG (FIGS. 5 and 6). Threshold elevations were greater when pure tones were presented to the ear (left) contralateral to the auditory cortex lesion. Thresholds for pitch change detection have not yet been tested.

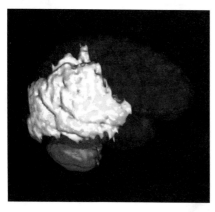

FIGURE 6. 3-D MRI reconstruction of the right cerebral hemisphere of case WKF. The *light area* marks the infarct.

Taken together, the effects of bilateral and unilateral auditory cortex lesions on pure-tone pitch perception in humans and macaques raise the following hypotheses: (1) frequency-selective neurons in both right and left A1 contribute to pitch change detection and pitch direction discrimination; (2) the contributions of right and left A1 neurons to pitch direction discrimination are additive; and (3) frequency-selective neurons in right A1 are necessary for normal pitch direction discrimination in most right-handed adults independent of musicality.

Effect of Auditory Cortex Lesions on Harmonic-Tone Pitch Perception

Bilateral lesions of auditory cortex elevate $\Delta F0$ thresholds for pitch direction discrimination whether F0 is present or missing in a harmonic complex tone.[37] The magnitude of the $\Delta F0$ threshold elevation appears to be proportional to the Δf threshold elevation obtained with pure-tone stimuli. This suggests that missing-F0 impairments may be attributable to derangements in "low-level" processes (e.g., frequency resolution) rather than higher-level processes (e.g., harmonic template matching).

Studies of unilateral lesion effects on F0 pitch perception were published by three laboratories between 1988 and 1990. Their results conflict with respect to the contribution of right auditory cortex when energy is present at F0. Sidtis and Volpe's harmonic-tone experiments in stroke patients used a dichotic-diotic match-to-sample pitch recognition task and the method of constant stimuli.[42] Deficits in F0 pitch perception were found in a population with right but not left hemisphere strokes. The site and size of the lesions were not reported. Zatorre's[38] harmonic-tone experiments in epilepsy patients with temporal lobectomies used a binaural, two-interval, two-alternative forced-choice task and the method of constant stimuli.[38] A dissociation between missing-F0 pitch (impaired) and F0 pitch (spared) was found in a population of patients with right-sided excisions involving anterior STG and TG. However, task difficulty differed for the two stimulus conditions. Harmonic tones in the F0 pitch task contained nine harmonics; those in the missing-F0 pitch task contained three or four. The resulting difference in stimulus pitch strength rendered the missing-F0 pitch task harder than the F0 pitch task and thus confounds a straightforward interpretation of the observed dissociation. Soon thereafter, Robin *et al.*[39] reported elevated $\Delta F0$ difference threshold for F0 pitch change detection in five stroke patients with right, but not left, STG and/or TG lesions.[39] Their experiment employed square-wave stimuli presented binaurally via loudspeaker (i.e., harmonic tones containing odd harmonics, including F0, that decrease in intensity with increasing f; duration = 200 ms, ISI = 500 ms, intensity = 70 dB SPL), an adaptive procedure, and a three-interval, two-alternative, forced-choice task that required listeners to judge which of the two comparison tones in the second and third stimulus intervals differed from the standard tone in the first interval. In summary: (1) the results of Robin *et al.* and those of Sidtis and Volpe agree that right-sided lesions impair F0 pitch perception in right-handed adults; (2) Zatorre's results indicate that right anterior STG and partial TG lesions impair missing-F0 pitch perception in right-handed adults; and (3) methodological differences hamper comparisons among the three studies.

The results of the most recent temporal lobectomy population study,[106] which employed a two-interval "same–different" forced-choice task and the method of constant stimuli, have been interpreted as evidence that right STG and TG excisions do *not* impair performance on same–different F0 pitch judgments of harmonic tones with 12 harmonics, even when there is little spectral energy at F0 relative to that at higher harmonics (binaural intensity = 75 dB SPL, tone duration = 500 ms, ISI = 100 ms). However, inspection of the population data (Ref. 106, FIG. 5, p. 1621, same timbre conditions) raises the possibility that patients with right excisions performed worse than all or almost all of those with left excisions and worse than most normal controls.

Using a two-interval, two-alternative, forced-choice task and an adaptive procedure, we measured F0 difference thresholds for pitch direction discrimination following unilateral infarction of right TG, STG, and adjacent gyri in Case WKF (FIG. 6). Our preliminary results obtained with missing-F0 tones raise the possibility that right auditory cortex mechanisms mediating the temporal processing of envelope periodicities generated by combinations of unresolvable high harmonics are dissociable from those mediating the spectral and/or temporal processing of individual, resolvable harmonics.

Acknowledging that the sum total of patients and nonhuman primates with focal lesions involving different sides and subdivisions of auditory cortex is small, and that replication of structure–function correlates by different laboratories remains in short supply, we tentatively interpret the results of lesion experiments employing tones or vocalizations with F0s, intensities, and durations typical of notes anchoring melodies in traditional Western vocal and instrumental music[30,33,35–38,40–43,106] as follows: (1) frequency-selective neurons in A1 and adjacent areas in right and left TG and STG play critical, additive roles in our ability to detect a small change in pitch between two successive pure tones or harmonic tones; (2) right-sided neurons play a critical role in our ability to perceive the direction of a pitch change between two successive pure tones or harmonic tones; (3) left-sided neurons play an additive role in our ability to perceive the direction of a pitch change between two successive pure tones or harmonic tones; (4) right auditory cortex mechanisms mediating perception of F0 pitch versus missing-F0 pitch are neurologically dissociable; and (5) right auditory cortex mechanisms mediating spectral versus temporal processing of harmonic tones are neurologically dissociable.

Gross Physiological Changes during F0 Pitch Perception

Measurements of cortical blood flow and metabolism during pitch perception also provide insights into structure–function relationships at the gross anatomical level.[48] Zatorre *et al.*[46] measured changes in cortical blood flow while ten young right-handed adults listened to pairs of consonant-vowel-consonant syllables differing in pitch and final consonant.[46] The size of the F0 difference between syllables was not given. When the volunteers completed 40 trials of a go/no-go two-interval, two-alternative forced-choice task that required them to respond differentially to the direction of the pitch change between syllables, blood flow was significantly higher in regions of right anterolateral frontal cortex, left posterodorsal frontal cortex, and medial occipital cortex than when the volunteers listened passively to the syllable pairs and performed a simple, repetitive task. Compared to lying still with minimal acoustic stimulation (ambient room noise), passive stimulation with syllables or amplitude-modulated noise during repetitive finger movements increased blood flow in portions of the left TG, right STG, left STG, and many other gyri. While large-scale differences in stimulus design, task requirements, and brain measures preclude a straightforward comparison of these structure–function correlates with those observed in lesion effect studies, the results (1) demonstrate that widespread regions of the cerebral cortex are active during acoustic stimulation and task performance; (2) neither support nor refute Zatorre's claim[38] that lesions of the right temporal lobe have no effect on complex-tone pitch perception when energy is present at F0, contrary to the authors' assertion that the observed right frontal activation during F0 pitch judgments was predicted on the basis of those results; and (3) suggest that supramodal systems involved in decision making and working memory during F0 pitch processing are lateralized to the right anterolateral frontal cortex.

Two recent fMRI studies suggest that the magnitude and spatial distribution of neural activity in A1 and adjacent auditory association cortex vary with pitch independent of stimulus physics.[49,50] For example, passive stimulation with harmonic tones that have a strong pitch evokes greater activation in left and right anterolateral TG and adjacent left STG than harmonic tones that have a weak pitch. Sequences with pitches that vary over time evoked greater activation in subregions of right STG than monotonic pitch sequences.[49] Magnetoencephalographic localization of responses evoked by tones with, versus without, spectral energy at F0 raise the possibility of a place representation of pitch in A1.[44,45]

NEURAL CODING OF PITCH

What are the neural representations and computations that give rise to pitch and harmony? An adequate theory needs to explain how the brain achieves high precision for pitch discrimination (pure-tone and complex-tone difference thresholds [~1% in naive listeners] and inharmonicity detection [~1% shift of one harmonic[51]]). It must also account for pitch equivalence despite differences in spectrum and level invariance over large dynamic ranges (within 1–2% across > 60 dB SPL).[4,52] In view of pitch equivalence and level invariance, the breakdown of tonotopy and nonmonotonic changes in discharge rates as a function of spectrum and intensity[53] confound a straightforward rate-place theory for pitch at the level of the auditory cortex.

Neurocomputational Mechanisms for Pitch Processing

Theories of pitch can be divided into two types: spectral-pattern models and temporal models. Spectral models first infer the pure-tone frequency components present in a sound from neural activity patterns, usually population rate-place profiles, and then carry out a harmonic analysis on that pattern. Connectionist networks can form harmonic templates if they have inputs that are highly frequency selective and well behaved in other ways.[54,55] Some evidence for the kinds of sharp frequency tuning and harmonic-combination selectivity that these models require has been observed in cortical neurons in barbiturate-anesthetized cats[56] and unanesthetized bats.[57–59] However, harmonic-combination selectivity has only been seen at best frequencies (BFs) well above those relevant for virtual pitches (BFs > 5 kHz). Another difficulty for spectral pattern models is that they cannot explain strong virtual pitches that can be evoked by sets of harmonics ($n > 6$) that are not resolved perceptually.[4,60]

Temporal models posit that neurons represent information about stimulus period-icities in the fine-timing of their discharges, specifically in the time intervals be-tween spikes ("interspike intervals"). Currently, the strongest neurocomputational models for pitch and harmony are based on population-wide distributions of all-order interspike intervals in the auditory nerve.[61-63] Features of these representa-tions correspond very well to many aspects of pitch perception: missing-F0 pitch; pitch equivalence; level and phase invariance; pitch strength (salience); unresolved harmonics, pitch shift of inharmonic tones (de Boer's rule); pitch ambiguity; the dominance region for pitch; Rameau's *basse fondamentale*; and the consonance and dissonance of musical intervals. Patterns of interval peaks in population-interval dis-tributions are similarly capable of representing vowel quality.[63-65] Since phase lock-ing and interspike interval information decline above 4 kHz, these models also

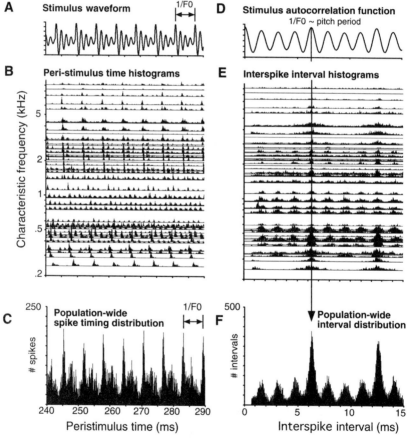

FIGURE 7. Temporal coding of pitch in the cat auditory nerve. (**Left**) Peristimulus time histograms of auditory nerve fibers in response to a vowel-like stimulus presented at moderate sound pressure level. (**Right**) Interspike interval distributions produced in single fibers and across the fiber population, whose features correspond closely to the pitches that are heard.[62]

explain the decline of pure-tone frequency discrimination at higher frequencies[66] and the existence region for musical tonality (up to about 4 kHz for octave matching and melodic recognition) (FIG. 7).

Coding Transformations in the Auditory Pathway

The spike timing information about pitch and harmony must either be analyzed in lower auditory stations in brain stem and midbrain, where it is abundantly available, or be transformed into some other form for processing in higher, thalamo-cortical centers.

The functional role of the auditory cortex is very different in these two models. In the first, fine-grained, temporal representations of pitch would exist only in lower stations. Here the cortex, through descending control paths, would dynamically organize lower centers to set up temporary, task-dependent circuits for pitch detection, discrimination, and recognition tasks. Cortical lesions would have the effect of degrading the control apparatus necessary for precise pitch distinctions. Relatively little is currently known about short-term bottom-up/top-down circuit dynamics in the auditory pathway, largely because most single-unit studies to date have been carried out in anesthetized animals. In the coding transformation conception, fine-grained pitch-related information is retained at higher levels, albeit in different form. A coding transformation would entail an orderly conversion of all-order interspike interval information into representations based on firing rate, spike latencies, more complex temporal patterns, interneural spike synchronies, and the like. In the transformational view, cortical lesions would have the effect of degrading neuronal populations that are involved in both conveying and processing pitch-related information.

Rate-based "periodicity detectors" situated in the midbrain have often been proposed as a plausible time-to-rate transformation.[67–71] While many neurons in the auditory pathway respond best to particular periodicity ranges (i.e., they have bandpass modulation-transfer functions), coding schemes based on periodicity-to-rate transformations are at odds with psychophysical observations: (1) modulation tuning is too coarse to support fine-grained pitch discrimination; (2) it is not level invariant;[72,73] (3) it cannot by itself account for pitch equivalence, since pure tones are unmodulated; and (4) it cannot account for the pitch of inharmonic tones, since modulation detectors, unlike pitch percepts, follow envelopes rather than fine structure when the two conflict.[1,74] Psychophysical data on the pitch of inharmonic tones strongly suggest a central temporal autocorrelation analysis rather than a modulation- or envelope-based one. Licklider's 1951 time-delay (TDNN) autocorrelation network could carry out the right time–place transformation, but no units with the requisite fine comb-filter tunings have been found at brain stem levels. Modulation tunings are seen in many units in the auditory thalamus and cortex,[75] but they are generally very coarse compared to pitch discrimination thresholds. While a few cortical units have best modulation frequencies (BMFs) in the periodicity pitch range (50–1000 Hz), the vast majority have BMFs well below this range (typically 4–16 Hz).

Central Temporal Codes

Within the central auditory system, evidence that pitch and harmony are encoded in the all-order interspike interval distribution is abundant in all three major divisions of the cochlear nucleus.[63,76–78] The existence of binaural periodicity pitches[79] suggests that mechanisms for pitch analysis must exist above the level of the superior olive, where timing information from both ears is preserved and combined. Interval-based representations of pitch and harmony at the level of the inferior colliculus are also possible.[80,81] Considerable phase-locking can be seen in the medial geniculate nucleus of lightly anesthetized animals: 10–20% of single units have synchronization indices of $\geq$0.3 to 1–2 kHz tones.[82] Neurons with these properties would almost certainly support fine pitch discriminations typical of human listeners.[66]

However, the extent to which these kinds of temporal representations can support pitch and harmony in the auditory cortex is much less clear. Averaged cortical potentials show periodicities up to several hundred hertz in response to click trains.[83–87] This, of course, is a pale remnant of the temporal information available in the auditory nerve and brain stem, where periodicities up to 4 or 5 kHz are seen. As one ascends the auditory pathway, both phase-locking and average discharge rates decline, and, as a consequence, it becomes increasingly difficult to find any units that produce a stimulus-related interspike interval pattern. If a cortical temporal code for pitch exists, it must be present in an asynchronous, covert, and/or sparse form, or it would have been observed by now with the methods at hand.

A sparse interval code is difficult to rule out completely. Since the numbers of neurons at higher stations increase dramatically, the same quantity of fine-timing information seen at lower stations might be more sparsely distributed across increasingly larger populations (sparse temporal code). De Ribaupierre and coworkers[88] found 2 of 179 neurons in A1 of unanesthetized cats that were capable of phase locking to click train F0s throughout the *entire* existence region for missing-F0 pitch (50–800 Hz). While < 2% is a small number in his sample, macaque A1 and surrounding areas contain over 10 million neurons;[89] consequently over 100,000 neurons may be available for interspike interval coding of missing-F0 pitch.

Could a central time code exist in covert form? For example, pitch-related timing information could be multiplexed with spike patterns that encode other aspects of the stimulus, such as timbre, roughness, location, loudness, and perceptual grouping, in which coding of each perceptual dimension relied on a different aspect of the population response (average rates, compactness of spike latency distributions, interval statistics, spike pattern correlations). Most of the common methods for cortical spike train analysis would miss temporal patterns of spikes embedded amid other spikes (multiplexed temporal codes), patterns that appeared at different times in the neuronal response (gated temporal code), patterns that involved interspike intervals be-

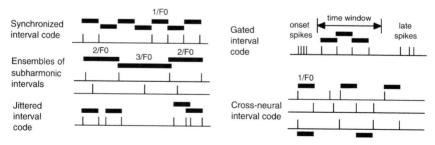

FIGURE 8. Possible synchronous and asynchronous coding schemes for encoding pitch-related periodicity information in spike correlation patterns.

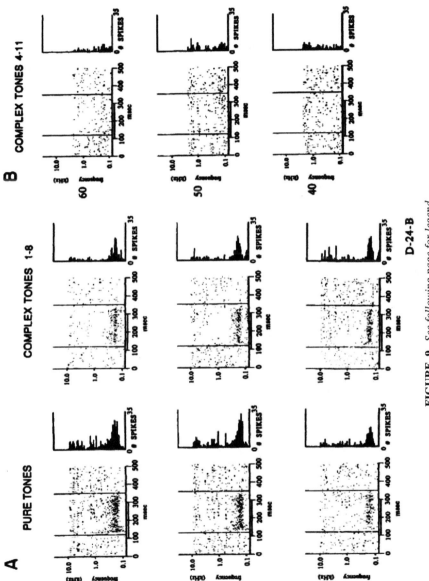

FIGURE 9. *See following page for legend.*

tween spike trains of multiple neurons (ensemble spike correlation code), patterns that were not rigidly time locked to the stimulus onset (asynchronous, jittered patterns), and periodicities related to subharmonics of F0 (Fig. 8). Although evidence for these and other alternative coding schemes has been found elsewhere,[90–100] most of these possibilities have yet to be explored in the auditory cortex.

Rate-Based Cortical Representations of Pitch

In recent decades most auditory neurophysiology has focused on characterizing input–output behavior of individual neurons rather than searching for response correlates of perceptual dimensions, distinctions, and invariances. In the latter strategy, one keeps the percept constant and analyzes for underlying commonalities in response patterns. Only one single-unit study in unanesthetized Old World monkeys has systematically investigated whether frequency-selective neurons change their firing rate as a function of pitch, independent of stimulus spectrum.[101] Stimuli included pure tones and harmonic tones with and without energy at F0, although Schwarz and Tomlinson found frequency-selective neurons in and around A1 that increased their discharge rates during stimulation with pure tones and harmonic tones with F0s near BF but not with harmonic tones missing F0 near BF (Fig. 9). Furthermore, no neurons displayed discharge rate profiles capable of resolving harmonic separations less than 300 Hz. In a much more limited study in anesthetized macaques, Riquimaroux and Hashikawa[102] reported that 15 of 15 neurons in A1 showed increases in firing rate to pure tones and harmonic tones with and without F0 at BF. It is not clear whether combination tones at BF or total stimulus spectral energy in the cell's frequency receptive field contributed to these results.

Evidence of a rate-based neural code for pitch in the auditory cortex of a New World monkey was recently reported by Bendor and Wang.[103] Peaks in the discharge rate profiles of 53 neurons tuned to low pure-tone frequencies in and around A1 were observed for harmonic-tone F0s and missing F0s near BF. Pitch equivalence might therefore be explained through the common activation of such units by sets of spectrally diverse sounds that have the same periodicity. Whether these neurons were, in fact, periodicity selective, as opposed to pitch selective, was not explicitly tested (e.g., with inharmonic tones, whose pitch does not match stimulus envelope periodicity). Moreover, inspection of the one published plot of cell discharge rate as a function of both pure-tone f and harmonic-tone missing F0 (i.e., the isointensity function in FIGURE 3a, p. 1163) finds that the neuron increased its discharge rate two or more standard deviations above spontaneous rate for most of the missing F0s tested—that is, the cell was *non*-selectively sensitive to harmonic-tone stimulation. By contrast,

---

FIGURE 9. Spike dot rasters and F0 vs. spike-count poststimulus time histograms showing responses of a frequency-selective neuron [best frequency (BF) = 173 Hz] to pure tones (*left column*), harmonic tones with energy at F0 (*middle column*), and harmonic tones missing F0 (**B**, *right column*) presented in quasi-free-field at 60 dB SPL (*top row*), 50 dB SPL (*middle row*), and 40 dB SPL (*bottom row*). This "F0 neuron" shows on-excitation during stimulation with pure tones and with harmonic tones when F0 is near BF (**A**), but there is no increase in firing during stimulation with harmonic tones missing F0, even when the missing F0 is near BF (**B**). (Reprinted by permission of the American Physiological Society.[101])

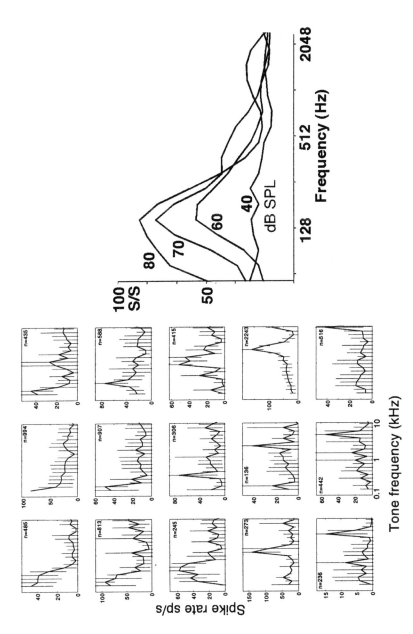

FIGURE 10. *See following page for legend*

the neuron responded to a much more restricted range of pure-tone frequencies. In addition, the peak in the rate versus missing-F0 isointensity function is much broader than psychophysical thresholds for missing-F0 discrimination predict.

It may be the case that this particular neuronal population has a special role to play in pitch perception, but there are reasons to doubt that the central representation for pitch is based per se on a profile of firing rates among these units, that is, on which of these detectors are firing the most frequently at any given instant or over a window of time between a few milliseconds to hundreds of milliseconds. One problem is that many periodicity-tuned units found by Bendor and Wang, like many neurons in A1, had nonmonotonic rate-level functions. These units tend to respond best to tones with the right periodicities that are presented at moderate levels, not far above their response thresholds (it should be noted that the periodicity tuning of these units was only demonstrated for low sound pressure levels, not high ones). They respond with lower firing rates when the same tones are presented at higher, sound pressure levels (e.g., > 60–80 dB SPL). In contrast, perception of musical pitch (low-frequency harmonic complex tones), does not break down at higher levels (e.g., for music listening in concert halls and clubs) and is almost completely invariant over the whole dynamic range of hearing. The prevalence of nonmonotonic rate-level functions means that different sets of these tuned periodicity units will respond maximally to the same tone presented at different sound levels. So, although such units may be able to give an economical explanation for why spectrally diverse stimuli presented at comparable levels have similar pitches, by the same argument, the same tones presented at different sound levels should have different pitches since different sets of units are activated. Although very slight shifts in pitch can be heard if levels are changed by 40 dB or more (maximally on the order of a few percent for pure tones and much less for complex tones), it is the extreme invariance of pitch with level that is striking. This invariance makes it possible for listeners to accurately match and discriminate pitches of tones even when their levels are randomly roved by tens of decibels.

The same problems arise when one attempts to account for the cortical representation of the pitches of low-frequency pure tones. In our single- and multiunit recordings of responses to pure tones in the core area of the alert macaques,[18] we found a variety of response types. For tones presented at moderate levels (70–80 dB SPL), we observed ON responders that responded to their range of preferred tone frequencies during the presentation of sustained pure tones, OFF responders that responded immediately after the tone ceased, and complex responders that gave ON responses to one set of frequencies and OFF responses to another. Even for ON responders that had one preferred range of tone frequencies (FIG. 10), frequency tuning was relative-

---

FIGURE 10. Frequency selectivity in the core area of alert macaques is intensity dependent. (*Left*) Mean spike rate vs. pure-tone frequency plots measured at 15 extracellular recording sites (7 from single neurons, 8 from clusters of 2–3 neurons). These isointensity functions analyze frequency selectivity during pure-tone stimulation (ON responses) at a fixed, moderate intensity (between 70–80 dB SPL near the tympanic membrane in the external auditory canal). Vertical bars in each plot show ±1 standard deviations from the mean spike rate computed across 10–20 repetitions of each frequency. (*Right*) Isointensity functions measured at four different tone intensities (40, 60, 70, 80 dB SPL) for a single neuron with a monotonic rate-level function. Note the increase in response bandwidth as tone intensity increases.

ly broad (typically 1–2 octaves), level dependent, and subject to a high degree of response variability (FIG. 11). These properties appear to confound simple representations of tone frequency that are based on profiles of average firing rates across neuronal populations.

In their systematic single-unit studies of the representation of pure-tone frequency in anesthetized cat auditory cortex, Phillips *et al.*[53] a half-decade earlier had found widespread nonmonotonic rate-level response functions and were forced to confront the problem of how such units could subserve a coherent rate-based representation of frequency (perceptually, pure-tone pitch). Their solution was to posit that the representation itself was level dependent, that is, a joint representation of

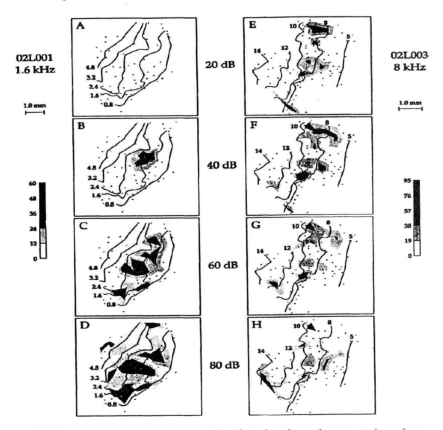

FIGURE 11. The topographic organization of rate-based neural representations of pure-tone frequency in A1 of anesthetized cats changes dramatically with tone intensity. (**A–H**) Flat maps of A1 (mostly its Layer IV) from two cats published by Phillips *et al.*[53] Lines demarcate isofrequency contours, i.e., longitudinal patches of neurons that are maximally sensitive to the same frequency near their response threshold. *Darker areas* on each map indicate penetration sites where single-unit discharge rates evoked by best-frequency tones were highest. For example, in **B**, stimulation with a best-frequency tone (1.6 kHz) at ≤20 dB above response threshold (**A**) caused neurons in a discrete 1 mm × 0.5 mm patch of A1 to fire. However, at higher intensities (**C** and **D**), the location of neurons maximally excited by the same tone frequency is markedly different. (Reprinted by permission of Springer, New York.[53])

both level and frequency. Unfortunately this solution simply intensifies the problem—now one must explain how level-invariant equivalence classes arise. Although connectionist learning models have been proposed that illustrate the general mechanisms by which pitch equivalence classes might arise through experience, such models have yet to show how robust equivalence classes could be formed using more realistic neuronal elements with highly variable responses coupled with nonmonotonic, level-dependent tuning.

In addition to meeting criteria related to perceptual acuity and constancy, viable neural codes for pitch must account for the functional dissociability of pitch change detection and pitch direction discrimination revealed by human lesion experiments.[30,35] The existence of A1 and belt neurons that increase their firing rate more when the second tone is higher than the first tone than when it is lower, and vice versa, has been firmly established by Brosch and colleagues in three anesthetized *Macaca fascicularis* monkeys.[108] To our knowledge, this remains the only systematic investigation of single- and multiunit responses to sequences of two pure tones in Old World monkeys with Δfs and ISIs similar to those used in psychoacoustic experiments. The prevalence of direction-sensitive units in different subdivisions of auditory cortex and their distribution in the left and right auditory cortices remain unknown.

Thus, an enigma persists concerning the nature of the neural codes and computations that support these universal and highly reliable perceptual invariances and equivalence classes at the cortical level. Whatever the code, it must behave in a highly reliable way under a huge variety of environmental conditions and for stimuli that are almost never present in the natural world. For the various reasons outlined here, we believe that the cortical representation of pitch is not likely to be based on either a simple temporal or rate-based code; nor is it likely to be based on idiosyncratic, special purpose assemblages of neural elements and interconnections. The neural representations and computations that subserve pitch perception may involve elegant and powerful information-processing principles and mechanisms that still elude us.

ACKNOWLEDGMENTS

This work was supported by NIH DC03328, DC006353, and DC00117; the Institute for Music and Brain Science; and the National Organization for Hearing Research. We thank the Pierfranco and Luisa Mariani Foundation, Dr. Maria Majno, Dr. Stefan Koelsch, and Mr. Steven Bohall; and our patients, colleagues, and students for their inspiration and contributions.

[Competing interests: The authors declare that they have no competing financial interests.]

REFERENCES

1. DE BOER, E. 1976. On the "residue" and auditory pitch perception. *In* Auditory System (Handbook of Sensory Physiology). W.D. Keidel & W.D. Neff, Eds.: 479–583. Springer-Verlag. Berlin.
2. LICKLIDER, J.C.R. 1951. A duplex theory of pitch perception. Experientia **VII:** 128–134.
3. TERHARDT, E. 1974. Pitch, consonance, and harmony. J. Acoust. Soc. Am. **55:** 1061–1069.
4. PLOMP, R. 1976. Aspects of Tone Sensation. Academic Press. London. p. 167.
5. MOORE, B.C.J. 2003. An Introduction to the Psychology of Hearing. Academic Press. San Diego.

6. RAMEAU, J.P. 1722. Treatise on Harmony. Dover (1971 reprint). New York.
7. PISTON, W. & M. DEVOTO. 1987. Harmony, 5th Ed. Norton. New York.
8. TOMLINSON, R.W.W. & D.W.F. SCHWARTZ. 1988 Perception of the missing fundamental in nonhuman primates. J. Acoust. Soc. Am. **84:** 560–565.
9. WRIGHT, B.A. *et al.* 2000. Music perception and octave generalization in rhesus monkeys. J. Exp. Psychol. Gen. **129:** 291–307.
10. IZUMI, A. 2000. Japanese monkeys perceive sensory consonance of chords. J. Acoust. Soc. Am. **108:** 3073–3078.
11. HEFFNER, H. & I.C. WHITFIELD. 1976. Perception of the missing fundamental by cats. J. Acoust. Soc. Am. **59:** 915–919.
12. CYNX, J. 1986. Periodicity pitch in a species of songbird, the European starling (*Sturnus vulgaris*). Assoc. Res. Otolaryngol. Abstr. p. 138.
13. CAPRIANCA, R.R., G.J. ROSE & E.A. BRENOWITZ. 1985. Time resolution in the auditory systems of anurans. *In* Time Resolution in Auditory Systems, A. Michelsen, Ed.: 28–57. Springer-Verlag. Berlin.
14. FAY, R.R. 1985. Time processing by the auditory system of fishes. *In* Time Resolution in Auditory Systems. A. Michelsen, Ed.: 28–57. Springer-Verlag. Berlin.
15. BORCHGREVINK, H.M. 1975. Musical chord preferences in humans as demonstrated through animal experiments. Tideskrit for den Norske Laegeforening (Norwegian) **95:** 356–358.
16. HULSE, H.S., D.J. BERNARD & R.F. BRAATEN. 1995. Auditory discrimination of chord-based spectral structures by European starlings. J. Exp. Psychol. **124:** 409–423.
17. JONES, E.G. *et al.* 1995. Subdivisions of macaque monkey auditory cortex revealed by calcium binding protein immunoreactivity. J. Comp. Neurol. **362:** 1–19.
18. TRAMO, M.J. 1998. Neural representations of acoustic information in relation to music and voice perception. Harvard University, Cambridge, MA. Ph.D. Dissertation.
19. HACKETT, T.A., T.M. PREUSS & J.H. KAAS. 2001. Architectonic identification of the core region in auditory cortex of macaques, chimpanzees, and humans. J. Comp. Neurol. **441:** 197–222.
20. MERZENICH, M.M. & J.F. BRUGGE. 1973. Representation of the cochlear partition on the superior temporal plane of the macaque monkey. Brain Res. **50:** 275–296.
21. PANDYA, D.N. 1995. Anatomy of the auditory cortex. Rev. Neurol. **151:** 486–494.
22. KAAS, J.H., T.A. HACKETT & M.J. TRAMO. 1999. Auditory processing in primate cerebral cortex. Curr. Opin. Neurobiol. **9:** 164–170.
23. RAUSCHECKER, J.F., B. TIAN & M.D. HAUSER. 1995. Processing of complex sounds in the macaque nonprimary auditory cortex. Science **268:** 111–114.
24. NEFF, W.D., I.T. DIAMOND & J.H. CASSEDAY. 1975. Behavioral studies of auditory discrimination: central nervous system. *In* Handbook of Sensory Physiology. W.D. Keidel & W.D. Neff, Eds.: 307–400. Springer-Verlag. Berlin.
25. THOMPSON, R.F. 1960. Function of auditory cortex of cat in frequency discrimination. J. Neurophysiol. **23:** 321–334.
26. ELLIOT, D.N. & C. TRAHOITIS. 1972. Cortical lesions and auditory discrimination. Psychol. Bull. **77:** 198–222.
27. EVARTS, E.V. 1952. Effect of auditory cortex ablation on frequency discrimination in monkey. J. Neurophysiol. **15:** 443–448.
28. JERGER, J. *et al.* 1969. Bilateral lesions of the temporal lobe: a case study. Acta Otolaryngol. Suppl. **252:** 1–51.
29. JERGER, J., L. LOVERING & M. WERTZ. 1972. Auditory disorder following bilateral temporal lobe insult: report of a case. J. Speach Hear Disord. **37:** 523–535.
30. TRAMO, M.J., G.D. SHAH & L.D. BRAIDA. 2002. The functional role of auditory cortex in frequency processing and pitch perception. J. Neurophysiol. **87:** 122–139.
31. SEASHORE, C.E., D. LEWIS & J.C. SAETVIT. 1960. Seashore Measures of Musical Talents Manual. The Psychological Corporation. New York.
32. TRAMO, M.J. 1990. Impaired perception of relative pure-tone pitch following bilateral lesions of auditory cortex in man. Soc. Neurosci. Abstr. **16:** 580.
33. TRAMO, M.J., A. GRANT & L.D. BRAIDA. 1994. Psychophysical measurements of frequency difference limens for relative pitch discrimination reveal a deficit following bilateral lesions of auditory cortex. Soc. Neurosci. Abstr. **20:** 325.

34. KAZUI, S. *et al.* 1990. Subcortical auditory agnosia. Brain Lang. **38:** 476–487.
35. JOHNSRUDE, I.S., V.B. PENHUNE & R.J. ZATORRE. 2000. Functional specificity in the right human auditory cortex for perceiving pitch direction. Brain **123:** 155–163.
36. DIVENYI, P.L. & A.J. ROBINSON. 1989. Nonlinguistic auditory capabilities in aphasia. Brain Lang. **37:** 290–326.
37. TRAMO, M.J., C.K. KOH, G.D. SHAH & L.D. BRAIDA. 2004 Functional role of auditory cortex in virtual pitch perception. Soc. Neurosci. Abstr. 30 apu.sfn.org/content/Publications/AnnualMeeting/.
38. ZATORRE, R.J. 1988. Pitch perception of complex tones and human temporal-lobe function. J. Acoust. Soc. Am. **84:** 566–572.
39. ROBIN, D., D. TRANEL & H. DAMASIO. 1990. Auditory perception of temporal and spectral events in patients with focal left and right cerebral lesions. Brain Lang. **39:** 539–555.
40. MASSOPUST, L.C. JR., *et al.* 1967. Changes in auditory frequency discrimination thresholds after temporal cortex ablations. Exp. Neurol. **19:** 245–255.
41. HARRINGTON, D.L., R.S. HEFFNER & H.E. HEFFNER. 2001. An investigation of sensory deficits underlying the aphasia-like behavior of macaques with auditory cortex lesions. Neuroreport **12:** 1217–1221.
42. SIDTIS, J.J. & B.T. VOLPE. 1988. Selective loss of complex-pitch or speech discrimination after unilateral lesion. Brain Lang. **34:** 235–245.
43. BHARUCHA, J.J., M.J. TRAMO & R.J. ZATORRE. 1993. Abstraction of the missing fundamental following bilateral lesions of auditory cortex. Soc. Neurosci. Abstr. **19:** p. 1687.
44. PANTEV, C. *et al.* 1996. Binaural fusion and the representation of virtual pitch in the human auditory cortex. Hear. Res. **100:** 164–170.
45. LANGNER, G. *et al.* 1997. Frequency and periodicity are represented in orthogonal maps in the human auditory cortex: evidence from magnetoencephalography. J. Comp. Physiol. **181:** 665–676.
46. ZATORRE, R. *et al.* 1992. Lateralization of phonetic and pitch discrimination in speech processing. Science **256:** 846–849.
47. TRAMO, M.J. *et al.* 2001. Neurobiological foundations for the theory of harmony in Western tonal music. Ann. N. Y. Acad. Sci. **930:** 92–116.
48. GAAB, N., C. GASER, T. ZAEHLE, *et al.* 2003. Functional anatomy of pitch memory—a fMRI study with sparse temporal sampling. Neuroimage **19:** 1417–1426.
49. PATTERSON, R.D. *et al.* 2002. The processing of temporal pitch and melody information in auditory cortex. Neuron **36:** 767–76.
50. PENAGOS, H., J.R. MELCHER & A.J. OXENHAM. 2004. A neural representation of pitch salience in nonprimary human auditory cortex revealed with functional magnetic resonance imaging. J. Neurosci. **24:** 6810–6815.
51. MOORE, B.C.J., R.W. PETERS & B.R. GLASBERG. 1985. Thresholds for the detection of inharmonicity in complex tones. J. Acoust. Soc. Am. **77:** 1985.
52. WIER, C.C., W. JESTEADT & D.M. GREEN. 1977. Frequency discrimination as a function of frequency and sensation level. J. Acoust. Soc. Am. **61:** 178–184.
53. PHILLIPS, D.P. *et al.* 1994. Level-dependent representation of stimulus frequency in cat primary auditory cortex. Exp. Brain Res. **102:** 210–226.
54. COHEN, M.A., S. GROSSBERG & L.L. WYSE. 1994. A spectral network model of pitch perception. J. Acoust. Soc. Am. **98:** 862–879.
55. BHARUCHA, J.J. 1991. Pitch, harmony and neural nets: a psychological perspective. *In* Connectionism and Music. P. Todd & G. Loy, Eds.: 84–99. MIT Press. Cambridge.
56. SUTTER, M.L. & C.E. SCHREINER. 1991. Physiology and topography of neurons with multipeaked tuning curves in cat primary auditory cortex. J. Neurophysiol. **65:** 1207–1226.
57. FITZPATRICK, D.C. *et al.* 1993. Combination-sensitive neurons in the primary auditory cortex of the mustached bat. J. Neurosci. **13:** 931–940.
58. SUGA, N. & W.E. O'NEIL. 1979. Harmonic-sensitive neurons in the auditory cortex of the Mustache bat. Science **203:** 270–274.
59. DEAR, S.P. *et al.* 1993. Tonotopic and functional organization in the auditory cortex of the big brown bat, *Eptesicus fuscus.* J. Neurophysiol. **70:** 1988–2009.

60. MOORE, B.C.J. 1997. An Introduction to the Psychology of Hearing, 4th Ed. Academic Press. London. p. 373.
61. MEDDIS, R. & L. O'MARD. 1997 A unitary model of pitch perception. J. Acoust. Soc. Am. **102:** 1811–1820.
62. CARIANI, P.A. & B. DELGUTTE. 1996. Neural correlates of the pitch of complex tones. I. Pitch and pitch salience. II. Pitch shift, pitch ambiguity, phase-invariance, pitch circularity, and the dominance region for pitch. J. Neurophysiol. **76:** 1698–1734.
63. CARIANI, P. 1999. Temporal coding of periodicity pitch in the auditory system: an overview. Neural Plast. **6:** 147–172.
64. PALMER, A.R. 1992. Segregation of the responses to paired vowels in the auditory nerve of the guinea pig using autocorrelation. *In* The Auditory Processing of Speech S.M.E.H., Ed.: 115–124. Mouton de Gruyter. Berlin.
65. HIRAHARA, T., P. CARIANI & B. DELGUTTE. 1996. Representation of low-frequency vowel formants in the auditory nerve. *In* Proceedings, ESCA Research Workshop on The Auditory Basis of Speech Perception, Keele University, United Kingdom. July 15–19, 1996.
66. GOLDSTEIN, J.L. & P. SRULOVICZ. 1977. Auditory-nerve spike intervals as an adequate basis for aural frequency measurement. *In* Psychophysics and Physiology of Hearing. E.F. Evans & J.P. Wilson, Eds. Academic Press. London.
67. LANGNER, G. 1992. Periodicity coding in the auditory system. Hear. Res. **60:** 115–142.
68. LANGNER, G. & C.E. SCHREINER. 1988. Periodicity coding in the inferior colliculus of the cat. I. Neuronal Mechanisms. J. Neurophysiol. **60:** 1799–1822.
69. ROSE, G.J. & R.R. CAPRIANICA. 1985. Sensitivity to amplitude modulated sounds in the anuran auditory nervous system. J. Neurophysiol. **53:** 446–465.
70. EPPING, W.J.M. & J.J. EGGERMONT. 1986. Sensitivity of neurons in the auditory midbrain of the grassfrog to temporal characteristics of sound. I Stimulation with acoustic clicks. Hear. Res. **24:** 37–54.
71. EPPING, W.J.M. & J.J. EGGERMONT. 1986. Sensitivity of neurons in the auditory midbrain of the grassfrog to temporal characteristics of sound. II. Stimulation with amplitude modulated sounds. Hear. Res. **24:** 55–72.
72. REES, A. & A.R. MØLLER. 1983. Responses of neurons in the inferior colliculus of the rat to AM and FM tones. Hear. Res. **10:** 301–330.
73. KRISHNA, M.B.S. & M.N. SEMPLE. 2000. Auditory temporal processing: responses to sinusoidally amplitude-modulated tones in inferior colliculus. J. Neurophysiol. **84:** 255–273.
74. DE BOER, E. 1956. On the "residue" in hearing. University of Amsterdam.
75. SCHREINER, C.E. & G. LANGNER. 1988. Coding of temporal patterns in the central auditory system. *In* Auditory Function: Neurobiological Bases of Hearing. G.M. Edelman, W.E. Gall & W.M. Cowan, Eds.: 337–362. Wiley. New York.
76. RHODE, W.S. 1995. Interspike intervals as correlates of periodicity pitch in cat cochlear nucleus. J. Acoust. Soc. Am. **97:** 2414–2429.
77. RHODE, W.S. & S. GREENBERG. 1992. Physiology of the cochlear nuclei. *In* The Mammalian Auditory Pathway: Neurophysiology. A.N. Popper & R.R. Fay, Eds.: 94–152. Springer-Verlag. New York.
78. YOUNG, E.D. 1984. Response characteristics of neurons in the cochlear nuclei. *In* Hearing Science. C. Berlin, Ed.: 423–460. College Hill. San Diego.
79. COLBURN, S. & N.I. DURLACH. 1978. Models of binaural interaction. *In* Handbook of Perception. E.C. Carterette & M.P. Friedman, Eds.: 467–518. Academic Press. New York.
80. DEPIREUX, D.A. & S.A. SHAMMA. 1998. Neuronal correlates of pitch in the inferior colliculus. Assoc. Res. Otolaryngol. Abstr. **20:** 182.
81. GREENBERG, S. 1980. Neural temporal coding of pitch and vowel quality: human frequency, following response studies of complex signals. UCLA Working Papers in Phonetics #52, Los Angeles.
82. DE RIBAUPIERRE, F. 1997. Acoustical information processing in the auditory thalamus and cerebral cortex. *In* The Central Auditory System. G. Ehret & R. Romand, Eds.: 317–397. Oxford University Press. New York.
83. MAKELA, J.P. *et al.* 1990. Steady-state responses from the cat auditory cortex. Hear. Res. **45:** 41–50.

84. GOLDSTEIN, M.H., N.Y.-S. KIANG & R.M. BROWN. 1959. Responses of auditory cortex to repetitive acoustic stimuli. J. Acoust. Soc. Am. **31:** 356–364.
85. GOLDSTEIN, M.H., JR. & N.Y.S. KIANG. 1958. Synchrony of neural activity in electric responses evoked by transient acoustic stimuli. J. Acoust. Soc. Am. **30:** 107–114.
86. STEINSCHNEIDER, M., J. AREZZO & H.G. VAUGHAN JR. 1980. Phase locked cortical responses to a human speech sound and low frequency tones in the monkey. Brain Res. **198:** 75–84.
87. STEINSCHNEIDER, M. *et al.* 1998. Click train encoding in primary auditory cortex of the awake monkey: evidence for two mechanisms subserving pitch perception. J. Acoust. Soc. Am. **104:** 2395–2955.
88. DE RIBAUPIERRE, F., M.H. GOLDSTEIN JR. & G. YENI-KOMSHIAN. 1972. Cortical coding of repetitive acoustic pulses. Brain Res. **48:** 205–225.
89. CHOW, K.L. 1951. Numerical estimates of the auditory central nervous system of the rhesus monkey. J. Comp. Neurol. **95:** 159–175.
90. VILLA, A.E.. 2000. Empirical evidence about temporal structure in multi-unit recordings. *In* Time and the Brain. R. Miller, Ed.: 1–52. Harwood Academic Publishers. Australia.
91. EGGERMONT, J.J. 1990. The Correlative Brain: Theory and Experiment in Neural Interaction. Springer-Verlag. Berlin. p. 307.
92. EGGERMONT, J.J. & G. SMITH. 1995. Synchrony between single-unit activity and local field potentials in relation to periodicity coding in primary auditory cortex. J. Neurophysiol. **73:** 227–245.
93. EGGERMONT, J.J. 1993. Functional aspects of synchrony and correlation in the auditory nervous system. Concep. Neurosci. **4:** 105–129.
94. SINGER, W. 1994. Time as coding space in neocortical processing. *In* Temporal Coding in the Brain. G. Buzsáki *et al.*, Eds.: 51–80. Springer-Verlag. Berlin.
95. ESPINOZA, I.E. & G.L. GERSTEIN. 1988. Cortical auditory neuron interactions during presentation of 3-tone sequences: effective connectivity. Brain Res. **450:** 39–50.
96. EMMERS, R. 1981. Pain: A Spike-Interval Coded Message in the Brain. Raven Press. New York.
97. MOUNTCASTLE, V. 1993. Temporal order determinants in a somatosthetic frequency discrimination: sequential order coding. Ann. N. Y. Acad. Sci. **682:** 151–170.
98. ABELES, M. *et al.* 1993. Spatiotemporal firing patterns in the frontal cortex of behaving monkeys. J. Neurophysiol. **70:** 1629–1638.
99. LESTIENNE, R. & B.L. STREHLER. 1987. Time structure and stimulus dependence of precise replicating patterns present in monkey cortical neuronal spike trains. Brain Res. **43:** 214–238.
100. DECHARMS, R.C. & M.M. MERZENICH. 1996. Primary cortical representation of sounds by the coordination of action-potential timing. Nature **381:** 610–613.
101. SCHWARZ, D.W.F. & R.W.W. TOMLINSON. 1990. Spectral response patterns of auditory cortical neurons to harmonic complex tones in alert monkey (*Macaca mulatta*). J. Neurophysiol. **64:** 282–298.
102. RIQUIMAROUX, H. & T. HASHIKAWA. 1994. Units in the primary auditory cortex of the Japanese monkey can demonstrate a conversion of temporal and place pitch in the central auditory system. Journal de Physique IV **4:** 419–425.
103. BENDOR, D. & X. WANG. 2005. The neuronal representation of pitch in primate auditory cortex. Nature **436:** 1161–1165.
104. AMERICAN NATIONAL STANDARDS INSTITUTE (ANSI). 1960, 1994. American National Standard Acoustical Terminology for Physiological and Psychological Acoustics. ANSI Suppl 1.1. American National Standards Institute. New York.
105. TRAMO, M.J., J.J. BHARUCHA & F.E. MUSIEK. 1990. Music perception and cognition following bilateral lesions of auditory cortex. J. Cog. Neurosci. **2:** 195–212.
106. WARRIER, C.M. & R.J. ZATORRE. 2004. Right temporal cortex is critical for utilization of melodic contextual cues in a pitch constancy task. Brain **127:** 1616–1625.
107. TRAMO, M.J. 2001. Music of the hemispheres. Science **291:** 54–56.
108. BROSCH, M., A. SCHULZ & H. SCHEICH. 1999. Processing of sound sequences in macaque auditory cortex:response enhancement. J. Neurophysiol. **82:** 1542–1559.

Aspects of Multisensory Perception

The Integration of Visual and Auditory Information in Musical Experiences

DONALD A. HODGES,[a] W. DAVID HAIRSTON,[b] AND JONATHAN H. BURDETTE[c]

[a]Music Research Institute, University of North Carolina at Greensboro, Greensboro, North Carolina 27402-6170, USA

[b]Department of Neurobiology and Anatomy, [c]Department of Radiology, Wake Forest University School of Medicine, Winston-Salem, North Carolina 27157, USA

ABSTRACT: One of the requirements for being a successful musical conductor is to be able to locate sounds instantaneously in time and space. Because this requires the integration of auditory and visual information, the purpose of this study was to examine multisensory processing in conductors and a matched set of control subjects. Subjects participated in a series of behavioral tasks, including pitch discrimination, temporal-order judgment (TOJ), and target localization. Additionally, fMRI scans were done on a subset of subjects who performed a multisensory TOJ task. Analyses of behavioral data indicate that, in the auditory realm, conductors were more accurate in both pitch discrimination and TOJs as well as in locating targets in space. Furthermore, these same subjects also demonstrated a benefit from the combination of auditory and visual information that was not observed in control subjects when locating visual targets. Finally, neural substrates in BA 37, 39/40 were identified as potential areas underlying the conductors' superior multisensory TOJs. Data collection and analyses are ongoing and will lead to an improved understanding of multisensory integration in a complex, musical behavior.

KEYWORDS: multisensory processing; auditory-visual discrimination; temporal-order judgments

INTRODUCTION

Music making is a prime example of multisensory processing in a complex form of human behavior. At a minimum, information from the ears, eyes, and motor systems must be fully integrated for successful performances. While a considerable amount has been learned in recent years about how the brain organizes musical behaviors,[1,2] little attention has been paid to how different systems integrate into a coherent whole. The purpose of this study was to examine multisensory processing in a select group of musicians.

Address for correspondence: Donald A. Hodges, School of Music, University of North Carolina at Greensboro, P.O. Box 26170, Greensboro, NC 27402-6170. Voice: 336-334-5176; fax: 336-334-5497.

dahodges@uncg.edu

Ann. N.Y. Acad. Sci. 1060: 175–185 (2005). © 2005 New York Academy of Sciences.
doi: 10.1196/annals.1360.012

Successful conductors have developed a myriad of skills, including reading a musical score (musical notation indicating the precise notes to be performed by each member of the ensemble), expressing musical information in precise physical gestures (e.g., tempo and dynamics), and retaining the idealized version of the sounds in auditory memory while monitoring the actual sounds produced in real time. Of particular interest is how visual information (e.g., from the score and from the players) is integrated with auditory information (both real and imagined). As just one example, consider that the conductor must be very adept at not only identifying errors, but also locating the errant sound in precise time and space. Thus, it is likely that experienced conductors have developed specialized skills at sound localization, instantly identifying exactly "who" played "what" wrong note. What are the neural substrates behind the visual and auditory processing required for such behaviors? Do experienced conductors possess enhanced neural processing for such tasks?

Multisensory integration is a natural brain function, as input from different sensory modalities is integrated into a coherent perceptual gestalt;[3] each sensory domain can receive input from other senses, and there are transitional multisensory zones between modality-specific cortical domains.[4] This type of integration between senses has been shown to enhance perception in a variety of realms, including speeding up responses,[5,6] increasing localization accuracy or the detectability of a stimulus,[7–9] and even enhancing perceptions of temporal-order judgments (TOJs).[10,11] However, it must be noted that enhancing effects are typically prominent only when the target stimuli are at or near a perceptual threshold or difficult to locate initially; this effect has become known as "inverse effectiveness,"[12] and suggests that the amount of cross-modal benefit observed is directly tied to the relative efficacy of each of the sensory channels involved.

While neuromusical research in general has increased significantly, few studies have investigated conductors specifically. In a study of error detection in conductors,[13,14] PET scans indicated different brain activations for the identification of melodic, harmonic, or rhythmic errors. In another study, event-related brain potentials and behavioral data suggested that conductors were superior to pianists and controls in focusing preattentively and attentively on auditory localization tasks.[15] Although it is clear that a conductor's perceptual gestalt is based on multimodal sensory input, many aspects of multisensory processing in such a complex, dynamic form of behavior remain poorly understood. Thus, it was the purpose of this experiment to investigate aspects of multisensory perception in the integration of visual and auditory information processing in conductors when compared to controls. Reported here are preliminary results from studies currently in progress.

SUBJECTS

Subjects for this experiment consisted of ten conductors and ten musically untrained controls. Conductors were between the ages of 30 and 40 (mean = 34.4, ± 4) and had from 6–18 (mean = 10.35, ± 3.9) years' experience as a middle or high school band or orchestra director. All were right handed, as determined by self-report. Seven of the subjects were male; three were female. These musicians had an average of 19.5 years of education (± 2.3). Control subjects were similar in age (mean = 33.4, ± 4.4), handedness, gender (6 males, 4 females), and general educational background (19 years, ± 3.7). However, they lacked formal music instruction, having minimal or no formal musical training.

METHODS FOR BEHAVIORAL TASKS

Subjects completed a series of behavioral tasks involving unisensory and multisensory processing. Some of these tasks were subsequently replicated during fMRI scanning.

Pitch Discrimination

Subjects reported which of two tones occurred first ("high first" versus "low first"), using two buttons on a response box. The order of presentation was always random. The two tones were played with a 500 ms pause between them, and they always differed in pitch. One tone was always 440 Hz and the other was always higher, but the specific frequency varied according to an adaptive staircase procedure, starting at 457 Hz (roughly two thirds the "tonal" distance between A and A#) and adjusting in "distance," and hence difficulty, until subjects were performing at roughly 71% accuracy; eventually the pitches were so close together that subjects could barely distinguish which of the two was the higher one. The end measurement was a frequency value that represented their threshold discrimination above the base tone (e.g., 442.8 Hz). Three such staircase procedures occurred simultaneously, and on each trial one was chosen at random, with the average of all three used as the final measure per subject.

A second task was identical, except that the base frequency was 500 Hz, and comparison tones started at 520 Hz. This was done to determine whether conductors were better with a well-known note (A) versus a tone not represented in the Western musical scale.

Temporal-Order Judgment

In the third task, auditory TOJ, the two tones did not change (440 and 660 Hz, A and E), but the onset time between them changed via an adaptive staircase; that is, they were presented closer together in time until they appeared to occur simultaneously. A determination was made of the time needed to discriminate the two pitches (e.g., 26 ms).

For the fourth task, visual TOJ, subjects determined which circle (of two) presented on a computer screen (200 Hz vertical refresh rate) was presented first—the top circle or the bottom circle. As in previous tasks, the stimulus-onset asynchronies (SOAs) varied according to an adaptive staircase to determine the threshold for visual discrimination.

Temporal-Order Judgment with Multisensory Cues

This TOJ task (visual circles with monotone beeps) was similar to the visual-alone TOJ task, in that the subjects determined which visual circle was presented first, but in this "multisensory" version, they were also presented with nonspatial 10 ms clicks with the presentation of the circles. The first beep was congruent with the first circle, but the second beep could be congruent with the second circle or delayed (50–350 ms). The delay between the auditory beeps was altered in an attempt to determine a window for multisensory integration. Measurements indicated how much change in accuracy occurred in the various sound-added conditions verses no sound, as well as the average time to respond.

Target Localization

Subjects were asked to locate visual (LED), auditory (broadband noise burst), or spatially congruent visual-auditory stimuli across a 180 arc. These stimuli were presented in a dark chamber, and the subjects were asked to locate either a briefly presented (50 ms) light, a sound, or both light and sound using a laser-pointing yoke device. Targets were presented 10, 20, 30, or 40 degrees to the left or right and were randomly interleaved across trials. Responses were measured with regard to both speed and accuracy.

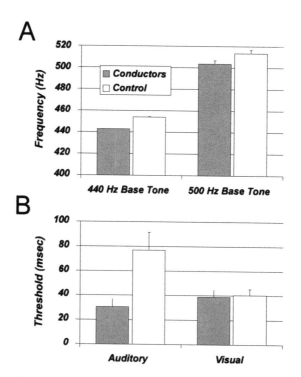

FIGURE 1. (A) Results for pitch discrimination tasks. **(B)** Results for auditory and visual temporal-order judgments. Error bars represent SEM for the group.

RESULTS FOR BEHAVIORAL TASKS

Pitch Discrimination

On average, conductors tested thus far are significantly more acute in their ability to discriminate between different pitches than musically untrained individuals. FIGURE 1A shows the average pitch that could be discriminated above two different baseline tones (440 and 500 Hz) for each of the two groups (error bars represent SEM). Because this trend was significant for both a baseline of 440 Hz ($t(18) = 3.66$, $P < .05$), A4 on the equal-tempered musical scale, and 500 Hz ($t(18) = 3.24, P < .05$), which is no particular note but would lay roughly between B4 and C5, it is not likely that this difference is selective to specific notes of the musical scale. As a matter of perspective, while control subjects require a tone to be roughly 2.95% of the base to discriminate it, conductors require a difference of only 0.67%.

Temporal Discrimination

Subjects also performed auditory and visual TOJs in order to ascertain their ability to discern between sounds over time. With regard to temporal discrimination, an interesting dichotomy appears, shown in FIGURE 1B. Specifically, in addition to discriminating between pitches, conductors are significantly more acute in the temporal domain with sounds as well. Control subjects require an average of 76.7 milliseconds between onsets to discriminate the two tones (440 and 660 Hz) at a threshold level, while conductors require only 33.7 ms in order to perform at the same level ($t(18) = 2.87, P < .05$). In contrast, both groups required approximately the same amount of time when the task involved visual stimuli (40.4 ms and 38.9 ms, respectively; $t(18) = 0.20, P > .05$), suggesting that these enhancements are limited to the auditory domain.

Temporal Discrimination with Multisensory Cues

For each subject, the threshold SOA determined from the above visual TOJ was used to fix the SOA to a single value within a set of visual TOJs that also included a task-irrelevant sound. Previous studies[10,11] have shown that the inclusion of this sound with a slight delay improves the accuracy with which subjects can perform the visual TOJ task. For both groups, this was indeed the case—subjects were significantly more accurate when the second sound was delayed by 100 ms than with either the sound synchronous (conductors: $t(9) = 4.32, P < .05$; control: $t(9) = 1.96, P = .05$) or no sound at all (conductors: $t(9) = 2.62, P < .05$; control: $t(9) = 9.93, P < .05$). Additionally, this effect showed a decline with larger delays, such that when the cross-modal SOA was as much as 350 ms, no benefit was observed (conductors: $t(9) = 1.0, P > .05$; control: $t(9) = 0.65, P > .05$); this is consistent with previous reports. FIGURE 2 shows the average time required to respond for each of the groups across all conditions.

Interestingly, response times are decreased (i.e., faster decisions) for both groups simply by including the sound synchronous with the onset of the two visual stimuli, even though it does not provide any task-relevant information; note that this does not quite reach statistical significance, although a trend is clear (conductors: $t(9) = 2.15$, $P < .016$; control: $t(9) = 1.95, P = .083$). Additionally, response times are fastest with

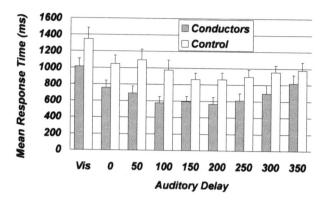

FIGURE 2. Results for visual TOJ when a task-irrelevant sound is added.

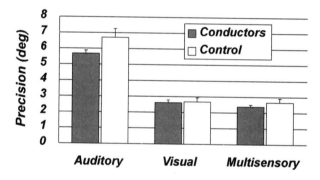

FIGURE 3. Results for localization tasks.

cross-modal SOAs of 150–200 ms between the latter two stimuli, and, as with accuracy, this effect weakens with an increased delay. Notice also that conductors are consistently faster than control subjects across all conditions tested ($F(1, 18) = 5.37$, $P < .05$).

Target Localization

After completion of all threshold testing, we assessed subjects' abilities to locate visual, auditory, and combined visual-auditory stimuli. FIGURE 3 shows the average localization precision for both groups for each target type.

In this case, "precision" is assessed as the average width, in degrees of angle, of the distribution of responses (standard deviation). Hence, low values (e.g., a small number of degrees) represent higher precision, whereas larger values denote less precision. First notice the pattern for control subjects. While their ability to locate the auditory target is relatively poor, it is significantly better when the target is visual in nature. However, the addition of an auditory stimulus to this visual target does not significantly improve this ability ($t(9) = 0.29$, $P > .05$). By contrast, conductors show a different trend. Primarily, their auditory localization ability is noticeably better than that of control subjects (while this does not quite reach statistical significance in a between-groups analysis ($t(18) = 1.8$, $P = .09$), this is likely due to the low number of subjects examined thus far). Additionally, unlike their control counterparts, conductors receive a significant benefit from the additional auditory signal, such that their localization ability is significantly better for multisensory than visual-only targets ($t(9) = 2.65$, $P < .05$), an effect that is likely due to their enhanced auditory performance. Finally, as with multisensory-mediated TOJs, conductors consistently responded faster than control subjects, showing significantly decreased response times with visual ($t(18) = 2.92$, $P < .05$), auditory ($t(18) = 3.08$, $P < .05$), and multisensory ($t(18) = 2.99$, $P < .05$) stimuli.

fMRI METHODS

We have begun an fMRI study in the same individuals on whom the behavioral tasks have been performed. We are first attempting to image differences in the patterns of activation between the conductors and nonconductors performing the multisensory TOJ task described previously. Specifically, we want to determine which underlying cortical networks are responsible for the speeding of response time in the conductors. Presented here are preliminary data for two conductors and four nonconductors.

fMRI Experimental Paradigm

During the fMRI session, subjects performed the identical multisensory TOJ task using an event-related paradigm. Stimuli were presented through MR-compatible goggles and headphones (<www.mrivideo.com>). Just as in the behavioral study discussed above, in each condition the subject had to state which of two circles appeared first. Two different visual SOAs were used for all conditions: one matched the individual threshold derived during the TOJ testing above, while the other was a constant value for all subjects (50 ms). Additionally, auditory delays were constrained to include only 100 and 300 ms. In addition to two visual-only control conditions, two auditory-only conditions were included, consisting of delays matching the two multisensory conditions but during which subjects did not respond. Together, each of these eight conditions (four multisensory, two visual, two auditory) were presented in a random order 14 times during each of five sessions, for a total of 70 stimulus "events" per condition.

fMRI Image Acquisition and Analysis

All imaging experiments were performed on a GE echo-speed Horizon LX MR scanner with a birdcage head coil (GE Medical Systems, Milwaukee, WI). Whole-brain activation was assessed by examining blood oxygenation level-dependent (BOLD) changes[16,17] by measuring changes in the T2*-relaxation rate caused by the changes in blood oxygenation that accompany cortical activation.[18,19] Functional imaging was performed in the axial plane using multislice gradient-echo echo-planar imaging (TR = 2500, TE = 40 ms) with a field of view of 24 cm (frequency) × 15 cm (phase), and an acquisition matrix of 64 × 40 (28 slices, 5 mm thickness, no skip). Statistical parametric maps (SPMs) were generated using SPM99[20,21] from the Wellcome Department of Cognitive Neurology (London, UK) implemented in Matlab (The Mathworks Inc., Sherborn MA, USA), with an IDL interface. Anatomic regions were defined using an anatomic MRI atlas[22] normalized to the same MNI-SPM template. SPMs were generated by means of the general linear model within SPM99. The data were analyzed using a fixed effects model for each group (conductors and nonconductors) and thresholded at $P < .001$, corrected for multiple comparisons at $P < .05$.

fMRI RESULTS

When analyzing the four multisensory conditions together, the conductors and nonconductors show, not unexpectedly, activity in the visual and motor cortex. However, the conductors showed greater activity in the visual cortex, especially in the higher-order visual cortices, and extending into the occipitotemporal regions in Brodmann's area (BA) 37 (FIG. 4). We probed the data using a region-of-interest (ROI) analysis constrained to known multisensory/heteromodal brain regions (BA37, 39/40), and the differences between the activation patterns for the conductors and nonconductors were more apparent (FIG. 5). Specifically, the conductors showed increased activity in bilateral occipitotemporal cortices in BA37. While the nonconductors did show activity in the higher parietal regions, they did not show increased activity in BA37 areas.

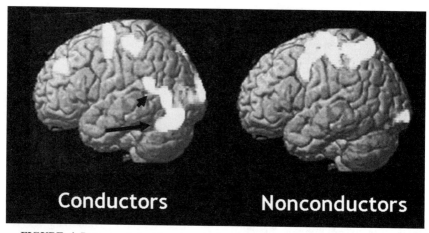

FIGURE 4. Patterns of significant cortical activation for conductors (*left*) and controls (*right*) on multisensory TOJs. Note the increased occipitotemporal activation (*black arrow*) and superior temporal sulcus activation (*arrow head*) in the conductors. Peak activity for the conductors was in the left occipital lobe (Montreal Neurological Institute (MNI) coordinates −8, −108, 10; peak t score = 9.61) and left occipitotemporal region (−20, −76, −15; peak t = 8.18). The peak activity for the controls was in the left parietal lobe (−32, −60, 60; peak t score = 7.36) and right occipital lobe (20, −80, −15; peak t = 6.65).

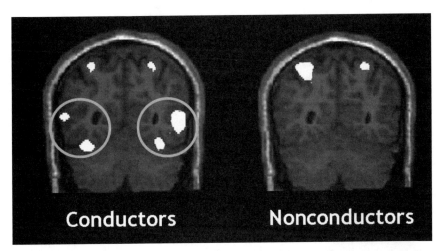

FIGURE 5. Increased activations in bilateral occipitotemporal cortices (*gray circles*) in BA37 for conductors (*left*) but not for controls (*right*). Peak activity for the conductors in the right BA37 was at MNI 56, −60, 10, with peak $t = 7.74$; and in the left BA37 was at −40, −64, −20, with peak $t = 6.22$. Peak activity for the controls was in the left parietal lobe (−32, −48, 45; peak $t = 6.82$).

CONCLUSIONS

Musical behaviors are primary examples of multisensory processing. Among musicians, conductors have a particular need for heightened spatial localization skills. It was the purpose of this experiment to examine multisensory processing in this select group of musicians. In the preliminary data gathered so far, ten experienced conductors and ten controls with very limited musical training participated in a series of behavioral tasks. Two conductors and four nonconductors subsequently underwent fMRI scans while performing the multisensory TOJ task.

Data analyses indicate that conductors have finely tuned auditory processing skills, including more refined pitch discrimination and shorter auditory temporal thresholds. In other words, they are more accurate in making pitch discriminations than control subjects and require less time between two sounds to be able to discriminate which of them occurred first. Furthermore, they demonstrated clearly improved response times to TOJs in multisensory conditions, when compared to visual alone. That is, besides the combination of auditory and visual information improving performance on the task, it also enabled them to provide a response more quickly. Although nonconductors also show such a benefit in their performance, the improvement in response time was less clear, and they were, in general, much slower to respond than were individuals with experience in both music and conducting.

The benefits of multisensory integration were also seen for conductors, but not nonconductors, in the target localization task. The integration of auditory and visual information enhanced performance, allowing conductors, once again, to respond faster and more accurately, while control subjects did not show such a benefit. These results fit nicely with the observation that conductors, while on the podium, must instantaneously be able to locate "who made what sound." Despite the typically seen highly accurate localization ability in the visual-alone condition (seen in both the conductors and nonconductors), the conductors still received a benefit in locating ability when the auditory stimulus was added. While such multisensory enhancement of behavior is usually present only with stimuli near threshold,[12] the conductors received a benefit from the added auditory stimulus even with both the auditory and visual stimuli substantially above threshold detection level. Perhaps the conductors have developed (or were born with?) neural ensembles or connections that allow the greater or more frequent integration of auditory and visual information.

In the first stages of the imaging component of this project, neural substrates underlying the conductors' superior multisensory TOJ performances have been identified. BA37, 39/40 are particularly implicated and are known as multisensory convergence areas involved in other behaviors. The cortex in BA37 at the occipito-temporal junction is a known heteromodal area of audiovisual convergence[23] and, interestingly, has been shown to be an important area in the acquisition of reading skills.[24] These early results may show a neural "signature" of multisensory binding and may show the brain network responsible for a more efficient processing of the simultaneously presented visual and auditory stimuli. Scanning additional subjects will determine whether these results are robust. If so, we will have the beginnings of an understanding of the neural mechanisms underlying multisensory processing in conductors.

REFERENCES

1. AVANZINI, G., C. FAIENZA, D. MINCIACCHI, et al., Eds. 2003. The Neurosciences and Music. Vol. 999. The New York Academy of Sciences. New York, NY.
2. ZATORRE, R.J. & I. PERETZ, EDS. 2001. The Biological Foundations of Music. Vol. 930. The New York Academy of Sciences. New York, NY.
3. WALLACE, M., G. ROBERSON, W. HAIRSTON, et al. 2004. Unifying multisensory signals across time and space. Exp. Brain Res. **158:** 252–258.
4. WALLACE, M., R. RAMACHANDRAN & B. STEIN. 2004. A revised view of sensory cortical parcellation. Proc. Natl. Acad. Sci. USA **101:** 2167–2172.
5. HUGHES, H., P. REUTER-LORENZ, G. NOZAWA & R. FENDRICH. 1994. Visual-auditory interactions in sensorimotor processing: saccades versus manual responses. J. Exp. Psychol. Hum. Percept. Perform. **20:** 131–153.
6. HARRINGTON, L. & C. PECK. 1998. Spatial disparity affects visual-auditory interactions in human sensorimotor processing. Exp. Brain. Res. **122:** 247–252.
7. HAIRSTON, W., P. LAURIENTI, G. MISHRA, et al. 2003. Multisensory enhancement of localization under conditions of induced myopia. Exp. Brain Res. **152:** 404–408.
8. LAURIENTI, P., R. KRAFT, J. MALDJIAN, et al. 2004. Semantic congruence is a critical factor in multisensory behavioral performance. Exp. Brain Res. **158:** 405–414.
9. LOVELACE, C., B. STEIN & M. WALLACE. 2003. An irrelevant light enhances auditory detection in humans: a psychophysical analysis of multisensory integration in stimulus detection. Brain Res. Cogn. Brain Res. **17:** 447–453.
10. MOREIN-ZAMIR, S., S. SOTO-FARACO & A. KINGSTONE. 2003. Auditory capture of vision: examining temporal ventriloquism. Brain Res. Cogn. Brain Res. **17:** 154–163.
11. HAIRSTON, W.D., J.H. BURDETTE, D.L. FLOWERS, et al. Abnormal temporal processing of cross-modal information in dyslexia. Exp. Brain Res. In press.
12. STEIN, B. & M. MEREDITH. 1993. In The Merging of the Senses. M. Gazzaniga, Ed. MIT Press. Cambridge, MA.
13. PARSONS, L., D. HODGES & P. FOX. 1998. Neural basis of the comprehension of musical harmony, melody, and rhythm. Society for Neuroscience Annual Meeting. J. Cogn. Neurosci. Abstracts.

14. PARSONS, L. 2001. Exploring the functional neuroanatomy of music performance, perception, and comprehension. Ann. N. Y. Acad. Sci. **930:** 211–230.
15. NAGER, W., C. KOHLMETZ, A. ALTENMÜLLER, *et al.* 2003. The fate of sounds in conductors' brains: an ERP study. Brain Res Cogn Brain Res. **17:** 81–93.
16. BUCHBINDER, B. & G. COSGROVE. 1998. Cortical activation MR studies in brain disorders. Magn. Reson. Imaging Clin. N. Am. **6:** 67–93.
17. TURNER, R., A. HOWSEMAN, G. REES, *et al.* 1998. Functional magnetic resonance imaging of the human brain: data acquisition and analysis. Exp. Brain Res. **123:** 5–12.
18. OGAWA, S., T. LEE, A. KAY, *et al.* 1990. Brain magnetic resonance imaging with contrast dependent on blood oxygenation. Proc. Natl. Acad. Sci. USA **87:** 9868–9872.
19. OGAWA, S., R. MENON, D. TANK, *et al.* 1993. Functional brain mapping by blood oxygenation level-dependent contrast magnetic resonance imaging. A comparison of signal characteristics with a biophysical model. Biophys. J. **64:** 803–812.
20. FRISTON, K., C. FRITH, R. FRACKOWIAK, *et al.* 1995. Characterizing dynamic brain responses with fMRI: a multivariate approach. Neuroimage **2:** 166–172.
21. FRISTON, K., C. FRITH, R. TURNER, *et al.* 1995. Characterizing evoked hemodynamics with fMRI. Neuroimage **2:** 157–165.
22. KIKINIS, R., P. GLEASON, T. MORIARTY, *et al.* 1996. Computer-assisted interactive three-dimensional planning for neurosurgical procedures. Neurosurgery **38:** 640–649; discussion 649–651.
23. BUCHEL C, C. PRICE & K. FRISTON. 1998. A multimodal language region in the ventral visual pathway. Nature **394:** 274–277.
24. PUGH, K., W. MENCL, A. JENNER, *et al.* 2001. Neurobiological studies of reading and reading disability. J. Commun. Disord. **34:** 479–492.

A Network for Sensory-Motor Integration

What Happens in the Auditory Cortex during Piano Playing without Acoustic Feedback?

SIMON BAUMANN, SUSAN KOENEKE, MARTIN MEYER, KAI LUTZ, AND LUTZ JÄNCKE

Department of Neuropsychology, University of Zurich, Zurich, Switzerland

ABSTRACT: Playing a musical instrument requires efficient auditory as well as motor processing. We provide evidence for the existence of a neuronal network of secondary and higher-order areas belonging to the auditory and motor modality that is important in the integration of auditory and motor domains.

KEYWORDS: sensorimotor integration; musicians; motor cortex; auditory cortex; plasticity; fMRI

INTRODUCTION

Playing a musical instrument poses enormous demands on the integration of sensory information and motor control.[1] Our data suggest that secondary auditory and motor areas play a critical role in the interaction of both modalities and that the intensity of piano training has a measurable influence on the neurophysiological basis of this process.

MATERIAL AND METHODS

Fourteen right-handed volunteers (seven pianists from the Zurich Conservatory of Music and seven participants with no history of musical training) took part in our study. Functional magnetic resonance imaging (fMRI) data of several piano-playing and piano-listening tasks were recorded. The piano-playing tasks consisted of the performance of a piano piece by Mozart ("Sonata Facile") and of scales. Participants had no auditory feedback since the movements were performed on a fiber glass board. The nonmusicians were instructed about how to play the scales and, instead of attempting the Mozart piece, performed random finger movements at a comparable speed. During the pure auditory tasks participants listened via earphones to recordings of the Mozart piece and to the scales. All playing and listening tasks were

Address for correspondence: Simon Baumann, Department of Neuropsychology, University of Zurich, Treichlerstrasse 10, CH-8032 Zurich, Switzerland. Voice: +41-44-634-1568.
s.baumann@psychologie.unizh.ch

Ann. N.Y. Acad. Sci. 1060: 186–188 (2005). © 2005 New York Academy of Sciences.
doi: 10.1196/annals.1360.038

subdivided further into subconditions controlling the level of attention to the modality not involved in the current task. In order to localize cortical structures that may play a role in audio-motor integration, we looked specifically for areas that were jointly activated in the pure motor and the pure auditory conditions. This procedure has been applied in previous studies in order to identify areas involved in multimodal integration in other tasks or modalities.[2,3]

fMRI data were recorded on a 3 T Siemens unit. Individual images were normalized into the standard stereotaxic space according to the Montreal Neurological Institute (MNI) convention. Individual activation maps for all conditions were calculated based on the "General Linear Model" statistic. A random effect statistic was applied for the group of fourteen subjects for all conditions. Overlapping areas of the activation maps for the pure auditory and the pure motor tasks were calculated by identifying voxels that were significantly active ($P < .001$, uncorrected) in the motor and the auditory conditions. Furthermore, activation in the relevant areas was compared between pianists and nonmusicians to identify changes due to intensive piano training.

RESULTS

The overlap maps for significantly activated areas during the combined pure motor tasks and the pure auditory tasks predominantly revealed secondary areas of the motor cortex and the auditory cortex (FIG. 1). These areas included the planum temporale (PT) and the planum polare (PP) in the auditory cortex, the supplementary motor area (SMA), and the ventral and dorsal part of the premotor cortex (PMv, PMd). All of these cortical structures showed overlapping areas in both hemispheres. Further recruited cortical areas were found in the superior parietal lobule and in the secondary somatosensory cortex (SII) in the left hemisphere. At the subcortical level, we found joint audio-motor activation in the nucleus lentiformis of the basal ganglia and the cerebellum in the right hemisphere. However, the jointly activated areas comprised neither primary auditory, primary somatosensory, nor primary motor areas.

A group comparison of activation maps between pianists and nonmusicians limited to the previously identified, jointly activated audiomotor areas revealed a

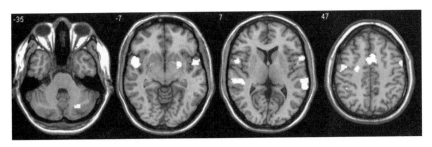

FIGURE 1. Transverse sections showing the relevant areas in MNI space (neurologic convention) active while listening to and performing a piano piece.

general trend to stronger activity among the musicians. Strongest differences were found in the PMd. However, significant between-group differences were evident only in the Mozart conditions, not in the scales conditions ($P < .05$, corrected for multiple comparisons).

Maximum t values and the number of significantly activated voxels were higher in all areas of the integratory network when attention was voluntarily directed to audiomotor coupling. In addition, the PMv, the inferior frontal gyrus (IFG), and the nucleus lentiformis showed significantly increased activation in the voluntary condition.

CONCLUSIONS

Our data indicate that in a multimodal task like piano performance, stimulation of one modality can lead to activation in another modality. In our case, piano-playing movements jointly activate areas (PT, PP) usually considered as belonging to the secondary auditory cortex. On the other hand, listening to piano sounds without moving evokes activity in secondary motor regions (PMd, PMv, and SMA). These results are considered clear evidence for information transfer between auditory and motor cortices, indicating that secondary areas of both modalities play a key role in audiomotor integration. This is in accordance with recent studies emphasizing the importance of secondary sensory areas for multisensory integration.[3]

Furthermore, our data suggest that audiomotor information flow consists of a voluntary and an involuntary component, which in part make use of the same structures. However, voluntary multimodal coupling activates additional cortical structures (PMv, IFG, and nucleus lentiformis).

The fact that musicians show increased activity in the dorsal premotor cortex while listening to music by Mozart suggests that cortical plasticity due to intensive training plays a role in transmodal activation and multimodal integration. However, the absence of significant differences between pianists and nonmusicians in the scales conditions in our data limits further speculation on potential transfer effects on tasks unrelated to music.

[Competing interests: The authors declare that they have no competing financial interests.]

REFERENCES

1. BANGERT M. & E.O. ALTENMÜLLER. 2003. Mapping perception to action in piano practice: a longitudinal DC-EEG study. BMC Neurosci. **4:** 26.
2. HICKOK G., B. BUCHSBAUM, C. HUMPHRIES, *et al.* 2003. Auditory-motor interaction revealed by fMRI: speech, music, and working memory in area Spt. J. Cogn. Neurosci. **15:** 673–682.
3. FOXE J.J., G.R. WYLIE, A. MARTINEZ, *et al.* 2002. Auditory-somatosensory multisensory processing in auditory association cortex: an fMRI study. J. Neurophysiol. **88:** 540–543.

The Power of Listening

Auditory-Motor Interactions in Musical Training

AMIR LAHAV,[a,b] ADAM BOULANGER,[c] GOTTFRIED SCHLAUG,[b]
AND ELLIOT SALTZMAN [a,d]

[a]*The Music, Mind and Motion Lab, Sargent College of Health and Rehabilitation
Sciences, Boston University, Boston, Massachusetts 02215, USA*

[b]*Department of Neurology, Beth Israel Deaconess Medical Center and
Harvard Medical School, Boston, Massachusetts 02115, USA*

[c]*The Media Lab, Massachusetts Institute of Technology,
Boston, Massachusetts 02139, USA*

[d]*Haskins Laboratories, New Haven, Connecticut 06511, USA*

ABSTRACT: We trained musically naive subjects to play a short piano melody
by ear in a fully monitored computerized environment and tested their potential to acquire a functional linkage between actions and sounds. Individual
notes that were simply acoustic pretraining signals became "physically meaningful" posttraining. In addition, we found preliminary evidence that passive
listening to a newly learned musical piece can enhance motor performance in
the absence of physical practice.

KEYWORDS: musical training; performance; auditory-motor; implicit learning

INTRODUCTION

When playing a musical instrument, movements are naturally paired with direct
auditory feedback. Musical training, therefore, can establish a multimodal functional mapping that links what we hear and what we play. Professional musicians often
report feeling a physical sensation when listening to music they know how to play.
Similarly, recent studies in the visual-motor domain have shown that when dancers
watched familiar and performable dancing movements, their brains automatically
simulated those movements.[1] While one may speculate on the underlying neural per-
ception-action mechanisms in the musician's brain,[2] it is unclear whether such an
audiomotor linking capability is an innate predisposition and even a prerequisite for
becoming a musician, or is simply experience dependent.[3] To address this, we
trained nonmusicians to play a novel piece of music and closely monitored their
learning process. We expected that subsequent listening to the acquired musical

Address for correspondence: Amir Lahav, The Music, Mind and Motion Lab, Boston University, 635 Commonwealth Ave., Boston, MA 02215. Voice: 617-353-7497.
lahav@bu.edu

Ann. N.Y. Acad. Sci. 1060: 189–194 (2005). © 2005 New York Academy of Sciences.
doi: 10.1196/annals.1360.042

piece might unconsciously prime the listener's corresponding motor repertoire, which in turn might lead to motor improvement.

METHODS

Experiment 1

Fifty-eight right-handed nonmusician college-level subjects (32 women, 26 men; mean age = 22.3 ± 2.8) attended a single piano session, in which they learned to play a novel musical piece (FIG. 1) on a piano keyboard, using their right hands and a set of five adjacent keys (F-G-A-B$^\flat$-C), in a fixed fingering position (i.e., the same finger always hit the same key). Motivated by previous training tools,[4] we designed innovative MIDI-based software for learning by ear with no sight-reading required. Subjects learned to play the piano role (solo) along with a prerecorded accompaniment (guitar, bass, and drums), while a computer didactically notified them when note (wrong key press) or timing ($\pm >1/16$th) errors occurred. To complete a piano

FIGURE 1. The musical piece: Subjects learned the piano part gradually, section by section, in the following measure order: 1–2; 3–4; 1–4; 5–6; 1–6; 7–8; 1–8.

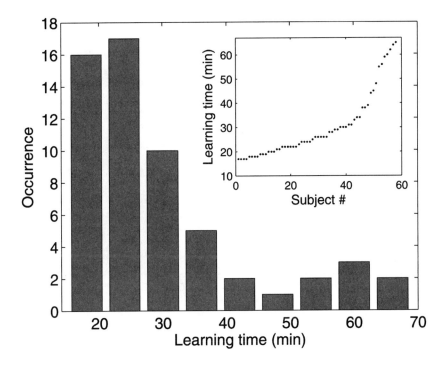

FIGURE 2. Learning times of the musical piece are shown for group distribution and for individual subjects (*inset*).

session, subjects went through a series of trials until reaching error-free performance. The minimum possible learning time was 12 minutes.

Pitch-Recognition-Production Test

Before and after their piano-training session subjects were presented with single notes (taken from the piano piece) in random order, and were asked after hearing each note, to press the corresponding piano key. Percent correct responses was measured.

Experiment 2

Following the initial piano session (Exp. 1), subjects ($n = 36$) were randomly assigned into three groups and underwent three additional 20-minute listening/practicing sessions over a period of one week. The *piano-listening group* passively listened to the same musical piece they had previously learned. During listening sessions, subjects lay supine with their palms facing up. Finger positions were monitored digitally using a motion-tracking system and a passive-marker glove to ensure that subjects followed instructions to stay as still as possible. The *nature-listening group*

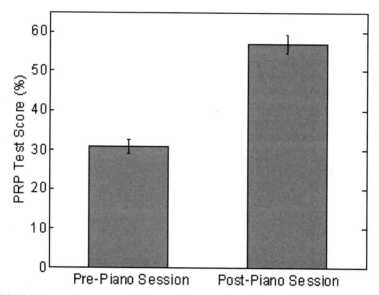

FIGURE 3. Mean PRP test score before and after piano training session. *Error bars* represent standard error of the mean.

listened to nature sounds, to control for treatment effects due merely to auditory exposure. The *practicing group* received additional practice sessions with the initial musical piece.

Performance Test

Subjects were asked to play the whole musical piece as accurately as possible, while their keyboarding was not accompanied by the original rhythmic-harmonic background instrumentation. Subjects' note/timing errors were measured.

RESULTS AND DISCUSSION

Nonmusicians Learn to Play by Ear

We first examined how nonmusicians learn to play a musical piece for the first time (Exp. 1). The average learning time to error-free performance was 28.5 minutes (STD = 12.9). Interestingly, learning times followed a bimodal distribution (FIG. 2), with most subjects learning relatively quickly and a small group taking considerably longer. A similar distribution pattern was found also for learning a different musical piece (data not shown). Classifying subjects by their learning times may open an experimental window for future brain imaging studies comparing neural activity of super learners versus poor learners.

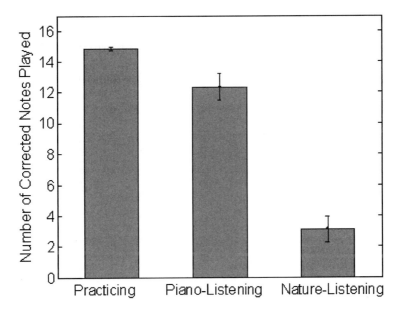

FIGURE 4. Mean correct notes for the musical piece played at the right time during performance test. *Error bars* represent standard error of the mean.

A Single Piano Session Facilitates Pitch Recognition

Here we ask whether learning to play a musical piece by ear could act to induce a pitch-to-key mapping, in which perceived notes can be identified independently of their original musical context. To this end, a pitch-recognition-production (PRP) test was performed before and after the initial piano session. To rule out learning effects during the PRP test, subjects did not receive knowledge of results (auditory feedback) when pressing the piano keys. Results showed that subjects improved their scores from approximately 30% pretraining to about 60% posttraining (FIG. 3, $P = 1 \times 10^{-15}$). These findings hint at a newly learned pitch recognition skill, consistent with the view that the ability to encode pitch develops with experience.[5]

Passive Listening to Music Improves Motor Performance

We tested the three listening/practicing groups (see METHODS, Exp. 2) on subjects' ability to play the previously learned musical piece. On average, the piano-listening group performed significantly better in the performance test than the nature-listening group ($P = .01$), but not as well as (though not far behind) the practicing group (FIG. 4, $P = 1 \times 10^{-7}$). Analysis of our digital motion-tracking system verified that subjects' fingers indeed did not move during listening sessions. These findings suggest that during passive listening, neural mechanisms linking sounds and actions may implicitly facilitate musical motor performance. We are now in the process of investigating further behavioral and functional imaging measures to determine in greater detail the processes underlying this motor improvement (manuscript in preparation).

[Competing interests: The authors declare that they have no competing financial interests.]

REFERENCES

1. CALVO-MERINO, B., D.E. GLASER., J. GREZES, *et al.* 2005. Action observation and acquired motor skills: an FMRI study with expert dancers. Cereb. Cortex **15:** 1243–1249.
2. HAUEISEN, J. & T.R. KNOSCHE. 2001. Involuntary motor activity in pianist evoked by music perception. J. Cogn. Neurosci. **13:** 786–792.
3. SCHLAUG, G. & C. GASER. 2003. Brain structures differ between musicians and nonmusicians. J. Neurosci. **23:** 9240–9245.
4. BANGERT, M. & E. ALTENMÜLLER. 2003. Mapping perception to action in piano practice: a longitudinal DC-EEG study. BMC Neurosci. **4:** 26–40.
5. TRAINOR, L.J. 2005. Are there critical periods for musical development? Dev. Psychobiol. **46:** 262–278.

The Mental Space of Pitch Height[a]

ELENA RUSCONI,[a,c] BONNIE KWAN,[a] BRUNO GIORDANO,[b]
CARLO UMILTÀ,[b] AND BRIAN BUTTERWORTH[a]

[a]University College London, London, United Kingdom
[b]Università degli Studi di Padova, Padova, Italy
[c]Institute of Cognitive Neuroscience,
17 Queen Square, WC1N 3AR London, United Kingdom

ABSTRACT: Through stimulus–response compatibility we tested whether sound frequency (pitch height) elicits a mental spatial representation. Musically untrained and, mostly, trained participants were shown a stimulus–response compatibility effect (Spatial–Musical Association of Response Codes or SMARC effect). When response alternatives were either vertically or horizontally aligned, performance was better when the lower (or leftward) button had to be pressed in response to a low sound and the upper (or rightward) button had to be pressed in response to a high sound, even when pitch height was irrelevant to the task.

KEYWORDS: pitch height; space; SMARC; stimulus-response compatibility

> The results are clear-cut and unequivocal. High tones are phenomenologically higher in space than low tones.
>
> C.C. PRATT, 1930

Pratt put forward the hypothesis of pitch–space correspondence after observing that the specific succession of tones in a musical phrase can generate a sensation of apparent movement (e.g., by successively presenting the notes of the diatonic scale from C3 to C4, almost everybody perceives an upward movement). Recent research in human cognition has shown that, in domains such as language and mathematics, discrete sets of unitary elements tend to be unconsciously mapped onto mental spatial positions. Manual key-press responses to large numbers, late months, letters at the end of the alphabet are faster with the right than the left key, whereas responses to small numbers, early months, letters at the beginning of the alphabet are faster with the left than the right key.[1,2] We predicted that, if spatial positions were also spontaneously assigned to the characteristic units of music, a better performance would result when pitch cognitive location corresponded to response location than when it did not. However, the vertical rather than the horizontal dimension could be predominant in the case of pitch

Address for correspondence: Elena Rusconi, University College London, London, UK. Voice: +44-207-6791127; fax: +44-207-9168517.
e.rusconi@ucl.ac.uk
[a]A more detailed discussion of this study can be found in Rusconi et al. Spatial representation of pitch height: the SMARC effect. Cognition. In press (available online at www.sciencedirect.com).

Ann. N.Y. Acad. Sci. 1060: 195–197 (2005). © 2005 New York Academy of Sciences.
doi: 10.1196/annals.1360.056

height, as suggested by its association with the vertical spatial dimension emerging consistently across languages and by early phenomenological reports (e.g., Ref. 3). Specifically, we expected to find responses to high pitches with the upper response key and low pitches with the lower response key (compatible condition) faster and more accurate than responses to high pitches with the lower and to low pitches with the upper response key (incompatible condition).

METHOD

Three experiments were conducted, in which participants (nonmusicians in experiments 1 and 2, and trained musicians in experiment 3) were asked to press one of two keys in response to a target sound. In experiment 1, their task was to determine whether the sound (E3 to G4, except C4) was higher or lower than the reference (C4) and, in experiments 2 and 3, whether the sound (F3 to A4#, except D4) was played by either a wind or a percussion instrument. For each trial participants fixed a cross in the center of a computer screen while the stimuli were presented. In experiment 1, the reference tone was played 1000 ms on each trial immediately before the target tone (lasting 1000 ms too). In experiments 2 and 3 a single tone was played on each trial for 1000 ms. The deadline for a response was fixed at 1300 ms and the intertrial interval at 1000 ms (inclusive of visual feedback on accuracy). Response keys were either horizontally (P and Q on a keyboard) or vertically[b] aligned (spacebar and 6); mappings and responding hands were varied orthogonally within participants (i.e., each possible mapping was performed with both uncrossed and crossed arms in different blocks, and the order of blocks was counterbalanced between participants).

RESULTS

Both latency and accuracy data were analyzed separately for the vertical and the horizontal alignments in each experiment. In experiment 1, two $2 \times 4 \times 2 \times 2$ repeated measures ANOVAs having pitch height (lower/higher), distance from the reference (1, 2, 3, 4 tones), responding hand (left, right) and response location (up/down or left/right) revealed, among the other effects, a significant interaction between pitch height and response location [RTs: $F(1,19) = 7.10$, $P =. 015$; errors: $F(1,19) = 7.10$, $P =. 015$] with vertically aligned response keys. The compatible condition was more accurate and faster than the incompatible condition. Two further ANOVAs on the data with horizontally aligned response keys revealed a marginally significant pitch height $\times$ response location interaction in latencies only [RTs: $F(1,19) = 4.19$, $P = .054$; errors: $F < 1$], showing an advantage for right responses to high, and left responses to low pitches. No main effects or interactions involving responding hand were significant.

Data of experiments 2 and 3 were analyzed in $2 \times 2 \times 4 \times 2$ mixed design ANOVAs having musical expertise (no/yes) as a between-participants factor, pitch

---

[b]In tasks with vertically aligned stimuli, stimulus–response compatibility effects are found irrespective of whether response keys are aligned along the frontal or the transverse plane.[4]

height (lower/higher), distance from the reference (1, 2, 3, 4 tones), and response location (up/down or left/right) as within-participants factors. In the vertical alignment, musical expertise was involved in a significant three-way interaction, with pitch height and response location in the reaction time (RT) analysis [$F(1,38)$ = 5.616, P =. 023] and t tests showing that the global compatibility effect was significant for musicians only ($t_{(38)}$ = 2.66, P = .011; nonmusicians: $t < 1$). However, the compatibility effect was present in either group at a distance of four tones from the implicit reference (D4) [pitch height × distance × response location significant interaction: $F(3,114)$ = 4.754, P = .004; 4-tone distance: $t_{(38)}$ = 3.46, P = .001]. The error analysis mirrored the RTs analysis, revealing a pitch height × distance × response location significant interaction (with stimulus–response compatibility at the four-tone distance), whereas musical expertise did not interact with any of the other factors. In the horizontal alignment, musical expertise was involved in a significant four-way interaction with pitch height, distance and response location in the error analysis [$F(3,114)$ = 3.423, P = .020], which approached significance in the RTs analysis [$F(3,114)$ = 2.535, P =. 060]. In both cases the compatibility effect was significant at large distances and for musicians only (errors: four-tone distance, $t_{(38)}$ = 2.64, P = .012; RTs: three-tone distance, $t_{(38)}$ = 2.13, P = .040, four-tone distance, $t_{(38)}$ = 3.07, P = .004).

CONCLUSIONS

Pitch height influenced performance consistently with vertically aligned responses irrespective of its relevance to the task, which suggests that our cognitive system maps pitch onto a mental representation of space. In turn this might shed some light "on the moot question of the apparent auditory movement which is set up by tones of different pitch when presented in succession."[3] We called this phenomenon the SMARC effect, analogous to the SNARC (Spatial-Numerical Association of Response Codes) effect, which was described for numbers.[2] Whether a directional spatial representation is epiphenomenal or plays an equally important role in the development, maintenance, and use of concepts in all these domains of knowledge is still a matter for debate (e.g., Ref. 5) and future investigation.

[Competing interests: The authors declare that they have no competing financial interests.]

REFERENCES

1. DEHAENE, S., S. BOSSINI & P. GIRAUX. 1993. The mental representation of parity and number magnitude. J. Exp. Psychol. Gen. **16:** 626–641.
2. GEVERS, W., B. REYNVOET & W. FIAS. 2003. The mental representation of ordinal sequences is spatially organized. Cognition **87:** B87–95.
3. PRATT, C.C. 1930. The spatial character of high and low tones. J. Exp. Psychol. **13:** 278–285.
4. VU, K.-P.L., R.W. PROCTOR & D.F. PICK. 2000. Vertical versus horizontal spatial compatibility: right–left prevalence with bimanual responses. Psychiatry Res. **64:** 25–40.
5. GALFANO, G., E. RUSCONI & C. UMILTÀ. Number magnitude orients attention but not against one's will. Psychon. Bull. Rev. Under revision.

Part IV: Developmental and Applied Perspectives on Music

Introduction

SANDRA E. TREHUB

Department of Psychology, University of Toronto at Mississauga, Mississauga, Ontario, Canada L5L 1C6

In recent years the burgeoning interest in music cognition has spawned intriguing questions about the ontogenesis of musical skills. It has also raised questions about the musical and nonmusical consequences of musical exposure or training. With respect to the first question, we have learned that human infants begin life with various musically relevant abilities, including fine-grained perception of pitch and rhythm patterns,[1,2] preferences for consonant over dissonant intervals,[3–5] cross-modal correspondences between sound and movement,[6] and heightened responsiveness to the expressively sung performances of mothers.[7–11] We have also learned that exposure to the music of their culture not only builds upon infants' initial biases but also reshapes them.[12–14] The collection of papers in this section focuses on a variety of issues related to short- or long-term musical exposure or training.

There is a long-standing belief that musical training early in life has a greater impact than comparable training later on, presumably because of the plasticity of young brains. The notion is analogous to critical, sensitive, or optimal periods that have been proposed for the acquisition of phonological and syntactic aspects of language.[15,16] Although there is clear evidence of age-related changes in the plasticity of the central auditory system,[17,18] evidence of sensitive periods for perceptual learning is much less clear. It is possible that enhanced learning at particular phases of life stems from lack of expertise or neural commitment in particular domains, which enables new learning to proceed without interference from prior learning.[19,20] For music in particular, there is little evidence that very early training is essential for high levels of ultimate musical achievement.[21,22] One phenomenon that implicates early training, however, is *absolute pitch*, which involves the ability to learn and retain arbitrary verbal labels for isolated musical pitches.[23–25] In any case, there is increasing interest in the neural consequences of early musical training.[26] Although it is of interest to document these consequences, it is of greater importance to determine whether the neural changes have functional correlates.

For obvious reasons, there is unabated interest in the general cognitive consequences of music lessons in childhood,[27] despite inconsistent findings across

Address for correspondence: Sandra E. Trehub, Department of Psychology, University of Toronto at Mississauga, Mississauga, Ontario Canada L5L 1C6. Voice: 905-828-5415; fax: 905-569-4850.

sandra.trehub@utoronto.ca

Ann. N.Y. Acad. Sci. 1060: 198–201 (2005). © 2005 New York Academy of Sciences.
doi: 10.1196/annals.1360.063

studies. For example, one study reported small IQ gains in children randomly assigned to one year of music lessons compared to drama lessons or no lessons.[28] Another study that involved random assignment to piano lessons for a full three years revealed no cognitive consequences but significant gains in self-esteem.[29] Schlaug, Norton, Overy, and Winner revisit the issue of music lessons by reporting preliminary neural and cognitive consequences of music lessons in 5- to 7-year-old and 9- to 10-year-old children. Conventional scanning protocols are inappropriate for use with young children. Accordingly, Overy, Norton, Cronin, Winner, and Schlaug outline a child-friendly protocol that can be administered while children are engaged in melody- and rhythm-processing tasks. Jentschke, Koelsch, and Friederici explore the consequences of music lessons by examining 10- and 11-year-old children's evoked responses to syntactic violations of linguistic and musical (chord) sequences. They also offer perspectives on reciprocal influences between music and language by exploring evoked responses to musical sequences in 5 year olds with normally developing language and 5 year olds diagnosed with language impairment.

Aside from the long-term consequences of systematic musical exposure or lessons, there are short-term effects of informal musical exposure on emotion and arousal.[30–32] Moderate elevations of mood or arousal can enhance adults' performances on cognitive tasks[33,34] and young children's performances on simple drawing tasks.[35] At times, such effects have been misinterpreted as direct cognitive consequences of exposure to specific kinds of music.[36] Schellenberg and Hallam extend this line of research by exploring 10- and 11-year-old children's performances on spatial problem-solving tasks after brief episodes of listening to popular music, a Mozart composition, or a discussion.

In addition to cognitive consequences of music that are mediated by mood or arousal, music may have unique advantages as a medium of information transmission. Historically, these advantages were exploited in oral cultures.[37] In the present section, Thaut, Peterson, and McIntosh make the case that verbal information that is sung rather than spoken induces greater neural synchrony, which enhances learning and memory beyond the benefits expected from temporal grouping or chunking.

Finally, Sloboda, Wise, and Peretz close this section by outlining differences between the small minority of individuals who are designated tone-deaf, or amusic, by virtue of their performance on the Montreal Battery for the Evaluation of Amusia,[38] and the larger minority of individuals whose tone deafness is self-declared. In contrast to the former group, whose problems are rooted in pitch-processing limitations,[39] problems of the latter group are rooted in real or imagined difficulties with singing.

REFERENCES

1. TREHUB, S.E. 2000. Human processing predispositions and musical universals. *In* The Origins of Music. N.L. Wallin, B. Merker & S. Brown, Eds.: 427–448. MIT Press. Cambridge, MA.
2. HANNON, E.E. & S.E. TREHUB. 2005. Metrical categories in infancy and adulthood. Psychol. Sci. **16:** 48–55.
3. TRAINOR, L.J. & B.M. HEINMILLER. 1998. The development of evaluative responses to music: infants prefer to listen to consonance over dissonance. Inf. Behav. Dev. **21:** 77–88.

4. TRAINOR, L.J., C.D. TSANG & V.H.W. CHEUNG. 2002. Preference for sensory consonance in two- and four-month-old infants. Mus. Percept. **20:** 187–194.
5. ZENTNER, M.R. & J. KAGAN. 1996. Perception of music by infants. Nature **383:** 29.
6. PHILLIPS-SILVER, J. & L.J. TRAINOR. 2005. Feeling the beat: movement influences
   infant rhythm perception. Science **308:** 1430–1430.
7. MASATAKA, N. 1999. Preference for infant-directed singing in two-day-old hearing
   infants of deaf parents. Dev. Psychol. **35:** 1001–1005.
8. NAKATA, T. & S.E. TREHUB. 2004. Infants' responsiveness to maternal speech and singing. Inf. Behav. Dev. **27:** 455–464.
9. SHENFIELD, T., S.E. TREHUB & T. NAKATA. 2003. Maternal singing modulates infant
   arousal. Psychol. Music. **31:** 365–375.
10. TRAINOR, L.J. 1996. Infant preferences for infant-directed versus noninfant-directed
    playsongs and lullabies. Inf. Behav. Dev. **19:** 83–92.
11. TREHUB, S.E. & L.J. TRAINOR. 1998. Singing to infants: lullabies and play songs. Adv.
    Infancy Res. **12:** 43–77.
12. HANNON, E.E. & S.E. TREHUB. 2005. Tuning in to musical rhythms: Infants learn more
    readily than adults. Proc. Natl. Acad. Sci. **102:** 12639–12643.
13. TRAINOR, L.J. & S.E. TREHUB. 1992. A comparison of infants' and adults' sensitivity to
    Western musical structure. J. Exp. Psychol. Hum. Percept. Perform. **18:** 394–402.
14. TRAINOR, L.J. & S.E. TREHUB. 1994. Key membership and implied harmony in Western
    tonal music:developmental perspectives. Percept. Psychophys. **56:** 125–132.
15. NEWPORT, E.L. 1990. Maturational constraints on language learning. Cogn. Sci. **14:**
    11–28.
16. WERKER, J.F. & R.C. TEES. 2005. Speech perception as a window for understanding
    plasticity and commitment in language systems of the brain. Dev. Psychobiol. **46:**
    233–234.
17. HARRISON, R.V., K.A. GORDON & R.J. MOUNT. 2005. Is there a critical period for
    cochlear implantation in congenitally deaf children? Analyses of hearing and speech
    perception performance after implantation. Dev. Psychobiol. **46:** 252–261.
18. SHARMA, A., E. TOBEY, M. DORMAN, *et al.* 2004. Central auditory maturation and bab-
    bling development in infants with cochlear implants. Arch. Otolaryn. Head Neck
    Surg. **130:** 511–610.
19. JOHNSON, M.H. & Y. MUNAKATA. 2005. Processes of change in brain and cognitive
    development. Trends Cognit. Sci. **9:** 152–158.
20. SEIDENBERG, M.S. & J.D. ZEVIN. 2005. Connectionist models in developmental cogni-
    tive neuroscience:insights about critical periods. *In* Attention and Performance, XXI:
    Processes of Change in Brain and Cognitive Development. Y. Munakata &
    M. Johnson, Eds. Oxford University Press. Oxford.
21. MOORE, D.G., K. BURLAND & J.W. DAVIDSON. 2003. The social context of musical suc-
    cess: a developmental account. Br. J. Psychol. **94:** 529–549.
22. TRAINOR, L.J. 2005. Are there critical periods for musical development? Dev. Psycho-
    biol. **46:** 262–278.
23. TAKEUCHI, A.H. & S.H. HULSE. 1993. Absolute pitch. Psychol. Bull. **113:** 345–361.
24. WARD, W.D. 1999. Absolute pitch. *In* The Psychology of Music. D. Deutsch, Ed.: 265–
    298. Academic Press. San Diego.
25. ZATORRE, R.J. 2003. Absolute pitch: a model for understanding the influence of genes
    and development of neural and cognitive function. Nat. Neurosci. **6:** 692–695.
26. SHAHIN, A., L.E. ROBERTS & L.J. TRAINOR. 2004. Enhancement of auditory cortical
    development by musical experience in children. Neuroreport **15:** 1917–1921.
27. SCHELLENBERG, E.G. Music and cognitive abilities. Curr. Dir. Psychol. Sci. In press.
28. SCHELLENBERG, E.G. 2004. Music lessons enhance IQ. Psychol. Sci. **15:** 511–514.
29. COSTA-GIOMI, E. 2004. Effects of three years of piano instruction on children's aca-
    demic achievement, school performance, and self-esteem. Psychol. Mus. **32:** 139–152.
30. KRUMHANSL, C.L. 2002. Music: a link between cognition and emotion. Curr. Dir. Psy-
    chol. Sci. **11:** 45–50.
31. SCHERER, K.R. & M.R. ZENTNER. 2001. Emotional effects of music: production rules.
    In Music and Emotion: Theory and Research. P.N. Juslin & J.A. Sloboda, Eds.: 361–
    392. Oxford University Press. New York.

32. SLOBODA, J.A. & P.N. JUSLIN. 2001. Psychological perspectives on music and emotion. *In* Music and Emotion: Theory and Research. P.N. Juslin & J.A. Sloboda, Eds.: 71–104. Oxford University Press. New York.
33. NANTAIS, K.M. & E.G. SCHELLENBERG. 1999. The Mozart effect: an artifact of preference? Psychol. Sci. **10:** 370–373.
34. THOMPSON, W.F., E.G. SCHELLENBERG & G. HUSAIN. 2001. Arousal, mood, and the Mozart effect. Psychol. Sci. **12:** 248–251.
35. SCHELLENBERG, E.G., T. NAKATA, P.G. HUNTER & S. TAMOTO. Exposure to music and cognitive performance: tests of children and adults. Psychol. Mus. In press.
36. RAUSCHER, F.H., G.L. SHAW & K.N. KY. 1993. Music and spatial task performance. Nature **365:** 611.
37. RUBIN, D.C. 1995. Memory in oral traditions: the cognitive psychology of epic, ballads, and counting-out rhymes. Oxford University Press. New York.
38. PERETZ, I., A.S. CHAMPOD & K. HYDE. 2003. Varieties of musical disorders: the Montreal battery of evaluation of amusia. Ann. N. Y. Acad. Sci. **999:** 58–75.
39. HYDE, K.L. & I. PERETZ. 2004. Brains that are out of tune but in time. Psychol. Sci. **15:** 356–360.

Music Listening and Cognitive Abilities in 10- and 11-Year-Olds: The Blur Effect

E. GLENN SCHELLENBERG[a] AND SUSAN HALLAM[b]

[a]Department of Psychology, University of Toronto at Mississauga, Mississauga, Ontario, Canada L5L 1C6

[b]Institute of Education, University of London, London, United Kingdom

ABSTRACT: The spatial abilities of a large sample of 10 and 11 year olds were tested after they listened to contemporary pop music, music composed by Mozart, or a discussion about the present experiment. After being assigned at random to one of the three listening experiences, each child completed two tests of spatial abilities. Performance on one of the tests (square completion) did not differ as a function of the listening experience, but performance on the other test (paper folding) was superior for children who listened to popular music compared to the other two groups. These findings are consistent with the view that positive benefits of music listening on cognitive abilities are most likely to be evident when the music is enjoyed by the listener.

KEYWORDS: music and cognition; Mozart effect; arousal and cognition; mood and cognition

The finding that listening to music composed by Mozart leads to improvements in spatial abilities[1] generated widespread interest among the media, policy makers, and the general public.[2,3] Interest among the scientific community in the so-called *Mozart effect* was scattered in comparison. One reason for the initial lack of scientific interest was that the mechanism said to be driving the effect was more or less miraculous. As articulated in the original authors' *trion model*,[4,5] cortical firing patterns arising from passive listening to complex music (such as that composed by Mozart) were said to be virtually identical to those that arise from tasks that require spatial–temporal reasoning. In other words, the model hypothesized intimate links—as exemplified by identical cortical activity—between domains that have no obvious connection.

It is not surprising, then, that many researchers failed to replicate the Mozart effect.[6] Nonetheless, there have also been many successful replications in independent laboratories,[7] which indicate that the effect is real but somewhat ephemeral. As such, the phenomenon needs a better explanation than that offered by the trion model. A reasonable alternative is provided by the *arousal and mood hypothesis*,[8] which considers the link between listening to Mozart and spatial–temporal abilities to be just one example of a pleasant stimulus that can improve a perceiver's emotional state, which can, in turn, affect cognitive performance. From this perspective, the

Address for correspondence: Glenn Schellenberg, Dept. of Psychology, University of Toronto at Mississauga, Mississauga, ON, Canada L5L 1C6. Voice: 905-828-5367; fax: 905-569-4326.
g.schellenberg@utoronto.ca

Ann. N.Y. Acad. Sci. 1060: 202–209 (2005). © 2005 New York Academy of Sciences.
doi: 10.1196/annals.1360.013

link between music and cognition is mediated by changes in listeners' arousal levels and moods. Accordingly, any pleasant or enjoyable musical or nonmusical stimulus that enhances arousal and mood could also enhance cognitive abilities. In contrast to the trion model, the arousal and mood hypothesis does not give special status to music composed by Mozart, music in general, or to spatial–temporal abilities.

In line with this perspective, enhancement in spatial–temporal abilities has been observed after participants listen to music other than Mozart, including Schubert,[9] Bach,[10] and Yanni.[11] In each instance, the comparison condition consisted of simply listening to nothing[9,10] or to relaxation instructions[11]—sometimes in groups[10,11]—which would be much less stimulating than listening to music. When the comparison condition involved listening to a nonmusical auditory stimulus of similar interest (e.g., a narrated story), the music advantage disappeared.[9] Instead, participants performed better after hearing the stimulus (music or story) they preferred. When the musical stimulus was a slow and sad-sounding classical piece (i.e., Albinoni's "Adagio"), the effect also disappeared, as one would expect if arousal and mood are the mediating factors.[8] Finally, when changes in arousal and mood from pre- to postlistening were measured and held constant, the cognitive benefits of listening to fast and happy sounding music composed by Mozart were greatly reduced in one instance,[12] and eliminated in another.[8]

Benefits of music listening also extend beyond measures of spatial–temporal ability, as one would expect from previous research on other stimuli (e.g., a cup of coffee or a small gift) that cause changes in arousal levels or moods and, consequently, changes in a variety of cognitive abilities.[13,14] In a recent study,[15] undergraduates listened to Mozart (i.e., up-tempo music in a major key) or to Albinoni (i.e., slow music in a minor key) before completing one of two subtests from the Wechsler Adult Intelligence Scale—Third Edition,[16] neither of which measured spatial–temporal (or spatial) abilities. When the two music-listening experiences elicited reliable differences in arousal and mood (favoring Mozart), a reliable difference on one of the subtests was also evident (favoring Mozart). Failure to find an effect on the second test indicates that changes in arousal and mood may be more influential for some cognitive tasks than for others, but a task's "spatial–temporal" status is irrelevant to this distinction.

In another experiment,[15] the creativity of Japanese 5-year-olds was tested after the children listened to Mozart, Albinoni, or familiar children's playsongs, or after they sang familiar songs. The prediction was that exposure to the children's music would be more enjoyable among these youngsters, such that their creativity would be enhanced compared to the children who listened to classical music. Indeed, the children who heard or sang familiar songs drew for longer periods of time, and their drawings were judged by adult raters to be more creative. In sum, much of the available evidence is consistent with predictions from the arousal and mood hypothesis. Music that is pleasant and enjoyed by a particular listener is the most likely to have positive impacts on the listeners' emotional states, and positive influences on emotional state can improve cognitive performance.[2,3]

In the present report, we sought to replicate and extend these findings by re-analyzing data collected previously from a large sample of 10- and 11-year-old children. In 1996, Hallam[17] tested over 8000 children residing in the United Kingdom. The study was conducted in collaboration with the British Broadcasting Corporation (BBC) a few years after the publication of the original Mozart-effect report. It was designed to test predictions of the trion model with children, specifically that their cognitive performance would be enhanced after listening to music composed by Mozart compared to control conditions that involved listening to popular music or to a discussion about the experiment. The children completed two spatial–temporal tasks after being assigned at random to one of three listening conditions. As if turns out, absolute levels of performance for the Mozart group were either lower than (on one test) or identical to (on another test) the comparison groups. This failure to replicate the Mozart effect was reported immediately (the day after) on BBC television. The null findings were also published (in 2000) in an outlet that is unavailable to the scientific community at large.[17]

From the perspective of the arousal and mood hypothesis, however, cognitive performance should be best for the children with optimal arousal levels and mood, which would be a likely consequence of the most pleasant and enjoyable listening experience. In our view, the popular music would undoubtedly be the most enjoyable listening experience for this particular age group. We also doubted that listening to Mozart would be particularly pleasing to the children's ears. Accordingly, we reanalyzed Hallam's data with two specific, orthogonal predictions: (1) performance on the spatial tasks would be better after listening to familiar popular recordings than after listening to a piece by Mozart or to a discussion about the experiment, and (2) performance would not differ between the Mozart and discussion groups. This reanalysis was motivated by the large sample size, an alternative hypothesis that emerged after the data were initially collected, and the fact that the earlier report, with its null findings, was published in a journal that is difficult for scholars to access. Although the collaboration with the BBC and the sheer scale of the project meant that the study was not as well controlled as it could have been if listeners had been tested individually in a laboratory, the huge sample size was expected to maximize the power to detect an effect if it existed.

METHOD

Participants

The participants were 8,120 10- and 11-year-olds recruited from schools in the United Kingdom. In March of 1996, the BBC undertook a large-scale publicity campaign aimed at recruiting schools to participate in a study that was designed to provide a test of the Mozart effect. The results were to be presented immediately afterward on the television program *Tomorrow's World* (BBC 1). More than 207 schools agreed to participate. These schools were distributed widely throughout the United Kingdom. The sample comprised all of the children in Year 6 (corresponding to fifth grade in the United States in terms of age) at each of the participating schools. The number of children participating from each school ranged from 6 to 142, with an average of 39 children from each school.

Measures

The outcome measures were two, 20-item paper-and-pencil tests of spatial abilities[18] obtained by the BBC from the National Foundation for Educational

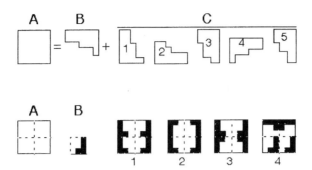

FIGURE 1. Examples of items from the square-completion test (*upper*) and the paper-folding test (*lower*). The correct answers are 3 (*upper*) and 2 (*lower*).

Research, a nonprofit, independent, research institution based in the United Kingdom. One test was *square completion* (FIG. 1, upper panel), a task that involves deciding whether two line drawings can be assembled to form a square (as in a jigsaw puzzle). On each trial, the participant sees a square (labeled A), a second enclosed line drawing (labeled B) with area less than A, and five possible options for C. The figures are positioned with mathematical signs indicating that A = B + C. The participant's task is to select the option for C that will form A when combined with B. The options for C can be rotated or flipped, or both.

The second test was *paper folding* (FIG. 1, lower panel), a task similar to the paper-folding-and-cutting task that has been used widely in previous research.[1,5,8,9,12] On each trial, participants view line drawings of a square piece of paper being folded in half vertically and horizontally, so that the folded square is one-quarter of its original size. Sections are then cut out of the folded square. The participants' task is to indicate which of four options represents the piece of paper when it is unfolded.

Procedure

At each school, all children in Year 6 were divided at random into three groups of approximately equal size. Each group was assigned to a different room where they had one of three 10-min listening experiences. The listening stimuli were broadcast simultaneously on three different BBC radio stations at 11:00 on the morning of Thursday, March 21, 1996. One of the groups listened to contemporary pop music on BBC 1, which included three recordings that were popular at the time: Blur, "Country House," Mark Morrison, "Return of the Mack," and PJ and Duncan, "Stepping Stone" (an updated recording of the Monkees' song from 1967). A second group heard the last 10 minutes of Mozart's String Quintet in D major, K593, on BBC 3, and a third group listened to the second author discussing the experiment with a journalist on BBC 5. After the listening experience, all of the children completed the square-completion test followed by the paper-folding test.

The teachers graded the tests immediately afterward and faxed the results to the BBC, who then forwarded them to the research team. Each child had two scores that could range from 0 to 20, based on the number of items answered correctly. The results were initially summarized and presented on television on Friday, March 22, the day after testing. Although the large sample and short time frame would undoubtedly involve some human error (e.g., in marking the tests and data entry), such errors should be distributed at random across the three conditions and should not affect the results in a systematic manner. Data from five children with impossible scores (i.e., >20 on either test) were excluded from analysis.

RESULTS

Because the two tests had a different number of response alternatives on each trial (i.e., square completion had 5, paper folding had 4), chance levels of responding differed between tests. Accordingly, scores were corrected for chance for each child separately for both tests by converting them to adjusted proportions, with the expected value of chance performance subtracted from both the numerator (number of items answered correctly) and the denominator (total number of items). After this transformation, scores on both tests were on the same scale, with a score of 0 corresponding to chance performance and a score of 1 indicating perfect performance. Means and standard errors are illustrated in FIGURE 1.

As one would expect, children who scored higher on one test also tended to score higher on the other test, $r = .50$, $N = 8115$, $P < .0001$. Nonetheless, 75% of the variance in either test was independent of variance in the other test. Differences between the three groups of children were analyzed initially with a 3×2 mixed-design analysis of variance that had one between-subjects variable (listening experience) and one within-subjects variable (spatial test). In general, the children found the square-completion task easier than the paper-folding task, $F(1, 8112) = 511.15$, $P < .0001$. A significant two-way interaction revealed that differences among the three groups of children varied across the two tests, $F(2, 8112) = 3.29$, $P = .0374$. Follow-up planned comparisons indicated that there were no differences among groups on the square-completion task, $Fs < 1$. For the paper-folding task, however, response patterns were consistent with predictions. The group that listened to popular music performed better than the other two groups of children, $F(1, 8112) = 5.22$, $P = .022$, who did not differ, $F < 1$. In sum, although the listening experience had no effect on performance for one of the spatial tests, the predicted "Blur effect" was evident for the other test.

DISCUSSION

We reanalyzed data from over 8,000 10- and 11-year-old children who were asked to complete two spatial tests after they had one of three 10-minute listening experiences. Whereas Hallam[17] concluded that these data provided no support for a Mozart effect (as predicted by the trion model), our reanalysis uncovered a Blur effect (as predicted by the arousal and mood hypothesis) for one of the tests. Children who listened to popular music by Blur and two other artists performed better on a subsequent paper-folding task compared to their counterparts who listened to Mozart or to a discussion about the experiment. On the square-completion task, however, mean levels of performance were virtually identical across the three groups.

These results provide additional evidence that is consistent with the arousal and mood explanation of the Mozart effect. In particular, the findings confirm that the type of music needed to generate cognitive benefits depends on the particular listener. Considering the literature as a whole, there is now evidence of a play-song effect for 5 year olds;[15] a Blur effect for 10 and 11 year olds; and Mozart,[1,7–9,11,12,15] Schubert,[9] and Yanni[11] effects for adults. The comparison condition also matters.[9] For example, many listening experiences—musical or nonmusical—would be more pleasant and engaging than sitting in silence. As such, a previous finding of Bach and Mozart effects among 10- to 12-year-old Australian children[10] is not likely to be a consequence of the children enjoying classical music to a great degree. Rather, the control condition (sitting in silence) was probably notable for being boring, possibly even unpleasant.

Why did we find a Blur effect for one outcome measure but not for the other measure? Schellenberg and his colleagues[15] reported a similar pattern of findings, namely, an advantage on one IQ subtest after participants listened to Mozart rather than Albinoni, but no such advantage on another subtest. Researchers interested in the interplay between emotion and cognition could explore this issue further in the hope of uncovering task and contextual factors that make some tests more susceptible than others to the emotional state of the participant. One possibility is that such effects are more likely when the task is particularly challenging (see FIG. 2). For the present sample of 10 and 11 year olds, the paper-folding test was more difficult than the square-completion test. Testing order could also have played a role in the present study because the square-completion test was always administered before the paper-folding test. Nonetheless, because effects of music listening on cognition are known to be temporary,[1] one would predict the exact opposite result if order were to matter (i.e., effects for the first test but not for the second).

In conclusion, our analysis provides further evidence that positive benefits of music listening on cognitive abilities are most likely to be evident when the music is enjoyed by the listener. In fact, although the arousal and mood hypothesis was formulated to explain cognitive benefits of music listening, links between enjoyable music, emotional state, and behavior extend well beyond cognitive abilities. For ex-

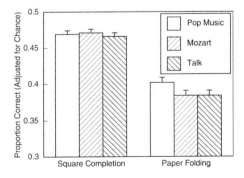

FIGURE 2. Children's performances on the square-completion and paper-folding tasks as a function of the prior listening experience.

ample, when patients select a piece of music to listen to while undergoing minor surgery, the pain they experience is less than that of patients who listen to white noise or the background sounds of the operating room, and their pain-medication requirements are reduced.[19] In short, positive effects of music listening are far reaching. Although music is not the only stimulus that has positive impacts on emotional state, it may be somewhat special in this regard because music does not have to be digested physically (unlike coffee or medication), no one is allergic to music, and music is easy (i.e., unobtrusive, noninvasive) to administer to oneself and others.

ACKNOWLEDGMENTS

This work was supported by the Natural Sciences and Engineering Research Council of Canada with cooperation from the British Broadcasting Corporation.

[Competing interests: The authors declare that they have no competing financial interests.]

REFERENCES

1. RAUSCHER, F.H., G.L. SHAW & K.N. KY. 1993. Music and spatial task performance. Nature **365:** 611.
2. SCHELLENBERG, E.G. 2005. Exposure to music: the truth about the consequences. *In* The Child as Musician: A Handbook of Musical Development. G.E. McPherson, Ed.: Oxford University Press. Oxford, UK. In press.
3. SCHELLENBERG, E.G. 2005. Music and cognitive abilities. Curr. Dir. Psychol. Sci. In press.
4. LENG, X. & G.L. SHAW. 1991. Toward a neural theory of higher brain function using music as a window. Concepts Neurosci. **2:** 229–258.
5. RAUSCHER, F.H., G.L. SHAW & K.N. KY. 1995. Listening to Mozart enhances spatial-temporal reasoning: towards a neurophysiological basis. Neurosci. Lett. **185:** 44–47.
6. CHABRIS, C.F. 1999. Prelude or requiem for the "Mozart Effect"? Nature **400:** 826–827.
7. HETLAND, L. 2000. Listening to music enhances spatial-temporal reasoning: evidence for the "Mozart effect." J. Aesthetic Edu. **34:** 105–148.
8. THOMPSON, W.F., E.G. SCHELLENBERG & G. HUSAIN. 2001. Arousal, mood and the Mozart effect. Psychol. Sci. **12:** 248–251.
9. NANTAIS, K.M. & E.G. SCHELLENBERG. 1999. The Mozart effect: an artifact of preference. Psychol. Sci. **10:** 370–373.
10. IVANOV, V.K. & J.G. GEAKE. 2003. The Mozart effect and primary school children. Psychol. Music **31:** 405–413.
11. RIDEOUT, B.E., S. DOUGHERTY & L. WERNERT. 1998. Effect of music on spatial performance: A test of generality. Percept. Mot. Skills **86:** 512–514.
12. HUSAIN, G., W.F. THOMPSON & E.G. SCHELLENBERG. 2002. Effects of musical tempo and mode on arousal, mood, and spatial abilities. Music Percept. **20:** 151–171.
13. ISEN, A.M. 2000. Positive affect and decision making. *In* Handbook of Emotions 2nd ed. M. Lewis & J.M. Haviland-Jones, Eds.: 417–435. Guilford. New York, NY.
14. SMITH, B.D., A. OSBORNE, M. MANN, *et al.* 2004. Arousal and behavior: biopsychological effects of caffeine. *In* Coffee, Tea, Chocolate, and the Brain. Nutrition, Brain, and Behavior. A. Nehlig, Ed.: 35–52. CRC. Boca Raton, FL.
15. SCHELLENBERG, E.G., T. NAKATA, P.G. HUNTER, *et al.* 2005. Exposure to music and cognitive performance: tests of children and adults. Psychol. Music. In press.
16. WECHSLER, D. 1997. Wechsler Adult Intelligence Scale—Third Edition. Psychological Corporation. San Antonio, TX.

17. HALLAM, S. 2000. The effects of listening to music on children's spatial task performance. Br. Psychol. Soc. Edu. Rev. **25:** 22–26.
18. ELIOT, J. & I.M. SMITH. 1983. An International Directory of Spatial Tests. NFER-Nelson. Windsor, UK.
19. AYOUB, C.M., L.B. RIZK, C.I. YAACOUB, *et al.* 2005. Music and ambient operating room noise in patients undergoing spinal anesthesia. Anesth. Analg. **100:** 1316–1319.

Examining Rhythm and Melody Processing in Young Children Using fMRI

K. OVERY,[a,b] A. NORTON,[a] K. CRONIN,[a] E. WINNER,[c] AND G. SCHLAUG[a]

[a]*Music and Neuroimaging Laboratory, Department of Neurology, Beth Israel Deaconess Medical Center and Harvard Medical School, Boston, Massachusetts 02215, USA*

[b]*Institute for Music in Human and Social Development, School of Arts, Culture and Environment (Music), University of Edinburgh, Edinburgh EH8 9DF, United Kingdom*

[c]*Department of Psychology, Boston College, and Harvard Project Zero, Cambridge, Massachusetts 02138, USA*

ABSTRACT: While it is often reported that musical experience can have positive effects on cognitive development in young children, the neural basis of such potential effects remains relatively unexplored. Employing functional magnetic resonance imaging (fMRI) for such research presents as many challenges as possibilities, not least of which is the fact that young children can find it difficult to remain still and attentive for long periods of time. Here we describe an fMRI scanning protocol designed specifically for young children using short scanning runs, a sparse temporal sampling data acquisition technique, simple rhythmic and melodic discrimination tasks with a button-press response, and a child-oriented preparation session. Children were recruited as part of a large-scale longitudinal study examining the effects of musical training on cognitive development and the structure and function of the growing brain. Results from an initial analysis of 33 children and from the first five children to be re-scanned after musical training indicate that our scanning protocol is successful and that activation differences can be detected both between conditions and over time.

KEYWORDS: music; fMRI; children; rhythm; melody; brain imaging

INTRODUCTION

Music plays an important part in the lives of young children. Traditional lullabies, nursery rhymes, singing games, and playground songs are perhaps increasingly being replaced by Disney musicals, TV theme songs, and pop music, but to no less enthusiastic response and indeed with great commercial success. This natural engagement with, and enjoyment of, music by children gives it enormous potential as an educational and therapeutic tool, particularly considering both the range of claimed benefits of music making (including enhanced language skills, motor skills, communication skills, and self esteem)[1–4] and the evidence suggesting that musical training may affect both brain function and brain structure.[5–10]

Address for correspondence: Dr. Katie Overy, Institute for Music in Human and Social Development, University of Edinburgh, Alison House, Edinburgh EH8 9DF, United Kingdom. Voice: +44-131-650-8248; fax: +44-131-650-2425.
k.overy@ed.ac.uk

Ann. N.Y. Acad. Sci. 1060: 210–218 (2005). © 2005 New York Academy of Sciences.
doi: 10.1196/annals.1360.014

In order to employ such potential benefits of music effectively, we must identify the specific mechanisms by which music can engage and develop specific perceptual, cognitive, motor, and personal skills. Some attempts at hypothesis development in this area have been made; for example, it has been proposed that temporal processing is a key mechanism underlying the potential of music lessons to support the language and literacy skills of dyslexic children,[11,12] while a range of studies have attempted to define the conditions and mechanisms under which listening to music can enhance spatial–temporal reasoning ability, usually in adults.[13–16] To date, however, there remains little understanding of the neural basis of such cognitive transfer effects, or indeed of the neural basis of music processing in young children. Such understanding of the parent domain of music processing remains crucial to an understanding of cross-domain transfer to other areas of learning.

In a large-scale, longitudinal study currently in progress,[17] the effects of musical instrument training are being monitored on the functional and structural brain development of children aged 5 to 7, along with a range of behavioral measures, including tests of language, mathematics, music, and motor skills. The use of functional magnetic resonance imaging (fMRI) in this study provides a powerful tool with which to examine potential changes in the localization of music processing after musical training, as well as any potential correlations with cognitive or structural development. However, the technique also presents a number of methodological challenges since it requires full concentration and an absence of physical movement in a noisy and potentially intimidating scanning environment.

Here we present the child-appropriate methods we have developed in order to address such challenges, which we have found to be successful. We also discuss preliminary data from the first five children to be retested after one year of musical training.

STIMULI AND TASK DESIGN

In order for neuroimaging data to be interpretable and useful, it is essential that participants be fully engaged in the behavioral tasks during scanning. This presents a particular challenge when working with young children since this population (1) can find it difficult to engage in a task unless they find it interesting, (2) can be easily distracted, and (3) generally take longer than adults to fully comprehend a task and perform it fluently. Thus, our aim was to design musical stimuli and a task paradigm that were as simple, quick to learn, and engaging as possible for children with or without musical training and on which performance would reflect a response to musical training.

The decision to focus on rhythm and melody skills was based on a number of factors. First, rhythm and melody are the two fundamental organizing principles of music[18] and are thus worthy of special attention. Second, most children, regardless of musical training, are familiar with the simple rhythms and melodies of children's songs. Third, rhythm and melody skills have often been found to be disassociated in tests of musical ability[19–21] and have also been found to show different hemispheric lateralization in nonmusicians, with melody tending toward a right-hemispheric dominance[22] and rhythm tending toward a left-hemispheric dominance.[23] While this effect appears to depend to some extent on the task paradigm, it is nevertheless worthy of examination in the developing brain. Fourth, it has sometimes been suggested that trained musicians show a reverse tendency for melody processing: a leftward dominance.[24–27] This indicates that musical training may have a particularly strong effect on the neural organization for melody-processing skills. Fifth, it has been hypothesized that rhythm skills play a key role in the transfer of musical abilities to other areas of cognitive ability,[12] making the neural basis of rhythm processing a particularly interesting area of study.

The choice of task paradigm was influenced by the fact that children of this age group (5–7) are particularly attuned to the concept of same/different. We thus presented pairs of rhythms or melodies to the children and asked them to determine whether the phrases in the pair were the "same" or "different." The phrases were composed using the first five notes of the C major scale (264, 297, 330, 352, and 396 Hz), since these lie in the natural vocal range for children and thus allow for comfortable and familiar listening. In order to avoid the potential experience bias of a real musical instrument sound, such as a piano or violin, a neutral, "marimba-like" sound was used (Cubase Universal Sound Module no. 13). The musical phrases were long enough to be musically interesting, but short enough to be memorable: each phrase was presented at 120 bpm, lasted for 5 beats, and consisted of 5 notes. Within the melody pairs, the pitches varied from c to g, while the durations remained constant at 500 ms;[a] within the rhythm pairs the durations varied from 125 ms to 1500

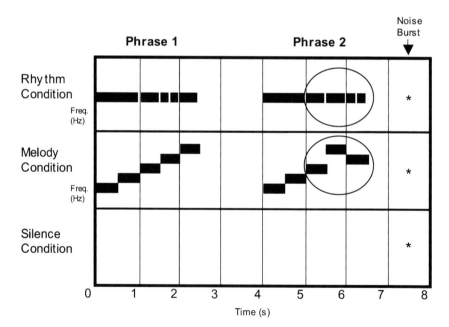

FIGURE 1. Task design. Diagram shows the three task conditions: rhythm discrimination, melody discrimination, and silence.

---

[a]We recognize that a melody usually includes both pitch and timing variation. Here, we use the term in its more limited sense of pitch variation.

ms, while the pitches remained constant (for both phrases) but were varied *between* trials using the same range of pitches from *c* to *g*. Thus, across all trials, the two experimental conditions were well matched acoustically, each with the same overall pitch content and number of notes.

The children gave their response using a button press (left button for same, with a soft-toy puppy held under the left arm as a reminder), allowing us to keep track of both their engagement with, and performance on, the task throughout the scanner session. The button press was cued by a short noise burst after the pair of musical phrases, thus limiting any motor activation to a specific temporal window after the cognitive discrimination had taken place. Similarly, the temporal window for potential cognitive discrimination was kept constant across trials: any difference between the two musical phrases occurred between beats 3 and 5 of the second phrase. Finally, a baseline condition of silence was included, in which the children simply heard a short noise burst cue and performed a bilateral button press (see FIG. 1).

SCANNING PROTOCOL

The high-volume noise of the fMRI scanner presents a number of potential difficulties for auditory studies, including masking of the auditory stimuli and causing unwanted auditory activation. Such noise might also be intimidating and distracting for young children. We overcame these potential difficulties with a sparse temporal sampling data acquisition technique that takes advantage of the natural delay in the cerebrovascular response to neural activity. Using this technique, single whole-brain images are acquired *after* each discrimination trial, thereby eliminating the possibility of either masking or distracting from the auditory stimuli, while considerably re-

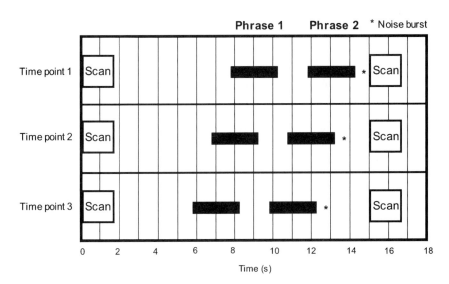

FIGURE 2. Sparse temporal sampling. Diagram shows the data acquisition method, with scans at three different time points relative to the stimuli, TR = 15.

ducing the amount of scanner noise during the test session. By "jittering" the relative position of the auditory stimuli to these scans between three different time points, we were also able to allow for differences in the cerebrovascular peak between brain regions and between individuals (see FIG. 2).

Two further considerations were the length of the scanning runs and the presentation order of the tasks. Since young children generally find it difficult to remain perfectly still and concentrate for long periods of time, we kept the scanning runs very short (3 min each), allowing the children some limited wiggling between runs. We also included only one experimental condition per run, with eight discrimination trials and four silence trials. This design avoided the potential confusion of switching tasks within a run and also created a clear, comprehensive structure of simple alternation between "rhythm" runs and "melody" runs.

PREPARATION SESSION

In order to help the children feel comfortable and confident during the fMRI scanning session, and in turn increase the chances of obtaining useful neuroimaging data, we included a practice session approximately one week before the scanning session, both to train the children on the rhythm and melody discrimination tasks and to familiarize them with the procedure of the scanning session.

Training on the discrimination tasks was a carefully staged process. Initially, the investigator asked the children to "be a listening detective" and determine whether pairs of sung melodies were the same or different. Next the children learned to press the correct buttons for *same*, *different*, and *silence* while keeping their eyes closed. The stimuli were then presented via computer, and the children learned to wait for the noise burst cue before giving their button press response. Last, the children heard recordings of the scanner noises after each trial and practiced keeping very still while listening and responding. Performance scores from the training session were monitored to compare with performance during the scanner session.

In addition to training on the task during the practice session, the children looked at a cartoon story about a boy having an MRI scan, were introduced to two soft-toy puppies, and were invited to choose the puppy they would take into the scanner with them. Finally, the children were shown examples of structural brain images of one of the investigators so that they would be familiar with some anatomical landmarks (e.g., eyes and teeth) when they saw their own "brain pictures" after the scanning session. All children and parents had previously given informed, written consent to take part in the study, which was formally approved by the Internal Review Board of the Beth Israel Deaconess Medical Center.

DATA ACQUISITION

Images were acquired on a 3T General Electric magnetic resonance imaging scanner. Prior to functional acquisition, a high resolution, strongly T1-weighted MR scan was performed, followed by a one-minute phase-encoded reference scan. Functional images were then acquired using a gradient-echo EPI sequence with an echo time of 25 ms and a 64×64 mm matrix. Using a mid-sagittal scout image, 26 slices were acquired over 1.75 s with a voxel size of $3.8 \times 3.8 \times 4$ mm. One volume set was acquired after each discrimination trial, taking advantage of the inherent delay in cerebrovascular response to neural activity. Scanning repetition time (TR) was kept constant at 15 s, while the musical stimuli were jittered between three different time points, such that the onset of the first axial slice varied between 1.25 and 3.25 s after the end of the musical stimuli.

DATA PROCESSING AND ANALYSIS

Preprocessing and analysis were conducted using SPM99 (Wellcome Dept. of Cognitive Neurology, London, UK; <www.fil.ion.ucl.ac.uk/spm>). Spatial normalization to a standard atlas was conducted by matching the T1-weighted images to a pediatric template created from 28 children's anatomical images. The identical transformation was applied to the functional data after realignment. Smoothing was applied with an 8-mm FWHM kernel. Condition effects were estimated according to the general linear model at each voxel in brain space.[28] The effect of global differences in scan intensity was removed by scaling each scan in proportion to its global intensity. Low-frequency drifts were removed using a temporal high-pass filter with a cutoff of 200 s, and no low-pass filter was applied. The data were not convolved with the hemodynamic response function (HRF), a box-car function was applied with an epoch length of 1 to the fMRI time series (12 acquisitions within each run), and no temporal derivatives were applied.[29]

Functional data from the three different time points (jitters) were combined for statistical analysis, thereby allowing for differences in the cerebrovascular peak between brain regions and between individuals.[30] Fixed effects analyses were conducted by combining the group data from all children and then contrasting the images from each condition using whole-brain, voxel-by-voxel *t* test comparisons.

PRELIMINARY RESULTS AND DISCUSSION

In an initial group analysis with 33 right-handed children,[30] mean performance scores were found to be approximately the same during the practice and scanner sessions (57% and 60%,[b] respectively; ns), demonstrating that the scanner environment did not have an adverse effect on performance. When images from each musical condition were contrasted with images from the silence condition, strong bilateral activation of the superior temporal gyrus (STG) was revealed during both melody and rhythm processing. No significant differences were found in direct contrasts between the rhythm and melody conditions, but in a region of interest analysis using the bilateral STG for a small volume correction, a small region in the right STG, slightly anterior and inferior to the primary auditory cortex, was found to show significantly higher activation for melody processing than rhythm processing (FWE, $P < .05$). Interestingly, this location has been identified as active during melodic tasks in several fMRI studies with adults.[32–34]

---

[b]The *different/same* ratio of the 8 phrase pairs per run was weighted so that a child detecting no differences and indicating *same* for every trial would score only 3/8 = 38%.

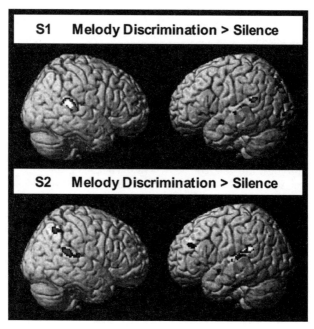

FIGURE 3. Example of images acquired during melody discrimination in scan 1 and scan 2 in one child; $P < .001$, uncorrected.

Preliminary data from the first five children to be rescanned after a year of musical training revealed that the children had improved considerably on the musical discrimination tasks (from a mean of 69% in scan 1 to a mean of 77% in scan 2, ns). Differences in neural activation between the two time points could also be identified, predominantly in the temporal lobes. No voxel-by-voxel statistical analyses were conducted with this small group, and thus no interpretation or conclusions can be made from these preliminary data. However, the results do indicate that our scanning protocol is effective: children were willing to return for a second scan, performance on the musical tasks improved, and differences in neural activation were revealed between scanning time points. For a visual example of the melody discrimination data acquired from one child at each time point, see FIGURE 3.

CONCLUSIONS

Our specially designed fMRI scanning protocol for examining music processing in young children proved extremely effective. The children were comfortable with the stimuli, understood the tasks, and did not become bored during the scanning session. A button press response was successfully used to monitor performance during scanning, and scores indicated that the MRI environment did not have an adverse effect on performance. The children particularly enjoyed certain aspects of the experience, such as the soft-toy puppies and receiving a CD-ROM with pictures of their own brain. The acquired fMRI data showed clear auditory activations and some evidence of differential specialization for melody and rhythm processing, while preliminary results from five children suggest that differences in neural activation patterns after a year of musical training can be revealed. Future analyses with larger numbers of both musically trained and untrained children will give the opportunity to differentiate normal developmental changes from those due to the effects of musical training.

In summary, we have described and demonstrated a useful methodology for examining the neural basis of musical processing in young children. It is hoped that these ideas and techniques will contribute toward providing a powerful tool with which to explore the enormous potential of music as an educational and therapeutic experience.

ACKNOWLEDGMENTS

This research was supported by grants from the National Science Foundation (BCS-0132508), the International Foundation for Music Research, and the Grammy Foundation. Dr. Overy also acknowledges support from the Harold Wingate Foundation.

[Competing interest: The authors declare that they have no competing financial interests.]

REFERENCES

1. SUTTON, J. 1995. The sound-world of speech- and language-impaired children: the story of a current music therapy research project. *In* Art and Music, Therapy and Research. A.L. Gilroy, Ed.: 153–163. Routledge. London.
2. THAUT, M.H., K.W. McINTOSH, G.C. McINTOSK, *et al.* 2001. Auditory rhythmicity enhances movement and speech motor control in patients with Parkinson's disease. Funct. Neurol. **16:** 163–172.
3. MALLOCH, S. 1999/2000. Mothers and infants and communicative musicality. Special Issue Mus. Sci. Rhythm Mus. Narrat. Origins Hum. Commun. 29–57.
4. KEEN, A.W. 2004. Using music as a therapy tool to motivate troubled adolescents. Soc. Work. Health Care **39:** 361–373.
5. SCHLAUG, G., L. JANCKE, Y. HUANG, *et al.* 1995. *In vivo* evidence of structural brain asymmetry in musicians. Science **267:** 699–701.
6. SCHNEIDER, P., M. SCHERG, H.G. DOSCH, *et al.* 2002. Morphology of Heschl's gyrus reflects enhanced activation in the auditory cortex of musicians. Nat. Neurosci. **5:** 688–694.
7. SLUMING, V., T. BARRICK, M. HOWARD, *et al.* 2002. Voxel-based morphometry reveals increased gray matter density in Broca's area in male symphony orchestra musicians. Neuroimage **17:** 1613–1622.
8. STEWART, L., R. HENSON, K. KAMPE, *et al.* 2003. Brain changes after learning to read and play music. **20:** 71–83.
9. GAAB, N. & G. SCHLAUG. 2003. The effect of musicianship on pitch memory in performance matched groups. Neuroreport **14:** 2291–2295.
10. KOELSCH, S., B.G. SCHMIDT, J. KANSOK, *et al.* 2002. Effects of musical expertise on the early right anterior negativity: an event-related brain potential study. Psychophysiology **39:** 657–663.
11. OVERY, K. 2000. Dyslexia, temporal processing, and music: the potential of music as an early learning aid for dyslexic children. Psychol. Mus. **28:** 218–229.

12. OVERY, K. 2003. Dyslexia and music: from timing deficits to musical intervention. *In* The Neurosciences and Music. G. Avanzini, C. Faienza, L. Lopez, *et al.*, Eds. Vol. 999: 497–505. Annals of the New York Academy of Sciences. New York.
13. RAUSCHER, F.H., G.L. SHAW & K.N. KY. 1993. Music and spatial task performance. Nature **365**: 611.
14. RAUSCHER, F.H. & G.L. SHAW. 1998. Key components of the "Mozart effect." Percept. Mot. Skills **86**: 835–841.
15. STEELE, K.M. 2000. Arousal and mood factors in the "Mozart effect." Percept Mot. Skills **91**: 188–190.
16. LINTS, A. & S. GADBOIS. 2003. Is listening to Mozart the only way to enhance spatial reasoning? Percept. Mot. Skills **97**: 1163–1174.
17. NORTON, A., E. WINNER, K. CRONIN, *et al.* 2005. Are there pre-existing neural, cognitive, or motoric markers for musical ability? Brain Cogn. In press.
18. KRUMHANSL, C.L. 2000. Rhythm and pitch in music cognition. Psychol. Bull. **126**:159–179. Review.
19. DENNIS, M. & T. HOPYAN. 2001. Rhythm and melody in children and adolescents after left or right temporal lobectomy. Brain Cogn. **47**: 461–469.
20. OVERY, K, R.I NICOLSON, A.J. FAWCETT & E.F. CLARKE. 2003. Dyslexia and music: measuring musical timing skills. Dyslexia **9**: 18–36.
21. HYDE, K.L. & I. PERETZ. 2004. Brains that are out of tune but in time. Psychol. Sci. **15**: 356–360.
22. ZATORRE, R.J. 2001. Neural specializations for tonal processing. Ann. N. Y. Acad. Sci. **930**: 193–210.
23. SAMSON, S., N. EHRLE & M. BAULAC. 2001. Cerebral substrates for musical temporal processes. Ann. N. Y. Acad. Sci. **930**: 166–178.
24. BEVER, T.G. & R.J. CHIARELLO. 1974. Cerebral dominance in musicians and nonmusicians. Science **185**: 537–539.
25. MESSERLI, P., A. PEGNA & N. SORDET. 1995. Hemispheric dominance for melody recognition in musicians and nonmusicians. Neuropsychologia **33**: 395–405.
26. EVERS, S., J. DANNERT, D. RODDING, *et al.* 1999. The cerebral haemodynamics of music perception: a transcranial Doppler sonography study. Brain **122**: 75–85.
27. OHNISHI, H., H. MATSUDA, T. ASADA, *et al.* 2001. Functional anatomy of musical perception in musicians. Cerebr. Cortex **11**: 754–760.
28. FRISTON, K.J., A.P. HOLMES., K.J. WORSLEY, *et al.* 1995. Statistical parametric maps in functional imaging: a general linear approach. Hum. Brain Mapp. **2**: 189–210.
29. GAAB, N., C. GASER, T. ZAEHLE, *et al.* 2003. Functional anatomy of pitch memory: an fMRI study with sparse temporal sampling. **19**: 1417–1426.
30. HALL, D.A., M.P. HAGGARD, M.A. AKEROYD, *et al.* 1999. "Sparse" temporal sampling in auditory fMRI. Hum. Brain Mapp. **7**: 213–223.
31. OVERY K., A.C. NORTON, K.T. CRONIN, *et al.* 2004. Imaging melody and rhythm processing in young children. Neuroreport **15**: 1723–1726.
32. ZATORRE, R.J., A.C. EVANS, E. MEYER, *et al.* 1994. Neural mechanisms underlying melodic perception and memory for pitch. J. Neurosci. **14**: 1908–1919.
33. OZDEMIR, E., N. GAAB, K. OVERY, *et al.* 2004. Shared neural substrates for singing, speaking, humming, and phonation. Program N. 595.5 2004 Abstract Viewer/Itinerary Planner. Society for Neuroscience. Washington, D.C.
34. OVERY, K., A. NORTON, E. OZDEMIR, *et al.* 2005. Activation of left inferior frontal gyrus after melodic intonation therapy in a Broca's aphasia patient. Program N. 595.7 2004 Abstract Viewer/Itinerary Planner. Society for Neuroscience. Washington, D.C.

Effects of Music Training on the Child's Brain and Cognitive Development

GOTTFRIED SCHLAUG,[a] ANDREA NORTON,[a] KATIE OVERY,[a] AND ELLEN WINNER[b]

[a]Department of Neurology, Music and Neuroimaging Laboratory, Beth Israel Deaconess Medical Center/Harvard Medical School, Boston, Massachusetts 02215, USA

[b]Department of Psychology, Boston College, Boston, Massachusetts 02215, USA

ABSTRACT: Research has revealed structural and functional differences in the brains of adult instrumental musicians compared to those of matched non-musician controls, with intensity/duration of instrumental training and practice being important predictors of these differences. Nevertheless, the differential contributions of nature and nurture to these differences are not yet clear. The musician–nonmusician comparison is an ideal model for examining whether and, if so, where such functional and structural brain plasticity occurs, because musicians acquire and continuously practice a variety of complex motor, auditory, and multimodal skills (e.g., translating visually perceived musical symbols into motor commands while simultaneously monitoring instrumental output and receiving multisensory feedback). Research has also demonstrated that music training in children results in long-term enhancement of visual–spatial, verbal, and mathematical performance. However, the underlying neural bases of such enhancements and whether the intensity and duration of instrumental training or other factors, such as extracurricular activities, attention, motivation, or instructional methods can contribute to or predict these enhancements are yet unknown. Here we report the initial results from our studies examining the brain and cognitive effects of instrumental music training on young children in a longitudinal study and a cross-sectional comparison in older children. Further, we present a comparison of the results in these children's studies with observations from our cross-sectional studies with adults.

KEYWORDS: music; instrumental music training; musicians; nonmusicians; brain plasticity; skill learning; longitudinal study; children; development morphometry; fMRI

BRAIN DIFFERENCES BETWEEN ADULT MUSICIANS AND NONMUSICIANS

Instrumental training is a multisensory motor experience, typically initiated at an early age. Playing an instrument requires a host of skills, including reading a com-

Address for correspondence: Gottfried Schlaug, Department of Neurology, Music and Neuroimaging Laboratory, Beth Israel Deaconess Medical Center/Harvard Medical School, 330 Brookline Avenue, Palmer 127, Boston, MA 02215. Voice: 617-632-8912; fax: 617-632-8920. gschlaug@bidmc.harvard.edu; www.musicianbrain.com

Ann. N.Y. Acad. Sci. 1060: 219–230 (2005). © 2005 New York Academy of Sciences. doi: 10.1196/annals.1360.015

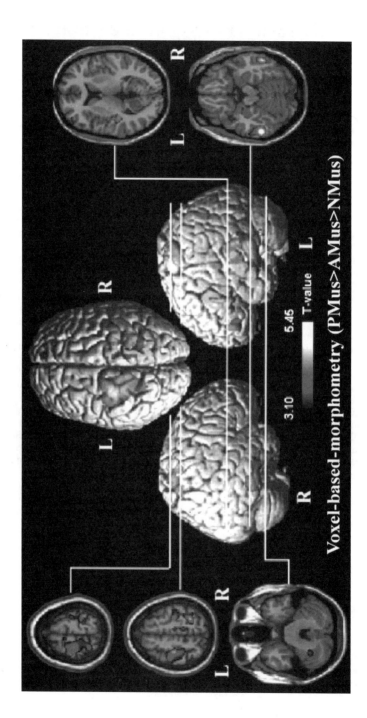

FIGURE 1. A voxel-based morphometric analysis of nonmusicians compared with amateur and professional musicians.

plex symbolic system (musical notation) and translating it into sequential, bimanual motor activity dependent on multisensory feedback; developing fine motor skills coupled with metric precision; memorizing long musical passages; and improvising within given musical parameters. Studies have explored the brain bases of these highly specialized sensorimotor,[1-5] auditory,[6-11] and auditory–spatial[12] skills. As shown in FIGURE 1 (a voxel-based morphometric analysis of nonmusicians compared with amateur and professional musicians), professional keyboard players, who reported approximately twice as much weekly practice time as the amateur musicians,[5] have significantly more gray matter in several brain regions, including the primary sensorimotor cortex, the adjacent superior premotor and anterior superior parietal cortex bilaterally, mesial Heschl's gyrus (primary auditory cortex), the cerebellum, the inferior frontal gyrus, and part of the lateral inferior temporal lobe, than either the amateur musicians or the nonmusicians.

While it may not be as surprising that structural differences are found in those brain regions that are closely linked to skills learned during instrumental music training (such as independent fine motor movements in both hands and auditory discrimination), structural differences outside of these primary regions (e.g., inferior frontal gyrus; see also Ref. 13) are of particular interest since this may indicate that plasticity can occur in brain regions that either have control over primary musical functions or serve as multimodal integration regions for musical skills. Functional correlates of music processing differences between musicians and nonmusicians typically show greater lateralization and stronger activation of auditory association areas in musicians, whereas nonmusicians may show stronger activation of primary auditory

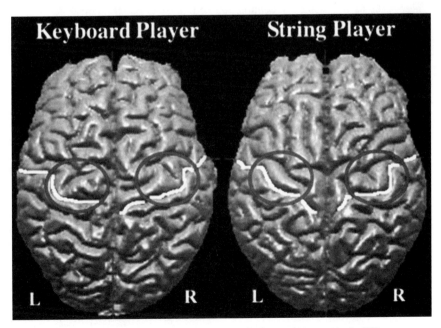

FIGURE 2. Within-musician, instrument-typical, gross-anatomical differences are seen in the precentral gyrus.

regions.[11] These effects have also been found in short-term training studies both in adult nonmusicians and in young children using auditory evoked potentials.[14–16]

Further support for the plasticity hypothesis comes from studies showing within-musician differences. Pantev and colleagues[17] found more pronounced cortical responses to trumpet and string tones in the respective players of those instruments, demonstrating that functional brain differences can be associated with the particular musical instrument played. Similarly, when comparing string and keyboard players, Bangert and colleagues[18] from our group have found within-musician differences in the omega sign (OS), an anatomical landmark of the precentral gyrus commonly associated with representation of hand/finger movement (see FIG. 2). The majority of the adult keyboard players had an elaborated configuration of the precentral gyrus on both sides, whereas most of the adult string players had this atypicality only on the left. There is evidence suggesting that these structural differences in musicians' brains are more pronounced in musicians who began study at a younger age[1,19,20] and who practiced with greater intensity.[5,10,21] Long-term motor training studies in animal studies also support the argument for training-associated brain plasticity.[22]

In order to determine whether the structural and functional differences seen in adult musicians reflect adaptations that occurred as a result of musical training during sensitive periods of brain development, or are instead, markers of musical interest and/or aptitude that existed prior to training, it is necessary to examine children and/or adults before the onset of instrumental music training and compare them to a group of control subjects not planning to study a musical instrument and practice regularly. Thus, we report here our baseline results and preliminary analyses after the first year of our pilot longitudinal study that aims to examine this hypothesis. These results are presented in conjunction with those of our cross-sectional studies of nine- to eleven-year-old children.

A CROSS-SECTIONAL COMPARISON OF FIVE- TO SEVEN-YEAR-OLD CHILDREN PRIOR TO INSTRUMENTAL MUSIC TRAINING

For the past two years we have been conducting a longitudinal study of the effects of music training on brain development and cognition in young children.[23–25] The major questions addressed were (1) whether there are pre-existing differences in brain structure/function and/or cognitive skills in children just beginning to study a musical instrument compared to those who are not; and (2) whether instrumental training initiated between the ages of five and seven leads to cognitive enhancement and stimulates regional brain growth in areas previously shown to be structurally different in adult musicians.[4,5,10,21] We have tested fifty, five- to seven-year-old children at baseline prior to beginning music lessons. Approximately two-thirds of those children chose to take piano, while the other third chose string lessons. We have also tested a smaller, untreated control group (currently $n = 25$) matched to the instrumental group in age, socioeconomic standard (SES), and verbal IQ. Each child underwent a battery of behavioral tests, including the Object Assembly, Block Design, and Vocabulary subtests from either the Wechsler Intelligence Scale for Children (WISC-III) (for children six years and older) or the Wechsler Preschool and Primary Scale of Intelligence (WPPSI-III) (for children under age six); the Raven's Colored Progressive Matrices (CPM) and Raven's Standard Progressive Matrices (SPM); the

Auditory Analysis Test[26] as a measure of phonemic awareness; Gordon's Primary Measures of Music Audiation (PMMA) as a measure of musical skill/aptitude; and two motor tests (an index finger tapping test and a motor sequencing task using four fingers) to measure speed and dexterity in both right and left hands.

Children also underwent structural and functional MR scans of their brains using a specially designed, child-appropriate protocol. MR images were acquired on a 3 Tesla General Electric Magnetic Resonance Imaging (MRI) Scanner. We found no pre-existing cognitive, music, motor, or structural brain differences between the instrumental and control groups at baseline,[25] thereby making it unlikely that children who choose to play a musical instrument do so because they have atypical brains, and suggesting that the brain atypicalities seen in adult musicians are more likely to be the product of intensive music training rather than pre-existing biological markers of musicality. The structural MR sequence had a spatial resolution of $1 \times 1 \times 1.5$ mm. We used a fully automatic technique for computational analysis of differences in local gray and white matter.[5,27] There were no differences in the absolute brain volume, gray matter volume, white matter volume, or the midsagittal corpus callo-

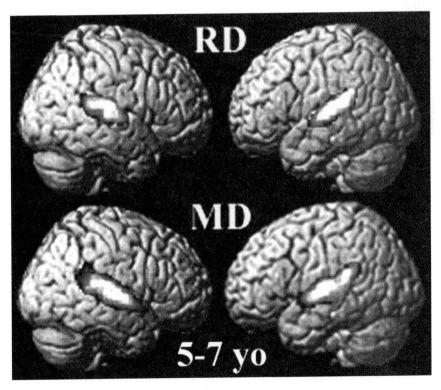

FIGURE 3. Statistical parametric images superimposed on surface renderings of a standardized anatomical brain depict significant group activations during rhythmic and melodic discrimination tasks in five- to seven-year-old children, naive for instrumental music training.

sum size (for more details, see Ref. 25). A voxel-based analysis[5,27] showed no significant differences in regional gray matter volume between the two groups.

Our fMRI scanning protocol, specifically for young children, uses short scanning runs, a sparse temporal sampling data acquisition technique,[28] simple rhythmic (RD) and melodic discrimination (MD) tasks with a button press response to indicate *same/different* judgments for pairs of short musical phrases, and a child-oriented MRI preparation session to overcome the challenges of scanning such young children. The functional images from each musical condition (RD and MD) were contrasted with the images from the silence condition (baseline) at a significance threshold of $P < .05$, using a family-wise error (FWE) correction for multiple comparisons. Both musical conditions led to strong bilateral activation of the superior temporal gyrus (STG; FIG. 3). A region in the right STG (slightly anterior and inferior to Heschl's gyrus) was found to show significantly higher activation during melodic discrimination than during rhythmic discrimination (for more details, see Ref. 23).

AFTER ONE YEAR OF INSTRUMENTAL MUSIC TRAINING

In our preliminary analyses (so far, only half of the children have completed their second round of testing) of the effects of one year of music training, we found significantly greater change scores in the instrumental group compared to the control group in behavioral tests directly linked to instrumental music training: fine motor skills (mean of 10% for the instrumental group compared to 5% for the control group) and auditory discrimination skills, as measured by Gordon's PMMA (1986) (9% vs. 6%). Although we have not yet found evidence for transfer effects in domains such as verbal, visual–spatial, and math after 14 months of observation, the instrumental group showed trends in the anticipated direction. Brain data also support this trend. In the groups used for the preliminary analyses, there was a nonsignificantly greater increase in gray matter volume in the instrumental group than in the control group, but as yet, no significant change in corpus callosum size has emerged. Since these between-group differences are likely to change as more subjects are added to the analyses, we are also investigating the influence of practice intensity on our behavioral outcomes and brain data within the instrumental music group. Preliminary analyses of the fMRI data suggest that functional changes during the melodic and rhythmic discrimination tasks occur after one year of instrumental music learning in both the right and left hemisphere, mainly in auditory association areas in the temporal lobe and temporal-parietal junction. No significant changes were seen when the control group's baseline was compared with their second set of results 14 months later.

A CROSS-SECTIONAL COMPARISON OF NINE- TO ELEVEN-YEAR-OLD CHILDREN: INSTRUMENTALISTS VERSUS NONINSTRUMENTALISTS

We recently added a new cross-sectional comparison between a group of nine- to eleven-year-old instrumentalists with an average of four years of training and a

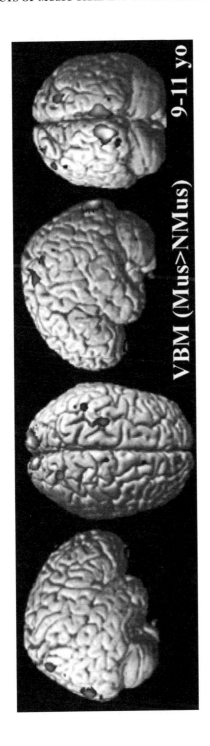

FIGURE 4. Voxel-based morphometry study comparing nine- to eleven-year-old instrumentalists with matched nonmusician controls.

group of noninstrumentalists (matched in age, handedness, and SES) to our ongoing longitudinal study. This group of children underwent the same battery of behavioral tests and imaging studies as the children in our longitudinal group of five to seven year olds did. The instrumental group performed significantly better than the well-matched control group on the Gordon's Intermediate Measures of Music Audiation (primarily due to their superior performance in the tonal subtest), the maximal left hand index finger tapping rate, and the Vocabulary subtest of the WISC-III. Strong, nonsignificant trends were seen in the phonemic awareness test (Auditory Analysis), the Raven's Progessive Matrices, and the Key Math test. We did not find any significant differences in the Object Assembly or Block Design tests. Because the tasks of reading music and playing an instrument call upon a wide variety of skills, there are plausible explanations for why music training could lead to transfer effects in other areas. For example, music training might enhance spatial reasoning because music notation itself is spatial. Mathematical skills may well be enhanced by music learning because understanding rhythmic notation actually requires math-specific skills, such as pattern recognition and an understanding of proportion, ratio, fractions, and subdivision (e.g., a half note is twice as long as a quarter note, and a quarter note can be evenly subdivided into four sixteenth notes). Phonemic awareness skills may be improved by music training because both music and language processing require the ability to segment streams of sound into small perceptual units.

The instrumentalists had significantly more gray matter volume (mean [SD] of 747 [75] cc compared to 661 [82] cc for the noninstrumental group) that was regionally pronounced not only in the sensorimotor cortex, but also in the occipital lobe bilaterally (FIG. 4). FIGURE 4 shows the regional distribution of gray matter volume differences when the instrumental group was compared with the noninstrumental group on a voxel-by-voxel basis.[29]

The nine- to eleven-year-old children participated in the same functional imaging experiments as the five to seven year olds in our longitudinal study (FIG. 5). Functional images from each musical condition (RD and MD) were contrasted with the images from the silence (control) condition at a significance threshold of $P < .05$, using a FWE correction for multiple comparisons. Preliminary analysis of all group comparisons revealed that both the instrumental and noninstrumental groups showed strong bilateral activation of the STG. However, the instrumental group showed more activation of the STG, particularly on the right, and also more activation of the posterior inferior and middle frontal gyrus in both hemispheres (more so in the MD than in the RD task). This trend of additional extratemporal lobe activation was found to be further increased in a group of adult subjects with long-term, intensive instrumental music training who also performed these functional tasks. This data is not reported in detail here, although FIGURE 6 shows the pattern of activation for the RD tasks in two adult groups (professional musicians vs. nonmusicians). Further, by comparing FIGURE 6 with FIGURES 5 and 3, the increase in extratemporal lobe activation with maturity and greater length of instrumental training becomes apparent.

The inferior and middle frontal regions that are activated by these rhythmic and melodic discrimination tasks may play a role in the integration of auditory events into larger units, or the sequential ordering of behaviorally relevant auditory events. The frontal and, in particular, the inferior frontal activations seen in auditory tasks should be considered in the context of the discussion on mirror neurons. "Mirror" neurons respond both when an action is observed and when that same action is per-

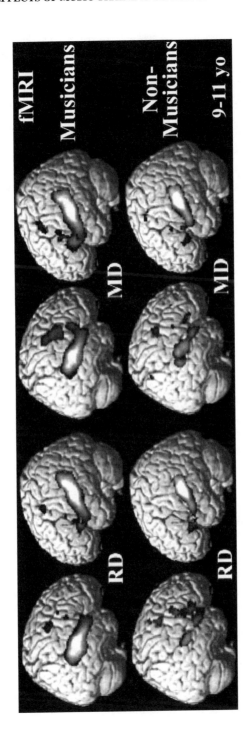

FIGURE 5. Statistical parametric images superimposed on surface renderings of standardized anatomical brains depict significant group activations during rhythmic and melodic discrimination tasks in nine- to eleven-year-old children with and without instrumental music training.

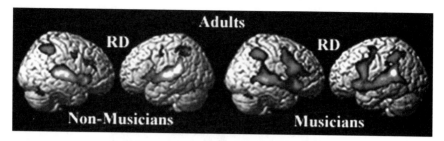

FIGURE 6. Statistical parametric images superimposed on standardized anatomical brains show significant activations during a melodic discrimination task in a group of professional musicians and a matched group of nonmusicians.

formed. In addition to the sight and performance mirror neurons, a subset of mirror neurons in monkeys also responds to the sound of an action.[30,31] These "auditory–visual" mirror neurons exemplify high-level abstraction in the representation of action—an identical neural system becomes activated regardless of whether a particular action is heard, seen, or performed. This may have implications for music learning over time. As musical skills are acquired, the same kinds of action–sound mappings occur.[32] The student learns by watching the teacher and/or conductor, by listening to the sounds that are produced by particular types of movement, by evaluating self-produced sounds either in isolation or in combination with sounds produced by other musicians, and by translating visual symbols into sound. Thus, it is likely that mirror neurons may play an important role in instrumental music learning. This notion is supported by the frontal activation that emerges in the nine- to eleven-year-old group and becomes more prominent in the adult musician group.

SUMMARY

Preliminary results of our longitudinal study in five- to seven-year-old children suggest that cognitive and brain effects from instrumental music training can be found. After 14 months of observation, these effects are still small and in domains such as fine motor and melodic discrimination that are closely related to the instrumental music training. Data from our cross-sectional study of nine- to eleven-year-old children with an average of four years of musical training suggest that the predicted effects become stronger, and that transfer effects begin to emerge in addition to those strong effects in closely related motor and auditory domains. Nevertheless, our nine- to eleven-year-old cross-sectional study is correlational, and although it supports the general trends seen across all three groups (from small, nonsignificant trends in five- to seven-year-olds after 14 months of observation, to prominent musician–nonmusician differences in adults), only an experimental study such as our longitudinal study can prove causality and test the role of other predictors such as intensity of training, skill at reading musical notation, and level of musical achievement.

[Competing interests: The authors declare that they have no competing financial interests.]

REFERENCES

1. ELBERT, T., C. PANTEV, C. WIENBRUCH, *et al.* 1995. Increased cortical representation of the fingers of the left hand in string players. Science **270:** 305–306.
2. AMUNTS, K., G. SCHLAUG, L. JANCKE, *et al.* 1997. Motor cortex and hand motor skills: structural compliance in the human brain. Hum. Brain Mapp. **5:** 206–215.
3. HUND-GEORGIADIS, M. & Y. VON CRAMON. 1999. Motor learning–related changes in piano players and nonmusicians revealed by functional magnetic resonance signals. Exp. Brain Res. **125:** 417–425.
4. SCHLAUG, G. 2001. The brain of musicians: a model for functional and structural adaptation. Ann. N. Y. Acad. Sci. **930:** 281–299.
5. GASER, C. & G. SCHLAUG. 2003. Brain structures differ between musicians and nonmusicians. J. Neurosci. **23:** 9240–9245.
6. BESSON, M., F. GAITA & J. REQUIN. 1994. Brain waves associated with musical incongruities differ for musicians and nonmusicians. Neurosci. Lett. **168:** 101–105.
7. PANTEV, C., R. OOSTENVELD, A. ENGELIEN, *et al.* 1998. Increased auditory cortical representation in musicians. Nature **392:** 811–814.
8. ZATORRE, R.J., D.W. PERRY, C.F. BECKETT, *et al.* 1998. Functional anatomy of musical processing in listeners with absolute pitch and relative pitch. Proc. Natl. Acad. Sci. USA **95:** 3172–3177.
9. KEENAN, J.P., V. THANGARAJ, A.R. HALPERN & G. SCHLAUG. 2001. Absolute pitch and planum temporale. Neuroimage **14:** 1402–1408.
10. SCHNEIDER, P., M. SCHERG, H.G. DOSCH, *et al.* 2002. Morphology of Heschl's gyrus reflects enhanced activation in the auditory cortex of musicians. Nat. Neurosci. **5:** 688–694.
11. GAAB, N. & G. SCHLAUG. 2003. The effect of musicianship on pitch memory in performance matched groups. NeuroReport **14:** 2291–2295.
12. MÜNTE, T.F., C. KOHLMETZ, W. NAGER & E. ALTENMÜLLER. 2001. Neuroperception: superior auditory spatial tuning in conductors. Nature **409:** 580.
13. SLUMING, V., T. BARRICK, M. HOWARD, *et al.* 2002. Voxel-based morphometry reveals increased gray matter density in Broca's area in male symphony orchestra musicians. Neuroimage **17:** 1613–1622.
14. TRAINOR, L.J., R.N. DESJARDINS & C. ROCKEL. 1999. A comparison of contour and interval processing in musicians and nonmusicians using event-related potentials. Aust. J. Psychol. **51:** 147–153.
15. BOSNYAK, D.J., R.A. EATON & L.E. ROBERTS. 2004. Distributed auditory cortical representations are modified when nonmusicians are trained at pitch discrimination with 40 Hz amplitude modulated tones. Cerebr. Cortex **14:** 1088–1099.
16. SHAHIN, A., L.E. ROBERTS & L.J. TRAINOR. 2004. Enhancement of auditory cortical development by musical experience in children. Neuroreport **15:** 1917–1921.
17. PANTEV, C., L.E. ROBERTS, M. SCHULZ, *et al.* 2001. Timbre-specific enhancement of auditory cortical representations in musicians. Neuroreport **12:** 169–174.
18. BANGERT, M., D. NAIR & G. SCHLAUG. 2005. Structural specialization of the specialized. Neuroimage **26** (Suppl. 1): 102.
19. SCHLAUG, G., L. JAENCKE, Y. HUANG & H. STEINMETZ. 1995. Increased corpus callosum size in musicians. Neuropsychologia **33:** 1047–1055.
20. LEE, D.J., Y. CHEN & G. SCHLAUG. 2003 Corpus callosum: musician and gender effects. Neuroreport **14:** 205–209.
21. HUTCHINSON, S., L.H.L. LEE, N. GAAB & G. SCHLAUG. 2003. Cerebellar volume of musicians. Cerebr. Cortex **13:** 943–949.
22. ANDERSON, B.J., P.B. ECKBURG & K.I. RELUCIO. 2002. Alterations in the thickness of motor cortical subregions after motor skill learning and exercise. Learn. Mem. **9:** 1–9.
23. OVERY K., A. NORTON, K. CRONIN, *et al.* 2004. Imaging melody and rhythm processing in young children. NeuroReport **15:** 1723–1726.

24. KOELSCH, S., T. FRITZ, K. SCHULZE, *et al.* 2005. Adults and children processing music: an fMRI study. Neuroimage **25:** 1068–1076.
25. NORTON, A., E. WINNER, K. CRONIN, *et al.* 2005. Are there neural, cognitive, or motoric markers for musical ability prior to instrumental training? Brain Cogn. In press.
26. ROSNER, J. & D. SIMON. 1971. Auditory analysis test. J. Learn. Disabil. **4:** 40–48.
27. ASHBURNER, J. & K.J. FRISTON. 2000. Voxel-based morphometry—the methods. Neuroimage **11:** 805–821.
28. GAAB, N., C. GASER, T. ZAEHLE, *et al.* 2003. Functional anatomy of pitch memory: an fMRI study with sparse temporal sampling. Neuroimage **19:** 1417–1426.
29. SCHLAUG, G., D.J. LEE, K. OVERY, *et al.* 2004. Does brain anatomy predict musicianship? Neuroimage **22:** 42.
30. KEYSERS, C., E. KOHLER, M.A. UMILTÀ, *et al.* 2003. Audiovisual mirror neurons and action recognition. Exp. Brain Res. **153:** 628–636.
31. KOHLER, E., C. KEYSERS, M.A. UMILTÀ, *et al.* 2002. Hearing sounds, understanding actions: action representation in mirror neurons. Science **297:** 846–848.
32. BANGERT, M. & E.O. ALTENMULLER. 2003. Mapping perception to action in piano practice: a longitudinal DC-EEG study. BMC Neurosci **4:** 26.

Investigating the Relationship of Music and Language in Children

Influences of Musical Training and Language Impairment

SEBASTIAN JENTSCHKE, STEFAN KOELSCH, AND ANGELA D. FRIEDERICI

Max Planck Institute of Human Cognitive and Brain Sciences, Leipzig, Germany

ABSTRACT: Language and music are human universals involving perceptually discrete elements organized in hierarchically structured sequences. The set of principles governing the combination of these structural elements into sequences is known as *syntax*. A violation of expectancies concerning syntactic regularities may be reflected by two ERP components: the ERAN (early right anterior negativity) and the ELAN (early left anterior negativity). The ERAN is evoked by a violation of musical regularities, whereas the ELAN is linked to syntax processing in the language domain. There is evidence from adult data to suggest that both ERAN and ELAN are, at least partly, generated in the same brain regions. Therefore, it seems plausible to expect transfer effects between music and language due to shared processing resources. Moreover, the ERAN is larger in adults with formal musical training (musicians) than in those without, indicating that more specific representations of musical regularities lead to heightened musical expectancies. The aim of this study is to investigate these issues in child development. We conducted two experimental sessions with the same participants and compared children with and without musical training (11 years old) and children with or without language impairment (5 years old). In a music experiment, the reactions to chord sequences ending either with a (regular) tonic or with an (irregular) supertonic were compared. For a language experiment we used syntactically correct and incorrect sentences. Preliminary results show that an ERAN is present in both groups and appears to have a larger amplitude in musically trained children. In addition, there are indications of an enhanced negativity in response to a syntactic violation in the musically trained children. The relationship between the ERP components is, moreover, manifested in the finding that an ERAN is present in linguistically nonimpaired children at the age of 5 years but not in children with language impairment of the same age.

KEYWORDS: processing of linguistic and musical syntax; development; musical training; language impairment; event-related brain potential (ERP); ERAN; ELAN

Address for correspondence: Sebastian Jentschke, Max Planck Institute of Human Cognitive and Brain Sciences, Stephanstr. 1A, D-04103 Leipzig, Germany. Voice: +49-341-35521712; fax: +49-341-35521730.
jentschke@cbs.mpg.de

Ann. N.Y. Acad. Sci. 1060: 231–242 (2005). © 2005 New York Academy of Sciences.
doi: 10.1196/annals.1360.016

INTRODUCTION

Music is one of the oldest and most basic sociocognitive domains of the human species. It is assumed that human musical abilities played a key phylogenetical role in the evolution of language and that music-making behavior covered important evolutionary functions such as communication, group coordination, and social cohesion.[1] Because the perception and, even more, the production of music involves practically every congnitve function, music has proven itself as a valuable tool for the investigation of the functional organization of the human brain.[1]

Natural environments contain highly structured systems, such as language and music, to which we are exposed in everyday life. The human brain internalizes regularities of these systems by passive exposure, and the acquired implicit knowledge influences perception and performance. Language and music provide two instances of highly structured systems that may be learned in an incidental manner and used in adults in an elaborate fashion. They consist of perceptually discrete elements organized in hierarchically structured sequences. The combination of these structural elements into sequences is governed by a set of principles that may be denoted as syntax.[2]

Few studies have examined the relationship between music and language processing. Musical expectancies are automatically generated during the perception of a musical context.[3–6] These expectancies are most presumably established with reference to a complex system of regularities (of major–minor tonal, or "classical" Western music) that build a musical structure and have been considered to represent part of a musical syntax.[7–9] A fruitful approach to investigate the processing of language and music are event-related brain potentials (ERPs). Processing of musical syntax was investigated in ERP studies that compared the responses evoked by harmonically inappropriate versus appropriate chords. A violation of musical regularities has been shown to be reflected in an early right anterior negativity (ERAN), that can be elicited preattentively.[10–12] Usually, the ERAN is followed by a late negativity, the N5 (maximal around 500–550 ms), which is supposed to reflect processes of musical integration.[10,11] Furthermore a late positive component has been described, which is thought to be evoked by detectional processes of a music-structural violation.[13]

With respect to effects of musical expertise on music perception in adults, a previous EEG study indicated that musicians react more sensitively to music-syntactic irregularities than nonmusicians.[14] The ERAN is larger in subjects with extensive formal musical training, suggesting that more specific representations of musical regularities lead to a stronger violation of musical expectancies.

Most previous ERP studies investigating music perception have been conducted with adults. Thus, the aim of this study is to specify in more detail how the neural correlates reflecting music processing develop during childhood. Previous studies investigating processing of musical structure in children with EEG[15] and fMRI[16] suggest that even five-year-old children show comparable processing mechanisms to adults. These studies indicate that children process chords according to their harmonic appropriateness, that is, according to a cognitive representation of the major–minor tonal system.

Research on language processing using ERPs found the early left anterior negativity (ELAN) and the late positivity (P600) as markers for syntactic processes.[17,18]

It is assumed that the ELAN reflects automatic initial structure building, which involves the identification of the incoming word's syntactic category upon which a local syntactic structure is built.[17–19] The syntactic structure not only provides the parsing system to build up structural hierarchies and relationships among various phrases, but it also allows more information to be kept in memory. The P600 is thought to reflect secondary parsing processes (under strategic control), including reanalysis and repair, depending on whether the sentence under consideration has a correct, but nonpreferred structure, or an incorrect structure.[18,19]

Behavioral and ERP studies show that, although children are able to comprehend sentences during early childhood, the mechanisms they use to achieve comprehension are different in early childhood and gradually develop through late childhood. The younger the age, the more children rely on contextual information. This is evidenced in a larger N400 in younger children[20] and in context-dependent monitoring times for function words in younger age groups.[21] Syntactic processes, by contrast, are not established as contextually independent (encapsulated), automatic processes before late childhood. This is supported by the context-dependent monitoring times for function words[21] and, moreover, by the ERP finding that only 8- to 10-year old nonimpaired children, but not developmentally delayed children, demonstrate a left lateralized negativity for function words.[22] These studies provide first indications regarding the developmental changes in the process of language comprehension and possibly in its neural basis. The age at which the ELAN can be observed depends upon the type of linguistic material used. For sentences with passive mode construction, also used in the present study, an ELAN appears at twelve to thirteen years of age; in younger children, however, a later, sustained negativity in response to a syntactic violation may be found.[23] For sentences with active mode construction, an ELAN can be found even in 32-month-old children.[24]

As stated previously, violations of the musical and linguistic syntax may be reflected in two ERP components: the ERAN, which is a response to a violation of the musical structure, and the ELAN, which reflects the detection of a linguistic syntax violation. These ERP components share several properties: both are negativities most prominent over frontal leads; they appear in a comparable time range with a maximum amplitude approximately 200 ms after stimulus onset; and they reflect very fast and automatic structure-building processes. They differ, however, in the laterality of the scalp distribution: processing of linguistic syntax is more lateralized to the left hemisphere, and processing of musical structure is more lateralized to the right hemisphere.

Most relevant to the present topic is that ERAN and ELAN are generated in comparable regions of the brain. Source localizations with MEG detected the sources of ERAN and ELAN in the inferior frontal cortex and suggest that neural correlates of music syntactic processing are, at least partly, located in the inferior frontolateral cortex (inferior pars opercularis)[25] and in the anterior superior temporal gyrus (STG).[7] These areas are also involved in the syntactic analysis of speech. Like the ERAN, the ELAN is thought to be generated in the inferior frontolateral cortex and in the anterior STG.[24] Comparable brain regions were found to be activated in experiments using fMRI when examining the neural bases of processing linguistic and musical syntax. Specifically areas in the inferior frontolateral parts of the brain were found to be activated in music as well as in language, but with a slightly different hemispheric weighting.[16,27–29] These studies indicate that both kinds of processes may share some common underlying neural substrate and that ERAN and ELAN are generated in comparable regions of the brain. Because of this and the close relationship between ERAN and ELAN, one might expect transfer effects between the two domains.

Taken together, this has led to three main topics of investigation for this study: First, we were interested in how violations of musical and linguistic syntax will be processed in different age groups. Second, we wanted to know whether there is a difference in ERAN and ELAN between children with and without musical training and with or without language impairments. Finally, we were interested in whether one could find a transfer due to additional musical training and if language impairment leads to a difference in the neural processing of musical structure. In order to address these issues, we conducted a within-subject comparison of ERAN and ELAN in children of different ages and took EEG measurements in experiments investigating either a violation of musical structure or of linguistic syntax.

MATERIALS AND METHODS

Children from two age groups participated in our studies. All were right-handed,[30] native speakers of German and had no known hearing or neurological deficits, attentional deficit disorders, or reading or learning disabilities.

The first group consisted of 28 children around 11 years of age. The 14 children who had received musical training were recruited from the Saint Thomas Boys Choir and the public music school (10 years, 2 months to 11 years, 6 months old, mean = 10 years, 10 months; 7 male, 7 female). The 14 children without musical training came from public schools in Leipzig (10 years, 4 months to 11 years, 11 months old, mean = 11 years, 2 months; 8 male, 6 female). There was no group difference in the results of the verbal part of the Wechsler Intelligence Scale for Children (HAWIK-III).[31] Even though the mean in the group of musically trained children (122.43, SE = 3.30) was slightly higher than in the group of the non–musically trained children (121.36, SE = 2.82), this difference was not significant ($t_{(26)} = 0.25$, $P = 0.807$). Moreover the two groups did not differ in duration with regard to their parents' education.

The second group consisted of 24 children around 5 years of age. Twelve of them had a specific language impairment (4 years, 2 months to 5 years, 11 months old, mean = 5 years, 3 months; 6 boys, 6 girls) and were evaluated at a kindergarten for special education. The 12 children (4 years, 3 months to 6 years, 3 months old, mean = 5 years, 4 months; 6 boys, 6 girls) without language impairment were recruited from public nursery schools in Leipzig. Children were only included if parents and teachers reported normal hearing and if they had at least 70 IQ points measured by the nonverbal part of the Kaufman Assessment Battery for Children.[32]

EEG data were recorded with Ag-AgCl electrodes from 25 scalp locations, referenced to the left mastoid, in an experiment that examined either a violation of musical syntax or of linguistic syntax. Processing of EEG data was conducted with EEGLab 4.512.[33] ERPs were evaluated statistically by computing two regions of interest (ROIs): left-frontal (F3, FC3) and right-frontal (F4, FC4). Because of a more posterior scalp distribution of the ERAN in the 5 year olds, different ROIs were used for that age group: left-anterior (F3, FC3, C3) and right-anterior (F4, FC4, C4).

Variances of ERPs were analyzed by general linear models (GLMs) for repeated measurements. These analyses are preliminary due to the small sample size and the need for additional subjects.

For the music experiment we employed two types of chord sequences that consisted of five chords. The first four chords were arranged according to the classical rules of harmony and established a musical context toward the end of the sequence. They were always the same: tonic, subdominant, supertonic and dominant. The fourth chord induced a strong expectancy for a tonic chord at the fifth position of a sequence,[3–5,34] as the dominant–tonic progression (i.e., the progression of chords built on the fifth and on the first scale tone) at the end of a chord sequence is a prominent marker for the end of a harmonic sequence and has been considered as a basic syntactic structure of major–minor tonal music.[35] The last chord was thus either a regular tonic, or an irregular supertonic, which in the case of the latter violated the expectancy of a regular musical structure.

The chord sequences were transposed to all twelve keys and were repeated eight times, leading to 96 sequences for each condition. Furthermore there were 18 sequences consisting of one chord played by another instrument. The task for the participants was to react to this different instrumental timbre with a keypress.

Stimuli of the language experiment were sentences in which four words had the same grammatical function, that is, each of the sentences consisted of an article, a noun, an auxilliary, and a past participle. Similar sentences had been used in several experiments before.[17,18,23] The correct sentences consisted only of these four words (e.g., Die Tante wurde geärgert. [The aunt was angered.]). A syntactic violation was introduced by sentences in which a preposition appeared after the auxiliary and was directly followed by a past participle (e.g., Die Mutter wurde im geärgert. [The mother was angered in.]) leading to a phrase structure error. Because the preposition indicates the beginning of a prepositional phrase, necessarily consisting of a preposition and a noun phrase, this sequence of words created a clear word category violation. Filler sentences that consisted of a whole prepositional phrase (i.e., preposition followed by a noun phrase) were introduced to ensure that participants would not necessarily anticipate a violation when encountering a preposition (e.g., Der Onkel wurde im Bett geärgert. [The uncle was angered in the bed.]). These sentences were not evaluated. In the experiment, we used 96 correct, 96 incorrect, and 48 filler sentences. Most of the sentences were read by a female speaker, but in 32 sentences one word was replaced by a word spoken by a male voice. The task here was to detect the change in the voice timbre.

RESULTS

The results from the music experiment with the 11 year olds are shown in FIGURE 1 (see figure caption for explanation). In the music experiment a considerable difference between the two conditions, maximal around 190 ms, became evident in musically trained as well as in non–musically trained children (see FIGS. 1A and 1B). The time window for statistical analyses was centered around that maximum (140 to 240 ms). The mean for the difference is -2.20 µV (SE = 0.57; mean of both frontal ROIs and for all subjects). Furthermore the difference in the amplitude of the ERAN is more pronounced in the group of the musically trained children (mean = -3.30 µV;

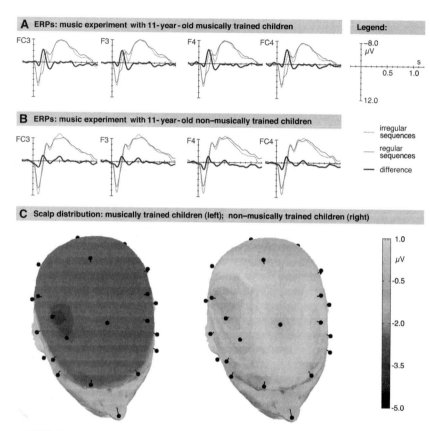

FIGURE 1. Results from the music experiment with 11-year-old children. (**A**) ERPs of the musically trained children. (**B**) ERPs of the non–musically trained children. (**A** and **B**) For the sake of simplicity, only the electrodes of the frontal ROIs are presented. The *solid gray lines* represent the responses to regular chords at the end of a sequence; *dotted gray lines* represent the responses to irregular chords. The *solid black lines* show the difference between the two conditions. (**C**) Scalp distributions of the differences between the two conditions (used time range is 180 to 200 ms). The results of musically trained children are shown on the *left*; the results of the non–musically trained children are shown on the *right*.

SE = 0.61) compared to the non–musically trained children (mean = −1.11 μV; SE = 0.90; see FIG. 1A vs. 1B). We observed a more right-lateralized scalp distribution of the ERAN (see FIG. 1C). The amplitude difference was more pronounced over the right hemisphere (mean = 2.41 mV; SE = 0.60) than over the left hemisphere (mean = 2.00 mV; SE = 0.57). Analyses that are preliminary due to the small sample size were carried out in a GLM with the within-subject-factors condition (supertonic vs. tonic) and hemisphere (left-frontal vs. right-frontal ROI) and the between-subjects-factor group (musically trained or nonmusically trained). The large amplitude difference between the two conditions is reflected in a significant main effect of condition ($F_{(1,26)} = 16.48$; $P < 0.001$). The interaction of condition and group is approaching significance ($F_{(1,26)} = 4.08$; $P = .054$).

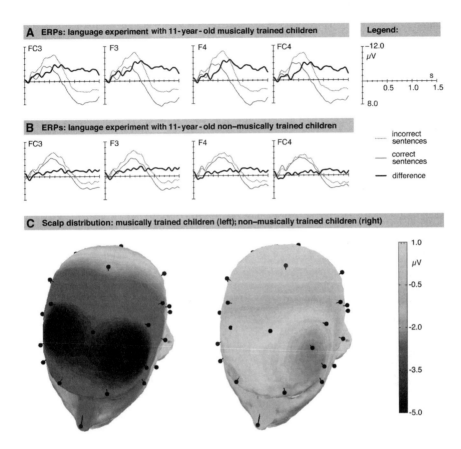

FIGURE 2. Results from the language experiment with 11-year-old children. (**A**) ERPs of the musically trained children. (**B**) ERPs of the non–musically trained children. (**A** and **B**) Responses to the syntactically correct sentences are plotted with *solid gray lines*; those to the syntactically incorrect sentences plotted with *dotted gray lines*. The *solid black lines* show the differences between the two conditions. (**C**) Scalp distributions of the differences between the two conditions (used time range is 500 to 1000 ms). The results of musically trained children are shown on the *left*; the results of the non–musically trained children are shown on the *right*.

The results from the language experiment for the 11 year olds are shown in FIGURE 2 (see figure caption for explanation). There is an amplitude difference between the two conditions (solid black lines) from around 200 ms to 350 ms, and from around 500 ms to 1500 ms. Time windows for statistical analyses were 220 ms to 320 ms for the earlier difference (which may be a precursor of the ELAN) and 500 ms to 1500 ms (for the later, sustained negativity in response to the syntactic violation). For the earlier time range the mean of the difference of the two conditions is -1.72 µV (SE = 0.65; mean of both frontal ROIs and for all subjects). Again the difference in the amplitude is more pronounced in the group of the musically trained children (mean = -2.24 µV; SE = 0.73) compared to the non–musically trained children (mean = −1.20 μV; SE = 1.08; see FIG. 2A vs. 2B). The amplitude difference is larger over the left hemisphere (mean = −2.17 μV; SE = 0.67) than over the right hemisphere (mean = −1.27 μV; SE = 0.68). Tested by means of a GLM with the within-subject-factors condition (syntactically correct vs. incorrect sentences) and hemisphere (left-frontal vs. right-frontal ROI) and the between-subjects-factor group (musically trained vs. non–musically trained children), this leads to a significant main effect of condition ($F_{(1,26)}$ = 6.94; P = 0.014) and an interaction of condtion and hemisphere ($F_{(1,26)}$ = 5.72; P = 0.024). Even though the two groups have distinct amplitude differences, this was not reflected in a significant interaction of condition by group. Future research will show these distinct amplitude differences, if such an interaction exists, when a larger group of subjects is used.

For the later time range (the negativity in response to the syntactic violation) the mean difference of the two conditions has a larger amplitude than that of the early negativity (−3.24 μV; SE = 0.69; mean of both frontal ROIs and for all subjects). The difference between the two groups is also larger than for the early negativity: it is −4.77 μV (SE = 0.92) for the musically trained children compared to −1.71 μV (SE = 0.87) in the children without musical training (see FIG. 2A and 2B). The scalp distribution of the difference is more bilateral (see FIG. 2C); it is only slightly larger over the left (−3.43 μV; SE = 0.70) compared to the right hemisphere (−3.05 μV; SE = 0.70). A GLM with the factors condition, hemisphere, and group revealed a significant main effect of condition ($F_{(1,26)}$ = 26.12, P < 0.001) and an interaction of condition by group ($F_{(1,26)}$ = 5.81; P = 0.023).

The results of the music experiment in the 5 year olds are shown in FIGURE 3 (see figure caption for explanation). An ERAN could be found in the children who are linguistically nonimpaired, whereas no such response can be seen for the language impaired children (see FIG. 3A vs. 3B). The amplitude of this response is maximal around 230 ms and therefore slightly later than in the older age group. Thus the time

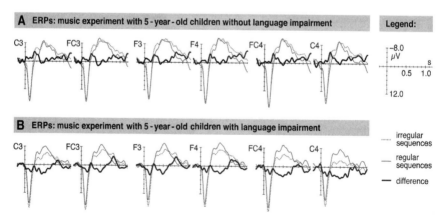

FIGURE 3. Results from the music experiment with 5-year-old children. (**A**) ERPs of the linguistically nonimpaired children. (**B**) ERPs of the language-impaired children. (**A** and **B**) The *solid gray lines* depict the responses to regular chords at the end of a sequence; the *dotted gray lines* depict responses to irregular chords. The *solid black lines* show the difference between the two conditions.

window was 180 to 280 ms after stimulus onset. In addition, the scalp distribution of the maximum amplitude for the difference of the responses on supertonics compared to tonics is weighted more posteriorly than in the older children. Therefore ROIs were computed of F3, FC3, C3 (left-anterior ROI), and F4, FC4, and C4 (right-anterior ROI). The mean amplitude of the difference between the two conditions is present in the linguistically nonimpaired (−3.10 µV, SE = 1.17) but not in the language-impaired children (1.14 µV, SE = 1.26). In the linguistically nonimpaired children the response is more prominent over the right-anterior ROI (mean = −3.50 µV; SE = 1.29) than over the left-anterior (mean = −2.69 µV; SE = 1.31). In the language-impaired children there is almost no response—none in the left-anterior (mean = 0.97 µV; SE = 1.19) and none in the right-anterior ROI (mean = 1.30 µV; SE = 1.48). In a GLM with the within-subject-factors hemisphere (left-anterior vs. right-anterior ROI) and condition (supertonic vs. tonic) and the between-subjects-factor group (language impaired vs. linguistically nonimpaired children), a condition-by-group-interaction is found ($F_{(1,22)} = 6.02$; $P = .023$), but no other main effects or interactions are significant.

DISCUSSION

In our ERP study we investigated processes during auditory sentence comprehension and music perception in children of two different age groups. We found that a violation of harmonic expectancies and linguistic syntax led either to an ERAN or to a later, sustained negativity in response to a syntactic violation. Furthermore we found differences between children with and without musical training and in linguistically nonimpaired compared to language-impaired children when carrying out these processes.

The results indicate that musical training facilitates the processing of musical structure. This is in accordance with earlier studies evidencing a comparable effect for adults.[14] Interestingly this difference was found as early as 11 years when children have not played an instrument for longer than 4 or 5 years. Despite such a relatively short period of musical training, these children may have acquired specific representations of music-syntactic regularities (e.g., more implicit and explicit knowledge about the theory of harmony underlying Western tonal music, and more specific representations of harmonic relatedness).[36]

A characteristic of the ELAN, of which a precursor was found in the language experiment, is a larger amplitude difference over the left hemisphere. This difference in ELAN amplitude between the group of the musically trained and the non–musically trained children is approaching significance, and is expected to become significant with a larger group of participants. Moreover, a later, sustained negativity was found in both groups with an enhanced amplitude in the group with musical training. This indicates a positive transfer from the music to the language domain. This finding was expected, since the neural resources underlying the processing of musical and linguistic syntax overlap to some extent, as outlined in the introduction. The fact that the negativity in response to the syntactic violation is distributed bilaterally is in accordance with an earlier study, which showed the same pattern of results: a bilateral anterior negativity between 600 and 1500 milliseconds.[23]

There are not many studies that investigated transfer effects from music to other cognitive domains. One of these studies investigated influences of musical training on the processing of prosody in language and of melody in music.[37] This study reported an increased N400 in nonmusicians, indicating that they rely more on semantic information when processing prosody. In contrast, there was an early negativity found for the musicians that might reflect the detection of incongruities in melodic and prosodic patterns. Other studies tested potential transfer effects in behavioral experiments (for an overview, see Schellenberg in this volume). Furthermore, there is a substantial body of evidence that suggests that sophisticated processing of musical elements of speech (e.g., prosody) is of considerable importance for the acquisition of language.[38–41]

Finally, no ERAN could be found for the 5 year olds with specific language impairment (SLI), whereas an ERAN could be seen in linguistically nonimpaired children. The finding that an ERAN is present even at that age is in accordance with an earlier study.[15] The difference between the studies is that the chord sequences used in the present study represent a more subtle violation than the Neapolitan chords used in that study. Both studies show that children at that age are able to process musical structure. This is particularly noteworthy because it was previously believed that the acquisition of the rules of "Western" music (which may be considered as a process of culture-specific attunement) is anything but rapid and that sensitivity to culture-specific details of tonal and harmonic structure seems to emerge between 6 and 7 years of age.[42,43] The difference between this assumption and our results might be due to two factors: First, it may be that chord sequences represent clearer irregularities than the melodies used in other studies. Second, the EEG might be a more sensitive method to investigate what processes children are capable of.

The finding that the amplitude of the ERAN is diminished in language-impaired children indicates that they have difficulties when processing musical syntax. There are studies reporting grammatical deficits as characteristic of specific language impairment (SLI).[44–47] Specifically, it seems that children with SLI have difficulties with the comprehension of specific types of syntactic relationships and with the processing of configurational aspects of grammar.[48] Our results fit well with the assumption of grammatical deficits, especially if one bears in mind that the neural correlates of syntax processing in language and music are (at least to some extent) shared by the two domains.

We observed indicators for an intricate relationship of syntax processing in language and music. Even though the analyses of our data are still preliminary, the data suggest that children can profit from musical training because of a more efficient processing of musical structure and because of its impact on the processing of linguistic syntax. This relationship might be especially important as well in therapy for language-impaired children.

ACKNOWLEDGMENTS

The work of S.J. was supported by the German Research Foundation (KO 2266/2-1/2, awarded to S.K.). We thank our participants and their parents. For their help in the acquisition of participants, we thank G.C. Biller, the leader of the St. Thomas Boys Choir; the staff of the Leipzig public music school; and the staff of the Leipzig kindergarten for special education. For their help in revising the manuscript, we thank Julia Grieser and Nikolaus Steinbeis.

[Competing interests: The authors declare that they have no competing financial interests.]

REFERENCES

1. ZATORRE, R.J. & I. PERETZ. 2003. The Cognitive Neuroscience of Music. Oxford University Press. Oxford.
2. PATEL, A.D. 2003. Language, music, syntax and the brain. Nat. Neurosci. **6:** 674–681.
3. BHARUCHA, J. & C. KRUMHANSL. 1983. The representation of harmonic structure in music: hierarchies of stability as a function of context. Cognition **13:** 63–102.
4. BHARUCHA, J. & K. STOECKIG. 1986. Reaction time and musical expectancy: priming of chords. J. Exp. Psychol. Hum. Percept. Perform. **12:** 403–410.
5. BHARUCHA, J. & K. STOECKIG. 1987. Priming of chords: Spreading activation or overlapping frequency spectra? Percept. & Psychophys. **41:** 519–524.
6. KRUMHANSL, C. & E. KESSLER. 1982. Tracing the dynamic changes in perceived tonal organization in a spatial representation of musical keys. Psychol. Rev. **89:** 334–368.
7. KOELSCH, S. & A.D. FRIEDERICI. 2003. Toward the neural basis of processing structure in music: comparative results of different neurophysiological investigation methods. Ann. N. Y. Acad. Sci. **999:** 15–28.
8. SLOBODA, J. 1985. The Musical Mind: The Cognitive Psychology of Music. Oxford University Press. New York, NY.
9. TILLMANN, B., J. BHARUCHA & E. BIGAND. 2000. Implicit learning of tonality: a self-organized approach. Psychol. Rev. **107:** 885–913.
10. KOELSCH, S., T. GUNTER, A.D. FRIEDERICI, *et al.* 2000. Brain indices of music processing: "non-musicians" are musical. J. Cogn. Neurosci. **12:** 520–541.
11. KOELSCH, S., E. SCHRÖGER & T.C. GUNTER. 2002. Music matters: preattentive musicality of the human brain. Psychophysiology **39:** 1–11.
12. KOELSCH, S., T. GUNTER, *et al.* 2001. Differentiating ERAN and MMN: an ERP-study. NeuroReport **12:** 1385–1389.
13. BESSON, M., F. FAITA & J. REQUIN. 1994. Brain waves associated with musical incongruities differ for musicians and non-musicians. Neurosci. Lett. **168:** 101–105.
14. KOELSCH, S., B. SCHMIDT & J. KANSOK. 2002. Influences of musical expertise on the ERAN: an ERP-study. Psychophysiology **39:** 657–663.
15. KOELSCH, S., T. GROSSMANN, *et al.* 2003. Children processing music: electric brain responses reveal musical competence and gender differences. J. Cogn. Neurosci. **15:** 683–693.
16. KOELSCH, S., T. FRITZ, *et al.* 2005. Adults and children processing music: an fMRI study. Neuroimage **25:** 1068–1076.
17. FRIEDERICI, A.D., E. PFEIFER & A. HAHNE. 1993. Event-related brain potentials during natural speech processing: effects of semantic, morphological, and syntactic violations. Cogn. Brain Res. **1:** 183–192.
18. HAHNE, A. & A.D. FRIEDERICI. 1999. Electrophysiological evidence for two steps in syntactic analysis: early automatic and late controlled processes. J. Cogn. Neurosci. **11:** 194–205.
19. FRIEDERICI, A.D. 2002. Towards a neural basis of auditory sentence processing. Trends Cogn. Sci. **6:** 78–84.
20. HOLCOMB, P.J., S.A. COFFEY & H.J. NEVILLE. 1992. Visual and auditory sentence processing: a developmental analysis using event-related brain potentials. Dev. Neuropsychol. **8:** 203–241.
21. FRIEDERICI, A.D. 1983. Children's sensitivity to function words during sentence comprehension. Linguistics **21:** 717–739.
22. NEVILLE, H.J., S.A. COFFEY, P.J. HOLCOMB & P. TALLAL. 1993. The neurobiology of sensory and language processing in language-impaired children. J. Cogn. Neurosci. **5:** 235–253.

23. HAHNE, A., K. ECKSTEIN & A.D. FRIEDERICI. 2004. Brain signatures of syntactic and semantic processes during children's language development. J. Cogn. Neurosci. **15:** 1302–1318.
24. OBERECKER, R., M. FRIEDRICH & A.D. FRIEDERICI. 2005. Neural correlates of syntactic processing in two-years-olds. J. Cogn. Neurosci. In press.
25. MAESS, B., S. KOELSCH, S., et al. 2001. Musical syntax is processed in the area of Broca: an MEG-study. Nat. Neurosci. **4:** 540–545.
26. FRIEDERICI, A.D., Y. WANG, et al. 2000. Localisation of early syntactic processes in frontal and temporal cortical areas: an MEG study. Hum. Brain Mapp. **11:** 1–11.
27. KOELSCH, S., T. GUNTER, et al. 2002. Bach speaks: a cortical "language-network" serves the processing of music. Neuroimage **17:** 956–966.
28. HEIM, S., B. OPITZ & A.D. FRIEDERICI. 2003. Distributed cortical networks for syntax processing: Broca's area as the common denominator. Brain Lang. **85:** 402–408.
29. FRIEDERICI, A.D., S-A. RÜSCHEMEYER, A. HAHNE, et al. 2003. The role of left inferior frontal and superior temporal cortex in sentence comprehension: localizing syntactic and semantic processes. Cereb. Cortex **13:** 170-177.
30. OLDFIELD, R.C. 1971. The assessment and analysis of handedness: the Edinburgh inventory. Neuropsychologia **9:** 97–113,
31. TEWES, U., P. ROSSMANN & U. SCHALLBERGER. 2000. HAWIK-III: Hamburg-Wechsler-Intelligenztest für Kinder III. Hogrefe. Göttingen, Germany.
32. KAUFMAN, A.S., N.L. KAUFMAN, et al. 2001. Kaufman Assessment Battery for Children (German version). Hogrefe. Göttingen, Germany.
33. DELORME, A. & S. MAKEIG. 2004. EEGLAB: an open source toolbox for analysis of single-trial EEG dynamics. J. Neurosci. Methods **123:** 9–21,
34. BIGAND, E., F. MADURALL, et al. 1999. Effects of global structure and temporal organization on chord processing. J. Exp. Psychol. Hum. Percept. Perform. **25:** 184–197.
35. RIEMANN, H. 1971. Musikalische Syntaxis: Grundriss einer Harmonischen Satzbildungslehre. (Reprint of the 1877 edition.) Saendig. Niederwalluf, Germany.
36. BHARUCHA, J.J. 1984. Anchoring effects in music: the resolution of dissonance. Cogn. Psychol. **16:** 485–518.
37. MAGNE, C., D. SCHÖN & M. BESSON. 2000. Prosodic and melodic processing in adults and children: behavioral and electrophysiologic approaches. Ann. N. Y. Acad. Sci. **999:** 461–476.
38. KRUMHANSL, C.L. & P.W. JUSCZYK. 1990. Infants' perception of phrase structure in music. Psychol. Sci. **1:** 70–73.
39. JUSCZYK, P.W., K. HIRSH-PASEK, et al. 1992. Perception of acoustic correlates of major phrasal units by young infants. Cogn. Psychol. **24:** 252–293.
40. JUSCZYK, P.W. & C.L. KRUMHANSL. 1993. Pitch and rhythmic patterns affecting infants' sensitivity to musical phrase structure. J. Exp. Psychol. Hum. Percept. Perform. **19:** 627–640.
41. FERNALD, A. 1989. Intonation and communicative intent in mothers' speech to infants: is the melody the message? Child Dev. **60:** 1497–1510.
42. KRUMHANSL, C.L. & F.C. KEIL. 1982. Acquisition of the hierarchy of tonal functions in music. Memory Cogn. **10:** 243–251.
43. TRAINOR, L.J. & S.E. TREHUB. 1994. Key membership and implied harmony in Western tonal music: developmental perspectives. Percept. & Psychophys. **56:** 125–132.
44. LEONARD, L.B. & J.A. EYER. 1996. Deficits of grammatical morphology in children with specific language impairment and their implications for notions of bootstrapping. *In* Signal to Syntax. J.L. Morgan & K. Demuth, Eds.: 233–248. Lawrence Erlbaum. Mahwah, NJ.
45. JOANISSE, M.F. & M.S. SEIDENBERG. 1998. Specific language impairment: a deficit in grammar or processing? Trends Cogn. Sci. **2:** 240–247.
46. TOMBLIN, J.B. & J. PANDICH. 1999. Lessons from children with specific language impairment. Trends Cogn. Sci. **3:** 283–285.
47. VAN DER LELY, H.K.J. 1996. Specifically language impaired and normally developing children: verbal passive vs. adjectival passive sentence interpretation. Lingua **98:** 243–272.
48. JOANISSE, M.F. & M.S. SEIDENBERG. 2003. Phonology and syntax in specific language impairment: evidence from a connectionist model. Brain Lang. **86:** 40–56.

Temporal Entrainment of Cognitive Functions

Musical Mnemonics Induce Brain Plasticity and Oscillatory Synchrony in Neural Networks Underlying Memory

MICHAEL H. THAUT, DAVID A. PETERSON, AND GERALD C. McINTOSH

Center for Biomedical Research in Music,
Molecular, Cellular, and Integrative Neuroscience Programs,
Colorado State University, Fort Collins, Colorado 80523, USA

ABSTRACT: In a series of experiments, we have begun to investigate the effect of music as a mnemonic device on learning and memory and the underlying plasticity of oscillatory neural networks. We used verbal learning and memory tests (standardized word lists, AVLT) in conjunction with electroencephalographic analysis to determine differences between verbal learning in either a spoken or musical (verbal materials as song lyrics) modality. In healthy adults, learning in both the spoken and music condition was associated with significant increases in oscillatory synchrony across all frequency bands. A significant difference between the spoken and music condition emerged in the cortical topography of the learning-related synchronization. When using EEG measures as predictors during learning for subsequent successful memory recall, significantly increased coherence (phase-locked synchronization) within and between oscillatory brain networks emerged for music in alpha and gamma bands. In a similar study with multiple sclerosis patients, superior learning and memory was shown in the music condition when controlled for word order recall, and subjects were instructed to sing back the word lists. Also, the music condition was associated with a significant power increase in the low-alpha band in bilateral frontal networks, indicating increased neuronal synchronization. Musical learning may access compensatory pathways for memory functions during compromised PFC functions associated with learning and recall. Music learning may also confer a neurophysiological advantage through the stronger synchronization of the neuronal cell assemblies underlying verbal learning and memory. Collectively our data provide evidence that melodic–rhythmic templates as temporal structures in music may drive internal rhythm formation in recurrent cortical networks involved in learning and memory.

KEYWORDS: multiple sclerosis; music; memory; EEG; brain plasticity; temporal entrainment

Address for correspondence: Michael Thaut, Center for Biomedical Research in Music, Molecular, Cellular, and Integrative Neuroscience Programs, Colorado State University, Fort Collins, CO 80523. Voice: 970-491-5529; fax: 970-491-7541.
Michael.Thaut@ColoState.edu

Ann. N.Y. Acad. Sci. 1060: 243–254 (2005). © 2005 New York Academy of Sciences.
doi: 10.1196/annals.1360.017

INTRODUCTION

The effect of temporal entrainment on motor function through rhythmic auditory stimuli has been well established in motor learning and therapeutic rehabilitation.[1] The physiological attractor function of auditory rhythm on the human motor system can facilitate the regulation of spatiotemporal and force parameters of movement by providing a stable anticipatory temporal structure onto which movement can be mapped and whose time cues drive movement into well-regulated motor patterns. The role of timing in human cognition is much less understood. There is a small body of research, however, that points to the importance for temporal organization in facilitating memory and learning.[2,3] Temporally structured learning templates (as inherent in most music) may enhance learning and may involve different plastic processes in the brain. Rhythm and music have been associated with providing temporal structure for information encoding.[4] Many clinical reports have emphasized the relative "survival" of musical memories as part of the cognitive functions in neurologic memory disorders.[5,6] Yet there is also considerable evidence that music can enhance memory for nonmusical material.[4,7–9] Music rehearsal has been shown to be more effective than verbal rehearsal in learning nonmusical materials with learning disabled and developmentally disabled students.[10–12]

Structured music listening has been shown to enhance a broad range of cognitive functions in autistic children.[13] Several researchers[7,14,15] have—based on Gestalt perception and learning principles—proposed as mechanisms for enhanced memory formation that music provides a highly effective mnemonic for learning by incorporating a temporal structure and redundancy that chunks information into more manageable units.[12,16] The metrical–rhythmic organization of text to enhance memory has been mentioned in the research literature as early as 1894.[17] Temporal structured learning research has demonstrated the significance for memory of how information is temporally grouped. The structure and redundancy in a musical template, for example, a song, can chunk material into more manageable units and increase relational processing that enhances interitem associations.[16] However, when conceptualizing the effect of rhythm and music on cognitive function, important differences between motor functions have to be considered. Behaviorally, time cues in auditory rhythm can be metrically understood in their effect on motor function and can be measured in phase and periodicity relationships. The temporal effect of rhythm on cognition has to be expanded to an understanding of rhythm not just in its metrical interpretation but as a structure for temporal distribution and patterned organization of elements in learning. In that broader, less chronometrically linear, sense one can conceptualize an effect of temporal structure—in a learning template—on intrinsic neural time coding of learning-related cell ensembles in the brain.

In a series of experiments we have begun to systematically investigate the effect of music as a mnemonic device on learning and memory and the underlying neural dynamics, that is, changes in the brain plasticity of oscillatory neural networks, during learning and recall. It has been shown that sensory cortical areas can be modified by musical training.[18,19] In the experiments we are reporting here, the overarching research question is whether external timing embedded in learning stimuli, via music, can modulate oscillatory synchrony in learning-related neural networks, that is, induce brain plasticity.

EVIDENCE FROM A MOTOR STUDY

In previous research we have shown that rhythmic motor synchronization involves changes in activation patterns of prefrontal and parietal networks, depending on the nature of the temporal tracking task.[20] During isochronous synchronization, prefrontal areas are activated nonsignificantly, whereas tracking of tempo-modulated rhythmic patterns shows a gradually expanding involvement of medio-, ventro-, and dorsolateral areas in the prefrontal cortex (PFC), contingent upon the modulation amplitude of the tempo changes and degree of conscious awareness of the tempo fluctuations.

EVIDENCE FROM A WORKING MEMORY STUDY

In a recent experiment we studied plasticity in electroencephalographic patterns associated with remembering the temporal structure of rhythmic patterns in an N-back working memory task.[21] During 2-s delays in working memory, spectral power analysis of electroencephalographic patterns showed increased high-frequency synchronizations in the right temporal cortex, with additional low-frequency synchronizations in the bilateral temporal and frontal cortices during 5-s delays. During the 10-s delays high-frequency synchronizations disappeared, and low-frequency desynchronizations emerged in bilateral temporal and frontal cortices. The data provided evidence that spatial patterns of neural synchronizations changed with duration of working memory, possibly indicating a progressive shift from phonological to semantic encoding.

MUSICAL MNEMONICS AND DIFFERENTIAL ELECTROENCEPHALOGRAPHIC TOPOGRAPHY

In the experiments we will report here, we used Rey's Auditory Verbal Learning Test (AVLT),[22] a supraspan verbal learning and memory test used in neuropsychological evaluations that uses repeated study/test learning. The single standard list of 15 words was presented either in spoken form or in a song. A number of learning trials was followed by an immediate recall (memory) trial and a second memory trial after a 20-min delay. In the first study,[23] twenty college-student subjects without musical background were each randomly assigned to either condition. The behavioral data showed no difference during the learning and recall trials. Both groups improved significantly over baseline, and the overall pattern of performance agreed with normative data. Performance returned to baseline levels during a distractor word list, indicating that the learning was not merely a time-dependent effect. It is important to note, however, that in the song condition subjects were asked to recall in spoken form the musically presented word list, creating modality incongruence between presentation and recall. Transfer-appropriate processing theory would actually predict in such a case that the incongruent learning group performs worse than the congruent group.

Spectral power of an electroencephalogram (EEG) was measured during the first learning trial a word was remembered in comparison to the preceding EEG when the

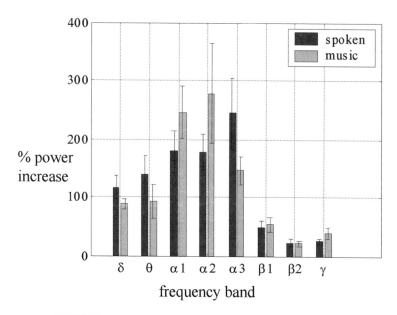

FIGURE 1A. Learning involves broadband synchronization.

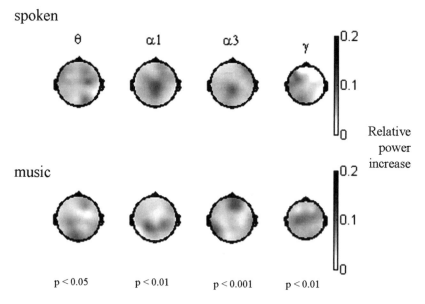

FIGURE 1B. Topography of network synchronization differs between verbal and musical learning.

word was not yet remembered. Learning was associated with significant increases in oscillatory synchrony, as evidenced by increased spectral power, across all frequency bands (after Bonferroni correction) (FIG. 1A). The largest increases in relative power were in low-frequency bands. Delta and theta power doubled, and power in the alpha subbands roughly tripled during the transition from unsuccessful to successful learning. No significant differences between the conditions were found. A significant difference, however, between the spoken and sung condition emerged in scalp distribution of the learning-related synchronization (LRS) in the theta, low alpha, high alpha, and gamma bands, using MANOVA analysis (FIG. 1B). During the spoken learning, the theta LRS was strongest in right central and right posterior regions. Learning with a musical template showed maximal theta LRS in right prefrontal and left occipital regions. Low-alpha in the midline centroposterior regions was slightly more bilateralized for the music condition. In music, high-alpha LRS was strongest in central prefrontal and left lateroposterior regions, while the spoken condition showed strongest activations in the centroposterior and right frontal regions. The highest gamma LRS emerged in the left hemisphere for the spoken condition and in the central bilateral regions for the music condition.

Although the two experimental conditions showed no behavioral differences in learning, the two learning modalities accessed different oscillatory brain networks. Previous studies have suggested that music-related processing in the brain involves cortical and subcortical networks that are more widely distributed than those used for nonmusical verbal function.[24,25] Thus, the topographically diverse cortical recruitment during music-related processing may produce the very different LRS topography seen in our study. Furthermore, there is neuropsychological evidence that the relative emphasis on bilateral activation during music processing may provide a compensatory role that could take advantage of the increasingly recognized lateral specialization of the PFC.[26] Differential patterns of frontal neural synchrony between spoken and sung conditions may reflect more interitem processing in music, contributing to a more organizational/chunking role of the learning items in the musical condition. Although the left frontal gyrus is associated with cue-based retrieval of words from long-term memory, patients with left frontal damage recruit the right inferior frontal gyrus for the same task.[27] Thus, the increased bilaterality in brain networks during musical template learning may also play a role in accessing compensatory pathways for memory functions during compromised PFC functions associated with learning and recall.

MUSIC LEARNING INDUCES INCREASED
BRAIN NETWORK COHERENCE

In a second experiment following the same design and task presented above, we analyzed electroencephalographic spectral power during the first successful learning trial for words related to their subsequent successful retrieval during the two memory trials.[28] These data would allow us to assess if neuronal synchronization patterns during learning could predict correct recall during subsequent memory tests. Again, behavioral results were not different between the two conditions. We studied network synchronization by measuring change in phase-locked oscillations among

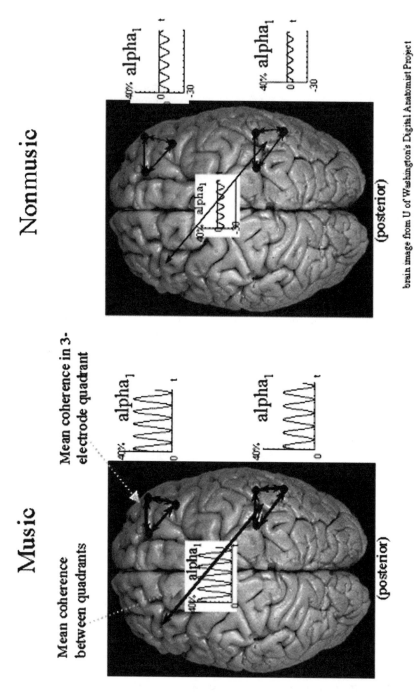

FIGURE 2. Learning related coherence and short-term recall.

three electrodes within four quadrants (left anterior and posterior/right anterior and posterior). Coherence C between each pair of electrodes i and j was defined as:

$$C_{ij} = \frac{|P_{ij}|^2}{P_{ii}P_{jj}},$$

where P_{ij} = the cross-spectral density between i and j; and P_{ii} and P_{jj} = the power spectral densities of electrodes i and j, respectively. Learning-related network coherence during short-term recall (1-min delay) showed significantly increased low alpha coherence in the music condition but not in the spoken condition (FIG. 2).

Local network coherence was evident in the right anterior and right posterior networks and globally between right posterior and left anterior networks. Long-term recall in music (memory trial after 20-min delay) showed additionally significant local network coherence in the gamma band in right anterior and posterior networks, while the global network coherence in the low-alpha band shifted to bilateral posterior regions. Network coherence results were consistent for music and nonmusic across both memory trials, in that only music facilitation of verbal learning was associated with increased neuronal synchronization. Thus, increased coherence in alpha and gamma bands in specific lateral and global brain networks predicted memory retention for words in the music condition but not in the spoken condition.

The results are important in several aspects related to the neurophysiology underlying learning and memory. Neuronal synchronization processes are involved in assembling coupled neuronal networks associated with effective learning and training. Brain plasticity associated with behavioral learning is thus critically dependent on precise temporal coding of neuronal responses. Long-term potentiation (LTP) is posited as a cellular mechanism for neuronal plasticity, and LTP in hippocampus and neocortex is influenced by phase synchrony of neuronal network oscillations.[29,30] It is therefore highly significant that music as a learning stimulus that is temporally overstructured on multiple levels of perception and stimulus properties—ranging from the time structure of its acoustical properties to the temporality in the melodic–rhythmic elements to its compositional macro-time structures of the phrasing and grouping—induces a statistically higher degree of temporal coherence in coupled neuronal cell assemblies than a spoken stimulus without explicit or implicit regulated temporal structure. Since the behavioral results are, however, similar regardless of differences in coherence of brain network oscillations, one may ask if the early induction of higher temporal synchrony in learning-related networks during music facilitation plays a role in memory formation in music that seems more resilient to neurological deficits (e.g., dementia or Alzheimer disease[31]) than nonmusical memory.

MUSICAL MNEMONICS FACILITATE VERBAL LEARNING IN PATIENTS WITH MULTIPLE SCLEROSIS

The final study reported here involved 40 patients with multiple sclerosis, randomly assigned to either a sung or spoken presentation of the AVLT, using the same experimental design as discussed above. Subjects, however, were instructed differently from the previous studies. In the music condition they were asked to sing back

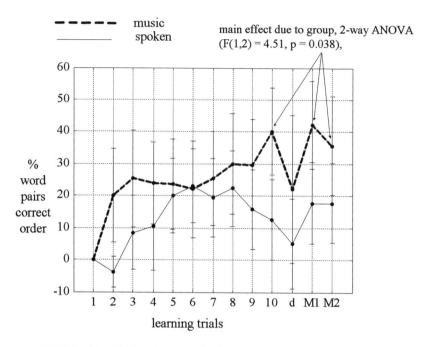

FIGURE 3A. Music enhances pair-wise word order learning and memory.

as many words as they recalled to create modality congruence between learning stimulus and recall. In the spoken condition they were additionally instructed to recall items in the order they were presented on the word list. Recall order in the sung condition was not explicitly asked for but assumed to be implicit in the stimulus structure. The behavioral analysis measured results for word chunks recalled in proper sequence, assessing chunk lengths of 2, 3, 4, 5, 6, and 7 words. In the analysis of pair-wise word order, learning a consistent trend emerged for better recall in music than spoken learning, reaching statistical significance at the last learning trial and the two subsequent memory trials ($F(1,2) = 4.51$, $P = .038$, 2-way ANOVA) (FIG. 3A).

In the longer word sequences the learning trials did not reach statistical significance; however, the percent change in correct recall from the last learning trial to the first memory trial was significant for all chunk lengths, most pronounced for the four- and five-word chunks ($P < .001$, two-tailed t test). This difference in better memory performance for the music condition was associated with a significant power increase in the low-alpha band over the spoken condition in bilateral frontal networks, indicating increased neuronal synchronization (FIG. 3B). Music learners also exhibited less posterior left beta-band synchronization than nonmusic learners. The finding of an increase in neuronal synchronization in cortical networks is remarkable in people with multiple sclerosis, since the demyelination process of the disease will affect and interrupt network dynamics of neuronal cell assemblies. The finding that

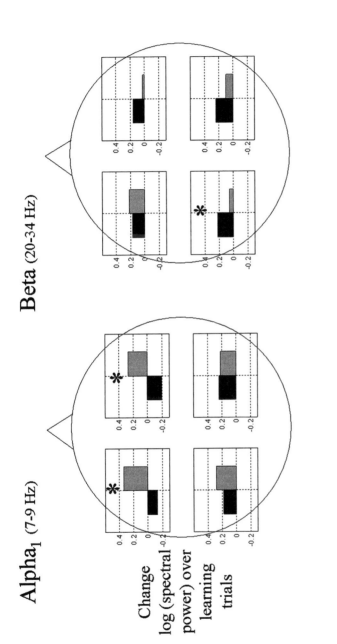

FIGURE 3B. Music enhances frontal electroencephalographic synchrony during learning in multiple sclerosis

music can improve order memory is also significant given the increasingly recognized cognitive deficits in multiple sclerosis.[32–34]

CONCLUSIONS

The presented studies investigate the neural dynamics of verbal memory using spoken and musical (singing) modalities. The data suggest that music, via melodic–rhythmic structures, enhances memory performance by mapping temporal order on learning information. Through the temporal order of music, a structure of units or chunks of learning items emerges that has two advantages for effective learning: (1) the units are separable and manageable in length and (2) the items within the units are interrelated through their temporal position in the melodic–rhythmic pattern. The temporally enhanced learning template leads to better memory performance, which specifically emerges in our current set of studies when temporal recall order and congruence between learning and recall are available. The time coding of learning materials in music also induces a neurophysiological advantage for the stronger formation of robust cell networks underlying increased memory performance: music increases the temporal sharpening of neuronal responses in localized brain networks, leading to tightly phase-locked synchronized neuronal cell assemblies. The cortical topography of these neuronal "groups" is also distributed differently in music than in spoken verbal learning, leading to music-induced spatial brain plasticity.

In summary, the following conclusions may be offered. Temporal synchrony is a prerequisite for efficient network formation in memory. A musical template (song) for verbal learning induces cortical plasticity characterized by higher synchrony in learning-related networks. Higher synchrony in learning related networks may produce more stable neural traces for long-term memory. Increased oscillatory synchrony in learning networks may be the neurophysiological basis for persistent memory for music despite severe memory loss, and improved access to verbal knowledge via music in neurologic conditions such as dementia and Alzheimer disease. Our data provide the first evidence that external rhythm as a temporal structure in music may drive internal rhythm formation in recurrent cortical networks for motor control and cognition.

[Competing interests: The authors declare that they have no competing financial interests.]

REFERENCES

1. HUMMELSHEIM, H. 1999. Rationales for improving motor function. Curr. Opin. Neurol. **12:** 697–701.
2. HITCH, G.J. *et al.* 1996. Temporal grouping effects in immediate recall: a working memory analysis. Quart. J. Exp. Psychol. A Hum. Exp. Psychol. **49:** 116–139.
3. KILGOUR, A.R., L.S. JAKOBSON & L.L. CUDDY. 2000. Music training and rate of presentation as mediators of text and song recall. Mem. Cognit. **28:** 700–710.
4. JAKOBSON, L.S., L.L. CUDDY & A.R. KILGOUR. 2003. Time tagging: a key to musicians' superior memory. Music Percept. **20:** 307–313.
5. HASLAM, C. & M. COOK. 2002. Striking a chord with amnesic patients: evidence that song facilitates memory. Neurocase **8:** 453–465.

6. BAUR, B. *et al.* 2000. Music memory provides access to verbal knowledge in a patient with global amnesia. Neurocase **6:** 415–421.
7. WALLACE, W.T. 1994. Memory for music: effect of melody on recall of text. J. Exp. Psychol. Learn. Mem. Cogn. **20:** 1471–1485.
8. RAINEY, D.W. & J.D. LARSEN. 2002. The effect of familiar melodies on initial learning and long-term memory for unconnected text. Music Percept. **20:** 173–186.
9. HO, Y.C., M.C. CHEUNG & A.S. CHAN. 2003. Music training improves verbal but not visual memory: cross-sectional and longitudinal explorations in children. Neuropsychology **17:** 439–450.
10. CLAUSSEN, D.W. & M.H. THAUT. 1997. Music as a mnemonic device for children with learning disabilities. Can. J. Music Ther. **5:** 55–66.
11. WOLFE, D.E. & C. HOM. 1993. Use of melodies as structural prompts for learning and retention of sequential verbal information by preschool students. J. Music Ther. **30:** 100–118.
12. GFELLER, K.E. 1983. Musical mnemonics as an aid to retention with normal and learning-disabled students. J. Music Ther. **20:** 179–189.
13. BETTISON, S. 1996. The long-term effects of auditory training on children with autism. J. Autism Dev. Disord. **26:** 361–374.
14. DEUTSCH, D. 1982. Organizational processes in music. *In* Music, Mind, and Brain. M. Clynes, Ed.: 119–131. Plenum Press. New York.
15. JANATA, P., B. TILLMANN & J.J. BHARUCHA. 2002. Listening to polyphonic music recruits domain-general attention and working memory circuits. Cogn. Affect. Behav. Neurosci. **2:** 121–140.
16. DEUTSCH, D. 1999. The Psychology of Music. Academic Press. San Diego. London.
17. MUELLER, G. & F. SCHUMANN. 1894. Experimentelle Beitraege zur Untersuchung des Gedaechtnisses. Z. Psychol. **6:** 81–190.
18. RAUSCHECKER, J.P. 1999. Auditory cortical plasticity: a comparison with other sensory systems. Trends Neurosci. **22:** 74–80.
19. PANTEV, C. *et al.* 2003. Music and learning-induced cortical plasticity. Ann. N. Y. Acad. Sci. **999:** 438–450.
20. STEPHAN, K.M. *et al.* 2002. Conscious and subconscious sensorimotor synchronization: prefrontal cortex and the influence of awareness. Neuroimage **15:** 345–352.
21. PETERSON, D.A. & M.H. THAUT. 2002. Delay modulates spectral correlates in the human EEG of nonverbal auditory working memory. Neurosci. Lett. **328:** 17–20.
22. LEZAK, M.D. 1995. Neuropsychological Assessment. Oxford University Press. New York.
23. PETERSON, D.A. & M.H. THAUT. 2003. Plasticity of alpha and theta synchronization during verbal learning with a musical template. Proc. Soc. Neurosci. **194:** 21.
24. PARSONS, L.M., D.A. HODGES & P.T. FOX. 1998. Neural basis of the comprehension of musical harmony, melody, and rhythm. Proc. Soc. Neurosci. **695:** 10.
25. ZATORRE, R.J. *et al.* 1996. Hearing in the mind's ear: a PET investigation of musical imagery and perception. J. Cogn. Neurosci. **8:** 29–46.
26. CABEZA, R., J.K. LOCANTORE & N.D. ANDERSON. 2003. Lateralization of prefrontal activity during episodic memory retrieval: evidence for the production–monitoring hypothesis. J. Cogn. Neurosci. **15:** 249–259.
27. BLASI, V. *et al.* 2002. Word retrieval learning modulates right frontal cortex in patients with left frontal damage. Neuron **36:** 159–170.
28. PETERSON, D.A., G.C. MCINTOSH & M.H. THAUT. 2004. Differential predictions in spectral EEG-plasticity associated with a verbal versus music-facilitated learning strategy. Proc. Soc. Neurosci. **766:** 16.
29. PAVLIDES, C. *et al.* 1988. Long-term potentiation in the dentate gyrus is induced preferentially on the positive phase of theta-rhythm. Brain Res. **439:** 383–387.
30. HOLSCHER, C., R. ANWYL & M.J. ROWAN. 1997. Stimulation on the positive phase of hippocampal theta rhythm induces long-term potentiation that can be depotentiated by stimulation on the negative phase in area CA1 *in vivo*. J. Neurosci. **17:** 6470–6477.
31. SON, G.R., B. THERRIEN & A. WHALL. 2002. Implicit memory and familiarity among elders with dementia. J. Nurs. Scholarsh. **34:** 263–267.

32. AMATO, M.P. *et al*. 2001. Cognitive dysfunction in early-onset multiple sclerosis: a reappraisal after 10 years. Arch. Neurol. **58:** 1602–1606.
33. PEYSER, J.M. *et al*. 1980. Cognitive function in patients with multiple-sclerosis. Arch. Neurol. **37:** 577–579.
34. RAO, S.M. 1990. Neurobehavioral Aspects of Multiple Sclerosis. Oxford University Press. New York.

Quantifying Tone Deafness in the General Population

JOHN A. SLOBODA,[a] KAREN J. WISE,[a] AND ISABELLE PERETZ[b]

[a]School of Psychology, Keele University, Staffordshire, ST5 5BG, United Kingdom

[b]Department of Psychology, University of Montreal, Montreal, Quebec, H3C 3J7, Canada

ABSTRACT: Many people reach adulthood without acquiring significant music performance skills (singing or instrumental playing). A substantial proportion of these adults consider that this has come about because they are "not musical." Some of these people may be "true" congenital amusics, characterized by specific and substantial anomalies in the processing of musical pitch and rhythm sequences, while at the same time displaying normal processing of speech and language. It is likely, however, that many adults who believe that they are unmusical are neurologically normal. We could call these adults "false" amusics. Acquisition of musical competence has multiple personal, social, and environmental precursors. Deficiencies in these areas may lead to lack of musical achievement, despite the fact that an individual possesses the necessary underlying capacities. Adults may therefore self-define as "unmusical" or "tone-deaf" for reasons unconnected to any underlying anomaly. This paper reports on two linked research studies. The first is an interview study with adults defining themselves as tone-deaf or unmusical. The interview schedule was designed to discover what criteria are being used in their self-definitions. Preliminary results suggest that performance criteria (e.g., judging oneself as unable to sing) play a major role, even for people who claim and demonstrate no perceptual deficits. The second study reports progress on the development of new subtests for a revised version of the Montreal Battery for the Evaluation of Amusia (MBEA, Peretz et al., 2003). This currently contains six tests that allow for the assessment of melodic perception: contour, intervals, scale, rhythm, meter, and recognition memory. The MBEA does not assess two capacities that are generally accepted as central to normal music cognition: harmony and emotion. The development and norming of the emotion subtest will be described. When completed, the MBEA(R) will form a robust screening device for use with the general population, whose purpose is to discriminate "true" from "false" amusics. Such discrimination is essential to achieve a better understanding of the variety of causes of low musical achievement.

KEYWORDS: amusia; tone-deaf; emotion; self-perception; Montreal Battery of Evaluation of Amusia

Address for correspondence: John A. Sloboda, School of Psychology, Keele University, Staffordshire, ST5 5BG, U.K. Voice: +44-1782-583381.

j.a.sloboda@keele.ac.uk

Ann. N.Y. Acad. Sci. 1060: 255–261 (2005). © 2005 New York Academy of Sciences.

doi: 10.1196/annals.1360.018

INTRODUCTION

In many cultures active participation in music is considered a birthright. Most take part and consider themselves equipped to do so. In industrialized Western societies, by contrast, a large proportion of people consider themselves unmusical, even tone-deaf, and avoid active participation in music.

What causes people to believe that they are musically impaired in one way or another? This paper addresses two lines of inquiry relevant to answering this question. One line of inquiry is focused on people's actual abilities. How can musical deficits be characterized in behavioral terms? Do people with self-perceptions of impairment possess shared behavioral or cognitive deficits? The other line of inquiry focuses on people's self-perceptions and their verbal accounts of their musical lives. How do people arrive at a self-concept of tone deafness and what does this mean to them? Could their perceptions be based on misattributions or misinterpretations of phenomena? Both of these approaches are necessary if we want to understand the nature and genesis of musical impairment. Research in this area may provide some pointers for how to improve the musical life of the self-defined "tone-deaf" or "unmusical" among us.

There is robust evidence for the existence of a musical learning disability, termed *congenital amusia*.[1] It emerges in early life and persists through adulthood. It occurs in the absence of any other identifiable cognitive or neurological difficulty and cannot be explained by lack of exposure to music. People with congenital amusia show impaired performance on a variety of basic musical tasks, including melodic discrimination and recognition, musical memory, metric discrimination, singing, and tapping with the beat.[1,2] The precise pattern of deficits varies across individuals. However, they have no problems processing speech prosody, voices, and other environmental sounds. The likely basis of the disorder is a deficit in processing fine-grained pitch variations.[3] It is thought to affect about 4% of the general population.[4] We can now reliably distinguish between the estimated 4% and the rest of the population by use of the Montreal Battery of Evaluation of Amusia (MBEA), which will be discussed later.

However it seems that around 15% of the population may self-label as tone-deaf.[5] Furthermore, many people who self-label as tone-deaf score within the normal range on the MBEA.[5,6] Some people therefore seem to mislabel themselves, believing they have a musical disability when they are in fact neurologically normal. How might we explain this apparent discrepancy? One possibility is that some self-declared tone-deaf people have deficits that the MBEA does not identify. For example, musical production skills, such as singing and playing, are not assessed by the MBEA. Neither are the receptive skills of emotional and harmonic recognition. We can also think of musical deficits as having broadly two levels of causation, namely neurological and environmental. Low perceived skill in producing music may be a result of, for example, lack of training but be misattributed to lack of capacity. A further possibility is that people think they have deficits when in fact they do not.

To begin to untangle some of the issues here, research needs to progress in two directions. First we need to probe exactly what it is people mean when they say they are tone-deaf. Second, we need to find out more about what they actually can or cannot do. The first question requires in-depth interviews, and the second question requires the gathering of behavioral data. In this paper we report preliminary moves in both directions.

CONCEPTIONS OF TONE DEAFNESS

Congenital amusia is a musical perceptual deficit that severely limits a person's capacity to be musical. We do not know, however, whether this corresponds to what is commonly understood as tone deafness among the general population. To investigate this, semistructured interviews were carried out with 15 participants aged between 18 and 70, with two broad research questions: (1) What do people think tone deafness is, and is it the same as being unmusical? and (2) What are the explanations people give for their own perceived tone deafness or lack of musicality?

Participants were recruited by means of a brief screening questionnaire with the questions "Do you consider yourself to be musical?" and "Do you consider yourself to be tone-deaf?" answered on a scale of 1 to 5, where 1 was "not at all" and 5 was "extremely." Participants invited for interview were chosen to reflect a range of age groups and responses. It is interesting to note that unmusicality and tone deafness were not always linked. So, some people considered themselves both musical and tone-deaf; others considered themselves unmusical but not tone-deaf.

The interview schedule included questions relating to participants' understanding of musicality and tone deafness in general, for example, "If you were told a person you had never met was musical, what would you expect?" and "If I introduce the idea of being tone-deaf, what comes to mind?" There were also questions about their musical self-perceptions, for example, "Do you consider yourself to be musical (or tone-deaf)?" and "How do you know you are musical/not musical/tone-deaf?"

When talking generally, people are very articulate about a tone-deaf stereotype, which is of someone singing very badly and often very loudly, either not knowing or not caring that they sound bad. Most people can recount stories of people they know:

"My music teacher was tone-deaf; oh god, she was terrible. She used to sing really loudly at the piano—absolutely dreadful...she wasn't getting any of the notes right at all" (Jenny, 19, self-declared tone-deaf and musical).

It was clear in every participant's account that whether applied to oneself or to others, *tone-deaf* meant *can't sing*:

"Mother...would have loved me to have joined the church choir so I went with John, my best friend, and I was told 'you can't sing' and I've never sung since. I've mimed all the way through my life really" (Paul, in his 60s, self-declared tone-deaf).

"I can't sing; I change key when I sing" (Susan, 46, self-declared tone-deaf).

By contrast, the role of perception in tone deafness is less well-defined in participants' accounts. People are sometimes aware of perceptual issues, but they do not usually think tone deafness affects the ability to listen to music. The word *hear* can have different meanings in participants' talk. On one level it means being able to appreciate music:

"I think if someone's tone-deaf they can still appreciate music sounds and...hear music the same as everybody else; they just can't express music verbally in the same way" (Jenny, 19, self-declared tone-deaf).

On another level to hear becomes "hearing yourself," which means being able to judge the accuracy of your own singing:

"I don't think she can hear herself because if she did...she'd match it more to the CD, and she'd be more in tune" (David, 19, self-declared unmusical).

Beyond this, ideas about perception are vague. People take basic musical perception for granted and focus on musical behavior. This is perhaps not surprising, as it is easier to tell whether one's behaviors are anomalous than whether one's perceptions are anomalous.

Two other observations are worth noting. First, tone-deaf does not necessarily mean unmusical. As far as the interviewed participants were concerned, being tone-deaf does not preclude someone from playing an instrument. However, it must also be said that people who consider themselves severely tone-deaf usually also consider themselves unmusical and do not play instruments. Second, most people believe that tone deafness is permanent and imposes limits on what a person can achieve, which means that the label is potentially restrictive.

Some of the main issues that arise in people's accounts of their negative musical self-perceptions have a social basis. First, accounts from people with negative musical self-concepts are full of unfavorable comparisons of their own abilities with the abilities of others. The social contexts of these comparisons, however, vary according to the individual's particular experiences. This means that in some cases, the comparator individual has very high levels of skill, and the actual skills of a person self-labeling as tone-deaf or unmusical may be average, or even above average. Second, many people think that musical people—either by training or talent—have access to privileged understanding about music that unmusical people do not, which allows them to perceive music differently. Third, people are very aware that one's own voice does not sound the same to others as it does to oneself. Lack of confidence in one's skills and the fear of a negative response from others are therefore salient. Self-defined tone-deaf people are sometimes not sure whether they sing badly or not, and fear keeps them from finding out. It remains to be seen, though, whether their self-perceptions correspond to their actual abilities.

The interviews show that while current research offers a description of musical deficits that is based on perception, the general population defines tone deafness primarily in terms of production (singing). It seems that perception may be difficult for people to reliably self-assess through everyday experience. People who falsely self-label as tone-deaf might therefore be helped by something to reassure them that they are musical, and to promote a clearer understanding of tone deafness.

Thus, we suggest two research directions. One involves the objective assessment of production skills, especially singing, and the other involves the development and application of a self-administrable objective test of perception. The remainder of this paper outlines how we are following the latter direction, by means of developing the Montreal Battery of Evaluation of Amusia, to which we referred earlier.

The battery currently comprises six tests assessing melodic discrimination, with regard to scale, interval, and contour; temporal discrimination, with regard to meter and rhythm; and memory. In each test, the same specially composed set of 30 tonal melodies are used as the target set, and most of the tasks require a "same–different" judgement between the target and an experimental manipulation of it. The memory test assesses recognition of melodies previously heard in the other tests, mixed with previously unheard foils.

A NEW EMOTION SUBTEST FOR THE MBEA

The MBEA is currently lacking measures for perception skills in a number of domains. It is important that a receptive test used for assessing and "reassuring" a normal population covers the main aspects of the response to tonal music. We have identified harmony and emotion as the two added elements needed to provide a comprehensive assessment. We will be focusing on emotion in this paper.

The key emotion-related skill is the ability to identify basic emotions in musical sequences (i.e., happy, sad, angry, or fearful). There is overwhelming evidence that this skill is well developed in musically untrained adults and emerges relatively early in childhood.[7] Research has demonstrated two different ways in which emotional competence can be demonstrated. First, people can classify different musical sequences on the basis of major emotional categories.[8] Second, people can emotionally classify different performances of the same melody.[9] Performers (both professional and amateur) are able to reliably project emotional character through their performance, by varying performance parameters such as speed, articulation, and timbre.[10] The existence of these reliable performance-emotion mappings allows us to construct a new MBEA test using the same target melodies as are used in the other subtests.

To generate potential stimuli for the emotion subtest we asked a professional violinist to record five performances of each target melody in the MBEA battery. These varied by emotional intention: happy, very happy, sad, very sad, and neutral. Piloting generated a set of 20 pairs of items that were consistently identified by normal adults as happy, sad, or neutral. Each pair consisted of two performances of the same tune, and the task was to judge whether the emotion conveyed was the same or different. The same–different format is used for two main reasons. First, it is consistent with the rest of the battery. Second, it controls for emotional cues that are communicated, for example, by tonality, melodic shape, and rhythmic patterns. We are therefore testing sensitivity to expressive variation in performance rather than global emotional response to music.

We ran the test with 9 individuals with amusia (aged 35–68, mean 59.4) who were recruited from a pool of amusic participants and had participated in many prior studies.[1,3] They had been previously screened to ensure that they met the criteria for a music-specific impairment. The 23 controls were a mixture of adults of similar age and education to the amusic individuals ($N = 4$, aged 45–65), and psychology undergraduates (aged 18–26, overall mean age of controls = 25.9).

As shown in TABLE 1, the amusic participants performed equally as well as the controls. However, the result on this test stands in stark contrast to the results on the existing tests of the MBEA. FIGURE 1 shows the performance of amusic individuals compared to controls on the existing tests of the MBEA,[6] plus the results of the new emotion test. On the existing tests, the controls score well, comparably to the scores we have observed on the emotion test. Amusic individuals, however, score considerably lower, in some cases no better than chance.

TABLE 1. Means and standard deviations of scores on the emotion test

	Amusic	Control
Mean score (out of 20)	18.89	18.61
SD[a]	1.27	1.37

[a]SD, standard deviation.

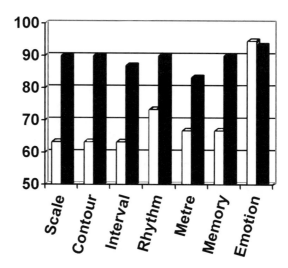

FIGURE 1. Percentage scores of amusic individuals and controls on the MBEA subtests. *Open bars*, amusic; *filled bars*, control.

How is the success of amusic participants in the emotion task to be explained? As shown by Juslin and Laukka,[11] the characteristics of music performances that communicate emotion are very similar to those used in speech and other human vocal prosody. Amusic individuals process speech intonation normally. They also retain the ability to process variations in articulation, tempo, and timbre,[1,12] which are the parameters identified by Juslin and colleagues as important in performers' projections of emotion in music. They can therefore use cues, such as tempo, to derive the happy–sad tone of musical selections.

The next stage for the development of the MBEA is the addition of a harmony test to assess understanding of conventional tonal language. We then plan to fashion a version of the entire battery that will take no longer than an hour to administer. This is a necessary precursor to developing a version of the battery to be used as a self-administered test, which may eventually be available for public use.

In conclusion, the results show that despite having severe difficulties with basic musical processing, people with congenital amusia are not hindered in all musical tasks. The emotion test is therefore a valuable addition to the MBEA, as it enriches the picture of musical skills. The MBEA is a potential means of reassuring people who mistakenly call themselves tone-deaf that there is nothing neurologically wrong with them. For those few who do have a perceptual disorder, it can allow insight into the nature of their difficulty. It appears from the interview study, however, that people think of tone deafness as a deficit of production rather than perception, and the role of singing skills with regard to musical impairments needs further investigation. In its current state, the MBEA cannot reassure people that they can sing. However, it may help remove some of the barriers to participation in music. If, as we hope, it is eventually made available to the public, it may prove to be a valuable means of dispelling myths and encouraging people to explore their hidden musical potential.

[Competing interests: The authors declare that they have no competing financial interests.]

REFERENCES

1. AYOTTE, J., I. PERETZ & K. HYDE. 2002. Congenital amusia: a group study of adults afflicted with a music-specific disorder. Brain **125:** 238–251.
2. PERETZ, I., J. AYOTTE, R.J. ZATORRE, *et al.* 2002. Congenital amusia: a disorder of fine-grained pitch discrimination. Neuron **33:** 185–191.
3. HYDE, K. & I. PERETZ. 2004. Brains that are out of tune but in time. Psychol. Sci. **15:** 356–360.
4. KALMUS, H. & D.B. FRY. 1980. On tune deafness (dysmelodia): frequency, development, genetics, and musical background. Ann. Hum. Genet. **43:** 369–382.
5. CUDDY, L.L., L-L. BALKWILL, I. PERETZ & R.R. HOLDEN. 2005. Musical difficulties are rare: a study of "tone deafness" among university students. Ann. N. Y. Acad. Sci. **1060:** 311–324. [This volume].
6. PERETZ, I., A.S. CHAMPOD & K. HYDE. 2003. Varieties of musical disorders: the Montreal Battery of Evaluation of Amusia. Ann. N. Y. Acad. Sci. **999:** 1–18.
7. KASTNER, M.P. & R.G. CROWDER. 1990. Perception of the major/minor distinction: IV. Emotional connotations in young children. Mus. Percept. **8:** 189–202.
8. GABRIELSSON, A. & E. LINDSTROM. 2001. The influence of musical structure on emotional expression. *In* Music and Emotion: Theory and Research. P.N. Juslin & J.A. Sloboda, Eds.: 223–248. Oxford University Press. New York.
9. SLOBODA, J.A. & A.C. LEHMANN. 2001. Performance correlates of perceived emotionality in different interpretations of a Chopin piano prelude. Mus. Percept. **19:** 87–120.
10. JUSLIN, P.N. 1997. Emotional communication in music performance: a functionalist perspective and some data. Mus. Percept. **14:** 383–418.
11. JUSLIN, P.N. & P. LAUKKA. 2003. Communication of emotion in vocal expression and musical performance. Different channels, same code? Psychol. Bull. **129:** 770–814.
12. FOXTON, J.M., J.L. DEAN, R. GEE, *et al.* 2004. Characterization of deficits in pitch perception underlying "tone deafness." Brain **127:** 801–810.

Does Music Instruction Improve Fine Motor Abilities?

EUGENIA COSTA-GIOMI

University of Texas–Austin, Center for Music Learning, Austin, Texas 78712-0435, USA

ABSTRACT: The fine motor abilities of children who participated in two years of piano instruction and those who had never received formal music training were compared before and after the instruction. A significant improvement in fine motor skills was found only for the children who received the lessons, and a significant difference in the speed of response was found between the two groups at the end of the two years of instruction. The innumerable opportunities to assess, refine, and time their motor responses to specific stimuli during musical practice and the availability of constant evaluative feedback (i.e., sound) may allow musicians to improve the accuracy and speed of perceiving and responding to relevant stimuli.

KEYWORDS: children; motor response; music instruction; music education; music practice

INTRODUCTION

Musicians outperform nonmusicians in certain perception tasks that require an accurate and immediate motor response to a visual stimulus.[1] These results have been taken as evidence that music training increases the speed and accuracy of the visual–motor association. Similarly, the anatomical differences in the sensorimotor cortex found between musicians and nonmusicians suggest that extensive music practice affects the organization of this cortical area.[2–5] Studies that focused on the cortical representation of hand fingers during intensive keyboard practice sessions have indeed shown clear changes over periods as short as five days and as long as two weeks.[5] No longitudinal research exists on the neurological changes that occur over longer periods of musical practice.

Although a few studies conducted with young children suggest that early music instruction improves performance in visual–motor tasks,[6,7] others showed no significant fine motor skills improvements after five months of violin instruction.[8] The effects of instrumental practice on fine motor skills in children have not been studied systematically despite the strong belief among parents and educators that learning to play an instrument improves dexterity and motor control.

Address for correspondence: Eugenia Costa-Giomi, University of Texas–Austin, Center for Music Learning, 1 University Station E3100, Austin, TX 78712-0435. Voice: 512-471-2495.
    costagiomi@mail.utexas.edu

Ann. N.Y. Acad. Sci. 1060: 262–264 (2005). © 2005 New York Academy of Sciences.
doi: 10.1196/annals.1360.053

METHOD

Children (n = 117) attending 16 public schools who had no formal music instruction, no piano at home, and family annual incomes below $30,000 were randomly
assigned to either the experimental group receiving two years of individual piano
lessons weekly or the control group receiving no formal music instruction. Children
in the experimental group received acoustic pianos and practiced, as an average, up
to 3.5 hours weekly.[9] The cognitive abilities, academic achievement, musical abilities, self-esteem, and motor proficiency of the two groups were comparable at the
start of the project as determined by standardized tests.[10]

Fifty-one children in the experimental group and 39 children in the control group
completed the fine motor components of the Bruinsky-Oseretsky Motor Proficiency
Test before and after the two years of instruction. The total scores of the fine motor
test and the scores in subtest 6: Response Speed, subtest 7: Visual-Motor Control,
and subtest 8: Upper-Limb Speed and Dexterity were analyzed through ANOVAs
with repeated measures.

RESULTS

Significant group (piano/non-piano group) × instruction (pre/posttest, repeated
measures) interactions were found for the total fine motor scores and the response
speed subtest scores $F(1,88) = 4.01$ $P < .05$ and $F(1,88) = 13.61$ $P < .001$, respectively. The fine motor skills of the piano group improved significantly more during
the two years of the project than did those of the control group, and the differences
in Subtest 6 scores between the two groups of children were significant only after
the two years of piano instruction. These results suggest that the improvement in motor proficiency was mainly caused by differential scores in the speed subtest which
required children to react quickly to catch a rod that was sliding down against a wall.
Scores in tasks that measured eye–hand coordination and dexterity were not affected
by the lessons.

CONCLUSIONS

Music performance requires accurate and quick motor reaction to visual, aural,
and kinesthetic stimuli. When performers practice their instruments, they receive
immediate and consistent aural feedback about their motor response to such stimuli.
The innumerable opportunities to assess, refine, and time their motor responses to
specific stimuli during musical practice and the availability of constant evaluative
feedback (i.e., sound) may allow musicians to improve the accuracy and speed in
perceiving the stimuli and responding to it.

[Competing interests: The authors declare that they have no competing financial
interests.]

REFERENCES

1. BROCHARD, R. *et al.* 2004. Effect of musical expertise on visuospatial abilities: evidence from reaction times and mental imagery. Brain Cogn. **54:** 103–109.
2. ELBERT, T., C. PANTEV, C. WIENDBRUCH, *et al.* 1995. Increased cortical representation of the fingers of the left hand in string players. Science **270:** 305–307.
3. JÄNCKE, L., N.J. SHAH & M. PETERS. 2000. Cortical activations in primary and secondary motor areas for complex bimanual movements in professional pianists. Brain Res. Cogn. Brain Res. **10:** 177–183.
4. PANTEV, C., A. ENGELIEN, V. CANDIA & T. ELBERT. 2001. Representational cortex in musicians. Plastic alterations in response to musical practice. Ann. N. Y. Acad. Sci. **930:** 300–314.
5. PASCUAL-LEONE, A. 2001. The brain that plays music and is changed by it. Ann. N. Y. Acad. Sci **930:** 315–332.
6. BROWN, J. *et al.* 1981. Effects of an integrated physical education/music program in changing early childhood perceptual–motor performance. Percept. Mot. Skills **53:** 151–154.
7. ORSMOND, G.I. & L.K. MILLER. 1999. Cognitive, musical, and environmental correlates of early music instruction. Psychol. Mus. **27:** 18–37.
8. ESSMA, S. 2005. Does music instruction increase children's fine motor abilities? University of Texas Austin. Unpublished manuscript.
9. COSTA-GIOMI, E. 2005. "I do not want to study piano!" Early predictors of student dropout behavior. Bull. Coun. Res. Music Educ. **161/162:** 57–64.
10. COSTA-GIOMI, E. 1999. The effects of three years of piano instruction on children's cognitive development. J. Res. Music Educ. **47:** 198–212.

The Effect of Early Musical Training on Adult Motor Performance

Evidence for a Sensitive Period in Motor Learning

VIRGINIA PENHUNE, DONALD WATANABE, AND TAL SAVION-LEMIEUX

Department of Psychology, Concordia University, Montreal, Quebec H4B 1R6, Canada

ABSTRACT: This experiment demonstrates that musicians who began training before age seven perform better on a rhythmic tapping task than musicians who began after the age of seven, when the two groups are matched for years of experience. These results support the idea that there may be a sensitive period in childhood for motor training, similar to that observed for language learning.

KEYWORDS: motor learning; sensitive period; brain plasticity; development

Mozart began piano training at the age of three years, Beethoven before the age of eight years. Many music programs for children emphasize starting training as early as possible in order to develop musical skill. However, very little is known about the real effects of early musical training on adult performance. Behavioral studies comparing early- and late-trained musicians have shown that early-trained musicians have better pitch discrimination and are more likely to develop absolute or "perfect" pitch.[1] More recently, studies using brain imaging have shown that there are structural and functional differences in auditory and motor regions of the brain associated with musical training and that these changes are greater in those who began training early in life.[2–6] Elbert[2] found that expert string players had enlarged cortical representations of the fingers of the left hand, and that this enlargement was greater for those who began training earlier.

Schlaug *et al.*[5] reported a larger anterior corpus callosum in musicians who commenced musical training before the age of seven years. These findings suggest that there may be a critical or sensitive period for motor skill training, similar to that observed for language acquisition. However, in these previous studies, it is also possible that differences in performance and brain structure can be accounted for by the amount of time spent in musical training. By definition, a musician who begins training early has more years of training and experience that one who begins later. There-

Address for correspondence: Dr. Virginia Penhune, Laboratory for Motor Learning and Neural Plasticity, Department of Psychology, SP-A 244, Concordia University, 7141 Sherbrooke W, Montreal, QC H4B 1R6, Canada. Voice: 514-848-2424 ext. 7535; fax: 514-848-4545.
vpenhune@vax2.concordia.ca

Ann. N.Y. Acad. Sci. 1060: 265–268 (2005). © 2005 New York Academy of Sciences.
doi: 10.1196/annals.1360.049

fore, the present experiment examined the effect of musical training on performance of a rhythmic tapping task in early- and late-trained musicians who were matched for years of musical experience.

The subjects tested in this experiment were currently practicing musicians who had an average of 14.3 years of musical training and experience. They were divided into two groups: early-trained musicians (ET) who began training before age seven and late-trained musicians (LT) who began training after age seven. Individual subjects were matched for years of training and experience as measured by an in-depth questionnaire. Years of formal musical training and hours of current weekly practice

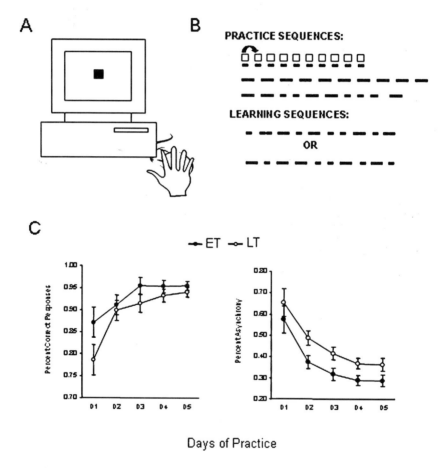

FIGURE 1. Experimental setup and results for ET and LT musicians on the TMST. Panel **A** shows how the subject taps in synchrony with a visual stimulus using a single button of the computer mouse. Panel **B** shows the temporal organization of the stimuli. Panel **C** shows the graphs for performance on measures of percent correct (*left*) and response asynchrony (*right*) for the two groups.

were also measured. The task used in this experiment was the temporal motor sequence task (TMST), which tests the subject's ability to learn and reproduce a complex sequence of finger taps similar to a musical rhythm.[7,8] In contrast with a typical musical rhythm, the TMST requires participants to reproduce the rhythmic sequence in synchrony with a visual stimulus. Therefore, performance on this task assesses the musician's ability to perform a familiar motor task in a novel context.

The visual stimuli are 10-element sequences made up of a series of white squares (3 cm^2) presented sequentially in the center of the computer screen (see Fig. 1A). Participants respond by tapping on a single key of the computer mouse with the index finger of the right hand. The sequences are made up five long (750 ms) and five short (250 ms) elements with a constant interstimulus interval (500 ms) and are constructed to have no more than two repeated elements and to have seven transitions from short to long (Fig. 1B). This results in sequences that are temporally regular but do not conform to a standard musical rhythm. Participants practice the sequence over five days and are tested again at four-week delayed recall. Key-on and key-off times are recorded and used to calculate three indices of learning: response accuracy, variance, and synchronization.

Results showed that the ET and LT musicians groups were well matched, with no significant differences in the total number of years of musical experience (ET = 14.8; LT = 13.7), the number of years of formal training (ET = 7.7; LT = 6.3), or the number of hours per week they currently practiced (ET = 11.8; LT = 13.4). As predicted, the groups were significantly different in terms of the age of start of musical training (ET = 5.9; LT = 11.4; $P < .01$) and in age (ET = 22.1; LT = 27.5). Analysis of performance on the TMST (see Fig. 1C) showed that ET and LT musicians were equally accurate in their reproduction of the order of short and long elements in the sequence, but that early-trained musicians were better able to synchronize their responses to the stimuli across the five days of practice ($F[1,27] = 3.95, P < .57$). Thus, the ET musicians showed an advantage in their performance of a novel motor task when compared with LT musicians with the same amount of musical experience. Most interestingly, this advantage was seen only for the measure of sensorimotor synchronization and was sustained across the five days of practice.

These findings show that ET musicians perform better on a novel rhythmic tapping task than LT musicians with similar levels of training and experience. This supports the idea that there may be a sensitive period in childhood for motor training, similar to that observed for language learning. Performance differences were greatest for the measure of response synchronization, suggesting that early training has its greatest effect on sensorimotor integration. Planned studies in our laboratory examining motor learning in children can shed light on developmental changes that might underlie this sensitive period. Importantly, however, whereas the two groups differed on average, there was considerable overlap in performance between the two groups. This indicates that early training is not the only factor affecting adult performance. Other potential contributors that are not controlled in this experiment are early ability, motivation, and family support for musical training. Future studies examining matched groups of early- and late-starting children undergoing the same type of musical training will shed light on the contributions of these factors.

[Competing interests: The authors declare that they have no competing financial interests.]

REFERENCES

1. BAHARLOO, S. *et al.* 1998. Absolute pitch: an approach for identification of genetic and nongenetic components. Am. J. Hum. Genet. **62:** 224–231.
2. ELBERT, T. *et al.* 1995. Increased cortical representation of the fingers of the left hand in string players. Science **270:** 305–307.
3. GASER, C. & G. SCHLAUG. 2003. Brain structure differences between musicians and nonmusicians. J. Neurosci. **23:** 9240–9245.
4. KOENEKE, S. *et al.* 2004. Long-term training affects cerebellar processing in skilled keyboard players. NeuroReport **15:** 1279–1282.
5. SCHLAUG, G. *et al.* 1995. Increased corpus callosum size in musicians. Neuropsychologia **33:** 1047–1055.
6. SCHNEIDER, P. *et al.* 2002. Morphology of Heschl's gyrus reflects enhanced activation in the auditory cortex of musicians. Nat. Neurosci. **5:** 688–694.
7. PENHUNE, V. & J. DOYON. 2002. Dynamic cortical and subcortical networks in learning and delayed recall of timed motor sequences. J. Neurosci. **22:** 1397–1406.
8. PENHUNE, V. & J. DOYON. 2005. Cerebellum and M1 interaction during early learning of timed motor sequences. NeuroImage **26:** 801–812.

Music Therapy: The Long Way to Evidence-Based Methods—Pending Issues and Perspectives

Introduction

LUISA LOPEZ

"Eugenio Litta" Rehabilitation Center, Grottaferrata and Child Neurology Unit, University of Rome "Tor Vergata," Rome, Italy

A roundtable on music therapy is definitely a challenge in a community of neuroscientists, even when their main interest is music. Unfortunately, typical articles on music therapy are characterized by small samples, lack of a proper control group, subjective measurement protocols, and huge inter-experimenter/therapist variability. There have been noticeable efforts to overcome these liabilities, and some results are appearing in selected literature databases.

As an example, I quote a systematic review by Gold *et al.,*[1] in which the authors analyzed "the effects of music therapy, or music therapy added to standard care, compared to placebo, standard care or no treatment for people with serious mental illnesses such as schizophrenia," including only randomized controlled trials. The interesting point is that music therapy, in addition to standard therapy, was found to improve the mental state and functioning, if a sufficient number of sessions was provided. Even more intriguingly, however, after excluding excessive drop-outs at follow-up and statistical heterogeneity, this metanalysis only included four studies.

Music therapy is used in a variety of applications, including communication and physical disability, pain control, relaxation during invasive or unpleasant procedures, modulation of cardiovascular function, improvement of adaptive and cognitive functions in both mental retardation and dementia, and psychiatric illness management. Many groups are involved in defining theoretical frameworks for this ever-growing discipline, and this roundtable was designed to provide a picture, albeit incomplete, of the effort that has been made in that direction. Robert Zatorre was chosen as the discussant and contributed to this very interactive session by giving input to the speakers and promptly responding to an audience that was alternatively skeptical and believing.

Eckart Altenmüller gave us examples of his research work on focal dystonia to introduce the issue of scientific methodology and replicability. The presentations that followed are all included in this section.

Address for correspondence: Dr. Luisa Lopez, Center for Developmental Disabilities "Eugenio Litta," Via Anagnina Nuova 13, 00046 Grottaferrata, Rome, Italy. Voice: +39-06-94315621; fax: +39-06-9411463.

lopez@uniroma2.it

Ann. N.Y. Acad. Sci. 1060: 269–270 (2005). © 2005 New York Academy of Sciences.
doi: 10.1196/annals.1360.064

The group from Heidelberg, represented by Thomas Hillecke and Anne Nickel presented two papers. The first describes the scientific method and the typical study designs implemented in music therapy research. The authors guide the readers through the development of their model comprising five factors they call "ingredients" that characterize music therapy in its current therapeutic applications: attention, emotion, cognition, behavior, and communication. Hillecke and collaborators very appropriately conclude their article with questions rather than answers. Anne Nickel, in a successful attempt to provide a few answers, continues by presenting three studies in which music therapy was used in patients with chronic, nonmalignant pain; children with migraine headache; and patients with tinnitus. All three studies were well designed, and this article, as the authors point out, is "a step on the long road to evidence-based methods."

Patricia Sabbatella, from the University of Puerto Real, Cadiz, takes a somewhat different point of view. Her analysis of the current literature selects two main areas of research in music therapy—the Spanish-speaking and the English-speaking ones. The author suggests that many studies also come from Asian countries; I would add that a good amount of research is also being published in Italy as well as several other countries. Sabbatella concludes her interesting literature review by acknowledging the evolution from an empirical approach to a more rigorously scientific one and she indicates clinical evaluation and assessment as a key factor in this process.

Michael Thaut, from Colorado State University, describes music in therapy as a mediator that influences brain function and behavior by arousing, guiding, organizing, focusing, and modulating perception, attention, and behavior in the affective, cognitive, and sensorimotor domains to obtain a therapeutic response. This paper focuses on the rhythmic properties of excitable cells and describes, for example, the entrainment of motor function used in the recovery of stroke. As a highly coherent set of sounds, music is an optimal oscillator in the auditory domain that can drive synchronization in specific brain areas, thus producing motor to cognitive facilitation effects.

This roundtable proved to be an interesting sequence of presentations, each with a different outlook on music therapy. We are probably still far from the goal of an evidence-based method in this discipline but so are many areas of rehabilitation and so-called alternative therapies. Probably the time window of our observation needs to be expanded, and another session on this topic might include the issue of preventive medicine as well.

In conclusion, we are always trying to work out a compromise between discrete measurable units and whole global meanings of music and therapy.

REFERENCE

1. GOLD, C., T.O. HELDAL, T. DAHLE & T. WIGRAM. 2005. Music therapy for schizophrenia or schizophrenia-like illnesses. Cochrane Database of Syst. Rev. 2: CD004025.

Scientific Perspectives on Music Therapy

THOMAS HILLECKE, ANNE NICKEL, AND HANS VOLKER BOLAY

German Center for Music Therapy Research, and Outpatient Department,
University of Applied Sciences Heidelberg, D-68123 Heidelberg, Germany

ABSTRACT: What needs to be done on the long road to evidence-based music therapy? First of all, an adequate research strategy is required. For this purpose the general methodology for therapy research should be adopted. Additionally, music therapy needs a variety of methods of allied fields to contribute scientific findings, including mathematics, natural sciences, behavioral and social sciences, as well as the arts. Pluralism seems necessary as well as inevitable. At least two major research problems can be identified, however, that make the path stony: the problem of specificity and the problem of eclecticism. Neuroscientific research in music is giving rise to new ideas, perspectives, and methods; they seem to be promising prospects for a possible contribution to a theoretical and empirical scientific foundation for music therapy. Despite the huge heterogeneity of theoretical approaches in music therapy, an integrative model of working ingredients in music therapy is useful as a starting point for empirical studies in order to question what specifically works in music therapy. For this purpose, a heuristic model, consisting of five music therapy working factors (attention modulation, emotion modulation, cognition modulation, behavior modulation, and communication modulation) has been developed by the Center for Music Therapy Research (Viktor Dulger Institute) in Heidelberg. Evidence shows the effectiveness of music therapy for treating certain diseases, but the question of what it is in music therapy that works remains largely unanswered. The authors conclude with some questions to neuroscientists, which we hope may help elucidate relevant aspects of a possible link between the two disciplines.

KEYWORDS: music therapy; therapy research; multidisciplinary approach; pluralistic point of view; working ingredients

INTRODUCTION

Music therapy and music therapy research currently represent heterogeneous but growing fields. In their clinical work, music therapists experience music as an effective tool in the treatment of various illnesses. Despite this clinical observation, it is necessary in modern societies and current health care systems to prove the effectiveness and efficacy of psychological as well as medical treatments. Therefore one of the major efforts in current music therapy research is to study effectiveness and efficacy of defined interventions for certain diseases. However, even if the corpus of outcome

Address for correspondence: Prof. Dr. Thomas Hillecke, German Center for Music Therapy Research, University of Applied Sciences Heidelberg, Outpatient Department, Maaßstraße 26, D-68123 Heidelberg, Germany. Voice: +49-6221-4154; fax: +49-6221-4152.
thomas.hillecke@fh-heidelberg.de

Ann. N.Y. Acad. Sci. 1060: 271–282 (2005). © 2005 New York Academy of Sciences.
doi: 10.1196/annals.1360.020

studies in music therapy were extensive and positive enough, the question would still remain about what it is in music therapy that works. There is an urgent need for the application of empirical research methods to studying the ingredients of music therapy. This application is needed especially because of the heterogeneous and often incommensurable theories music therapists use to describe and explain their work. One consequence of this is that communication between music therapy centers nationally as well as internationally can be described as, at best, unsatisfactory.

The different approaches also offer advantages in an evolutionary and epistemological sense. They reflect the broad way to study and explain relevant aspects as well as the complexity of music therapy work. However—as Darwin said—only the fittest survive. These different approaches reveal that music is associated with many biological, psychological, and sociocultural phenomena in human life. Therefore a single explanation, such as a great unifying music therapy theory, seems nothing more than a utopian vision. The major disadvantage of theoretical heterogeneity makes the study of working ingredients of music therapy more difficult and a theoretical agreement among representatives of the field unlikely. Concerning these aspects one can conclude that knowledge of working factors is far from evidence based, whereas the field of outcome studies is growing, supporting the clinical experience of music therapists and allied professions that music represents a useful tool in the treatment of different mental and somatic diseases.[1–10]

MUSIC THERAPY RESEARCH STRATEGIES

The combination of the two terms *music* and *therapy* implies the application of both music research and therapy research methods. One possibility is to use therapy research methodology in music therapy. This perspective opens the door to different research designs to study specific questions and draw respective conclusions. TABLE 1 distinguishes basic research, single-case research, group research, and reviews.

For example, the description of phenomena as well as experimental research can help to identify relevant aspects of music therapy. It often leads to new ideas of interventions and underlying mechanisms of music as a therapeutic means. The observation and clear description of a phenomenon is very important for all scientific work. It represents one first step to reducing complexity and to enhancing objectivity (understood as intersubjectivity), also for music therapy.

Single-case research facilitates the understanding of relevant phenomena in a defined therapy process. With single-case course studies we comprehend the process of change of clients or patients with regard to the complexity of music therapy. Especially in this area, qualitative and quantitative designs complement each other. Nowadays the field of single-case studies with a different purpose is growing. The goal is to guarantee process quality and probably external validity of a given intervention. One important project of the German Center for Music Therapy Research was to develop a quality assurance system for the Music Therapy Outpatient Department of the University of Applied Sciences, Heidelberg,[11] which is unique in the music therapy field. Every therapy in this department is videotaped and evaluated for clinical significance.[12,13]

TABLE 1. Therapy research strategies

Type	Design	Conclusions
Basic research	Description of phenomena	Identification of relevant aspects (generation of hypotheses)
	Experimental research on psychological and physiological effects of music	Identification of relevant aspects (explanation of effects)
Single-case research	Descriptive case studies	Understanding of specific relevant aspects (generation of hypotheses)
	Qualitative single-case course studies	Understanding of specific process aspects (generation of hypotheses)
	Quantitative single-case course studies	Objective illustration of specific process aspects (generation of hypotheses)
	Comparative case studies	Identification of similarities and differences (generation of hypotheses)
	Single-case studies as a measure of quality assurance	Inspection of process and outcome standards
Group research	Descriptive qualitative research on group comparisons	Identification of different effects by different interventions in comparable patient groups. Identification of differences between different patient groups in comparable interventions (generation of hypotheses).
	Effectiveness (efficacy) research • pre–post comparison • controlled course analysis • randomized comparisons with waitlist controls • randomized comparisons with alternative treated group	Identification of change during the therapeutic process (post hoc) Identification of change by the therapeutic process (propter hoc) (hypotheses testing)
	Follow-up effectiveness (efficacy) research • uncontrolled • controlled • randomized	Identification of stable change after the therapeutic process (post hoc). Identification of stable change by the therapeutic process (propter hoc) (hypotheses testing)
	Research on working factors (micro-analysis, event research, comparison of different treatment conditions)	Identification of working ingredients (generation of hypotheses, hypothesis testing, and explanation of effects)
	Multicenter studies	Identification of differences and similarities by different institutional frameworks
Reviews	Studies of literature	Identification of relevant literature (generation of hypotheses and theories)
	Metanalysis	Comprehensive presentation of empirical studies (testing of hypotheses and theories)

Most common are group research designs. These designs are of great value in every clinical research area. Different group research designs can have various intentions. The main interest is to test outcome and to find out what works. Effectiveness studies and efficacy studies are the main work of the German Center of Music Therapy Research. With these designs the outcome of music therapy in the field of chronic nonmalignant pain (1), children with migraines (2), tinnitus (3), and heart catheter examinations (4) was tested. The results show clinically significant positive change in the first three areas and poor effects in the last one. In an ongoing multicenter study, the outcome of neurological music therapy (NMT) for hemiparetic stroke patients is being tested in collaboration with the Center for Biomedical Research in Music of Colorado State University.

Another growing field is the review method. This method offers two major possibilities. One is to study the literature by summarizing contents. This is a necessary tool to comprehend the state of the art. Since the 1950s, researchers have carried out what is called metanalysis. This is a debatable review integrating and combining statistical results. In the field of music therapy, metanalytical studies have existed since 1986.[7] Currently quite a few of them reflect the effect sizes of music therapy in different clinical fields.[1-5,7-10]

This systematology (TABLE 1) highlights the important and growing role neuroscience can play in music therapy research. Neurocognitive research has the potential as a basic research approach to identify and explain relevant effects of music in therapy by the use of experimental research designs and neurophysiological investigation methods. It is also a useful tool to identify working ingredients, to generate new hypotheses, and especially to test and explain the correlation between music therapeutic intervention techniques and empirically observed outcome.

FIGURE 1. Music therapy (research): a multidisciplinary field.

THE NECESSITY OF PLURALISM IN MUSIC THERAPY

Music therapy is a multidisciplinary field in which the researchers can learn from others. The field overlaps with a wide spectrum of scientific areas, including mathematics, natural sciences, behavioral and social sciences, as well as the arts (FIG. 1). The word *music* stands for a multiplicity of human events that are difficult to analyze from a reductionist point of view. Therefore the study of music and music therapy needs to be multidisciplinary as well as theoretically and scientifically pluralistic.

Some examples may illustrate this requirement: (1) Physics may be a useful tool to study psychophysical aspects of music as acoustical phenomena. It describes physical aspects of music, such as sound waves, volume, and acoustic pressure. (2) Biological and biomedical aspects are relevant in understanding how music is processed by the nervous system and how its effects reach other organic structures of the body. Biological background is necessary for the explanation of how music leads to physiological changes. From this point of view music as a personal experience is a result of physiological information processing. (3) Psychotherapy research is currently a very systematic research field (see above). (4) Also psychology with psychological experiments and psychological diagnostics, such as questionnaires and tests, including the paradigm of cognition, are very important tools in analyzing what happens in music therapy. Music psychology is mainly very important to understanding how music influences behavior and experience. (5) Sociological as well as ethnological aspects play an important role in music therapy research. For example, do the social and ethnic backgrounds play important roles in the reaction of patients to music therapy interventions? It could also be asked whether music therapy is a development of industrialized Western culture, or does it exist in (all) other cultures? What are the differences in music therapy interventions among these cultures? (6) Another relevant aspect is musicology. This field, for example, contributes different possibilities for describing music in symbols and interpreting music as art and creativity.

PROBLEMATIC ASPECTS OF APPLYING THERAPY RESEARCH TO MUSIC THERAPY

The first problem, of relevance for music therapy, especially in the treatment of psychological disorders, can be called the specificity problem and is a result of more than 50 years of psychotherapy research. Psychotherapy theorists have often been frustrated by this problem and therefore mainly ignored it. This problem was first formulated by the psychiatrist Jerome D. Frank[14] and empirically first observed by the psychoanalyst Lester Luborsky.[15] It is labeled "the dodo bird verdict" because the dodo bird says to Alice in Wonderland: "Everybody has won and all must have prizes." Its consequences are best summarized by Michael Lambert,[16] who came to the conclusion that extratherapeutic aspects determine 40%, therapeutic relationship 30%, expectancy and placebo effects 15%, and specific therapeutic techniques 15% of the observed outcome variance in psychotherapy studies. This reveals that unspecific factors play a major role, and specific therapeutic techniques seem to be almost negligible.

The dodo bird verdict characterizes empirical results of outcome studies and comparative studies.

- The differences in outcome between various approaches, such as depth-oriented psychotherapy, humanistic therapy, behavior therapy, and cognitive therapy can almost be ignored. All of them produce comparable effects.

- One consequence is the evidence-based assumption that common factors—as ingredients that are shared by all these approaches—are of much more significance than specific factors, which reflect the specific assumptions of these different theoretical approaches.

For music therapy as a treatment of psychological disorders, the problem leads to the serious question of whether the observed outcome of clinical studies depends on music as the specific ingredient, or on common factors. The only possibility for coping with that challenge is to use comparative therapy studies or working factor studies, which correlate music therapy techniques to observed outcome.

The second problem is of more practical relevance. It can be called the *eclectic problem*. It points out that traditional theoretical frameworks are often obviously not satisfying for music therapy clinicians. This is also comparable to the situation regarding psychotherapy. An extensive corpus of modalities and treatment theories exist in the music therapy world. Some of them are psychoanalytic music therapy,[17] humanistic music therapy,[18] behavioral music therapy,[19,20] Nordoff-Robins music therapy,[21] and music medicine.[22] Often music therapists refer to one of them as theoretical background; nevertheless these traditional approaches can be considered as more or less belief systems. In clinical reality most of the music therapists mix techniques and theories, creating their own blend and personal music therapy theory. Additionally they also combine music therapy with other therapeutic modalities. To handle this eclecticism problem, the development of treatment manuals, including defined musical interventions, is a promising possibility. The best way, however, would be to develop theories that are testable as well as practical, and that would contain empirical knowledge of etiology, pathology, working factors, and expected outcome. These theories should then be tested by clearly defined and suitable empirical methods. If falsification is the outcome, they should be dropped. Unfortunately the philosophy of falsification[23] is currently not common enough in the field of music therapy. Therefore epistemology should be more emphasized in music therapy training programs.

The question that we are asking is whether neuroscience or neurocognitive approaches constitute new upcoming paradigms for the music therapy field. Since the publication of *The Structure of Scientific Revolutions*,[24] the idea of a scientific paradigm shift is often used to proclaim new perspectives. However, for a new scientific paradigm to exist as a result of a scientific revolution, an old paradigm is necessary. The situation in the field of music therapy is very different. There is a lack of specific evidence-based theories on the one hand, and there is therapeutic eclecticism in practical music therapy work on the other. The situation may rather be described as a continuous search for adequate theoretical frameworks that help practitioners in their everyday work. The transference of theories from other disciplines, especially psychology, was and is predominant (for example, psychodynamic music therapy, humanistic music therapy, behavioral music therapy, and cognitive music therapy). A music-specific therapeutic paradigm itself does not exist. If a paradigm changes, it concerns basic research areas (such as physics), in general, not applied sciences, like music therapy or medicine. If the upcoming or current paradigm (or scientific matrix) of medicine or psychology is naturalism containing neurocognition, it would be better to ask what music therapy can learn from that current perspective. In psychotherapy research, different developments can be detected. Pure naturalism in this field has been analyzed and fundamentally criticized by Slife.[25] This author points out the limitations of a naturalistic point of view. Naturalism, like other theoretical frameworks, is based on implicit and often unexamined assumptions. In the case of naturalism these are the following:

- Objectivism: The objective world of therapy occurs outside our subjectivity, and thus in a value-free world without meaning and morality. Alternatively, music therapeutic relationships occur between (two or more) subjects, who interpret musical experiences individually and by interaction.

- Materialism: Matter is what is important and sufficient for understanding. Hence, nonobservational constructs are operationalized, and psychotherapy is increasingly biologized. Alternatively, social contexts are of significance in music therapy settings. Especially psychological disorders and symptoms, as well as reactions to music, are often culture bound and cannot be understood as mere biological phenomena.

- Hedonism: All living things seek pleasure and avoid pain, with all higher animals ultimately concerned with benefits to the self (well-being as major outcome). Alternatively, outcome may include other different values, such as altruism. Musical experiences that moderate change can initially be emotionally disturbing and may not always lead to pleasure.

- Atomism: The natural world comprises self-contained atoms, each with unique properties and qualities contained therein. Therefore the individual is of relevance. Alternatively, the relevant focus of music therapy could also be on a family or other social groups.

- Universalism: The most fundamental things are the things that do not change. This ideal is a matching of diagnostic and treatment "universals." Alternatively, every music therapy may be described as a unique phenomenon that cannot be repeated or experienced by unique human beings.

To do justice to the complex matter of music therapy, different research approaches, such as quantitative, qualitative, biological, psychological, or sociological are necessary. Despite the complexity of music therapy it is probably better to work with specific microtheories, and specific operational hypotheses as often used in current medicine and psychotherapy (research).[26] These specific theories should be tested using adequate and modern scientific methods, techniques, and approaches. Neurophysiological investigation methods represent especially important new tools that are relevant in music therapy research and should be integrated into the pluralistic corpus of significant music therapy research methods.

In addition to the two problems characterized above, others exist. One is the gap between theory, research, and practice. Here the main question is whether general objectivistic theories and empirical results are adequate for music therapy research. In empirical studies there is a strong emphasis on homogeneity of groups and interventions, but the clinical work often consists of heterogeneous single cases.

Awareness of these problems, combined with the epistemological insight that there is no final conclusion, means that knowledge will advance. In addition to coincidental findings, insights that happen by chance, inductive generalizations, or deductive conclusions, keeping an open mind that respects other approaches is especially important to the growth of a field of applied science like music therapy. Therefore there is a reasonable chance that music therapy will profit from the quickly advancing field of neuroscience in music by means of interdisciplinary cooperation and discussion by experts.

A HEURISTIC WORKING FACTOR MODEL FOR MUSIC THERAPY AS A THEORETICAL FRAMEWORK

Because of the often unsatisfactory heterogenous theoretical situation in music therapy, the team at the University of Applied Science in Heidelberg has started to develop a specific model for the most effective ingredients in the field of music therapy. One provisional result is a heuristic model of working factors. The emphasis in this model is on a systematic collection of such ingredients seen as significant in music therapeutic work. It focuses on pragmatic therapeutic aspects and should be more specific if applied in special clinical fields (e.g., a therapy manual). Currently it consists of five factors.

(1) The first is called *attention modulation* or the *attentiveness factor*: The basic assumption is that music as an auditive quality has the power to attract attention (more than other sensual modalities). This factor includes phenomena associated with musical experience, like distraction, the relaxational use of music, and the so-called anxiolytic or algolytic effect that often is discussed in music medicine.[16] This musical ingredient is commonly used as an auditive signal for waking up (alarm clock effect). Additionally, many people use music for distraction from stressful events in everyday life. An impressive example for the power of auditive experiences to attract attention is the tinnitus experience. Patients suffering from such symptoms are often unable to concentrate on other sensory stimuli.[27] Attention modulation is supported by the phylogenetic function of audition as an early warning system. Anesthesiological studies of Schwender[28] and others imply that the auditive system is the sensory system that switches off last. A typical clinical use of that working factor is pain therapy, and it is very useful in music therapy with autistic and ADHS children.

(2) The second ingredient is called *emotion modulation* or the *emotional factor*: The basic assumption is that music (more than other human experiences) has the power to modulate emotions. Not only basic but also complex emotions, like national sentiments, can be stimulated by music. This factor is most relevant in active music therapy and includes direct emotional activation as well as the recall of emotional events associated with musical and auditive experiences. Emotion modulation is discussed in nearly all music therapy approaches, but empirical knowledge is still rather limited.[29] Emotion modulation is often a component in film music or love

songs. Neuroscientific research indicates that music is processed in the emotional brain (e.g., limbic system, gyrus cinguli, and the paralymbic cortical regions), and highly dissonant music tends to be unpleasant. In recent years research has been conducted on the so-called thrill or chill effect, which is an emotional and physiological reaction to music often associated with getting goose pimples.[30–32] Emotion modulation is common, especially in the psychotherapeutic use of music to directly evoke emotions, to recall emotional memories, and to learn more flexible emotional reactions.[33,34] It is very important to study the emotion modulation potential of music precisely because most psychological disorders and mental illnesses can also be described as emotional disturbances. Eliciting emotions by music may be a relevant topic and useful tool in the future treatment of these emotional disorders.

(3) The third factor is called the *cognition modulation* or *cognitive factor*: The basic assumption is that music represents a neurocognitve capacity. Music is produced by the human brain. The neuronal processing of music as well as its complexity is comparable to speech.[35,36] Therefore it is obvious that understanding music means thinking and creating (subjective) significance and experience. This factor implies subjective and cultural meanings for music as well as psychological, cognitive associations connected with specific musical experiences. In the same way the evocation of synaesthetic experiences is often used in music therapy. Musically induced visual imagery contributes to that factor as well as musical imagery itself.[29] Music always has a subjective meaning for humans. It is often observed that subcultures define themselves by specific musical styles, which are coded as some kind of group-specific language. Additionally music is used to alter states of consciousness in different native cultures.[37] Music is also known to facilitate recall of episodic memories.[38] Cognition modulation is clinically used to change subjective cognitions and meaning patterns, and is also important in music guided imagery techniques.[39,40] Some music therapists use music to induce a hypnotic trance and to alter states of consciousness.[41]

(4) The fourth factor is called *behavior modulation* or *motoric behavioral factor*: The basic assumption is that music represents a useful possibility to evoke and condition behavior, such as movement patterns, without the necessity of conscious will. The association of music and dance is well known. Marching songs are common, and the military offers a great variety of military marches. Neuroscientists, like the team of Michael Thaut, point out that rhythmic stimulation influences timing processes in the frontal brain and associated neural structures (neurologic music therapy, NMT). This factor is used therapeutically in gait rehabilitation of stroke patients and in the treatment of movement problems, for example, in Parkinson patients.[42,43] Music and auditive stimulation—known since the time of Pavlov—is a useful tool in behavioral conditioning in general. The analysis of the behavioral component of patients' performing music is of central interest in active music therapy and important in facilitating the learning of new behaviors. It is used as a theoretical framework in behavioral music therapy.[44,45]

(5) The fifth and last factor we call *communication modulation* or *interpersonal factor*: The basic assumption is that psychotherapeutic music therapy represents a complex paradigm of nonverbal communication.[46] This factor should be more emphasized in the future. It is often pointed out that music therapy uses nonverbal interpersonal interaction, but there is still little empirical knowledge of the patient–therapist interaction in music therapy settings. This phenomenon can be observed as a kind of community-building effect of music, for example, songs and music of sport fans, or music at the campfire. Many cultures use music to evoke the communal spirit and to enhance group cohesion. Humans use music to communicate simultaneously and not alternately, as in verbal dialogue. Improvisational music activities in therapeutic contexts are expecially seen as a form of nonverbal communication. Communication modulation is often used in a clinical setting for the learning of interpersonal competencies, especially in group music therapy. The reality of many music therapy interventions (active music therapy) is realized when therapist and patient are sharing a joint music activity and experience.

CONCLUSION

We conclude that music therapy does not need a new paradigm but may profit from new research methods. In clinical observation the therapeutic use of music often seems adequate and beneficial, but the empirical knowledge in our field is rare and limited, yet growing. If we want music therapy to have a more respected and defined role in modern health care systems, enhanced efforts will be needed. Neuroscientific research can help to support the long road toward evidence-based music therapy. We have started to walk this road, but the end is still not in sight. In conclusion, we pose several questions for neuroscientists:

• Do specific neurocognitive theories have the potential to explain the connections between music therapy interventions and the pathophysiology of mental and somatic diseases?

• Are specific neurocognitive theories and methods relevant for the explanation of observed short-term effects of music therapy interventions?

• Are neurocognitive theories and methods relevant for the explanation of observed positive outcome of effectiveness studies in music therapy (long-term effects)?

• Do neurocognitive theories and methods deliver new intervention strategies and techniques for music therapy?

• Do neurocognitive theories and methods help us to get a clearer view of possible working factors in music therapy?

[Competing interests: The authors declare that they have no competing financial interests.]

REFERENCES

1. GOLD, C. *et al.* 2004. Effects of music therapy for children and adolescents with psychopathology: a meta-analysis. J. Child Psychol. Psychiatry **45:** 1054–1063.
2. STANDLEY, J.M. 2002. A meta-analysis of the efficacy of music therapy for premature infants. J. Pediatr. Nurs. **17:** 107–112.
3. STANDLEY, J.M. 1996. A meta-analysis on the effects of music as reinforcement for education/therapy objectives. J. Res. Music. Edu. **44:** 105–133.
4. PELLETIER, C.L. 2004. The effect of music on decreasing arousal due to stress: a meta-analysis. J. Music. Ther. **41:** 192–214.
5. WHIPPLE, J. 2004. Music in intervention for children and adolescents with autism: a meta-analysis. J. Mus. Ther. **41:** 90–106.
6. EVANS, D. 2002. The effectiveness of music as an intervention for hospital patients: a systematic review. J. Adv. Nurs. **37:** 8–18.
7. STANDLEY, J.M. 1986. Music research in medical/dental treatment: meta-analysis and clinical applications. J. Mus. Ther. **23:** 56–122.
8. KOGER, S.M. *et al.*1999. Is music therapy an effective intervention for dementia? A meta-analytic review of literature. J. Music. Ther. **36:** 2–15.
9. SMEIJSTERS, H. 1997. Musiktherapie bei Alzheimerpatienten. Eine Metaanalyse von Forschungsergebnissen. Musikther. Umsch. **18:** 268–283.
10. STANDLEY, J.M. 1996. A meta-analysis on the effects of music as reinforcement for education/therapy objectives. J. Res. Mus. Edu. **44:** 105–133.
11. WORMIT, A.F. 2003. Qualitätssicherungsinstrumentarien in der Ambulanten Musiktherapie sind möglich! Musiktherapeut. Umsch. **24:** 291–292.
12. JACOBSON, N.S. *et al.* 1984. Psychotherapy outcome research: methods for reporting variability and evaluating clinical significance. Behav. Ther. **15:** 336–352.
13. JACOBSON, N.S. & P. TRUAX. 1991. Clinical significance: a statistical approach to defining meaningful change in psychotherapy research. J. Consult. Psychol. **59:** 12–19.
14. FRANK, J.D. & J.B. FRANK. 1991. Persuasion & Healing: A comparative Study of Psychotherapy. John Hopkins University Press. Baltimore.
15. LUBORSKY, L. *et al.* 1975. Comparative studies of psychotherapies: is it true that "everyone has won and all must have prizes?" Arch. Gen. Psychiatry **32:** 995–1008.
16. LAMBERT, M.J. 1992. Psychotherapy outcome research: implications for integrative and eclectic therapists. *In* Handbook of Psychotherapy Integration. J.C. Norcross & M.R. Goldfried, Eds.: 94–129. Basic Books. New York.
17. PRIESTLEY, M. 1994. Essays on Analytical Music Therapy. Barcelona Publishers. Phoenixville, PA.
18. SALAS J. & D. GONZALES. 1991. Like singing with a bird: improvisational music therapy with a blind four-year-old. *In* Case Studies in Music Therapy. K.E. Brucia, Ed.: 17–27. Barcelona Publishers. Phoenixville, PA.
19. MADSEN, C.K. *et al.* 1968. A behavioral approach to music therapy. J. Mus. Ther. **5:** 69–71.
20. HANSER, S.B. 1983. Music therapy: a behavioral perspective. Behav. Therap. **6:** 5–8.
21. NORDOFF, P. & C. ROBBINS. 1975. Music in Special Education. The John Day Company. New York.
22. SPINTGE, R. & R. DROH. 1992. Musik-Medizin. Gustav Fischer. Stuttgart.
23. POPPER, K. 1994. Logik der Forschung. 45–46. Mohr. Tübingen.
24. KUHN, T.S. 1996. The Structure of Scientific Revolutions. University of Chicago Press. Chicago.
25. SLIFE, B.D. 2004. Theoretical challenges to therapy practice and research: the constraint of naturalism. *In* Handbook of Psychotherapy and Behavior Change. Bergin & Garfield, Eds.: 44–83. John Wiley & Sons. New York.
26. LAMBERT, M.J. *et al.* 2004. Overview, trends, and future issues. *In* Handbook of Psychotherapy and Behavior Change. Bergin & Garfield, Eds.: 805–821. John Wiley & Sons. New York.
27. CUNY, C. *et al.* 2004. Reduced attention shift in response to auditory changes in subjects with tinnitus. Audiol. Neurotol. **9:** 294–302.
28. SCHWENDER, D. *et al.* 1991. Bewusste und unbewusste akustische Wahrnehmung während der Allgemeinanästhesie. Anaesthesist **40:** 583–593.

29. ZATORRE, R. 2003. Music and the brain. Ann. N. Y. Acad. Sci. **999:** 4–14.
30. PANKSEPP, J. 1995. The emotional sources of "chills" induced by music. Mus. Percep. **13:** 171–207.
31. PANKSEPP, J. 2002. Emotional sounds and the brain: the neuro-affective foundations of musical appreciation. Behav. Processes **60:** 133–155.
32. GABRIELSON, A. & W. LINDSTRÖM. 2003. Strong experiences related to music: a descriptive system. Musicae Scientiae **7:** 157–217.
33. BOLAY, H.V. *et al.* 1998. Musiktherapeutische Handlungsstrategien in der Behandlung von Schmerzpatienten. Musiktherapeut. Umsch. **19:** 268–277.
34. HILLECKE T. & H.V. BOLAY. 2000. Musiktherapie bei chronischen Schmerzen—theoretische Grundlagen—das Heidelberger Modell. Anesthesiol. Intensivmed. Notfallmed. Schmerzther. **35:** 394–400.
35. KOELSCH, S. *et al.* 2002. Bach speaks: a cortical "language-network" serves the processing of music. NeuroImage **17:** 956–966.
36. KOELSCH, S. *et al.* 2000. Musical syntax is processed in the area of Broca: an MEG study. NeuroImage **11:** 56.
37. ROUGET, G. 1985. Music and Trance: A Theory of the Relations between Music and Possession. The University of Chicago Press. Chicago.
38. SLOBODA, J.A. & P. JUSLIN. 2001. Psychological perspectives on music and emotion. *In* Music and Emotion: Theory and Research. P. Juslin & J.A. Sloboda, Eds.: 71–104. Oxford University Press. Oxford.
39. BONNY, H. 1978. Facilitating G.I.M. Sessions. G.I.M. Monograph, Vol. 1. ICM Books. Baltimore.
40. BONNY, H. 1989. Sound as symbol: guided imagery and music in clinical practice. National Association for Music Therapy California Symposium on Clinical Practices. Music. Ther. Perspect. **6:** 7–10.
41. DITTRICH, A. 1987. Bedingungen zur Induktion außergewöhnlicher Bewusstseinszustände. *In* Ethnopsychotherapie. A. Dittrich & C. Scharfetter, Eds.: 7–34. Enke. Stuttgart.
42. THAUT, M.H. *et al.* 1997. Rhythmic facilitation of gait training in hemiparetic stroke rehabilitation. J. Neurol. Sci. **15:** 207–212.
43. THAUT, M.H. *et al.* 1999. The connection between rhythmicity and brain function: implications for therapy of movement disorders. IEEE. Eng. Med. Biol **18:** 101–108.
44. MADSEN, C.K. *et al.* 1968. A behavioral approach to music therapy. J. Music. Therap. **5:** 69–71.
45. HANSER, S.B. 1983. Music therapy: A behavioral perspective. Behav. Therap. **6:** 5–8.
46. BOLAY, H.V. 1983. Musiktherapie. *In* Handbuch der Psychotherapie. R. Corsini, Ed.: 279–754. Psychologie Verlags Union. München.

Outcome Research in Music Therapy

A Step on the Long Road to an Evidence-Based Treatment

ANNE KATHRIN NICKEL,[a,b] THOMAS HILLECKE,[a,b] HEIKE ARGSTATTER,[a] AND HANS VOLKER BOLAY[a,b]

[a]German Center for Music Therapy Research (Viktor Dulger Institute), Heidelberg, Germany

[b]Outpatient Department of the Music Therapy Department, University of Applied Sciences, Heidelberg, Germany

ABSTRACT: Music therapy is the therapeutic use of music and musical activities in the treatment of somatic and mental diseases. In the last decades it has developed from a quasi-professional working field into an increasingly evidence-based treatment for various diseases. Selected outcome studies that were carried out in order to give music therapy a scientific and empirical base are presented in this article. Results show that music therapy is an effective intervention for patients with chronic pain, children with migraine, and patients suffering from chronic tinnitus. This positive outcome, in combination with the observed moderate to large effect sizes in different metanalyses, provides evidence for the use of music therapy in specific clinical fields. Future research should focus on studies that compare well-defined music therapy interventions to standard treatment.

KEYWORDS: evidence-based music therapy; music therapy outcome research; effectiveness of music therapy

INTRODUCTION

The use of music for preventing and curing mental and physical diseases is one of the oldest, and at the same time youngest, therapeutic interventions in human history.[1] Since ancient times music has been used for healing, but it was not until the twentieth century that music therapy established itself as an independent profession within the modern health care systems. Only in the last decades has it developed from a quasi-professional working field into a profession with a growing corpus of evidence-based interventions for various diseases.

Address for correspondence: Dr. Anne Kathrin Nickel, Deutsches Zentrum fur Musiktherapieforschung (Viktor Dulger Institut) DZM e.V., Maaßstr. 26, 69123 Heidelberg, Germany. Voice: +49-6221-884150; fax: +49-6221-884152.
anne.nickel@fh-heidelberg.de

Ann. N.Y. Acad. Sci. 1060: 283–293 (2005). © 2005 New York Academy of Sciences.
doi: 10.1196/annals.1360.021

DEFINITION OF MUSIC THERAPY

Music therapy is the therapeutic use of music and musical activities for patients suffering from somatic and mental diseases. The term *music therapy* contains two main, but overlapping, directions: psychotherapeutic music therapy and music medicine. Clinical work is completed by the use of music in the field of special education and consulting.

Typical work sites of music therapists include in-patient medical care, such as psychiatric clinics, psychotherapeutic clinics, neurological clinics and hospitals, but music therapists also offer out-patient care in counseling centers, pain centers, special kindergartens/schools, or in private practices.

Modern evidence-based music therapy needs to integrate current methods and theories of neuroscience, biology, psychotherapy, and music psychology, and should combine empirical research approaches with quantitative and qualitative research designs.[2] It integrates effectiveness and efficacy studies and standards of modern psychotherapy research.

In order to comprehend the state of the art, metanalytical studies need to be performed, some of which already exist for music therapy in medical applications,[3-5] music therapy with autistic children,[6] music therapy for people with dementia,[7] and for effects of music therapy in children and adolescents with psychopathology.[8]

THE HEIDELBERG MODEL OF MUSIC THERAPY

Health care systems are undergoing profound social and historical transformation processes. The rising demand for evidence-based medical and psychotherapeutic care has to be met by music therapy. Theoretically founded treatment manuals that are scientifically and empirically evaluated need to be developed. This is the main aim of the German Center for Music Therapy Research (Viktor Dulger Institute). The benefit lies not only in the fact that the proof of effectiveness may lead to funding of music therapy by health insurance, but also within the training context, where practical modules of the curriculum can be based on transparent learnable and replicable treatment techniques for different diagnosis groups. It also contributes to fewer school-oriented, personalized approaches within the professional field, inasmuch as communication (for example in the form of multicenter studies and study replications) across different centers and national borders becomes possible.

At the German Center for Music Therapy Research the efforts of the last ten years have centered around clinical effectiveness and efficacy studies, in which manualized music therapy concepts for defined patient populations have been evaluated. Studies were carried out within an interdisciplinary treatment context, using randomised and controlled designs and modern outcome data analysis methods, such as the method of clinical significance.[9,10]

The patient group most focused upon is that with chronic, psychosomatic diseases. For this area of work, music therapy is considered to be a form of artistic psychotherapy. Multimodal, symptom-specific treatment concepts are developed, addressing the symptom itself as well as comorbid psychological distress.

Lately the German Center for Music Therapy Research has begun to advance more into the field of music medicine or so-called functional music therapy. More recent research efforts have examined the effects of music therapy (or music as a relaxing/distracting stimulus) on patients undergoing heart or cataract examination, and in a multicenter study in cooperation with the Center for Biomedical Research in Music (Colorado State University) on *rhythmic-auditory stimulation*[11] in gait rehabilitation of hemiparetic stroke patients.

EFFECTIVENESS OF MUSIC THERAPY WITH CHRONIC, NONMALIGNANT PAIN PATIENTS

Introduction

Interdisciplinary therapy is most effective in the treatment of chronic pain.[12] In this project pain patients are treated by physicians trained in special pain therapy and in the experimental group additionally by music therapists using the Heidelberg model of music therapy for patients suffering from chronic pain.[13]

Methods

A randomized controlled design was carried out. The analyzed sample consisted of 40 patients randomly assigned to an experimental group ($n = 21$; age 51 ± 11 years, 67% female) receiving music therapy in addition to standard pharmacological pain treatment, and a waiting group ($n = 19$; age 52 ± 10 years, 74% female) receiving pharmacological treatment only. Patients were referred from the Pain Center of the Department of Anesthesiology (University Hospital Heidelberg) if they (1) were older than 18 years, (2) qualified for a diagnosis of one or several chronic nonmalignant pain disorders, (3) suffered from psychological distress but not to a psychiatric extent (except somatoform pain disorder) and (4) did not take part in other psychological intervention programs. Most patients suffered from headache or back pain. Statistical methods included parametric and nonparametric tests, pre- and postanalysis, group comparison, as well as the method of clinical significance.[14] Respecting the subjective experience of chronic pain, the focus of the measures was on self-rating data of patients using reliable and valid psychological questionnaires.

Results

Significant results in the group comparison condition can be demonstrated in pain measures (visual analogue scale, Schmerzempfindungs-Skala)[15] and psychological measures (OQ45.2)[16] (see TABLE 1).

By integrating the main measures (VAS: pain in the last 4 days; SES: affective pain; OQ: symptom distress) using the reliable change index,[14] one can conclude that about 70% of the patients improve after music therapy and medical treatment, whereas about 35% improvement in the waiting list control group is observed. The Mantel-Haenszel chi^2 test reveals a tendential superiority of the music therapy group. Analysis of patients with stable medication (experimental group, $n = 12$; waitlist, $n = 11$) results in statistically significant superiority of music therapy (see FIG. 1).

TABLE 1. Results of the postgroup comparison

Measure	Scale	EX n	EX M ± SD	WL n	WL M ± SD	P value
Pain intensity score (VAS)	Momentary pain	21	4.7 ± 2.4	19	5.5 ± 2.3	.092 ns
	Pain in the last 4 days, including today	21	5.1 ± 2.0	19	7.0 ± 1.6	.014*
Pain sensation scores (SES)	Affective pain	21	50 ± 8.4	19	53 ± 10.3	.355 ns
	Sensory pain	19	56 ± 10.5	19	54 ± 10.9	.832 ns
Psychotherapeutic outcome (OQ45.2)	Total score	21	65 ± 20.8	19	72 ± 20.1	.042*
	Symptom distress	21	40 ± 13.9	19	45 ± 13.3	.063 ns
	Interpersonal relations	19	12 ± 5.4	19	12 ± 6.8	.552 ns
	Social role	21	13 ± 4.9	19	14 ± 3.8	.027*

NOTE: Analysis of covariance (premeasure = covariate; F value = influence of group). The figure shows means (M) and standard deviations (SD) after therapy. EX = experimental group; WL = waitlist; ns = not significant; * = significant ($P < .05$).

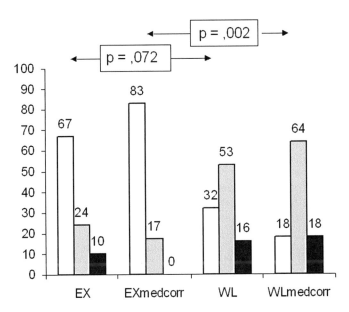

FIGURE 1. Percent of reliable change. Reliable improvement means reliable improvement in at least one of the defined criteria and no change in others. No change means no reliable change in all criteria or mixed reliable positive and negative changes. Reliable deterioration means reliable deterioration in at least one of the defined criteria and no change in the others. *Open bars,* reliable improvement; *light gray bars,* no change; *dark gray bars,* reliable deterioration. medcorr, corrected for medication: Only those patients were included whose medication intake was not changed during the course of the study.

Conclusion

Results of the study indicate that music therapy is an effective adjuvant intervention for patients suffering from chronic nonmalignant pain, doubling the effects of pharmacological treatment. Significant results were documented in pain measures and in psychological measures. Music therapy not only reduces pain but also addresses associated psychological distress in a positive way.

The results of the present study support findings concerning the effectiveness of active music therapy in patients with chronic pain in a personalized setting.[17,18]

EFFECTIVENESS OF MUSIC THERAPY IN THE TREATMENT OF CHILDREN WITH MIGRAINE HEADACHE

Introduction

Migraine in childhood is a serious health problem with a tendency toward chronification.[18,19] According to the biopsychosocial model, migraine is considered a disorder that is generated by multiple factors and that requires an interdisciplinary treatment concept, consisting of medical and psychological intervention. The aim of this study is to evaluate music therapeutic treatment and drug versus placebo treatment in children with migraine headaches, the strength and/or frequency of which requires preventive interval treatment.

An evaluated music-therapeutic concept for adults with chronic pain has been adapted to the requirements of child treatment.[18–21] As a preventive drug, we used Petadolex®, which has proved effective in adult migraine interval treatment.

Methods

A randomized, placebo-controlled, three-armed parallel group design was used; the two drug arms were double blind. Fifty-eight children (aged 10.4 ± 1.4 yr) were included in the trial (after diagnosis, baseline, and informed consent); because of drop out, group sizes decreased in postline to $n = 18$ music therapy, $n = 19$ Petadolex, and $n = 19$ placebo; and in follow-up to $n = 17$ music therapy, $n = 15$ Petadolex, and $n = 18$ placebo.

Inclusion criteria according to the guidelines for controlled trials of drugs in migraine of the International Headache Society:[21] ages 8–12 years, 2–6 attacks/month, 24 h between attacks, migraine Code 1.1, 1.2, and migraines for more than 1 year. Exclusion criteria according to the International Headache Society:[22] acute headache medication taken more than 10 d/month, antipsychotic or antidepressant medication in last 3 months, ongoing psychotherapy, and other severe illness.

Treatment duration was 3 months, including 12 weekly sessions of music therapy or 2 × 2 capsules (25 mg) per day. Outcome variables were headache parameters obtained through a headache diary, but also psychological factors obtained through interviews and psychological questionnaires. Data have been collected through pre- and postmeasurements, process measurement daily (+8 weeks baseline and postline), as well as follow-up one year after inclusion (see FIG. 2).

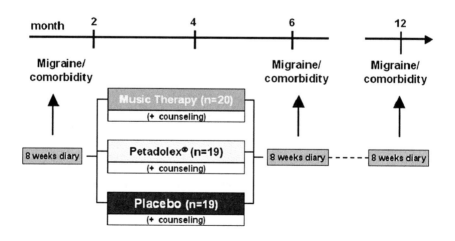

FIGURE 2. A migraine design study.

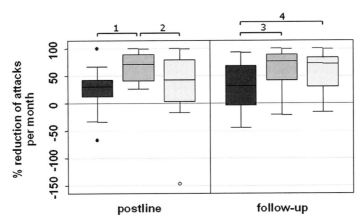

FIGURE 3. Relative reduction of migraine per month. Relative reduction pre- and post- (**1**) music therapy (med = 71%) vs. placebo (med = 31%), $P = .003$. (**2**) Music therapy vs. Petadolex® (med = 43%), $P = .097$. Relative reduction pre-follow-up (**3**) music therapy (med = 77%) vs. placebo (med = 31%), $P = .018$. (**4**) Petadolex (med = 73%) vs. placebo, $P = .044$.

Results

Seventy-one percent of the music therapy group profited reliably from treatment (their headache frequency was reduced by a minimum of 50% after treatment) and are thus therapy responders, according to guidelines of the International Headache Society.[22] Therapy effects are stable until follow-up. Music therapy patients are significantly more satisfied with treatment compared to the other conditions.

In the postmeasurement only the music therapy condition shows a significant placebo superiority with regard to relative reduction of headache frequency versus baseline. In the follow-up measure, both treatment groups (music therapy and Petadolex) supersede the placebo condition significantly (see FIG. 3).

Conclusion

Results suggest that music therapy and Petadolex are effective prophylactic treatments for pediatric migraine, while music therapy has the additional benefit of an earlier symptom reduction, higher therapy satisfaction, and patient compliance. Results are comparable or superior to other non–drug treatment methods for children with migraine.[23,24] Another advantage of music therapy compared to drug or drug-placebo treatment is that music therapy encourages active coping.[25] It tends to prevent passive coping strategies, which may lead to both legal and illegal drug abuse often observed in adolescents and adults suffering from chronic pain.[26,27]

EFFECTIVENESS OF MUSIC THERAPY WITH TINNITUS PATIENTS

Introduction

Tinnitus is one of the most common disorders in ENT medicine/otorinolaryngology. Patients suffering from chronic tinnitus experience such psychiatric distress as sleeping disturbance, scant attention, anxiety, and depression.[28] Because of the complexity of tinnitus, a comprehensive and interdisciplinary treatment is required. The music therapy concept aims at integrating the tinnitus sound into a musically controllable acoustic process.[29] The aim of the present study is to evaluate the effectiveness of this innovative concept.

Methods

This study was prospective, controlled, and randomized.

Sample: Twenty patients (10 male, 10 female; mean age 51 ± 7 yr) were chosen who were suffering from decompensated chronic tinnitus (mean score in the Tinnitus Questionnaire[30] TQ = 46.8 ± 9.6). All subjects underwent a comprehensive medical and psychological checkup in order to rule out organic and psychiatric diseases and were randomly allocated to the groups; therapy consisted of 12 50-min sessions of individual music therapy ($n = 10$); the control group ($n = 10$) did not receive any treatment.

Outcome variables were tinnitus parameters as well as coping styles and psychological factors. The outcome variables were obtained through interviews and psychological questionnaires.

Data collection consisted of pre- and postmeasurements, process measurement every 4 weeks, and follow-up after 24 weeks (see FIG. 4).

Results

Tinnitus pathology decreases continuously in the course of music therapy. Regression analyses indicate a highly statistical and clinical significant decline in mean TQ scores ($B = -3.8$; $F(1) = 26.38$; $P = .000$). Group comparison shows that TQ scores diminish in the music therapy group in the course of the therapy by 24.9 points or 53%, and in the control group by 2.4 points or 5%. Results of ANOVA reveal significant group differences pre-post ($F(1) = 14.190$, $P = .001$), with a large effect size of $d = 1.73$. Analysis of the follow-up results in the music therapy group shows that results remain almost unchanged. A decline pre-follow-up by 23 points or 49% has been observed (univariate repeated measures ANOVA: $F(5) = 5.68$, $P = .000$).

Because of the small sample size, the method of clinical significance[10] does lend itself to interpreting individual changes, as displayed in TABLE 2.

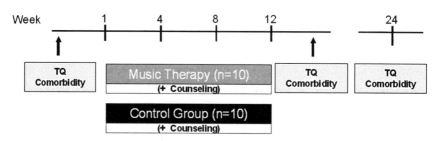

FIGURE 4. A tinnitus design study.

TABLE 2. Therapy results of the music therapy group

Category	Absolute number	TQ change absolute	TQ change relative
Clinically significant reduction (winner) (≤ −16)	n = 8	−30 points	−63%
Reliable reduction (responder) (−6 to −15)	n = 1	−8 points	−20%
No change (nonresponder) (−5 to +15)	n = 1	±0 points	±0%
Aggravation (loser) (≥ +16)	n = 0	—	—

Conclusion

Despite the small sample size, the innovative music therapy concept yields statistically and clinically significant results. Further research would seem to be beneficial. Brain imaging (PET, fMRI, and MEG) reveals cortical plasiticity in the auditory cortex similar to reorganization in chronic phantom pain.[31–33] This indicates that upcoming studies should correlate therapy outcome with neurophysiological investigation methods. A replication of the above-described project, integrating a bigger sample and more treatment conditions as well as neuroimaging, is projected in cooperation with the Max Planck Institute, Leipzig.

CONCLUSIONS

Music therapy seems to be an effective treatment for chronic pain, migraine in children, chronic tinnitus, as well as conditions in other clinical fields. Music therapy research is a growing field, but more outcome studies of effectiveness and efficacy of treatment are necessary. These studies need to be designed as comparative studies, testing music therapy versus standard treatment. Multicenter studies and replications are necessary to generalize observed effects. In order to achieve these aims, intraprofessional communication has to be enhanced. According to advances in the neurosciences, neurophysiological investigation methods should be increasingly integrated into outcome and working factor research.

Taking into account that nearly all established therapy approaches in the medical as well as in the psychological field have had to present an extensive corpus of well-designed outcome studies in order to achieve recognition in health care systems, music therapy should focus further efforts on this widely neglected field.

[Competing interests: The authors declare that they have no competing financial interests.]

REFERENCES

1. HERZOG, I. & L. BERGER. 1997. 5000 Jahre Musik in der Medizin. *In* Musik, Magie, und Medizin. Berger, Ed.: 49–52. Jungfermann. Paderborn.

2. HILLECKE, T., A. NICKEL & H.V. BOLAY. 2005. Scientific perspectives on music therapy. Ann. N. Y. Acad. Sci. **1060:** 271–282. [This volume].
3. STANDLEY, J.M. 1986. Music Research in medical/dental treatment: meta-analysis and clinical applications. J. Mus. Ther. **23:** 56–122.
4. BUNT, L. 1997. Clinical and therapeutic uses of music. *In* The Social Psychology of Music. D.J. Hargreaves & A.C. North, Eds.: 249–267. Oxford University Press. New York.
5. DILEO, C. 2003. Music therapy entrainment: a meta-analysis of the literature in medical music therapy and music medicine with an agenda of future research. Book of abstracts of the VIIIth. Symposium for Music in Medicine of the International Society of Music in Medicine (ISMM). Hamburg. June 24–28, 2003.
6. WHIPPLE, J. 2004. Music in intervention for children and adolescents with autism: a meta-analysis. J. Music. Ther. **26:** 90–106.
7. VINK, A.C., J.S. BIRKS, M.S. BRUINSMA & R.J. SCHOLTEN. 2004. Music therapy for people with dementia. Cochrane Database Syst. Rev. **3:** CD003477.
8. GOLD, C. *et al.* 2004. Effects of music therapy for children and adolescents with psychopathology: a meta-analysis. J. Child. Psychol. Psychiatry **45:** 1054–1063.
9. JACOBSON, N.S. & P. TRUAX. 1991. Clinical significance: a statistical approach to defining meaningful change in psychotherapy research. J. Consult. Clin. Psychol. **59:** 12–19.
10. KORDY, H. & W. HANNOVER. 2000. Die Evaluation von Psychotherapie und das Konzept der klinisch bedeutsamen Veränderung. *In* Diagnostik in der Psychotherapie. A. R. Laireiter, Ed.: 477–495. Springer-Verlag. Wien.
11. THAUT, M.H. *et al.* 1997. Rhythmic facilitation of gait training in hemiparetic stroke rehabilitation. J. Neurol. Sci. **151:** 207–121.
12. FLOR, H. *et al.* 1992. Efficacy of multidisciplinary pain treatment centers: a meta-analytical review. Pain. **49:** 221–230.
13. HILLECKE, T. & H.V. BOLAY. 2000. Musiktherapie bei chronischen Schmerzen—theoretische Grundlagen—das Heidelberger Modell. Anästhesiol. Intensivmed. Notfallmed. Schmerzther. **35:** 394–400.
14. JACOBSON, N.S. & P. TRUAX. 1991. Clinical significance: a statistical approach to defining meaningful change in psychotherapy research. J. Consult. Clin. Psychol. **59:** 12–19.
15. GEISSNER, E. 1996. Die Schmerzempfindungs-Skala-Handanweisung. Hogrefe. Gottingen.
16. LAMBERT, M.J. *et al.* 1996. Administration and Scoring Manual for the OQ(TM)-45.2 (Outcome Questionnaire). M.D. Stevenson, American Professional Credentialing Services, LLC.
17. MULLER-BUSCH, H.C. & P. HOFFMANN. 1997. Aktive Musiktherapie bei chronischen Schmerzen: eine prospektive Untersuchung. Schmerz **11:** 91–100.
18. NICKEL, A.K. *et al.* 2002. Musiktherapie mit Kindern mit Migräne. Psychotherapeut **47:** 285–290.
19. OELKERS-AX, R. & F. RESCH. 2002. Kopfschmerzen bei Kindern: auch ein kinder- und jugendpsychiatrisches Problem? Pathogenese, Komorbidität, Therapie. Z. Kinder. Jugendpsychiatr. **30:** 281–293.
20. WINNER, P. 1999. Pediatric headaches: What's new? Curr. Opin. Neurol. **12:** 269–272.
21. NICKEL, A.K. *et al.* 2003. Music therapy in the treatment of children with migraine. Mus. Ther. Today 4. Available at http://musictherapyworld.net.
22. INTERNATIONAL HEADACHE SOCIETY CLINICAL TRIALS SUBCOMMITTEE. 2000. Guidelines for controlled trials of drugs in migraine. Cephalalgia **20:** 765–786.
23. MCGRATH, P.J. *et al.* 1990. Help Yourself: A Treatment for Migraine Headaches. University of Ottawa Press. Ottawa.
24. UBERALL, M.A. *et al.* 2001. Kopfschmerztherapie im Kindes- und Jugendalter. *In* Schmerztherapie bei Kindern. B. Zernikow, Ed.: 238–257. Springer. Berlin.
25. BANDURA, A. 1977. Self-efficacy: toward a unifying theory of behavioral change. Psychol. Rev. **84:** 191–215.
26. HAAG, G. *et al.* 2004. Prophylaxe und Therapie des medikamenteninduzierten Dauerkopfschmerzes. Nervenheilkd. In press.

27. GOEBEL, H. 1997. Die Kopfschmerzen. Springer. Berlin.
28. TAYLOR, R.S. 2000. Tinnitus Handbook. Singular Publishing Group. San Diego.
29. ARGSTATTER, H. *et al.* 2005. Musiktherapie bei chronischem Tinnitus: Pilotstudie zur Entwicklung und Uberprüfung einer neuartigen Behandlungsmethode. Z. Mus. Tanz. Kunstther. **16:** 1–6.
30. GOEBEL, G. & W. HILLER. 1998. Tinnitusfragebogen: ein Instrument zur Erfassung von Belastung und Schweregrad bei Tinnitus. Hogrefe. Göttingen Bern.
31. MUHLNICKEL, W. *et al.* 1998. Reorganization of auditory cortex in tinnitus. Proc. Natl. Acad. Sci. **95:** 10340–10343.
32. MIRZ, F. *et al.* 2000. Cortical networks subserving the perception of tinnitus: a PET study. Acta. Oto. Laryng. Suppl. **543:** 241–243.
33. PENAGOS, H. *et al.* 2004. A neural representation of pitch salience in nonprimary human auditory cortex revealed with functional magnetic resonance imaging. J. Neur. Sci. **24:** 6810–6815.

Music Therapy Research in Ibero-American Countries

An Overview Focused on Assessment and Clinical Evaluation

Evaluation is a new discipline,
but an ancient practice.
—MICHAEL SCRIVEN

PATRICIA L. SABBATELLA

University of Cádiz, Faculty of Sciences Education,
Campus Universitario de Puerto Real, 11519 Puerto Real, Cádiz, España

ABSTRACT: The aim of this article is to present an overview of some contemporary ideas concerning the status of music therapy research in Ibero-American countries, with a focus on assessment and clinical evaluation of music therapy clinical practice.

KEYWORDS: international music therapy; music therapy in Ibero-America; music therapy research; music therapy assessment

INTRODUCTION

Music therapy practice, research, and theory differ from one another in purpose and process. Yet they are integrally linked to one another,[1] as are applications of music therapy around the world. Several important factors, based on social, cultural, political, geographic, linguistic, and professional elements, have influenced and contributed to the development of music therapy. Contemporary music therapy is diverse in theoretical and methodological approaches to clinical practice and research. From an international perspective, we differentiate between two professional perspectives: the Anglo-Saxon one (English scientific language), and the Ibero-American[a] (Spanish scientific language). Within these contexts, the theoretical and methodological orientations of music therapy clinical practice are different, with implications for research interests and scientific publications.[2] Nowadays, the international perspective of music therapy is expanding, and it is possible to differentiate a third emerging professional context: Asian music therapists are communicating their findings to Western music therapists.[3]

Address for correspondence: Patricia L. Sabbatella, Ph.D., University of Cádiz, Faculty of Sciences Education, Campus Universitario de Puerto Real, 11519 Puerto Real, Cádiz, España. Voice: +34-956-016225; fax: +34-956-016253.

patricia.sabbatella@uca.es

[a]Ibero-American countries include Latin American countries, Spain, and Portugal.

Ann. N.Y. Acad. Sci. 1060: 294–302 (2005). © 2005 New York Academy of Sciences.
doi: 10.1196/annals.1360.022

MUSIC THERAPY RESEARCH IN IBERO-AMERICAN COUNTRIES

Since the 1960s Latin American countries, especially Argentina, Brazil, and Uruguay, have contributed widely to the development of music therapy in South America and influenced other countries like Chile, Peru, Colombia, Venezuela, Mexico, Spain, and Portugal.[4] In 1993 the Latin American Music Therapy Committee was founded to promote the development of music therapy in the region, and four Latin American conferences were organized (in 1996, 1998, 2002, and 2004). Nowadays, music therapy as practiced in Latin America is represented in international seminars, congresses, and conferences, and many music therapists from these countries work as representatives in international institutions.

From an international perspective, political and financial issues determine the status of research, and the current socioeconomic situation in Latin America does not permit financial support for these kinds of research projects. Within this framework, the status of research funds for Spain and Portugal is different than for funds in Latin America.

At the beginning of the twenty-first century, a survey of the literature shows that music therapy research in the Ibero-American countries is increasing.[5–9] Research activities are coordinated by music therapists who establish partnerships with professionals from medical and educational institutions, some of them receiving financial support from public or private organizations. At the present time, there are not enough music therapists trained in research methods working in Latin American, Spanish, and Portuguese universities, and no Ph.D. programs in music therapy are yet offered.

ASSESSMENT AND EVALUATION IN MUSIC THERAPY RESEARCH

The field of assessment and clinical evaluation in music therapy is broad. Within music therapy theory, attention to the importance of assessment has been less than rigorous. In this paper, my aim is to introduce some contemporary ideas regarding the status of research on this topic, and to discuss (1) descriptive results of an extensive literature review of articles published in Spanish, English, and Portuguese music therapy journals between 1985 and 2004, and in proceedings of world, European, and Latin American music therapy congresses and conferences held between 1993 and 2004; and (2) results of a research survey conducted by the author about the status of music therapy clinical practice in Ibero-American countries, focusing on assessment and evaluation of clients and covering the main areas of clinical work from 1998 to 2001.

LITERATURE REVIEW

Nowadays there is a sizable body of literature discussing a wide variety of topics that support clinical application, research, and teaching. Literature is a referent, allowing one to observe the philosophical, theoretical, and methodological orientation covered by authors and the way these ideas influence the clinical practice of music therapy. This review examined a total of 11 music therapy journals and 12 conference proceedings.

TABLE 1. Articles about assessment and evaluation in music therapy journals (1985–2004)

Journals Reviewed	Number of Articles
Spanish journals	
Música, Terapia y Comunicación	1
Música, Arte y Proceso	1
Revista Española de Musicoterapia	No reference
Revista Internacional Latinoamericana de Musicoterapia	1
Icmus, Investigación y Clínica Musicoterapéutica	No reference
English journals	
Journal of Music Therapy	19
Music Therapy	1
Music Therapy Perspectives	15
Bristish Journal of Music Therapy	9
Nordic Journal of Music Therapy	1
Portuguese journals	
Revista Brasileira de Musicoterapia	No reference

A total of 94 articles about assessment and evaluation in music therapy were published between 1985 and 2004: 48 in music therapy journals, and 46 in conferences proceedings and abstract books from world, European, and Latin American music therapy congresses. Most of the articles were published in English language music therapy journals ($n = 45$), and only 3 references were located in Spanish language music therapy journals (TABLE 1).

Forty-six papers were published in conferences proceedings. The number of presentations at music therapy congresses was low, ranging from 0 to 4, except during the VIII World Congress (7 papers), the IX World Congress (13 papers), the X World Congress (7 papers), and the VI European Congress (7 papers). Only 2 papers were presented at the II Latin American Music Therapy Meeting (TABLE 2).

Although there were articles published covering topics about assessment and evaluation during the 1980s and early 1990s, publications in this area started to increase in the late 1990s, with another increase at the beginning of the twenty-first century.

During the 1980s, articles published emphasized the importance of the assessment process in music therapy, the necessity of designing specific instruments, implementation of effective approaches to client assessment,[10–11] and a description of assessment instruments for different client populations.[12–17]

During the 1990s, publications focused on assessment for specific diagnosis areas,[18–35] methods for improvisation analysis,[36–40] data collection,[41–42] treatment effectiveness,[43–45] and evaluation research.[46,47]

Beginning in 2000, publications focused on theoretical aspects of assessment and evaluation,[48–54] methods for improvisation analysis,[55–57] assessment used with different client populations,[58–71] and service evaluation.[72,73]

TABLE 2: Papers about assessment and clinical evaluation published in conference proceedings and abstract books of music therapy congresses (1993–2004)

	Congresses	Number of Papers
World congresses	VII (1993)	2
	VIII (1996)	7
	IX (1999)	13
	X (2002)	7
European congresses	III (1995)	2
	IV (1998)	2
	V (2001)	4
	VI (2004)	7
Latin-American congresses	I (1996)	–
	II (1998)	2
	III (2002)	–
	IV (2004)	–

A comprehensive review of the literature shows that the area of assessment and evaluation in music therapy is still weak and, as a topic of research, still in progress. There is not much literature dealing specifically with theory, methods, research, assessment, and clinical evaluation, and several gaps can be found in the literature:

- Most of the articles focus on client assessment but not on the process and practice of music therapy assessment for any clinical setting.

- Authors do not specifically address music therapy evaluation as a methodological process, and there is not enough information about objectives, criteria, data collection, measurement, categorization of data, standardization of instruments, areas of evaluation, relationships between assessment and evaluation, report style, interpretation of data and reports, and evaluation of treatment effectiveness.

- Informal assessment tools are abundantly available in music therapy practice, but few reliable assessment tools have been designed. There is a lack of standardized or formal music therapy assessment instruments. Some authors adapt test or assessment tools from other disciplines (e.g., music education, psychotherapy, or physical rehabilitation).

- There are no publications about assessment research.

- Concerning authors, a review confirms that most of them belong to the Anglo-Saxon perspective, with articles published in English language journals. There is an absence of specific literature and studies about assessment and clinical evaluation in Ibero-American countries and few papers published in Spanish or Portuguese. Literature available on assessment and clinical evaluation in music therapy remains sparse and usually appears fragmented.

RESEARCH SURVEY

Research about the status of music therapy clinical practice in Ibero-American countries was conducted by the author between 1998 and 2001.

The method of research was based on a survey design, and data collection was made through the Práctica Profesional de la Musicoterapia (Survey of Music Therapy Clinical Practice). The survey had 25 open and closed questions organized in three sections: demographic and professional issues; methods and approach to clinical practice, and methods of assessment and evaluation of clients.

The criterion used to delimit the target audience and select participants was membership in the Comité Latinoamericano de Musicoterapia (CLAM-Latin American Music Therapy Committee). To overview the context, Spain and Portugal were included. A total of 119 responses was received, and no responses were obtained from Mexico, Venezuela, and Cuba.

Descriptive analysis of data collected summarizes that the ability of the music therapist to assess clients depends on his/her clinical experience and the theoretical orientation of the clinical practice. A flexible approach to client assessment and evaluation is used. The approach to clinical evaluation is not formal and can be categorized as nonstructured assessment organized in a number of sessions (between 1 and 6), where the music therapist assesses different aspects of the client's case using various methods. Assessment purposes vary according to client population, clinical settings, type of assessment tool used, and the theoretical orientation of the therapy process.

Instruments used for data collection vary widely according to the clinical population, the level of clinical practice, institutional needs, and the theoretical orientation of the therapist. Improvisation and listening to music are the assessment tasks more frequently used.

The recording methods used most frequently to collect and preserve data in music therapy assessment are (1) written reports (85%), (2) audiotape recording (60%), (3) music notation (37.4%), (4) behavioral checklists (30%), (5) and videotape recording (16.5%).

The most frequently evaluated skill areas in different client populations are:

- nonmusical: communication skills (94.8%), emotional responsiveness (94%), sensorimotor skills (90.6%), and cognitive skills (55.6%).

- musical: participation in musical activities (80.5%), general musical skills (70.5%), use of musical instruments (61.2%), and voice skills (66.1%).

Descriptive reports are the model used to present assessment and evaluation data.

CONCLUSIONS

Music therapy theory has developed out of empirical practice, and attention to the importance of assessment has been limited, and less rigorous and standardized than in other disciplines. Nowadays, this trend seems to be changing, and more and more music therapists are looking for studies reporting relevant information on the topic because of its importance in the treatment process and its role in furthering the credibility of the profession. Research and a review of the literature demonstrate (1) that music therapy as a profession needs to develop assessment and clinical evaluation tools on a higher scientific level, and (2) that music therapy as a discipline needs to increase the number and quality of studies related to evaluation in different music therapy areas in order to provide a more reliable and substantive body of knowledge.

In clinical evaluation, it is necessary (1) to increase the volume and quality of assessment research, connecting theoretical backgrounds from such fields as psychological assessment, evaluation theory, statistical analysis, musical tests, and musical disorders; (2) to promote the study of clinical evaluation in music therapy from an evaluative perspective, and (3) to develop assessment tools and clinical evaluation protocols on a higher scientific level and based on evidence research methods.

Conducting assessments and evaluation is seen as a way of seeking credibility and professional recognition. As professionals, we should be able to produce high-quality service in our clinical practice. More research and writing is needed in this emerging area.

[Competing interests: The author declares that she has no competing financial interests.]

REFERENCES

1. Bruscia, K. 1998. Defining Music Therapy. Second Edit. Barcelona Publishers. Gilsum, NH.
2. Sabbatella, P. 2001. El desarrollo profesional de la musicoterapia: un estudio en el contexto iberoamericano. Rev. Bras. Musicoter. **IV:** 9–28.
3. Rika, I. 2005. Reflections on the first Asian music therapy symposium. Voices: Main Issues. Vol. 5, no. 1, Art. #166, available at <http://www.voices.no/mainissues/mi40005000166.html.>
4. Sabbatella, P. 2004. Music therapy in Latin-America: an overview of the First International Summer Congress: development in music therapy. Mus. Ther. Today (online) Vol. V, Issue 3, available at <http://www.musictherapyworld.net.>
5. Sabbatella, P. 2001b. La investigación en musicoterapia: ecos de una identidad en desarrollo. Rev. Española Musicoter. **4:** 32–41.
6. Smith, M. 2003. Music therapy in Brazil. [online] Voices: a world forum for music therapy. Retrieved from <http://www.voices.no/country/monthbrazil_april2003.html.>
7. Ferrari, K. & V. Snchez. 2003. Music therapy in Argentina [online]. Voices: a world forum for music therapy. Available at <http://www.voices.no/country/monthargentina_july2003.html.>
8. Leite, T. 2002. Music therapy in Portugal [online] Voices: a world forum for music therapy. Available at <http://www.voices.no/country/monthportugal_sept2002.html.>
9. Sabbatella, P. 2004. Music therapy in Spain. [online] Voices: a world forum for music therapy. Available at <http://www.voices.no/country/monthspain_march2004.html.>
10. Bruscia, K. 1988. Standards for clinical assessment in the arts therapies. Arts Psychoter. **15:** 5–10.
11. Isenberg-Grezda, I. 1988. Music therapy assessment: a reflection of professional identity. Mus. Ther. **3:** 156–169.
12. Braswell, C. et al. 1986. Development and implementation of music: activity therapy intake assessment for psychiatric patients. J. Mus. Ther. **23:** 126–141.
13. Gfeller, K. & A. Baumann. 1988. Assessment procedures for music therapy with hearing impaired children. J. Mus. Ther. **25:** 192–205.
14. Hunter, L. 1989. Computer-assisted assessment of melodic and rhythmic discrimination skills. J. Mus. Ther. **26:** 79–87.
15. Jones, R. 1986. Assessing developmental levels of mentally retarded students with the musical-perception assessment of cognitive development. J. Mus. Ther. **23:** 166–173.

16. JAMES, M. 1986. Utilization of motor skill assessments in clinical practice. Mus. Ther. Perspect. **3:** 9–12.
17. SANDROCK, D. & M. JAMES. 1989. Assessment instruments for music-assisted relaxation training. Mus. Ther. Perspect. **7:** 44–50.
18. BROWNING, A., J. COLEMAN & S. MURRAY. 1990. Music therapy assessment of the cerebrovascular accident patient. Mus. Ther. Perspect. **8:** 23–31.
19. CASSITY, M. & J. CASSITY. 1994. Psychiatric music therapy assessment and treatment in clinical training facilities with adults, adolescents, and children. J. Mus. Ther. **31:** 2–30.
20. DARROW, A. 1991. An assessment and comparison of hearing impaired children's preference for timbre and musical instruments. J. Mus. Ther. **28:** 48–59.
21. GRIGGS-DRANE, E. & J. WHEELER. 1997. The use of functional assessment procedures and individualized schedules in the treatment of autism: recommendations for music therapists. Mus. Ther. Perspect. **15:** 87–93.
22. HOOPER, J. 1996. The Somatron: evaluating and understanding the initial reactions of clients with learning disability. *In* Book of Abstracts: VIII World Congress of Music Therapy. H. Smeijsters & F. Mecklenbeck, Eds. Trialog. Bremen.
23. LECOURT, E. 1999. Proposition for a scale evaluating the sonorous and musical world of persons with autism. Abstr. IX World Congress Mus. Ther.
24. LIPE, A. 1995. The use of music performance tasks in the assessment of cognitive functioning among older adults with dementia. J. Mus. Ther. **32:** 137–151.
25. MAGEE, W. 1995. Case studies in Huntington's disease: music therapy assessment and treatment in the early to advanced stages. Br. J. Mus. Ther. **9:** 13–19.
26. METZNER, S. 2000. Evaluación de la improvisación grupal en musicoterapia. Rev. Int. Latinoamericana Musicoter. **6:** 2.
27. OLFIELD, A. 1999. Music therapy assessment and short-term treatment in child psychiatry. Abstr. IX World Congr. Mus. Ther.
28. SCHUMACHER, K. 1998. Evaluation of interpersonal relationship through the use of instruments in music therapy with autistic children. Abstr. IV Eur. Mus. Ther. Congr.
29. STIGE, B. 1996. Music and social interaction. Music therapy assessment, intervention and evaluation in a transactional perspective. *In* Book of Abstracts: VIII World Congress of Music Therapy. H. Smeijsters & F. Mecklenbeck, Eds. Trialog. Bremen.
30. SLOBODA, L. 1996. Individual music therapy as a tool for the assessment and treatment of mentally ill offenders in a secure psychiatric unit. *In* Book of Abstracts: VIII World Congress of Music Therapy. H. Smeijsters & F. Mecklenbeck, Eds. Trialog. Bremen.
31. TOWSE, E. 1997. Group analysis and improvisation: a musical perspective. Br. J. Mus. Ther. **11:** 51–55.
32. YORK, E. 1994. The development of a quantitative music skills test for patients with Alzheimer's disease. J. Mus. Ther. **31:** 280–296.
33. WERGER, C. 1995. Assessment. Abstr. III Eur. Mus. Ther. Congr.
34. WIGRAM, T. 1995. Rett syndrome therapy clinics: a multidisciplinary approach to therapeutic assessment, and the place of music therapy in this process. Abstr. III Eur. Mus. Ther. Congr.
35. WIGRAM, T. 1996. El diagnóstico de niños con trastornos comunicativos a través de la musicoterapia. Mus. Arte Proceso **1:** 18–32.
36. ANSDELL, G. 1991. Mapping the territory. J. Br. Mus. Ther. **5:** 18–27.
37. HAKES, R. 1999. An electroacoustic analysis of musical improvisation in group psychotherapy: a case study. Abstr. IX World Congr. Mus. Ther.
38. LEE, C. 1996. A method of analysing improvisations in music therapy. *In* Book of Abstracts: VIII World Congress of Music Therapy. H. Smeijsters & F. Mecklenbeck, Eds. Trialog. Bremen.
39. NOLAN, P., C. ROBBINS & B. SCHEIBY. 1999. Clinical musical models of tape analysis. Abstr. VIII World Congr. Mus. Ther.
40. TOCANTIS, R. 1999. Semiotical analysis of musical discourse in music therapy. Abstr. IX World Congr. Mus. Ther.
41. BERGSTROEM-NIELSEN, C. 1993. Graphic notation as a tool in describing and analyzing music therapy improvisations. Mus. Ther. **12:** 40–58.

42. LOUREIRO, C. 1998. Coleta de dados na prática musicoterápica. Abstr. II Encuentro Latinoamericano Musicoter.
43. BEHRENS, G. 1999. Methods for evaluating the results of single-subject research. Abstr. IX World Congr. Mus. Ther.
44. CODDING, P. 1999. Music and art therapy in the correctional setting: qualitative research as an evaluative tool for client progress and program effectiveness. Abstracts IX World Congress of Music Therapy.
45. HEANY, C. 1992. Evaluation of music therapy and other treatment modalities by adult psychiatric inpatients. J. Mus. Ther. 29: 70–86.
46. OLFIELD, A. 1993. A study of the way music therapists analyze their work. Br. J. Mus. Ther. 7: 14–22.
47. SABBATELLA, P. 1998. How to Evaluate Music Therapy? Abstracts IV European Music Therapy Congress.
48. BRUNT, B. & K. COLEMAN. 2000. Development of a special education music therapy assessment process. Mus. Ther. Perspect. 18: 59–68.
49. BRUSCIA, K. 2001. A qualitative approach to analyzing client improvisations. Music Therapy Perspectives 19: 7–21.
50. DE BATISTA, R. 2001. Taking note: working with children with communication disorders and observing our observation of them in assessment. Abstracts V European Music Therapy Congress.
51. GREGORY, D. 2000a. Test instruments used by Journal of Music Therapy authors from 1984–1997. J. Mus. Ther. 37: 79–94.
52. GREGORY, D. 2000. Information sharing: technological assistance in client assessment: implications for music therapy. Mus. Ther. Perspect. 18: 69–71.
53. LOEWY, J. 2000. Music psychotherapy assessment. Mus. Ther. Perspect. 18: 47–58.
54. GANTT, L. 2000. Assessments in the creative arts therapies: learning for each other. Mus. Ther. Perspect. 18: 41–46.
55. ARNASON, C. 2002. An eclectic approach to the analysis of improvisations in music therapy sessions. Mus. Ther. Perspect. 22: 4–12.
56. BRUSCIA, K. 2001. A qualitative approach to analyzing client improvisations. Mus. Ther. Perspect. 19: 7–21.
57. LEE, C. 2000. A method of analysing improvisations in music therapy. J. Mus. Ther. 37: 147–167.
58. ALA-RUONA, E. 2004. Nonstructured assessment of psychiatric client in music therapy: therapist as a hands-on researcher. Paper presented at the VI European Music Therapy Congress.
59. BOYLE, M. 2004. An overview of measurement strategies for music therapists in general medical settings. Paper presented at the VI European Music Therapy Congress.
60. CHASE, K. 2004. Music therapy assessment for children with developmental disabilities: a survey study. J. Mus. Ther. 41: 28–54.
61. DE BATISTA, R. 2001. Taking note: working with children with communication disorders and observing our observation of them in assessment. Abstr. V Eur. Mus. Ther. Congr.
62. EDWARDS, J. 2004. Issues in testing and measurement of responses to music therapy in research with children receiving hospital care following injury or illness. Paper presented at the VI European Music Therapy Congress.
63. MATILA, L. 2001. Assessing developmentally retarded patients. Overall assessment with music therapy, a study of the therapist's own clinical work. Abstracts V European Music Therapy Congress.
64. OLDFIELD, A. & J. HOLMES. 2004. An investigation into how music therapy diagnostic assessments can help the psychiatric team diagnose children suspected of having autistic spectrum disorders. Paper presented at the VI European Music Therapy Congress.
65. SAGI, N. & R. SAGI. 2001. A study of qualitative assessment on music therapy for senile dementia with familiar music. Abstracts V European Congress of Music Therapy.
66. SCALENGE, R. & K. MURPHY. 2000. Music therapy assessment in the managed care environment. Mus. Ther. Perspect. 18: 23–30.

67. YORK, E. 2002. Musical tasks in clinical assessment: dialogue with the residual music skill test. Paper presented at the X World Congress of Music Therapy.
68. VARKILA, L. 2004. Improvisation as a method in music therapeutical assessment with developmentally retarded clients. A study of the therapist's own clinical work. Paper presented at the VI European Music Therapy Congress.
69. WHEELER, B. 2004. Issues in measuring the effects of music therapy with patients with cardiac problems. Paper presented at the VI European Music Therapy Congress.
70. WILSON, B. & D. SMITH. 2000. Music therapy assessment in school settings: a preliminary investigation. J. Mus. Ther. **37:** 95–117.
71. WIGRAM, T. 2000. A method of music therapy assessment for the diagnosis of autism and communication disorders in children. Mus. Ther. Perspect. **18:** 13–30.
72. MOSS, H. 2003. Service evaluation: music therapy and medicine for the elderly. Br. J. Mus. Ther. **17:** 76–89.
73. ROBB, S. 2000. The effect of therapeutic music interventions on the behavior of hospitalized children in isolation: developing a contextual support model of music therapy. J. Mus. Ther. **37:** 118–146.

The Future of Music in Therapy and Medicine

MICHAEL H. THAUT

Center for Biomedical Research in Music,
Molecular, Cellular, and Integrative Neuroscience Programs,
Colorado State University, Fort Collins, Colorado 80523, USA

ABSTRACT: The understanding of music's role and function in therapy and medicine is undergoing a rapid transformation, based on neuroscientific research showing the reciprocal relationship between studying the neurobiological foundations of music in the brain and how musical behavior through learning and experience changes brain and behavior function. Through this research the theory and clinical practice of music therapy is changing more and more from a social science model, based on cultural roles and general well-being concepts, to a neuroscience-guided model based on brain function and music perception. This paradigm shift has the potential to move music therapy from an adjunct modality to a central treatment modality in rehabilitation and therapy.

KEYWORDS: music; neuroscience; aesthetics; neurologic music therapy

FOUNDATIONS IN AESTHETICS AND NEUROSCIENCE

The study of the neurobiological basis of music has intrinsically linked music to a role in influencing and shaping brain function. The brain—one may state—that engages in music is changed by engaging in music. This reciprocal relationship in music and brain function has been discovered over the past 10 years by connecting the fields of neuroscience, music cognition, music therapy, and rehabilitation. The connection has unfolded within the larger context of a very fascinating line of research demonstrating the experience-dependent plasticity of the brain, which is one of the most powerful motors of change in the understanding of learning, cognition, and therapeutic rehabilitation. In the modern history of music therapy—encompassing roughly the past 60 years—social science concepts of music's influence and role in human life and society have dominated music therapy. The therapeutic value of music was explained mainly by music's cultural role in facilitating social learning and emotional well-being. However, more recently—under the influence of new data in brain research in music—new findings suggest that music can stimulate complex cognitive, affective, and sensorimotor processes in the brain, which can then be generalized and transferred to nonmusical therapeutic purposes.

Address for correspondence: Michael Thaut, Center for Biomedical Research in Music, Molecular, Cellular, and Integrative Neuroscience Programs, Colorado State University, Fort Collins, CO 80523. Voice: 970-491-5529; fax: 970-491-7541.
Michael.Thaut@ColoState.edu

Ann. N.Y. Acad. Sci. 1060: 303–308 (2005). © 2005 New York Academy of Sciences.
doi: 10.1196/annals.1360.023

These research developments fit surprisingly well with developments in aesthetics and psychobiology that help our understanding of the role of art works in perception. One of the key concepts suggested in aesthetic perception is the function of the mediating response, as described for the first time comprehensively by Berlyne in 1971.[1] Berlyne contends that aesthetic (sensory) stimuli can have a facilitating, clarifying, organizing, and amplifying function in the perception and analysis of nonaesthetic objects and behavior experiences. Furthermore, they can facilitate and enhance the development of adequate and adaptive responses to such objects and experiences.

The concept of music as a mediating stimulus is a useful one for music in therapy, because it locates the musical response and musical stimulus as a mediator among current brain and behavior function, the aims of therapy, and the desired therapeutic response. In experimental aesthetics the mediating response in music is caused by the meaningful perception of the intrinsic patterns in the music itself, as well as by other symbolic, iconic, or behavioral meanings that have become linked to the music through associative learning processes. As a mediating stimulus, music—based on its uniquely ordered structure of sensory patterns in aesthetic forms—initially engages human behavior and brain function meaningfully by arousing, guiding, organizing, focusing, and modulating perception, attention, and behavior in the affective, cognitive, and sensorimotor domains. From these premises we can build research models that show how music can influence human behavior and brain function in general as well as in a therapeutically meaningful way. Interestingly, a renewed and complex focus on the aesthetic foundations of music perception and music production has provided the key to a firm rooting of the future of music therapy in the neurosciences.

HOW RESEARCH TRANSFORMS CLINICAL PRACTICE

A brief review will illustrate how these concepts have served since the early 1990s as a foundation for a new research agenda that is scientifically explanatory as well as translational and foundational to a new clinical practice in music therapy.

Studies have shown impressively over the past 15 years that rhythmic entrainment of motor function can actively facilitate the recovery of movement in patients with stroke,[2–7] Parkinson disease,[8–14] cerebral palsy,[15] or traumatic brain injury.[16] There is strong physiological evidence that rhythmic sounds act as sensory timers, entraining brain mechanisms that control the timing, sequencing, and coordination of movement. Recovery of speech functions can also be facilitated with music.[17–22] Music's strong timing mechanisms are thought to entrain oscillatory circuits in the speech centers of the brain.[23] Recognizing the importance of temporal organization in cognitive functions, new frontiers in research have investigated the effect of music and rhythm on critical aspects of timing in learning, attention, executive function, and memory.[24–33]

In a fascinating analogy, it may be suggested that music—written in the time code of rhythm, creating meaningful sound patterns in time—simulates or resembles the oscillatory "rhythmic" synchronization codes of neural information processing in the brain, thus becoming a powerful stimulus to communicate sensory and cognitive-perceptual information to the brain.[34] Music's temporal-based grammar may be in fundamental parallel to how the brain processes information. Neurophysiological studies have shown that sound can arouse and excite the spinal motor neurons mediated by auditory-motor connections at the brain stem and spinal cord level.[35] This priming effect sets the motor system in the brain in a state of readiness, facilitating the execution of movements. However, rhythmic sounds also entrain the timing of the muscle activity, thereby providing a physiological template for cueing the timing of movements. Patients with neurological movement disorders do benefit from this effect of music and rhythm to retrain their motor functions. Thus, music provides a stimulus that substitutes for compromised internal functions, accesses compensatory networks in the brain, and may help build new pathways shaping the plasticity of the brain. The rhythmic patterns of music can help patients with Parkinson disease overcome bradykinesia and episodes of "freezing" of movement because the music acts as a sensory sequencer that provides critical neural "movement" command signals that are not generated reliably in time by brain areas affected by the disease.[18,36]

In cognition, we may consider as an example the effect of music on memory. The organizing element of chunking, a critical element in memory coding, is always present in all music as a necessary component to build musical forms through melodic, harmonic, and rhythmic phrasing.[37] Studies have shown that music can function as an excellent memory template for nonmusical declarative or procedural learning.[38–40] Studies with memory disorders, such as Alzheimer disease (AD), frequently show retention of musical information in patients that is preserved longer and out of proportion with their concurrent state of memory loss.[41] Such data suggest that neuronal memory traces built through music are deeply ingrained and more resilient to neurodegenerative influences.[27,28] Findings that in memory tasks people with AD access prefrontal-amygdaloid networks rather than prefrontal-hippocampal networks may make music a useful modality to access and enhance memory function in AD, based on its highly emotional saliency.[42] The organizational basis of music as a temporally "overstructured" language of sound patterns may play a critical role in such effective memory formation.[43] Recent research has shown that neuronal oscillations, which build rhythmically synchronized firing patterns in network ensembles of neurons, form the neurobiological basis of perception and learning. The precise synchronization of neuronal activation patterns is a crucial element in building the tightly coupled networks that physiologically underlie the process of effective learning.[44] Thus, music's temporality, expressed in its rhythmic nature, may optimize the formation of such rhythmic neuronal networks, because music, as the learning stimulus that drives the physiological activations in the brain, is already tightly organized within temporal structures.[45]

CONCLUSION

Music can communicate information to the brain that has profound effects on learning, development, recovery of function, and aesthetic engagement. Research into the neurobiology as well as biomedical effects of music has made great and unprecedented progress in the last decade due to a fruitful merger of lines of investigation from neuroscience, psychology, medicine, psychophysics, and musicology, supported by other disciplines, such as mathematics, physics, and engineering.[46,47]

A large number of clinical studies have shown striking evidence that auditory rhythm and music can be effectively harnessed for specific therapeutic purposes. The emerging research base has guided the establishment of neurologic music therapy as a comprehensive new clinical model of music therapy practice that has found recognition and acceptance as an evidence-based rehabilitation discipline.

[Competing interests: The authors declare that they have no competing financial interests.]

REFERENCES

1. BERLYNE, D.E. 1971. Aesthetics and Psychobiology. Appleton, Century, and Croft. New York.
2. THAUT, M.H., R.R. RICE, G.C. MCINTOSH & S.G. PRASSAS. 1993. The effect of auditory rhythmic cueing on temporal stride parameters and EMG patterns in hemiparetic gait of stroke patients. J. Neurol. Rehabil. **7:** 9–16.
3. THAUT, M.H., G.C. MCINTOSH & R.R. RICE. 1997. Rhythmic facilitation of gait training in hemiparetic stroke rehabilitation. J. Neurol. Sci. **151:** 207–212.
4. THAUT, M.H., G.P. KENYON, C.P. HURT, et al. 2002. Kinematic optimization of spatiotemporal patterns in paretic arm training with stroke patients. Neuropsychologia **40:** 1073–1081.
5. HUMMELSHEIM, H. 1999. Rationales for improving motor function. Curr. Opin. Neurol. **12:** 697–701.
6. WHITALL, J., W.S. MCCOMBE, K.H. SILVER & R.F. MACKO. 2000. Repetitive bilateral arm training with rhythmic auditory cueing improves motor function in chronic hemiparetic stroke. Stroke **31:** 2390–2395.
7. MAURITZ, K.H. 2002. Gait training in hemiplegia. Eur. J. Neurol. **9** (Suppl. 1): 23–29; discussion 53–61.
8. MILLER, R.A., M.H. THAUT & J. AUNON. 1996. Event-related brain wave potentials in an auditory–motor synchronization task. In Music Medicine, Vol. 2. R. Pratt & R. Spintge, Eds.: 76–84. MMB Music. St. Louis.
9. THAUT, M.H., G.C. MCINTOSH, R.R. RICE, et al. 1996. Rhythmic auditory training in gait training with Parkinson's disease patients. Mov. Disord. **11:** 193–200.
10. MCINTOSH, G.C., S.H. BROWN, R.R. RICE & M.H. THAUT. 1997. Rhythmic auditory–motor facilitation of gait patterns in patients with Parkinson's disease. J. Neurol. Neurosurg. Psychiatry **62:** 122–126.
11. HOWE, T.E., B. LOVGREEN, F.W. CODY, et al. 2003. Auditory cues can modify the gait of persons with early-stage Parkinson's disease: a method for enhancing Parkinsonian walking performance. Clin. Rehabil. **17:** 363–367.
12. FREEDLAND, R.L., C. FESTA, M. SEALY, et al. 2002. The effects of pulsed auditory stimulation on various gait measurements in persons with Parkinson's disease. Neurorehabilitation **17:** 81–87.
13. PACCHETTI, C., F. MANCINI, R. AGLIERI, et al. 2000. Active music therapy in Parkinson's disease: an integrative method for motor and emotional rehabilitation. Psychosom. Med. **62:** 386–393.
14. FERNANDEZ DEL OLMO, M. & J. CUDEIRO. 2003. The timing in Parkinson's disease: effects of a rehabilitation programme based on rhythmic sound cues. Proc. Soc. Neurosci. **734:** 2.
15. THAUT, M.H., C.P. HURT, D. DRAGAN & G.C. MCINTOSH. 1998. Rhythmic entrainment of gait patterns in children with cerebral palsy. Dev. Med. Child Neurol. **40:** 15.
16. HURT, C.P., R.R. RICE, G.C. MCINTOSH & M.H. THAUT. 1998. Rhythmic auditory stimulation in gait training for patients with traumatic brain injury. J. Mus. Ther. **35:** 228–241.
17. PILON, M.A., K.W. MCINTOSH & M.H. THAUT. 1998. Auditory versus visual speech timing cues as external rate control to enhance verbal intelligibility in mixed spastic-dysarthric speakers: a pilot study. Brain Injury **12:** 793–803.

18. THAUT, M.H., G.C. MCINTOSH, K.W. MCINTOSH & V. HOEMBERG. 2001. Auditory rhythmicity enhances movement and speech motor control in patients with Parkinson's disease. Funct. Neurol. **16:** 163–172.
19. BOUCHER, V., J.L. GARCIA, J. FLEURANT & J. PARADIS. 2001. Variable efficacy of rhythm and tone in melody-based interventions: implications of the assumption of a right-hemisphere facilitation in nonfluent aphasia. Aphasiology **15:** 131–149.
20. BONAKDARPOUR, B., A. EFTEKHARZADEH & H. ASHAYERI. 2003. Melodic intonation therapy in Persian aphasic patients. Aphasiology **17:** 75–95.
21. GLOVER, H., J. KALINOWSKI, M. RASTATTER & A. STUART. 1996. Effect of instruction to sing on stuttering frequency at normal and fast rates. Percept. Motor Skills **83:** 511–522.
22. BELIN, P., P. VAN EECKHOUT, M. ZILBOVICIUS, et al. 1996. Recovery from nonfluent aphasia after melodic intonation therapy. Neurology **47:** 1504–1511.
23. WARREN, J.D., J.E. WARREN, N.C. FOX & E.K. WARRINGTON. 2003. Nothing to say, something to sing: primary progressive dynamic aphasia. Neurocase **9:** 140–153.
24. HOMMEL, M., B. PERES, P. POLLAK & B. MEMIN. 1990. Effects of passive tactile and auditory stimuli on left visual neglect. Arch. Neurol. **47:** 573–576.
25. BAUR, B., I. UTTNER, J. ILMBERGER, et al. 2000. Music memory provides access to verbal knowledge in a patient with global amnesia. Neurocase **6:** 415–421.
26. CHAN, A.S., Y.C. HO & M.C. CHEUNG. 1998. Music training improves verbal memory. Nature **396:** 128.
27. FOSTER, N.A. & E.R. VALENTINE. 2001. The effect of auditory stimulation on autobiographical recall in dementia. Exp. Aging Res. **27:** 215–228.
28. HASLAM, C. & M. COOK. 2002. Striking a chord with amnesic patients: evidence that song facilitates memory. Neurocase **8:** 453–465.
29. HO, Y.C., M.C. CHEUNG & A.S. CHAN. 2003. Music training improves verbal but not visual memory: cross-sectional and longitudinal explorations in children. Neuropsychology **17:** 439–450.
30. KILGOUR, A.R., L.S. JAKOBSON & L.L. CUDDY. 2000. Music training and rate of presentation as mediators of text and song recall. Mem. Cogn. **28:** 700–710.
31. RAINEY, D.W. & J.D. LARSEN. 2002. The effect of familiar melodies on initial learning and long-term memory for unconnected text. Mus. Percept. **20:** 173–186.
32. SON, G.R., B. THERRIEN & A. WHALL. 2002. Implicit memory and familiarity among elders with dementia. J. Nurs. Scholarship **34:** 263–267.
33. OVERY, K. 2003. Dyslexia and music: from timing deficits to musical interventions. Ann. N. Y. Acad. Sci. **999:** 497–505.
34. MERZENICH, M.M., C. SCHREINER, W. JENKINS & X. WANG. 1993. Neural mechanisms underlying temporal integration, segmentation, and input sequence representation: Some implications for the origin of learning disabilities. Ann. N. Y. Acad. Sci. **682:** 1–22.
35. ROSSIGNOL, S. & G. MELVILL JONES. 1976. Audiospinal influences in man studied by the H-reflex and its possible role in rhythmic movement synchronized to sound. Electroencephalogr. Clin. Neurophysiol. **41:** 83–92.
36. HAUSDORF, J.M., J.D. SCHAASMA, Y. BALASH, et al. 2003. Impaired regulation of stride variability in Parkinson's disease subjects with freezing of gait. Exp. Brain Res. **149:** 187–194.
37. WOLFE, D.E. & C. HOM. 1993. Use of melodies as structural prompts for learning and retention of sequential verbal information by preschool students. J. Mus. Ther. **30:** 100–118.
38. WALLACE, W.T. 1994. Memory for music: effect of melody on recall of text. J. Exp. Psychology Learn. Mem. Cogn. **20:** 1471–1485.
39. GFELLER, K.E. 1983. Musical mnemonics as an aid to retention with normal and learning-disabled students. J. Mus. Ther. **20:** 179–189.
40. CLAUSSEN, D.W. & M.H. THAUT. 1997. Music as a mnemonic device for children with learning disabilities. Can. J. Mus. Ther. **5:** 55–66.
41. COWLES, A., W.W. BEATTY, S.J. NIXON, et al. 2003. Musical skill in dementia: a violinist presumed to have Alzheimer's disease learns to play a new song. Neurocase **9:** 493–503.
42. ROSENBAUM, R.S., M.L. FUREY, B. HORWITH & L.L. GRADY. 2004. Altered communication between emotion-related brain regions support short-term memory in Alzheimer's disease. Proc. Soc. Neurosci. **203:** 8.

43. DEUTSCH, D. 1982. Organizational processes in music. *In* Music, Mind and Brain. M. Clynes, Ed.: 119–131. Plenum Press. New York.
44. KLIMESCH, W., M. DOPPELMAYR, H. RUSSEGER, *et al.* 1998. Induced alpha band power changes in the human EEG and attention. Neurosci. Lett. **244:** 73–76.
45. THAUT, M.H., D.A. PETERSON & G.C. MCINTOSH. 2005. Temporal entrainment of cognitive functions: musical mnemonics induce brain plasticity and oscillatory synchrony in neural networks underlying memory. Ann. N.Y. Acad. Sci. **1060:** 243–254 [This volume.]
46. AVANZINI, G., C. FAIENZA, D. MINCIACCHI, L. LOPEZ & M. MAJNO, EDS. 2003. The Neurosciences and Music. Vol. 999. The New York Academy of Sciences. New York.
47. THAUT, M.H. 2005. Rhythm, Music, and the Brain: Scientific Foundations and Clinical Applications. Routledge. New York and London.

Part V: The Neurology of Music

Introduction

TIM GRIFFITHS

Auditory Group, University of Newcastle Medical School,
Newcastle upon Tyne, NE2 4HH, United Kingdom

Remarkable progress in the study of the neurology of music has made since the first discussion meeting on music and the brain was organized by the New York Academy of Sciences five years ago in New York. At that time the systematic study of deficits in musical analysis due to acquired brain disorder had been facilitated greatly by the introduction of systematic measures of musical cognition by Peretz: these measures have allowed a better characterization of acquired disorders within a theoretical framework.[1] At that meeting the first efforts to apply this approach to a developmental musical disorder were described, taking the description of tone deafness beyond the level of the previous anecdotal reports going back more than a century. Later contributions further developed these efforts.[2]

Here, Cuddy describes a systematic survey of tone deafness in a large population of undergraduates. The data are interesting in suggesting that individuals who label themselves as tone-deaf do not necessarily have the condition based on objective testing of their musical perception. I do not think this need alarm us. Consider, for example, the relationship between symptomatic memory deficits and objective testing using batteries such as the Wechsler Memory Scale.[3] I frequently see people in the clinic with profound symptomatic memory loss and no deficit on formal memory testing, or with no symptoms and a clearcut memory disorder such as Alzheimer disease. That dissociation does not invalidate the memory scale or the concept of Alzheimer disease.

Evidence in favor of a biological basis for developmental deficits in musical perception is emerging from a number of fields. Drayna's group has shown a strong genetic predisposition to abnormal performance in the Distorted Tunes Test (similar to the scale test in the Montreal battery). Further genetic studies are presented here that extend the pioneering work on twins.[4] Other work discussed at this meeting considers structural brain changes using the MRI technique of voxel-based morphometry and functional changes measured with evoked potentials. These early studies are consistent with a disorder of cortical processing or cortical connectivity, although the details are still far from clear.

An interesting contrast to these studies of disordered musical development is provided by the work of Levitin's group, who have pioneered the systematic study of

Address for correspondence: Tim Griffiths, Auditory Group, University of Newcastle Medical School, Framlington Place, Newcastle-upon-Tyne, NE2 4HH, United Kingdom. Voice: +44-191-222-6648; fax: +44-191-222-5227.
   T.D.Griffiths@ncl.ac.uk

Ann. N.Y. Acad. Sci. 1060: 309–310 (2005). © 2005 New York Academy of Sciences.
doi: 10.1196/annals.1360.065

Williams syndrome, a rare genetic disorder where auditory and musical perception are preserved as a "island" of ability in the presence of often profound intellectual deficits in other domains. Study of this disorder has the potential to unlock the essential elements for musical cognition.

The neurology of musical performance is evolving as rapidly as the neurology of musical perception. At the earlier meeting in New York, Pantev presented data on the normal organization of motor maps in musicians. Here, Candia presents data on motor disorganization in the career-threatening condition of focal dystonia and describes one of the first studies to evaluate a program of motor retraining. The work combines the twenty-first-century technique of magnetoencephalography to study motor maps with the use of a splinting device reminiscent of a contraption used by Robert Schumann in the early nineteenth century. In motor as in perceptual aspects of music, robust assessment measures are allowing evaluation of both the conceptual bases for disorders and their clinical progression.

REFERENCES

1. PERETZ, I. & M. COLTHEART. 2003. Modularity of music processing. Nat. Neurosci. **6:** 688–691.
2. PERETZ, I., A-S. CHAMPOD & K. HYDE. 2003. Varieties of musical disorders: the Montreal Battery of Evaluation of Amusia. Ann. N. Y. Acad. Sci. **999:** 58–75.
3. WECHSLER, D. 1987. Wechsler Memory Scale-Revised Manual. The Psychological Corporation. San Antonio, TX.
4. DRAYNA, D. et al. 2001. Genetic correlates of musical pitch recognition in humans. Science **291:** 1969–1972.

Musical Difficulties Are Rare

A Study of "Tone Deafness" among University Students

LOLA L. CUDDY,[a] LAURA-LEE BALKWILL,[a] ISABELLE PERETZ,[b]
AND RONALD R. HOLDEN[a]

[a]Department of Psychology, Queen's University, Kingston, Ontario, K7L 3N6, Canada

[b]Département de Psychologie, Université de Montréal,
Montréal, Québec, H3C 3J7, Canada

ABSTRACT: This study was concerned with self-reported "tone deafness" and
its possible relationship to congenital amusia. Nearly 17% of over 2000 first-
year psychology students at Queen's University self-reported tone deafness.
Two hundred students were recruited from this pool of students, comprising
100 who reported tone deafness and 100 who reported that they were not tone-
deaf (NTD). The study contained two parts. In part 1, participants completed
the six tests of the Montreal Battery of Evaluation of Amusia (MBEA) developed by Peretz and collaborators. In part 2, participants completed an extensive questionnaire designed to elicit details about musical experiences, abilities,
training, and interests. Twenty-eight questionnaire items allowing a quantitative response were subjected to factor analysis. Four orthogonal components
emerged from the analysis. The components reflected self-report of (1) vocal
production, (2) music instruction, (3) listening attitudes, and (4) childhood
memories of musical environment. Results for each of the MBEA tests and
composite scores for all tests were regressed on participants' factor scores. The
best and significant predictors of the MBEA scores were factor I and factor II,
followed by factor III. Factor scores accounted for a higher percentage of the
variance in MBEA composite test results (27%) than the self-report of tone
deafness alone (7%). The musical difficulties revealed by the MBEA test results
for some participants warrant further attention and study. However, an
encouraging conclusion from the MBEA results is that many individuals who
consider themselves "tone-deaf" may not, in fact, have perceptual difficulties,
and these individuals should be supported in any of their efforts to proceed
with music enjoyment and instruction.

KEYWORDS: neuropsychology; tone deafness; congenital amusia; MBEA battery

Tone deafness is a popular term in Western culture, but the referent is not clear.
The term may indicate lack of musical instruction, or lack of interest, or it may
alternatively refer to true musical deficiencies with possibly a number of manifestations or subcategories of the difficulty. The purpose of the present study was to

Address for correspondence: Lola L. Cuddy, Department of Psychology, Queen's University,
Kingston, Ontario, K7L 3N6, Canada. Voice: 613-533-5013; fax: 613-533-2499.
cuddyl@post.queensu.ca

Ann. N.Y. Acad. Sci. 1060: 311–324 (2005). © 2005 New York Academy of Sciences.
doi: 10.1196/annals.1360.026

investigate the use of tone deafness as a descriptor by young adults—to explore what personal characteristics lead to a self-report of tone deafness and to determine whether perceptual difficulties might accompany a self-report of tone deafness.

The further motivation was to clarify the relationship between tone deafness and a musical disorder known as *congenital amusia*. Congenital amusics display a life-long selectively musical handicap that is not explained by low intelligence, impaired hearing, or lack of music instruction.[1-3] Language skills, general memory, including memory for song lyrics, and recognition of speech intonation, are unimpaired. More-over, no history of brain assault or injury is present; thus the disorder appears to be of the genuine developmental type rather than an acquired type of amusia. It is prob-ably hereditary in origin.[4] The prevalence is assumed to be around 4–5% of the population[5]—although the reliability of this estimate has been questioned.[6]

Yet, as reported below, the prevalence of self-reported tone deafness in two large samples of young adults is almost 17%. This relatively large estimate leads to the suspicion that many of the self-reported "tone-deaf" (TD) may not be amusic. They may undervalue their abilities, but may, in fact, function normally. Furthermore, they may, with encouragement, be capable of acquiring and appreciating many of the social and personal benefits of the musical world.[7] On the other hand, other self-reported TD individuals may be truly amusic and may function abnormally. Behav-ioral identification of true amusics is important for the future study of the neural and genetic correlates of musical deficiencies and of learning disabilities in general.

A variety of behavioral tests have confirmed that self-declared amusics are defi-cient in pitch discrimination and pitch pattern recognition (see, e.g., Refs. 1–3, 6, and 8), timing tasks that engage pitch as well as temporal processing,[1,2] and discrim-ination of nonlinguistic analogues of speech intonation.[1,9] The presence of amusia is typically verified by low scores, relative to matched controls, on the Montreal Battery of Evaluation of Amusia.[10] The Montreal Battery of Evaluation of Amusia (MBEA) is a comprehensive behavioral test validated against brain damage of various types. A strong feature of the MBEA battery is that the individual tests have been developed to isolate theoretical processing components of a model of music recognition.[11]

Amusics in the above studies were typically selected from volunteers who met the criteria of high general education, childhood music instruction, a history of musical failure, and absence of neuroaffective cause. The criterion of a history of musical failure requires the selection of older participants with extensive personal experience of failure. To what extent amusia may occur in younger populations, and may be re-flected in their musical self-assessments, is unknown. The purpose of the present study was to contribute relevant information toward this issue, using the MBEA as a diagnostic device.

METHOD

Participants

Prescreening

A prescreening questionnaire was administered to a pool of psychology under-graduate students entering the first year of study in the fall of 2003 and spring of 2004. The questionnaire included, among other items, questions addressing demographic background, years of music instruction, and self-assessment of tone deaf-ness. Responses were obtained from 1589 students (398 males and 1191 females, mean age = 19.3; range = 17–51 years). Of these students, 269, or 16.9%, self-reported tone deafness (males, 15%; females, 17.6%, a nonsignificant difference, $z = 1.15$, $P > .20$). The (point biserial) correlation between self-report of tone deaf-ness (yes = 1, no = –1) and years of music training was –.15, a significant ($P < .0001$) but very weak relation. To assess the reliability of the above figures, responses from 1222 students in the introductory class were obtained in the fall of 2004 (312 males and 910 females, mean age 18.6; range = 16–68 years). Of these students, 205 or 16.8% reported tone deafness (males, 15.4%; females, 17.3%, a nonsignificant difference, z = .76, $P > .50$). The correlation between self-reported tone deafness and years of music training was again –.15, a significant ($P < .0001$) but weak relation.

Recruitment Sample

A sample of 200 participants was recruited from the 2003/2004 prescreening pool. Recruitment was subject to five constraints. The first constraint was to obtain two groups of equal size ($n = 100$), with the first group comprising individuals who self-reported tone deafness, and the second comprising individuals who self-report-ed no tone deafness. The second constraint was to ensure that both groups represent-ed the full range of music instruction on the primary instrument reported in the prescreening questionnaire. Mean years of music instruction for the TD group was 3.04 (range 0 to 20) and for the not tone-deaf (NTD) group 3.89 (range 0 to 21). The correlation between self-report of TD and years of music training was not signifi-cant, $r(198) = -.05$, $P = .46$. The third constraint was to restrict the age range so that the sample represented the typically young students in the pool. Mean age for the TD group was 19.7 years, range 18–26 years, and for the NTD group 19.4 years, range 17–23 years. The fourth constraint was to obtain a male/female ratio similar to the ratio of males/females in the prescreening pool. For the TD group the proportion of males to females was 29:71, and for the NTD group the proportion of males to females was 28:72. The fifth constraint was that participants must not have reported hearing loss in either ear.

Materials

Montreal Battery of Evaluation of Amusia (MBEA)

Each participant was tested on the battery of six subtests of the MBEA. The tests are named the scale, contour, interval, rhythm, meter, and memory tests. Examples of musical stimuli are shown in FIGURE 1, and the test construction is described in detail in Ref. 10.

The first three tests address melodic organization. They each contain 30 same–different trials in which a standard melody (A in FIG. 1) is followed by a comparison melody. For trials where the comparison melody differs from the standard, one note is altered in one of the following ways: it is out of the scale of the melody while pre-serving the contour (B in FIG. 1), it changes the contour of the melody while preserv-

Stimuli   Responses

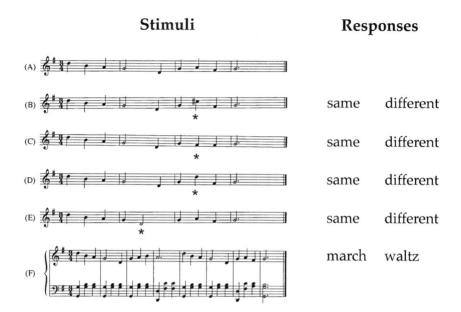

FIGURE 1. Example of the stimuli used in each of the six tests of the Montreal Battery of Evaluation of Amusia (redrawn from Peretz et al.[10] FIG. 2).

ing the scale (C in FIG. 1), or it changes the intervallic distance between notes while maintaining the contour and the scale (D in FIG. 1).

The fourth and fifth tests address temporal organization. The fourth test is also a same–different paradigm with 30 trials. For trials where the comparison melody differs from the standard, the duration of two adjacent notes is changed so that the rhythmic grouping of the melody is altered (E in FIG. 1). For the metric test, half the 30 trials are written in duple meter and half in triple meter (F in FIG. 1 is an example of triple meter), and the participant is asked to classify each trial as either a march or a waltz.

The sixth test is a memory recognition test in which the participants must discriminate "old" melodies that they heard in the previous five tests from "new" melodies similar in style to the old melodies but never heard before.

Testing was conducted in one of two sound-attenuated booths. A CD recording of the tests was reproduced in one booth by a Panasonic CD player (4.4 PDS MASH) through Sennheiser headphones (HD 480) and in the other booth by a Panasonic CD player (RX-ED 77) through Precision headphones (HD 580).

Self-Assessment Questionnaire

Each participant answered an extensive eight-page pencil-and-paper questionnaire designed to elicit details about childhood exposure to music, music instruction and involvement, further self-assessment of musical perception and production, and suspected etiology of musical difficulties.

Procedure

Participants were tested individually and were randomly assigned to one of the two sound-attenuated booths. A test administrator conducted the session; the test administrator was not informed of the participant's group assignment (TD or NTD). The MBEA was administered first, followed by the questionnaire. The MBEA tests were presented in a fixed order from test 1 to test 6. Printed sheets with test instructions and response forms were distributed one test at a time. For each test, the test administrator repeated the written instructions, presented the two examples preceding each test, and answered questions from participants. Each test lasted approximately ten minutes, with a short rest break between tests.

Participants were allowed unlimited time to write responses to the questionnaire. The test administrator was available to clarify items, if requested, and also ensured that there were no unintentionally missing answers. At the end of the session, participants were debriefed and allowed to ask questions about the nature of the study. The entire session lasted approximately 90 minutes.

RESULTS

MBEA Test Results

FIGURE 2 shows the distribution of mean composite test scores for the TD and NTD groups. (Composite test scores are the average across the six tests.) For comparison, the inset in FIGURE 2 shows the distribution for 120 normal controls (age 14–79 years) and a group of 27 self-reported amusics (age 20–89 years) described in Peretz *et al.*[10] Three findings to note in the present data: first, the test score dis-

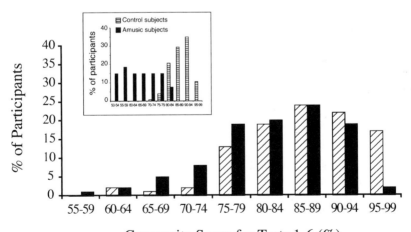

FIGURE 2. Distribution of composite scores on the MBEA battery for self-reported tone-deaf (TD) and self-reported not tone-deaf (NTD) participants. The insert (from Peretz *et al.*[10] FIG. 5) shows the distribution of composite scores on the MBEA battery for self-declared amusics and a normative sample. *Striped columns*, NTD; *filled columns*, TD.

TABLE 1. MBEA test means with standard deviations (in parentheses) for 100 TD and 100 NTD participants, t tests of mean differences, and point biserial correlation between self-reported tone deafness and test scores

Test	TD	NTD	t	$r_{(pbi)}$
Scale	24.3 (2.9)	25.5 (2.7)	−3.16***	−.22***
Contour	23.8 (3.5)	25.6 (3.1)	−3.92****	−.27****
Interval	23.7 (3.2)	25.4 (3.4)	−3.63****	−.25****
Rhythm	25.5 (3.4)	26.4 (3.1)	−1.94+	−.136+
Meter	24.3 (5.3)	25.9 (5.4)	−2.00*	−.14*
Incidental memory	26.8 (2.6)	27.7 (2.5)	−2.42*	−.17*
Composite	24.7 (2.5)	26.1 (2.3)	−3.98****	−.27****

NOTE: **** $P < .001$; *** $P < .005$; * $P < .05$; + $P < .10$.

tributions for both groups were skewed; second, there was considerable overlap between the TD and NTD distributions; and third, there was, nonetheless, a tendency for NTD participants to obtain higher scores than TD participants. These findings held generally for the individual tests. The mean and standard deviation for each group and each test is given in TABLE 1, along with the results of t tests of the difference between group means, and the point biserial correlation between self-report of tone deafness and test scores. TABLE 1 reports significant differences between NTD and TD mean scores; the point biserial correlations in TABLE 1 were also significant (but accounted for less than 7% of the variance).

The TD distribution in FIGURE 2 does not seem to match the distribution for congenital amusics in Peretz et al.[10] (see the inset in FIGURE 2) where perceptual abilities of the amusics appear clearly disadvantaged and separable from the normative sample. Peretz et al.[10] found that 89% of their self-declared amusic volunteers obtained a composite score below a criterion of 77% correct (the mean of controls minus 2 SD). By contrast, only 11% of TD participants scored below a comparable criterion obtained from the distribution for 100 NTD participants. This criterion score for the present sample was 72%. (We have no immediate explanation for the slight difference between the distribution for the Peretz et al.[10] controls and the NTD participants; we leave the question to future research.)

Prediction from the Self-Assessment Questionnaire

Subsequent analysis sought to determine whether greater prediction of MBEA results would result from consideration of the extended self-assessment of musical skills and experience. The analysis proceeded in two steps. First, the responses to quantifiable questionnaire items were entered into a factor analysis. Second, the resulting factor scores were used as predictors of self-report of tone deafness and as predictors of the MBEA scores.

For the first step, item information was summarized through a principal components analysis of the correlations among item responses. Although six eigenvalues exceeded unity, this heuristic is noted for yielding an overextraction of components.[12] A scree test, Velicer's[13] minimum average partial criterion, and both mean

TABLE 2. Items with highest factor loadings greater than .50 for four independent components

Factor I. Vocal Production
How well can you match notes when humming along to favorite recorded music?
How easy is singing in general?
How easy is it to sing a note to match one played on the piano?
When singing, can you tell when you are out of tune?

Factor II. Music Instruction
Years of playing a musical instrument
Number of instruments played
Type(s) of music education—e.g., private, group, self-taught, conservatory examinations
At the peak of interest, number of hours per week practicing the main instrument

Factor III. Listening Attitudes
How often do you purposefully listen to music?
How often do you get a tune stuck in your head?
When music is played on the radio, in a store, or on TV, how frequently can you recognize a familiar song in the first two or three notes?
Number of positive responses to music (from a checklist of items)

Factor IV. Childhood Memories
Was singing encouraged in your childhood environment?
How many members of your family sang to you when you were a child?
How often did you hear music in your childhood?

and 95th percentile values of Horn's[14] parallel analysis criterion all converged in indicating four components that were subsequently rotated to a varimax solution. Items with factor loadings above .50 on each factor are summarized, in descending order of weight, in TABLE 2. Inspection of the items led to the following interpretations. Factor I is primarily assessment of vocal production ability—whether singing is easy and accurate. Factor II reflects music instruction—primarily the amount of formal training and practicing. Factor III reflects listening attitudes. A critical term in the items is *often* or *frequently*, with high self-assessment related to a positive searching for, and engagement with, music. Factor IV reflects childhood memories—the extent of singing and musical exposure during childhood. The four factors accounted for 54% of the variance in questionnaire responses. The full list of questionnaire items and factor loadings for each item are shown in APPENDICES 1 and 2.

Next, the self-report of tone deafness was regressed on individual factor scores. For the regression, TD was coded "1" and NTD as "−1." As shown in TABLE 3, factors I and III, but not II and IV, contributed significantly to predictions of self-report of tone deafness. TD participants rated their vocal abilities more negatively than the NTD and rated attraction to, and engagement with, music less positively than the NTD.

Finally, MBEA scores for the composite scores and for individual test scores were regressed on individual factor scores. Results of the regression are shown in

TABLE 3. Results of regression analysis predicting self-report of tone deafness from factor scores—beta weights for each factor and the multiple correlation

Factor	Beta weight
I. Vocal production	−.41**
II. Music instruction	−.09
III. Listening attitudes	−.23**
IV. Childhood memories	.03
	Multiple correlation
R (n = 200)	.48***

NOTE: **P < .01; ***P < .001

TABLE 4. Factors I, II, and III contributed significantly to prediction of composite MBEA scores; the variance accounted for by the factor scores approached 30%. Thus, high MBEA scores were associated with more positive ratings of vocal ability, more extensive musical instruction, and more positive attitudes toward music than low MBEA scores. Results for the individual test scores were generally similar, with factors I and II the strongest predictors for all tests.

Tone Deafness and Amusia

We will consider two sets of TD participants scoring at the two different extremes of the TD distribution. The two sets of TD participants provide examples of discrepancies from the general pattern of results. The regression results yield the expectation that low MBEA scores will be generally accompanied by low factor scores, and high MBEA scores will generally be accompanied by high factor scores. However, some participants did not follow this expectation.

The first set contains those participants with very low MBEA scores, the 11% noted above whose scores were below criterion. Although seven had low factor scores on music instruction (a finding consistent with the regression analysis), four had average or above average scores on the music instruction factor. In other words, despite music instruction, the four remain handicapped. Their pattern of low ability on the MBEA combined with low self-assessment on factors I and III (see the questions in TABLE 2) suggests that they are likely candidates for the diagnosis of amusia (see Ref. 10, p. 68).[a]

The second set contains TD participants who obtained high MBEA scores. Of the top 11% of TD scores, nine had below-average scores for factors I and/or III. Their negative feeling about singing ability and/or music interest is not consistent with

---

[a]As well, two of the TD participants who scored below criterion responded that they could not tell if they sang out of tune unless someone told them, and six responded that if asked to repeat a song recently sung to them, they could remember the lyrics but would have problems remembering the tune. These items are similar (the second is not identical) to the items used to select potential amusics by Peretz.[10] In the present study, one individual below criterion responded affirmatively to both items, and another responded affirmatively to the first but with respect to the second said neither lyrics nor tunes could be remembered. Both individuals could be considered particularly likely candidates for the diagnosis of amusia.

TABLE 4. Results of regression analysis predicting MBEA scores from factor scores

Test	Scale	Contour	Interval	Rhythm	Meter	Memory	Composite
Factor							
I. Vocal production	.21**	.31***	.28***	.19**	.14*	.30***	.32***
II. Music instruction	.28***	.33***	.23***	.17**	.34***	.22***	.39***
III. Listening attitudes	.04	.10	.07	.06	.16**	.17**	.15*
IV. Childhood memories	.02	−.07	−.04	−.07	−.02	.004	−.04
R (n = 200)	.36***	.47***	.37***	.27**	.40***	.41***	.52***

NOTE: Factor cell entries are beta weights. Cell entries for R are multiple correlations.
*P < .05; **P < .01; ***P < .001

their good scores on the MBEA battery. According to the battery scores, these, and in fact most other, TD participants in the present study are not candidates for a diagnosis of amusia.

DISCUSSION

Young adult students who self-declare tone deafness are reporting some aspects of their musical identity. They feel they cannot sing (a finding urging the development of a battery for production skills in line with the MBEA for perception) and they are not particularly drawn to music. Formal music instruction, at least as presently recovered from questionnaire responses, does not play a major role in self-assessment of tone deafness. The relationship between self-report of tone deafness and amount of music instruction is weak at best.

Self-report of tone deafness alone was not a strong indicator of amusia, as measured by the MBEA battery. Although the TD group scored slightly but significantly below the NTD group on all tests, there was considerable overlap in the distribution of scores for both groups. The majority of participants in both groups scored above the distribution of scores for amusics[10] (reproduced in FIG. 2).

A number of directions could be followed to explain why individuals report tone deafness when there is no evidence of amusia. One would be to introduce in the questionnaire items probing the sociocultural context of music lessons and musical activities. Contexts could include cultural pressures for success, the amount of competition in music classes, and the degree of peer group evaluation. Another direction would be to reconsider the factor of childhood memories, a factor with no predictive value in the current study. These questions are the furthest removed from the participant's current experience; memories may be generally inaccessible or may be biased toward providing an inaccurate account. Information from family members could usefully verify or contradict early memories and perhaps explain an individual's negative self-assessment.

The predictive factors of vocal production and listening attitude deserve two additional comments. First, with respect to vocal production, it makes sense that production deficits will arise from a deficient perceptual/memory system.[1] What is not clear, and deserving further investigation, is why some participants with above-average MBEA scores rated their vocal production as poor. A reliability check on the accuracy of self-report is needed, along with further exploration of vocal difficulties. Second, with respect to listening attitude, we may note that individuals who actively seek out music in their environment and are thereby constantly exposed to the regular patterns and grammar of music may have internalized mental schemata for music. These schemata, it has been proposed, guide perception, recognition, and memory.[15] Supportive evidence from the present study is that individuals scoring high on the listening attitude factor are privileged on the MBEA tests. As a corollary, it may be proposed that increasing musical exposure will lead to improved test scores.

In conclusion, we have learned that the self-report of tone deafness is not a reliable indicator of musical difficulty. The label *tone-deaf* reflects different referents, two of which were uncovered here—self-assessment of poor singing and lack of musical interest and exposure. True amusia among the TD, according to our MBEA and factor score results, may be very rare. On a complementary note, the MBEA can help to reveal false amusics—individuals who think they are unmusical but are capable of normal function.[7] In revealing an absence of musical difficulty, test results can be used to promote encouragement of further musical activity.

ACKOWLEDGMENTS

Support from the Natural Sciences and Engineering Council of Canada (L.L. Cuddy) and the Canadian Institutes of Health Research (I. Peretz) is gratefully acknowledged. We thank C.G. Trimmer, J. Li, and the team of undergraduates trained to administer the MBEA, for research assistance. Data for the 100 NTD students are available on the Web site (<http://www.brams.umontreal.ca/peretz>).

[Competing interests: The authors declare that they have no competing financial interests.]

REFERENCES

1. AYOTTE, J., I. PERETZ & K. HYDE. 2002. Congenital amusia: a group study of adults afflicted with a music-specific disorder. Brain **125:** 238–225.
2. PERETZ, I. *et al.* 2002. Congenital amusia: a disorder of fine-grained pitch discrimination. Neuron **33:** 185–191.
3. PERETZ, I. & K.L. HYDE. 2003. What is specific to music processing? Insights from congenital amusia. Trends Cogn. Sci. **7:** 362–367.
4. DRAYNA, D. *et al.* 2001. Genetic correlates of musical pitch recognition in humans. Science **291:** 1969–1972.
5. KALMUS, H. & D.B. FRY. 1980. On tune deafness (dysmelodia): frequency, development, genetics, and musical background. Ann. Hum. Genet. Lond. **43:** 369–382.
6. HYDE, K.L. & I. PERETZ. 2004. Brains that are out of tune but in time. Psychol. Sci. **15:** 356–360.
7. SLOBODA, J.A., K.J. WISE & I. PERETZ. 2005. Quantifying tone deafness in the general population. Ann. N. Y. Acad. Sci. **1060:** 255–261. [This volume.]
8. FOXTON, J.M. *et al.* 2004. Characterization of deficits in pitch perception underlying "tone deafness." Brain **127:** 801–810.
9. PATEL, A., J.M. FOXTON & T.D. GRIFFITHS. 2004. Musically tone-deaf individuals have difficulty discriminating intonation contours extracted from speech. Brain Cogn. **59:** 310–313.
10. PERETZ, I., S. CHAMPOD & K. HYDE. 2003. Varieties of musical disorders: the Montreal Battery of Evaluation of Amusia. Ann. N. Y. Acad. Sci. **999:** 58–75.
11. PERETZ, I. 1993. Auditory agnosia: a functional analysis. *In* Thinking in Sound: The Cognitive Psychology of Human Audition. S. McAdams & E. Bigand, Eds.: 199–230. Oxford University Press. New York.
12. NUNNALLY, J.C. & I.H. BERNSTEIN. 1994. Psychometric Theory, 3rd ed. McGraw-Hill. New York.
13. VELICER, W.F. 1976. Determining the number of components from the matrix of partial correlations. Psychometrika **41:** 321–327.
14. HORN, J.L. 1965. A rationale and test for the number of factors in factor analysis. Psychometrika **30:** 179-185.
15. KRUMHANSL, C.L. 1990. Cognitive Foundations of Musical Pitch. Oxford University Press. New York.

APPENDICES

APPENDIX 1.

Summary of Questionnaire Items Entered into Factor Analysis

The code for each question indicates the section and order within section in the questionnaire, followed by the type of response entered in the factor analysis (R = numerical rating; C = numerical count; Y/N = yes [1], no [0]). Thus, C10-R refers to the tenth question in section C with a rating entered into the analysis. Ratings were always oriented so that the high end of the scale reflected a positive response (e.g., very easy, very frequently, and so forth). Not all questionnaire items appear below; some were qualitative and not appropriate for the analysis. The full questionnaire may be obtained from the contact author.

The order of questions below corresponds to the order of questions with factor loadings in Appendix 2.

C10-R	I find it hard/easy to sing or hum along with my favorite recorded music (rate ease/difficulty of matching notes).
C2-R	How difficult do you find singing in general (rate ease/difficulty)?
C11-R	Singing a note to match one played on the piano is a task I find... (rate ease/difficulty).
C13-R	When I sing, I can tell when I'm out of tune (rate ability to tell).
C14-R	When I sing, I perform best when... (rate solo vs. group situations).
C6-R	I find it hard/easy to repeat a tune someone else has recently sung to me (rate ease/difficulty).
C1-R	When you listen to music, how difficult is it to hear the difference between the notes (rate ease/difficulty)?
C7-R	If I imagine the tune *Happy Birthday*, I can hear the melody in my head (rate accuracy).
B3a-C	Years of training on primary instrument
B3-C	Number of instruments played
B1-C	Types of musical education (e.g., private, group, self-taught)
B6-C	At the peak of your interest, how many hours per week did you play/practice this instrument?
A8-Y/N	Were musical instruments played in your childhood environment?
B11-R	Given the opportunity, my interest in participating in future musical instruction is... (rate very high to nonexistent).
B7-C	Regarding your peak of interest (B6), how long did you maintain this peak?
B15-R	How often do you purposely listen to music, as opposed to music in your environment that you had no part in choosing, e.g., music in stores, elevators, restaurants (rate frequency)?
C15-R	How often do you get a tune stuck in your head (rate frequency)?
C9-R	When music is being played in my environment (e.g., on the radio, in a store, on TV), I can recognize familiar songs by the first two or three notes (rate frequency).

APPENDIX 1. (*continued*)

	Summary of Questionnaire Items Entered into Factor Analysis
B16-C	Which of these statements do you agree with (number of positive statements regarding music)?
C5-R	Rate your ability to memorize a short song (rate excellent to nonexistent).
C12-R	If someone played two notes on the piano, separately, and asked me which was higher in pitch, I would find this task…(rate ease/difficulty).
A2-Y/N	Was choral or individual singing encouraged in your childhood environment?
A1-C	Did any members of your family sing to you when you were a child (number of people)?
A3-R	How often did singing occur in your childhood environment (rate frequency)?
A5-R	How often did you hear music in your childhood environment (rate frequency)?
A6-C	Were any of your family members particularly fond of music (number of people)?
B13-R	I sing in public (as part of a group or solo: e.g., a choir, carols, a sing-a-long, hanging out with friends) (rate frequency).
B12-R	I sing in private (e.g., in my car, in the shower, in my environment) (rate frequency).

APPENDIX 2. Rotated component matrix with factor loadings

	Component			
Item	1	2	3	4
C10	.80	.18	.14	.07
C2	.76	.16	.14	.25
C11	.73	.40	.18	.05
C13	.66	.23	.14	.12
C14	.63	−.15	.09	.24
C6	.49	.29	.35	.09
C1	.45	.40	.43	.00
C7	.42	.27	.40	.01
B3a	.13	.81	.10	.19
B3	.12	.80	.08	.16
B1	.17	.76	.17	.12
B6	.15	.71	.14	.07
A8	−.01	.49	.09	.45
B11	.30	.48	.29	.10
B7	.14	.46	−.02	.22
B15	−.04	.00	.71	.27
C15	.19	.14	.69	.19
C9	.40	.12	.64	.04

APPENDIX 2. (*continued*) **Rotated component matrix with factor loadings**

Item	Component			
	1	2	3	4
B16	.09	.09	.62	.28
C5	.36	.04	.58	.02
C12	.13	.39	.49	−.06
A2	.21	.29	.04	.69
A1	.13	−.06	.05	.69
A3	.23	.27	.14	.67
A5	.05	.17	.41	.56
A6	−.07	.29	.24	.55
B13	.41	.15	.06	.48
B12	.42	.07	.36	.47

NOTE: Extraction method: principal component analysis. Rotation method: varimax with Kaiser normalization. Rotation converged in 7 iterations. Rows are ordered based on the magnitude of a loading within a component.

Musical Behavior in a Neurogenetic Developmental Disorder

Evidence from Williams Syndrome

DANIEL J. LEVITIN

Department of Psychology and Program in Behavioural Neuroscience,
McGill University, Montreal, QC H3A 1B1, Canada

ABSTRACT: This paper reviews a series of studies performed to assess the musical abilities and behaviors of individuals with Williams syndrome, a neurogenetic developmental disorder, in the hope of eventually being able to link genes, neurodevelopment, and cognition. Two questionnaire studies addressing the role of music in everyday life, and unusual reactions to sound, are described. Additionally, the findings from two empirical behavioral studies and a neuroimaging study are reviewed. The findings show that individuals with Williams syndrome tend to be more engaged in musical activities than others, and I report a possible neuroanatomical correlate of this engagement, with increased activation in the right amygdala to music and to noise. Williams syndrome represents a compelling model of the relationship between genes, brains, and such complex cognitive behaviors as music.

KEYWORDS: Williams syndrome; Williams-Beuren syndrome; neurodevelopmental disorders; neurogenetic disorders; music cognition

The study of distinct, well-defined, and atypical populations is of increasing importance to cognitive neuroscientists because it offers a unique window into specific aspects of cognition, and it is useful in establishing the degree to which various cognitive abilities are correlated with, or can be decoupled from, one another.[1-3] Williams syndrome (WS, also referred to as Williams-Beuren syndrome), a neurogenetic developmental disorder, offers one of the most compelling human models of the links between genes, neurological function, cognition, and behavior.[4] This article does not present any new data, nor does it attempt to provide an integrative review or synthesis of the excellent work done by colleagues on WS. Rather, the purpose of this paper is to summarize a series of studies my collaborators and I have performed over the past 10 years on music and WS. Excellent work is being conducted in other laboratories as well, as attested to in a number of recent papers.[5-8]

A diagnosis of WS is generally made in one of three ways: a physician's diagnosis; a score of 3 or more points on the Williams syndrome diagnostic score sheet;[9,10] or confirmation of a microdeletion on chromosome 7, including the gene

Address for correspondence: Dr. Daniel J. Levitin, 1205 Avenue Penfield, Montreal, QC H3A 1B1, Canada.
daniel.levitin@mcgill.ca

Ann. N.Y. Acad. Sci. 1060: 325–334 (2005). © 2005 New York Academy of Sciences.
doi: 10.1196/annals.1360.027

for elastin (ELN), following application of the fluorescent *in situ* hybridization (FISH) test. It was recently discovered that some individuals lack the full, previously documented deletion associated with WS.[11] With these cases of partial deletion, the diagnostic situation has become somewhat muddled; some individuals with partial deletions present the WS phenotype and some do not. In the work reported here, we have followed the more conservative and exclusive criterion, and we report results from only those individuals who have met the stricter criterion of the hemizygous deletion of 17–20 genes on the long arm of chromosome 7, in region 7q11.23 and including the ELN gene and approximately 50,000 base pairs.[12]

The physical manifestations of WS include supravalvular aortic stenosis (narrowing of the aorta), a deficit in the elastin production, hypercalcemia, scoliosis, and elfin or "pixie-like" facial features.[13–15] Cognitive manifestations include generally impaired cognitive function (mean full scale IQ = 58–61); poor spatial, quantitative, and reasoning abilities; distractibility; poor attention span; poor eye–hand coordination; and delayed acquisition of reading (if reading is acquired at all).[16–18] What has made the study of WS so interesting is the finding of relatively spared ability in four cognitive domains: face processing, sociability, language, and music. However, there exists a great deal of variability within the group. IQs can range from near 40 to over 100, and competencies in all domains can vary from one individual to another. Although it is a characteristic of WS that they tend toward being hypersocial and hypermusical, this is a general tendency, and individuals do show variation.

With respect to their oral language skills, individuals with WS tend to have fluent speech, good phonology, preserved morphology and syntax, and a good vocabulary. However, we have encountered many persons in which their expressive vocabulary tends toward low-frequency and unusual words. For example, on the name-as-many-animals-as-you-can test, one child responded with "newt, saber-tooth tiger, ibex, antelope" as her first responses. Individuals with WS are also prone toward using exaggerated prosody in their speech, and attention getters during story-telling, such as "all-of-a-sudden" or "lo-and-behold!" We've also encountered children with linguistic preservation problems. For example, when I first met child CB, all he could say was Tchaikovsky's *1812 Overture* over and over again, with different prosody and emotional presentation. The following day, he could speak more fluently but could only speak about the musical instrument steel drums: "I'd like to play—to play steel drums. Do you have any steel drums? Steel drums? Are *those* steel drums over there? I wanna get some steel drums."

Anecdotal reports over the past decade suggested that individuals with WS might be more musical than others, although what is meant by "musical" can vary from description to description. On the basis of our own observations, claims of musicality involve a range of proclivities, including being drawn toward frequent music listening, music performance (in spite of a general eye–hand coordination deficit), a deep emotional engagement with music, or an above-average musical memory. Alongside such anecdotal reports we also heard many reports of unusual sensitivity to sound, including being able to hear sounds other can't, being able to categorize or label sounds that other can't (such as the make and model of a vacuum cleaner, based on the sound of its motor), or being fearful of sounds that others don't find aversive. Collectively, these sound sensitivities were being referred to under an umbrella term, *hyperacusis*, in a way that was at odds with its precise medical definition. A fundamental task, as my collaborators and I saw it, would be to attempt to quantify these anecdotal claims of musicality and of sound sensitivities, and document them in a systematic fashion.

What can the uneven cognitive profile in WS teach us about intelligence and independence of mental faculties? Although recent years have seen most cognitive neuroscientists shying away from claims of strong modularity, it nevertheless goes against traditional neurological understanding that motor action plans could be so domain specific: how can we account for individuals who can play the clarinet or piano but cannot button their shirts or tie their shoes? What does it say about the motor system and about motor action planning circuits that they function in a musical context, but not in others? How can we quantify "musicality" in WS? How can we assess perception and performance in this population? Because the genetic profile in WS is well known, WS can help us to better understand the links between genes, brain, and musical behaviors. It has further been speculated that their hypersociability and lack of social inhibitions might be related to their musicality, a notion that I will take up later.

CHARACTERIZING THE MUSICAL PHENOTYPE IN WILLIAMS SYNDROME

To better understand and to systematically document the nature of musical behaviors in individuals with WS, we administered a questionnaire to the caregivers of 130 individuals with WS (age = 5–50, M = 20.4, S.D. = 10.4), and to comparison groups of individuals with Down syndrome (DS, $n = 30$, age = 5–51, M = 17.2, S.D. = 9.2), individuals with autism (AUT, $n = 40$; age = 9–39, M = 18.2, S.D. = 7.7), and a group of typically developing normal controls (NC, $n = 130$, age = 5–44, M = 20.9, S.D. = 7.4), all matched for chronological age.[19] The questionnaire contained 46 items: 33 multiple choice items (including Likert-like scales) and 13 free-response items. The questionnaires gathered information about physical variables (age, sex, handedness, hearing loss, physical deficits), interest in music, emotional responses to music, musical training, the amount of time engaged in various musical activities, and the age of onset of musical activities. The reliability of the questionnaire was established using split-subjects analysis, and we found no significant differences between the halves ($F (1,305) = 0.16, P \sim .69$).

Individuals with WS showed a significantly younger age of onset of musical interest, spent a greater number of hours per week listening to and playing music, and were reported to experience higher levels of emotion while listening to music (by ANOVA, all $P < .05$ and adjusted for multiple comparisons).

A principal components analysis revealed seven underlying orthogonal factors (FIG. 1) that contributed to the profile we obtained from the questionnaire. A recent reanalysis of our data by Goldberg[20] shows the derivational tree for the seven factors. At the seven-factor level, the components include content related to musical complexity, reproduction, sensitivity, musical theory and achievement, listening habits, positivity, and emotions. The reproduction factor splits off at the second level, staying virtually unchanged all the way down the hierarchy. What is particularly important about this representation is that factors at different levels are able to differentiate between the four populations studied: Williams syndrome, autism, Down syndrome, and the normal controls. A discriminant function analysis predicted

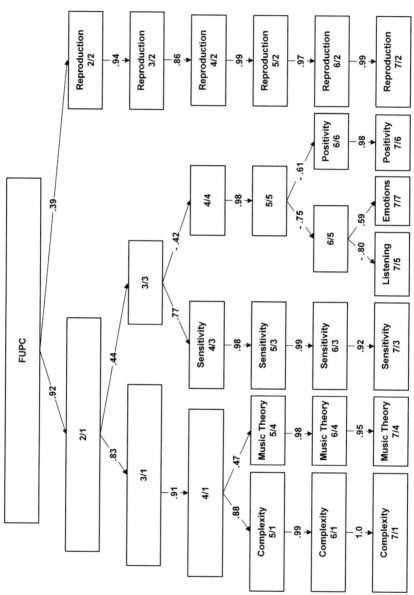

FIGURE 1. Varimax-rotated components derived from 33 kinds of musical behaviors and experiences.

group membership for 70% of the cases ($P < .01$). Next, to control for age and sex, we performed stepwise linear regressions of age and sex against the seven factors, and found that neither factor resulted in a statistically significant improvement in the model.

UNDERSTANDING CLAIMS OF HYPERACUSIS IN WS

A careful reading of the literature suggested that both caregivers and medical/research professionals were using the term *hyperacusis* in an inconsistent fashion. The medical definition of hyperacusis (also known as oxyacusis) is that it is an "abnormal sensitivity to sound,"[21,22] where sensitivity is meant in its psychophysical, not psychiatric connotation; in other words, as an ability to hear soft sounds that others cannot. Yet published and anecdotal accounts were reporting a host of other auditory abnormalities, including people with WS who report that some normal sounds are uncomfortable for them, some normal sounds are simply annoying or aversive to them, and some sounds are especially attractive, what Bellugi has referred to as "auditory fetishes." Amy Bihrle referred to this cluster of behaviors as "aversion, awareness, and attraction," which became the title of the paper in which we investigated this phenomenon.[23]

On the basis of a questionnaire that we administered to the same caregivers as for the musical phenotype study, we found that the incidence of *true hyperacusis*, or lowered-hearing thresholds, was just under 5% for people with WS, and there were no reports of it among our sample of people with AUT, DS, or NC. Three other categories of auditory anomalies emerged: *odynacusis* (a lowered-pain threshold for loud sounds, also known as lowered uncomfortable loudness levels, or LULLs[24]); *auditory allodynia*, also referred to as *phonophobia* (an aversion to sounds not normally found aversive), and *auditory fascinations* (a substantial attraction to certain sounds). Persons with WS were significantly more likely to experience all three of these symptoms, or behaviors, than the other groups. Interestingly, we discovered that many WS children outgrew their fear of certain sounds, and those same sounds subsequently became objects of intense fascination. We heard many stories of children who would sit for hours listening to leaf blowers outdoors, or who loved the sounds of vacuum cleaners. One child had a collection of vacuum cleaners, and every year at Christmas he would ask for a new one.

THE NEURAL CORRELATES OF AUDITORY PERCEPTION IN WS

In trying to understand the pattern of auditory anomalies we observed, as well as the intense involvement with music and sound experienced by many people with WS, we hypothesized that we would find differences in brain activation between people with WS and NCs. On the basis of cytoarchitectonic studies, Galaburda and his colleagues[25–27] have shown morphological and neurophysiological differences between people with WS and normals, including differences in cell-packing density, cortical layering, and gray matter to white matter ratios. We hypothesized in particular that individuals with WS would show a wider and more diffuse pattern of activation to music and noise stimuli than NCs, and that they would show a greater amygdaloidal activation, indexing their heightened emotional reactions to music and noise.

We were initially pessimistic about being able to perform any neuroimaging studies, however, based on the high incidence of LULLs and overall aversion to loud noises: the noisy environment of the scanner and its relatively confined space would have made it frightening to most individuals with WS. Positron emission tomography (PET), while quieter than functional magnetic resonance imaging (fMRI) would have required injection of radioactive tracers, and individuals with WS are known to hate needles even more than normal children. In response to these concerns, Allan Reiss developed a desensitization program that involved a professionally produced video introduction to the fMRI scanning procedure, using a child's-eye-view of the facility and a child's narration. This was followed by a visit to an fMRI simulator in which the participants could become acclimated to the noises and enclosed space. In the end we were able to recruit five participants with WS for an fMRI study of differential processing of music and noise, and five age- and sex-matched controls.[27]

Participants listened to excerpts from familiar and unfamiliar classical music, as well as the types of noisy sounds that individuals with WS are often sensitive to, such as fans, motors, and leaf blowers. For our analyses, we examined brain activations from the blood oxygenation level–dependent (BOLD) signal for music compared to rest, noise compared to rest, and music compared to noise. Our hypotheses were confirmed. Comparing music to noise, WS individuals showed a significantly lower voxel intensity bilaterally in the superior temporal cortex, middle temporal gyri, and superior temporal sulcus. In a comparison of responses to music-minus-rest versus noise-minus-rest, control participants showed significantly higher temporal lobe activations to the music than the noise, while the WS participants showed virtually indistinguishable activation levels. Persons with WS are apparently unable to modulate neural activity in the temporal cortex in response to music and noise in a manner similar to that of controls. We also observed marked differences between WS patients and controls in the right amygdala, with WS patients exhibiting far greater activation intensity in the music-minus-noise contrast. This amygdala result points to a possible neural basis for the unusual acoustical and musical sensitivities observed in affected individuals. Overall, WS participants displayed more variable and diffuse activations throughout the brain, and they showed increased activation in the amygdala and cerebellum, thus providing new and converging evidence that their neural organization may differ from that of normal individuals.

RHYTHMIC PRODUCTION ABILITY

In an effort to better understand the music production facility of individuals with WS, we separately tested rhythmic production, pitch production, and song production. To test rhythm, we presented eight WS individuals and eight mentally age-matched controls with a set of clapped rhythms in increasing complexity.[28] The task of the participant was to clap back the rhythm as accurately as possible. Independent coders, blind to hypothesis, and group membership, analyzed audio tapes of the test sessions and scored each trial as correct or incorrect. Part of the way through the coding process, the two coders, both professional musicians, independently reported that some of the trials contained responses that were clearly not replications of the presented rhythm, but that they felt bad marking them "incorrect" because they seemed to be musically compatible with the target phrase; that is, a subset of the incorrect answers struck the coders as completing a musical phrase, and as "musical," though clearly not straight repetitions. The coders were instructed to then go back and assign trials to one of three categories: "right," "wrong," or "wrong, but very musical nonetheless."

The results showed that the WS and NC participants obtained an equal number of correct trials, approximately 66%. However, WS individuals were three times more likely when incorrect to supply a musically compatible rhythm. We interpreted this as a marker of rhythmic ability or creative rhythmicity among the WS participants.

MELODIC PRODUCTION ABILITY

We presented 12 WS individuals, 12 chronologically age-matched normal controls, and 12 individuals with DS a set of melodies increasing in complexity, to assess their melodic reproduction ability. WS and NC were statistically better at melodic repetition than the DS, and not significantly different from one another. We then presented all participants with a set of melodic fragments and instructed them to complete the melodies. As FIGURE 2 shows, the WS individuals were *not* as good at melodic completion as the NCs. Thus, WS individuals are better at rhythmic production than melodic production.

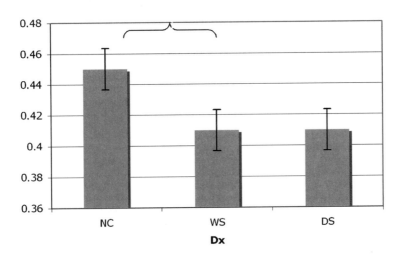

FIGURE 2.

TABLE 1. An opposite pattern of phenotypic traits and brain volumes obtained for individuals with autism and Williams syndrome

	Autism	Williams
Sociability	low	high
Musical engagement	typically low	high
Empathy	low	high
Cerebral volume	normal	small
Paleocerebellar volume (vermal I–IV)	normal	small
Neocerebellar volume (vermal V–VI)	small	large

RHYTHMIC AND MELODIC PERCEPTION

In a study currently underway, we presented 20 individuals with WS and 20 Juilliard students with the Gordon Primary Measures of Musical Audiation rhythm and tonal tests. These tests probe the ability of participants to detect differences in the rhythm or pitch in pairs of sequences that are either the same or different. We found that the WS and NC participants performed equivalently. Thus, the disparity between rhythm and melodic *production* does not carry over to the *perception* of rhythm and production.

HYPERSOCIABILITY AND MUSICALITY: A POSSIBLE CONNECTION?

David Huron[29] has noted that autism and Williams syndrome present some interesting double dissociations, diagrammed in TABLE 1. Huron suggests this as evidence for a possible genetic link between sociability and musicality, speculating that during our evolutionary history, music played a role in social bonding and social communication, and thus those same genes that were selected for sociability were those involved in musical behaviors.

CONCLUSIONS

We have documented that individuals with WS are more engaged with music than members of other groups. Their music perceptual abilities are equivalent to those of typical developing normal individuals, as is their rhythmic (but not their melodic) production abilities. Neural activations in individuals with WS are marked by distinctive differences from normal individuals, including more widespread and diffuse activation to music and noise, and greater right lateralized amygdala activation.

ACKNOWLEGMENTS

I am grateful to the Fondazione Pierfranco e Luisa Mariani and the Max Planck Institute for supporting my participation in the Neurosciences and Music II meeting, at which a version of this paper was originally presented, and at which I received helpful feedback from many colleagues, including Angela Friederici, Stefan Koelsch, Sandra Trehub, and Dennis Drayna. I am grateful also to the following collaborators for their participation in the work reported herein: Ursula Bellugi, Albert Galaburda, Julie Korenberg, Vinod Menon, and Allan Reiss. Thanks also go to the following for helpful comments on previous versions of this paper: Jamshed Bharucha, Carolyn Drake, Francesca Happé, Pamela Heaton, Donald Hodges, Carolyn Mervis, Helen Neville, Isabelle Peretz, Mike Posner, John Sloboda, and William F. Thompson. The preparation of this report was supported by a grant from the Social Sciences and Humanities Research Council of Canada (SSHRC) to DJL.

[Competing interests: The author declares that he has no competing financial interests.]

REFERENCES

1. BELLUGI, U., E.S. KLIMA & P.P. WANG. 1996. Cognitive and neural development: clues from genetically based syndromes. *In* The Life-Span Development of Individuals: A Synthesis of Biological and Psychological Perspectives. (Proceedings of the Nobel Symposium, Stockholm, Sweden, June 19–22, 1994.). D. Magnusson, Ed.: 223–243. Cambridge University Press. New York.
2. BURACK, J.A. 1997. The study of atypical and typical populations in developmental psychopathology: The quest for a common science. *In* Developmental Psychopathology: Perspectives on Adjustment, Risk, and Disorder. S.S. Luthar, *et al.*, Eds.: 139–165. Cambridge University Press. Cambridge.
3. KARMILOFF-SMITH, A. *et al.* 1995. Is there a social module? Language, face processing, and theory of mind in individuals with Williams syndrome. J. Cogn. Neurosci. **7:** 196–208.
4. TASSABEHJI, M. 2003. Williams-Beuren syndrome: a challenge for genotype–phenotype correlations. Hum. Mol. Genet. **12:** R229–237.
5. SMOOT, L. *et al.* 2005. Medical overview and genetics of Williams-Beuren syndrome. Prog. Pediatr. Cardiol. **20:** 195–205.
6. NAMIHARA, T., Y. HIRAYASU & Y. KOGA. 2004. The assessment of cognitive function in a Williams syndrome patient: a case report. Psychiatry Clin. Neurosci. **58:** 99.
7. MEYER-LINDENBERG, A. *et al.* 2004. Neural basis of geneically determined visuospatial construction deficit in Williams syndrome. Neuron **43:** 623–631.
8. LANDAU, B. & J.E. HOFFMAN. 2005. Parallels between spatial cognition and spatial language: evidence from Williams syndrome. J. Mem. Lang. **53:** 163–185.
9. AMERICAN ACADEMY OF PEDIATRICS. 2001. Health care supervision for children with Williams syndrome (RE0034). Pediatrics **107:** 1192–1204.
10. KORENBERG, J. 2005. Williams Syndrome Diagnostic Scoresheet. Accessed from <http://www.csmc.edu/csri/korenberg/forms.html>, Nov. 27, 2005.
11. HELLER, R. *et al.* 2003. Partial deletion of the critical 1.5 Mb interval in Williams-Beuren syndrome. J. Med. Genet. **40:** E99.
12. FRANCKE, U. 1999. Williams-Beuren syndrome: genes and mechanisms. Hum. Mol. Genet. **8:** 1947–1954.
13. FANCONI, G. 1952. Textbook of Pediatrics. Wm. Heinemann. London.
14. WILLIAMS, J.C.P., B.G. BARRATT-BOYES & J.B. LOWE. 1961. Supravalvular aortic stenosis. Circulation **24:** 1311.
15. LENHOFF, H.M. *et al.* 1997. Williams syndrome and the brain. Sci. Am. **277:** 68–73.
16. MERVIS, C.B. *et al.* 1999. Williams syndrome: findings from an integrated program of research. *In* Neurodevelopmental Disorders: Contributions to a New Framework from the Cognitive Neurosciences. H. Tager-Flusberg, Ed. MIT Press. Cambridge, MA.
17. BELLUGI, U., J.R. KORENBERG & E.S. KLIMA. 2001. Williams syndrome: an exploration of neurocognitive and genetic features. J. Clin. Neurosci. Res. **1:** 217–229.

18. BELLUGI, U. *et al.* 2000. The neurocognitive profile of Williams syndrome: a complex pattern of strengths and weaknesses. J. Cogn. Neurosci. **12** (Suppl.): 7–29.
19. LEVITIN, D.J. *et al.* 2004. Characterizing the musical phenotype in individuals with Williams syndrome. Child Neuropsychol. **10**: 223–247.
20. GOLDBERG, L.R. Doing it all bass-ackwards: the development of hierarchical factor structures from the top down. Submitted for publication.
21. DIRCKX, J.H. 2001. Stedman's Concise Medical Dictionary for the Health Professions. Lippincott Williams & Wilkins. Philadelphia.
22. VENES, D., C.L. THOMAS & C.W. TABER. 2001. Taber's Cyclopedic Medical Dictionary. F.A. Davis Company. Philadelphia.
23. LEVITIN, D.J. *et al.* 2005. Aversion, awareness, and attraction: understanding hyperacusis in Williams syndrome. J. Child Psychol. Psychiatry Allied Discip. **46**: 514–523.
24. PHILLIPS, D.P. & M.M. CARR. 1998. Disturbances of loudness perception. J. Am. Acad. Audiol. **9**: 371–379.
25. GALABURDA, A. & U. BELLUGI. 2000. Multilevel analysis of cortical neuroanatomy in Williams syndrome. J. Cogn. Neurosci. **12** (Suppl.): 74–88.
26. GALABURDA, A.M. *et al.* 2002. Williams syndrome: neuronal size and neuronal-packing density in primary visual cortex. Arch. Neurol. **59**: 1461–1467.
27. LEVITIN, D.J. *et al.* 2003. Neural correlates of auditory perception in Williams syndrome: an fMRI study. NeuroImage **18**: 74–82.
28. LEVITIN, D.J. & U. BELLUGI. 1998. Musical abilities in individuals with Williams syndrome. Mus. Percept. **15**: 357–389.
29. HURON, D. 2001. Is music an evolutionary adaptation? *In* Biological Foundations of Music, Vol. 930. pp. 43–61. Annals of the New York Academy of Sciences. New York.

Changing the Brain through Therapy for Musicians' Hand Dystonia

VICTOR CANDIA,[a,b] JAUME ROSSET-LLOBET,[b] THOMAS ELBERT,[c] AND ALVARO PASCUAL-LEONE[d]

[a]Collegium Helveticum, ETH-Zentrum/STW, Schmelzbergstrasse 25, CH-8092 Zürich, Switzerland

[b]Institut de Fisiologia i Medicina de l'Art-Terrassa, Ctra. de Montcada 668, E-08227 Terrassa, Barcelona, Spain

[c]Department of Psychology and Lurija-Institute for Neurorehabilitation, University of Konstanz, Fach D25, D-78457 Konstanz, Germany

[d]Center for Non-invasive Brain Stimulation, Beth Israel Deaconess Medical Center, Harvard Medical School, Boston, Massachusetts 02115, USA, and Hospital de Rehabilitación Institut Guttmann, Barcelona, Spain

ABSTRACT: Focal hand dystonia is a disorder in which sensory and motor anomalies emerge that appear to be grounded in maladaptive routes of cortical plasticity. Remodeling cortical networks through sensory-motor retuning (SMR), we achieved long-term reduction in the symptoms of focal hand dystonia. Magnetoencephalography confirmed that SMR modified the representational cortex of the fingers, whereby the representation of the affected hand was reorganized so that it resembled more the organization of the non-affected side. Furthermore, we observed differences in abnormal tactile acuity between patients with musician's cramp and those with writer's cramp: Using two-point finger discrimination, dystonic musicians showed perceptual asymmetry between hands, while writer's cramp patients did not. To further evaluate the occurrence of collateral disturbances in focal dystonia, we assessed the clinical histories of 101 affected musicians. An important finding from this study was that dystonic musicians who play a similar first and second instrument reported a continuous worsening of their symptoms. In addition, collateral disturbances appeared with a shorter delay when more than one instrument was played. Taken together, these studies suggest that (1) neurological dysfunction can be reversed by context-specific training protocols, (2) specific symptomatic and etiological differences among various forms of focal hand dystonia might result from different behavioral experiences and their central representation, and (3) the spread of symptoms might be prevented by avoiding training that implies movement patterns similar to the main affected task, and by reducing the amount of task-associated movement behavior.

KEYWORDS: focal hand dystonia; sensory discrimination; cortical plasticity; hand rehabilitation

Address for correspondence: Dr. Victor Candia, Collegium Helveticum ETH-Zentrum/STW, Schmelzbergstrasse 25, CH-8092 Zürich, Switzerland. Voice: +41-44-632-54-04; fax: +41-44-632-12-04.

candia@collegium.ethz.ch

Ann. N.Y. Acad. Sci. 1060: 335–342 (2005). © 2005 New York Academy of Sciences.
doi: 10.1196/annals.1360.028

INTRODUCTION

Focal hand dystonia in musicians, also termed "musician's cramp," is a greatly feared condition that leads to reduced performance levels[1] and usually the termination of a musician's career. The illness is a sensorimotor disorder characterized by a loss of control over individual finger movements, especially during specific actions related to playing a given musical instrument. Hand dystonia seems to be caused or at least related to the excessive performance of repetitive activities[2] and has been treated in different ways, including physical therapy, prolonged rest, anticholinergic drugs, and botulinum toxin.[3-7] Nevertheless, these treatments have not induced long-term improvements[1,8] even though isolated observations suggest that electromyographically guided botulinum toxin injections may have led to lasting reductions in symptoms in some of the musicians treated.[9]

Animal models have shown that repeated and prolonged use of the contralateral hand results in changes in the functional organization of the motor[10] and the primary sensory cortex.[11] Similarly, changes have been observed in the sensory cortical organization of violinists,[12] and in the motor cortex of piano learners.[13] Changes in auditory cortex of pianists[14] and trumpeters[15,16] have also been reported. It is thought that these changes underlie musical ability. However, it also seems possible to drive practice-induced changes in brain organization into the maladaptive range, a phenomenon that has been associated with focal hand dystonia.[17-20] Observed sensory anomalies suggest that focal dystonia is not simply a motor problem,[21] because abnormalities extend beyond the particular task affected[22-24] and can be measured even during passive tactile stimulation.[25-28] Abnormal sensory processing, for instance, has been repeatedly demonstrated,[27-32] and it seems likely that these changes might contribute to the emergence of the disease.

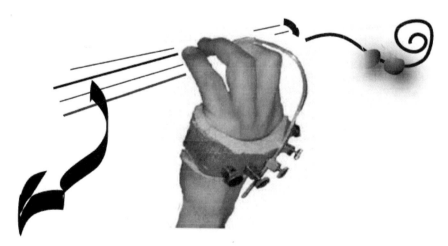

FIGURE 1. In sensory-motor retuning (SMR), a special device is used to splint fingers in such a way that dystonic movements can be avoided while practicing. It is important that the fingers can be splinted in positions that are similar to those adopted during normal playing. (Modified from Candia *et al.*[35])

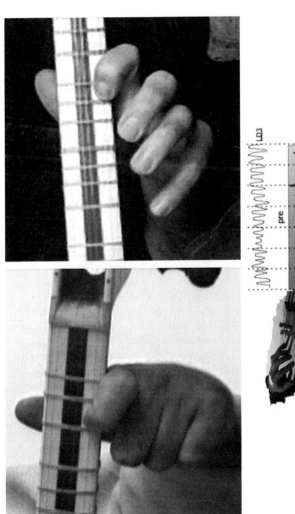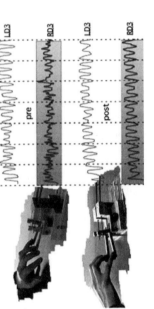

FIGURE 2. Focal hand dystonia is a condition in which voluntary control of the fingers is lost (*left*). After SMR, performance levels clearly increased (*right*). These changes correlated with the movement smoothness of the affected fingers, as measured with a displacement dexterity device (*lower panel*).[3]

RESULTS

To evaluate the occurrence of collateral disturbances in focal dystonia, we assessed clinical histories of 101 affected musicians.[33] A prominent finding was that dystonic musicians playing a similar first and second instrument (e.g., guitar and electric bass) reported continuous worsening of symptoms, whereas musicians playing only one or two rather different instruments (e.g., guitar and clarinet) did not show such a deterioration. In addition, generalized disturbances appeared with a longer delay (75% of cases) when only one instrument was played.

Assuming that (a) deviant brain organization contributes to focal hand dystonia and (b) motivated training can retune such brain abnormality, we developed a context-specific behavioral intervention, which we called sensory-motor retuning or SMR.[34,35] All subjects treated so far presented with a chronic condition and had received diverse prior treatments for their symptoms, which, however, had resulted in little or no relief. In SMR therapy, a hand splint immobilizes one or more finger(s), allowing different permutations of finger movements on the musical instrument for short periods of time (FIG. 1).

Repertoire practice without a splint is also added in order to accomplish the transfer of the exercise-induced improvement into the real-world environment. Supervised treatment is administered for eight consecutive days for 1.5 to 2.5 hours per session, depending on the patient's fitness.[35] We assessed the treatment outcomes with a device that continuously measures finger displacements and a subjective dystonia evaluation scale (DES).[36] Pianists and guitarists visibly improved from pre- to posttreatment (see FIG. 2). Follow-up proceeded for 3 to 25 months, suggesting that these results are long term. By contrast, the wind players did not improve. The subjective ratings and the clinical evaluation correlated with the findings obtained from the dexterity and displacement device. Thus, the movements of the dystonic fingers were smoother after SMR, indicating enhanced motor control.

To assess whether SMR would also induce observable alterations in the organization of the somatosensory cortex of the treated musicians, we studied the finger representations in somatosensory cortex in 10 patients pre- and posttreatment, using magnetoencephalography (MEG). We correlated post- and pretreatment differences between the displacement dexterity device and the patients' subjective ratings. We also correlated the difference pre- to posttreatment for the data collected with the displacement dexterity device with the difference pre- to posttreatment of the MEG-recorded dipole moment for the dystonic fingers. In addition, we calculated the cortical area of a triangle comprising the dystonic finger and its right and left neighboring fingers. These calculations showed that (1) prior to treatment, somatosensory relationships of the individual fingers differed between hands; (2) following treatment, the finger representations contralateral to the dystonic side were similar to the representation of the less affected side; (3) somatosensory finger representations were more ordered according to homuncular laws following treatment (FIG. 3); and (4) the predefined dystonic area was significantly smaller for the nontreated hand before treatment and was reduced for the dystonic hand after treatment.[37] These physiological changes correlated well with the behavioral data.

To assess tactile perceptual symmetry between hands, we measured two-point finger discrimination in musicians affected in the right hand and compared the find-

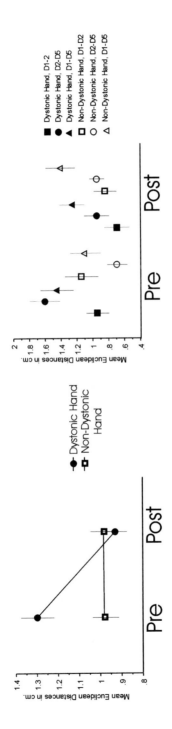

FIGURE 3. The areas of the brain that represent different fingers had centers that were further apart on the affected hand before treatment (*left*). After treatment this relationship and even the amount of brain activity responding to stimulation of the most affected finger became more normal (*right*).[37] These changes were associated with better performance on the instrument (see also *right panel* of FIG. 2).

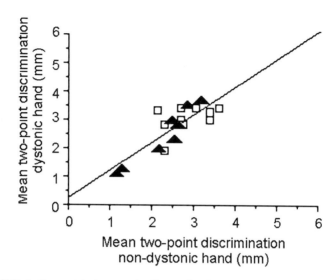

FIGURE 4. Symmetrical two-point finger discrimination was observed in the musicians' control group (*filled triangles*). In the dystonic group, no linear trend was evidenced (*open squares*). Thresholds are shown in mm. Single values represent the average across all fingers of one hand.

ings with results in patients with writer's cramp.[38] Only the dystonic musicians showed perceptual asymmetry between hands (FIG. 4).

The spatial discrimination within a defined dystonic area following the same criteria used in the MEG study revealed perceptual asymmetry within but not out of this area. Writer's cramp patients and their control group did not show such changes.

DISCUSSION

The reported data are consistent with a variety of studies that suggest that cortical organization may be modified through extensive use.[39–41] Our results confirm that cortical changes, together with emergent neurological dysfunction, can be redressed by context-specific treatment. In addition, the use of a movement intervention capable of producing measurable changes in the cortical organization of sensory areas underscores the tight relationship between sensory and motor systems. This might explain the diverse anomalies reported at different levels of the sensorimotor system of the affected patients by different research groups.

It has been suggested that changes in the sensorimotor system of focal hand dystonia patients may not be exclusively limited to task-relevant stimulation but may instead be a more generalized phenomenon.[22–24,32] In agreement with this notion, our data for musicians suffering from focal hand dystonia showed that changes in the symmetry of the static two-point threshold appear to be part of a more general profile of sensory anomalies: for the instruments being played by our patients, simultaneous two-point discrimination does not seem to be a task of relevance.

Conspicuously, the assessments of the 101 affected musicians revealed that those playing similar instruments (containing similar kinematics) experience continuous aggravation of their symptoms. Moreover, those playing more than one musical instrument developed collateral disturbances faster, suggesting that increasing the degree of task similarity and the amount of practice devoted to such tasks may be crucial for worsening of symptoms. We, therefore, speculate that manual activities containing similar kinematics may lead to dystonic disturbances at some point. It is not the general reduction of muscular afferents from the affected limb[42] but the specific reduction of activities containing similar kinematics that will most probably contribute to a limitation of the symptoms to a particular task, making them more amenable to context-specific interventions, such as sensory-motor retuning.

ACKNOWLEDGMENTS

Research was supported in part by the Cogito Foundation, Fundació Ciència I Art, the Deutsche Forschungsgemeinschaft, the Harvard-Thorndike General Clinical Research Center (NCRR MO1 RR01032), the National Institutes of Health (RO1MH60734, RO1EY12091, K24 RR018875), and the Goldberg Foundation.

[Competing interests: The authors declare that they have no competing financial interests.]

REFERENCES

1. ALTENMÜLLER, E. 1996. Fokale Dystonien bei Musikern: eine Herausforderung für die Musiker-Medizin. Musikphysiol. Musikermed. **3:** 29–40.
2. HALLETT, M. 1998. The neurophysiology of dystonia. Arch. Neurol. **55:** 601–603.
3. FRY, H.J. 1993. The treatment of overuse injury syndrome. Md. State Med. J. **42:** 277–282.
4. NEWMARK, J. & F.H. HOCHBERG. 1987. Isolated painless manual incoordination in 57 musicians. J. Neurol. Neurosurg. Psychiatry **50:** 291–295.
5. LEDERMAN, R. 1988. Occupational cramp in instrumental musicians. Med. Prob. Performing Artists **3:** 45–51.
6. LEVINE, W.R. 1983. Behavioral and biofeedback therapy for a functionally impaired musician: a case report. Biofeedback Self-Regul. **8:** 101–107.
7. FAHN, S. & C.D. MARSDEN. 1987. The Treatment of Dystonia. Butterworths. London.
8. BEJJANI, F.J., G.M. KAYE & M. BENHAM. 1996. Musculoskeletal and neuromuscular conditions of instrumental musicians. Arch. Phys. Med. Rehabil. **77:** 406–413.
9. SCHUELE, S. *et al.* 2005. Botulinum toxin injections in the treatment of musicians' dystonia. Neurology **64:** 341–343.
10. NUDO, R.J. 2003. Functional and structural plasticity in motor cortex: implications for stroke recovery. Phys. Med. Rehabil. Clin. N. Am. **14:** S57–76.
11. WANG, X. *et al.* 1995. Remodelling of hand representation in adult cortex determined by timing of tactile stimulation. Nature **378:** 71–75.
12. ELBERT, T. *et al.* 1995. Increased cortical representation of the fingers of the left hand in string players. Science **270:** 305–307.
13. PASCUAL-LEONE, A. *et al.* 1995. Modulation of muscle responses evoked by transcranial magnetic stimulation during the acquisition of new fine motor skills. J. Neurophysiol. **74:** 1037–1045.
14. PANTEV, C. *et al.* 1998. Increased auditory cortical representation in musicians. Nature **392:** 811–814.
15. PANTEV, C. *et al.* 2001. Timbre-specific enhancement of auditory cortical representations in musicians. Neuroreport **12:** 169–174.

16. PANTEV, C. *et al.* 2001. Representational cortex in musicians. Plastic alterations in response to musical practice. Ann. N. Y. Acad. Sci. **930:** 300–314.
17. BYL, N.N., M.M. MERZENICH & W.M. JENKINS. 1996. A primate genesis model of focal dystonia and repetitive strain injury. I. Learning-induced dedifferentiation of the representation of the hand in the primary somatosensory cortex in adult monkeys. Neurology **47:** 508–520.
18. BYL, N.N. & M. MELNICK. 1997. The neural consequences of repetition: clinical implications of a learning hypothesis. J. Hand Ther. **10:** 160–174.
19. BYL, N.N. *et al.* 1997. A primate model for studying focal dystonia and repetitive strain injury: effects on the primary somatosensory cortex. Phys. Ther. **77:** 269–284.
20. MCKENZIE, A.L. *et al.* 2003. Somatosensory representation of the digits and clinical performance in patients with focal hand dystonia. Am. J. Phys. Med. Rehabil. **82:** 737–749.
21. PUJOL, J. *et al.* 2000. Brain cortical activation during guitar-induced hand dystonia studied by functional MRI. Neuroimage **12:** 257–267.
22. TINAZZI, M. *et al.* 2002. Deficits of temporal discrimination in dystonia are independent from the spatial distance between the loci of tactile stimulation. Mov. Disord. **17:** 333–338.
23. LIM, V.K. *et al.* 2003. Perceptual differences in sequential stimuli across patients with musician's and writer's cramp. Mov. Disord. **18:** 1286–1293.
24. TINAZZI, M., T. ROSSO & A. FIASCHI. 2003. Role of the somatosensory system in primary dystonia. Mov. Disord. **18:** 605–622.
25. SANGER, T.D. *et al.* 2002. Nonlinear sensory cortex response to simultaneous tactile stimuli in writer's cramp. Mov. Disord. **17:** 105–111.
26. ELBERT, T. *et al.* 1998. Alteration of digital representations in somatosensory cortex in focal hand dystonia. Neuroreport **9:** 3571–3575.
27. BARA-JIMENEZ, W. *et al.* 1998. Abnormal somatosensory homunculus in dystonia of the hand. Ann. Neurol. **44:** 828–831.
28. SANGER, T.D., D. TARSY & A. PASCUAL-LEONE. 2001. Abnormalities of spatial and temporal sensory discrimination in writer's cramp. Mov. Disord. **16:** 94–99.
29. BARA-JIMENEZ, W., P. SHELTON & M. HALLETT. 2000. Spatial discrimination is abnormal in focal hand dystonia. Neurology **55:** 1869–1873.
30. BARA-JIMENEZ, W. *et al.* 2000. Sensory discrimination capabilities in patients with focal hand dystonia. Ann. Neurol. **47:** 377–380.
31. GRUNEWALD, R.A. *et al.* 1997. Idiopathic focal dystonia: a disorder of muscle spindle afferent processing? Brain **120:** 2179–2185.
32. MOLLOY, F.M. *et al.* 2003. Abnormalities of spatial discrimination in focal and generalized dystonia. Brain **126:** 2175–2182.
33. ROSSET-LLOBET, J. *et al.* 2005. Collateral disturbances in 101 cases of musician's dystonia. Submitted for publication.
34. ELBERT, T. & B. ROCKSTROH. 2004. Reorganization of human cerebral cortex: the range of changes following use and injury. Neuroscientist **10:** 129–141.
35. CANDIA, V. *et al.* 2002. Sensory motor retuning: a behavioral treatment for focal hand dystonia of pianists and guitarists. Arch. Phys. Med. Rehabil. **83:** 1342–1348.
36. CANDIA, V. *et al.* 1999. Constraint-induced movement therapy for focal hand dystonia in musicians. Lancet **353:** 42.
37. CANDIA, V. *et al.* 2003. Effective behavioral treatment of focal hand dystonia in musicians alters somatosensory cortical organization. Proc. Natl. Acad. Sci. USA **100:** 7942–7946.
38. CANDIA, V., A. PASCUAL-LEONE & T. ELBERT. 2005. Two-point discrimination in the fingers is different in patients with musician's or writer's cramp. Submitted for publication.
39. STERR, A. *et al.* 1998. Changed perceptions in Braille readers. Nature **391:** 134–135.
40. STERR, A. *et al.* 1998. Perceptual correlates of changes in cortical representation of fingers in blind multifinger Braille readers. J. Neurosci. **18:** 4417–4423.
41. PASCUAL-LEONE, A. *et al.* 1995. The role of reading activity on the modulation of motor cortical outputs to the reading hand in Braille readers. Ann. Neurol. **38:** 910–915.
42. PRIORI, A. *et al.* 2001. Limb immobilization for the treatment of focal occupational dystonia. Neurology **57:** 405–409.

Music to Electric Ears: Pitch and Timbre Perception by Cochlear Implant Patients

DANIEL PRESSNITZER,[a] JULIE BESTEL,[b] AND BERNARD FRAYSSE[c]

[a]Equipe Audition, LPE-CNRS UMR 8581, Département d'Etudes Cognitives, ENS, 75230 Paris cedex 05, France

[b]Advanced Bionics Europe, Clinical Research Department, 68170 Rixheim, France

[c]Service ORL, Hopital Purpan, 31059 Toulouse, France

ABSTRACT: The sounds of music play with many perceptual dimensions. We devised a set of psychophysical procedures to better understand how recipients of cochlear implants perceive basic sound attributes involved in music listening.

KEYWORDS: psychophysics; hearing impairment; melody; pitch memory; cochlear implant

INTRODUCTION

The technology of cochlear implants (CIs) consists of directly stimulating the auditory nerve of profoundly deaf patients to restore some aspects of auditory function. Much research and progress has been made toward improving speech perception by cochlear implant recipients. However, music perception has not received the same amount of interest so far. Music is, however, an integral part of the auditory experience of each and every individual, and there is a growing demand from implant recipients for improvement of their musical experiences. Subjective assessments of music appreciation indicate that whereas rhythm cues are satisfactorily transmitted by the implant, other aspects of music may not be.[1] A common method of evaluating pitch perception consists of testing the recognition of familiar tunes; other performance measures use simple pitch discrimination tasks (for recent reviews, see Refs. 2 and 3). For timbre perception, which is also part of musical listening, performance measures address the recognition of musical instruments.[4] Recognition tasks rely on the familiarity of the listeners with the stimulus to be recognized, and a successful recognition can be attributed to many different cues. We aimed to address pitch and timbre perception by cochlear implant recipients using a set of psychophysical measures of performance that focus on distinct aspects of these musical attributes and that do not rely on the previous musical experience of listeners.

Address for correspondence: Daniel Pressnitzer, Equipe Audition, LPE-CNRS UMR 8581, Département d'Etudes Cognitives, ENS, 29 rue d'Ulm, 75230 Paris cedex 05, France. Voice: +33-1-4432-2673.

daniel.pressnitzer@ens.fr

Ann. N.Y. Acad. Sci. 1060: 343–345 (2005). © 2005 New York Academy of Sciences.
doi: 10.1196/annals.1360.050

PSYCHOPHYSICAL PROCEDURES TO TEST
PITCH AND TIMBRE PERCEPTION

The first pitch task was a higher–lower pitch judgement. It is more complex than a discrimination task, because a ranking judgement is requested; however, it still addresses a basic level of pitch processing. The second task involved the comparison of two short random melodies.[5] Listeners first heard a four-note chromatic melody, and then the melody repeated with a pitch difference introduced in one of the notes at random. Their task was to indicate which note had been altered. This task is easy when presented to normal hearing listeners, even if they are nonmusical, provided that a pitch is clearly heard for each of the notes.[5]

Stimuli were bandpass-filtered harmonic complex tones delivered acoustically. The repetition rate of the tones was varied to control pitch, whereas the spectral region remained fixed. This meant that the fundamental frequency of the sounds (~100 Hz) was absent for most trials. Presentation level was roved to reduce cues unrelated to pitch.

The timbre tasks measured just-noticeable differences on two important dimensions of musical timbre: attack time and spectral center of gravity. These dimensions repeatedly appear in multidimensional scaling studies of timbre of musical instruments (e.g., McAdams et al.[6]), indicating that they underlie the more salient perceptual differences among instruments. Bandpass-filtered complex tones were again used. For the attack time, the spectral content of the sound was fixed, whereas the temporal envelope was varied. For the spectral center of gravity, the temporal envelope was fixed, whereas the rate of decay between successive harmonics was varied.

All tasks, pitches, and timbres were adaptive procedures with feedback and an N-alternative unforced choice.[7] A training period was included when listeners could ask for feedback before giving their answers.

PITCH, MELODIES, AND TIMBRE

We now describe the first results obtained with these objective tasks, in the framework of a larger clinical study. Subjects were postlingually implanted patients using the Advanced Bionics HiRes™ stimulation scheme.

Listeners had higher–lower pitch-ranking thresholds ranging between 2 and 7 semitones. This is higher than what is observed with a non–age-matched control group (0.2 semitones), but it is in line with previous results.[3] The melody task, however, proved impossible for most implant recipients. Note that all intervals between the notes of the melodies were chosen to be larger than the pitch-ranking threshold for each individual listener. The control group, as expected, displayed ceiling performance for the melody task.

On both dimensions of timbre tasks that we investigated, implant recipients performed almost as well as the control group. The dimensions we investigated were not necessarily familiar to our listeners, and we could not describe in the experimental instructions how they would sound after being processed by the implant. After the feedback received during the training period, however, all listeners could perform the task to a good level of performance.

CUES TO MUSICAL SOUNDS PERCEPTION

Models of pitch perception show that pitch is correlated with spectral cues, fine structure temporal cues, and envelope cues.[5] The processing schemes of the implants as well as inherent limitations of electrical stimulation of the auditory nerve reduce both spectral and fine-structure cues. This could explain the reduced performance of implant recipients for the pitch-ranking task. The melody task raises another question. Memory for pitch is different from memory for other auditory attributes, in that it allows for the short-term encoding of more items.[8] Pitch, thus, is especially suited for conveying melodies that extend in time. If the percepts associated with repetition rate for implant recipients do not have the memorization characteristics of pitch, specific impairments in the perception of melodies would be predicted.

Discrimination along dimensions of musical timbre proved to be accurate in our group of implant recipients. We thus hypothesize that the processing scheme used in the implant provided informative timbre cues, even though they might be different from what listeners were used to hearing before implantation. The fact that these cues can subserve discrimination should allow for musical instrument recognition after some training.

ACKNOWLEDGMENTS

These data were collected as part of a multicentric clinical trial, with the support of Advanced Bionics. We thank Pr. Michel Mondain, Guy de Chauliac Hospital, Montpellier; Pr. Bruno Frachet, Avicenne Hospital, Bobigny; Pr. J.P. Bébéar, Pellegrin Hospital, Bordeaux; and Pr. Olivier Sterkers, Beaujon Hospital, Clichy.

[Competing interests: The authors declare that they have no competing financial interests.]

REFERENCES

1. GFELLER, K. *et al.* 2000. Musical backgrounds, listening habits, and aesthetic enjoyment of adult cochlear implant recipients. J. Am. Acad. Audiol. **11:** 390–406.
2. MCDERMOTT, H.J. 2004. Music perception with cochlear implants: a review. Trends Amplif. **8:** 49–82.
3. MOORE, B.C.J. & R.P. CARLYON. 2005. Perception of pitch by people with cochlear hearing loss and by cochlear implant users. *In* Pitch: Neural Coding and Perception. C.J. Plack *et al.*, Eds.: 234–277. Springer. New York.
4. LEAL, M.C. *et al.* 2003. Music perception in adult cochlear implant recipients. Acta Otolaryngol. **123:** 826–835.
5. PRESSNITZER, D., R.D. PATTERSON & K. KRUMBHOLZ. 2001. The lower limit of melodic pitch. J. Acoust. Soc. Am. **109:** 2074–2084.
6. MCADAMS, S. *et al.* 1995. Perceptual scaling of synthesized musical timbres: common dimensions, specificities, and latent subject classes. Psychol. Res. **58:** 177–192.
7. KAERNBACH, C. 2001. Adaptive threshold estimation with unforced-choice tasks. Percept. Psychophys. **63:** 1377–1388.
8. MCFARLAND, D.J. & A.T. CACACE. 1992. Aspects of short-term acoustic recognition memory: modality and serial position effects. Audiology **31:** 342–352.

Part VI: Music Performance

Introduction

DIEGO MINCIACCHI

Degree Course in Sciences of Movement, Faculty of Medicine, University of Florence, Florence, Italy

> *Not only science but art, also, shows us that reality, at first incomprehensible, gradually reveals itself, by the mutual relations that are inherent in things.*
> PIET MONDRIAN[1]

Music performance is a stunning topic for scientists to explore, because of the multiplicity of sensory channels involved, the huge intra- and intermodal central integration, and the actualization and monitoring of motor output through external sensory and internal "musically associated" feedback.

Even with this tiny sketch in mind, we can grasp, at first glance, a discouraging consciousness of incapacity. However, several investigators approached courageously, from different standpoints and for different purposes, the world of music performance and its constellation of inherent evils. They extended accessible techniques and developed thorough experimental designs to move toward the unraveling of one of the most complex human activities. The collection of five papers in this section gives a short though assorted palette of relevant recent achievements in this mostly unexplored field of neuroscience.

The opening contribution by Lim and coworkers concentrates on one of the most annoying and invalidating problems that afflict performers, the musician's cramp, a special family of "action" or "training-induced" dystonia. The work is a paradigmatic illustration of a study where music performance is used to ultimately deepen our knowledge of the pathophysiology of a prominent class of movement disorders, the focal dystonia. Using an interesting experimental design the authors provide evidence supporting the idea that subjects with musician's cramp have multiple sensory deficiencies and thus that focal dystonia is an expression of extensive multilevel changes in the central nervous system. It should be mentioned, however, that a recent paper by Rosenkranz and coworkers suggests a different pathophysiology for musician's cramp with respect to other forms of focal dystonia, such as writer's cramp, considering the differential role played by the hand sensory input in the pathological change of motor commands.[2]

Address for correspondence: Prof. Diego Minciacchi, University of Florence, Faculty of Medicine, Degree Course in Sciences of Movement, Viale Pieraccini 6, I-50134 Florence, Italy. Voice: +39-055-4279788.

diego@unifi.it.

Ann. N.Y. Acad. Sci. 1060: 346–348 (2005). © 2005 New York Academy of Sciences.
doi: 10.1196/annals.1360.066

The contribution by Palmer discusses how memory retrieval function and movement preparation combine in music performance. The author moves from a personal model for temporal activation of memory operations during sequence production, a model that would be predictive of the time course for retrieval processes. Taking into consideration the possibilities of serial or cascade arrangements for the combination of item retrieval and movement preparation, Palmer concludes that the cascade model, where movement preparation starts before item retrieval is completed, is the best candidate to be involved in music performance. Ordering events into a cascade sequence has many advantages and is definitely the choice when we deal with behaviors requiring fast neural operations.

Information on strategies for controlling voluntary timed motor sequences is also offered in the elegant study by Ullén and colleagues, where the opportunity for an independent codification of the temporal sequence and the ordinal structure is discussed. It is shown that temporal sequences and ordinal structures are learned and processed by distinct sets of brain regions. In addition, a group of cortical areas including the supplementary motor area, the superior temporal gyrus, and the inferior frontal cortex, is distinctively designated as relevant for processing movement-independent temporal sequences. This highly specialized regional organization appears to be flexible for the management of voluntary motor sequential tasks and well suits the needs of music performance where several brain functions are hard-pressed to their temporal limits.

An additional dissociation of the brain apparatus relevant for music performance is illustrated in the contribution by Stewart, which explores and discusses the spatial sensorimotor mapping of scores from a set of vertically organized cues into a horizontally organized set of responses. The author ultimately emphasizes that the "what" and "when" aspects of music reading appear to involve the what and when of occipitoparietal and occipitotemporal cortical streams. A concurrent report, based on piano playing from score reading, highlights the same question, that is, a dissociation between the involvement of the dorsal visual stream, for spatial pitch processing, and the ventral visual stream, for temporal movement preparation.[3]

Finally, the remarkable contribution by Schneider and colleagues explores possible correlations between the structure and functioning of the Heschl's gyrus and the preference for particular instruments and musical performance. A huge number of musicians and nonmusicians was subdivided into groups of fundamental or spectral pitch listeners, that is, subjects that perceive sounds in synthetical or analytical mode, respectively. After study of the auditory cortex with magnetic resonance imaging and magnetoelectroencephalography, the authors were able to correlate the asymmetries of the Heschl's gyrus with the perceptual attitude of subjects and with their predilection for instruments and music performance, inferring that the mode of listening and the specific music preference for performance may be predicted by the structural and functional asymmetry of the auditory cortex.

Musical performance is the realm in which humans produce the most elaborate integration processes, involving perceptual, cognitive, emotional, and motor skills. Music production can be regarded as the peak of human central nervous system performance, and the scientific effort to disentangle its brain machinery and tasks is far from complete.

Even future studies based on more sophisticated experimental designs and techniques of analysis than those used in the present volume, however, will not provide a definitive explanation for musical genius. We will end with the enlightening recent words by Baader, Kazennikov, and Wiesendanger, who, in a paper dealing with motor control in violin playing, affirm that the neuroscientific approach to music performance "deals with motor (or psychological) mechanisms and technical (or cognitive) skills, that is, the prerequisites of musical proficiencies. However, it does not necessarily encompass true musicality."[4]

REFERENCES

1. MONDRIAN, P., H. HOLTZMAN & M.S. JAMES, Eds. 1993. The New Art—The New Life: The Collected Writings of Piet Mondrian. Da Capo Press.
2. ROSENKRANZ, K., A. WILLIAMON, K. BUTLER, et al. 2005. Pathophysiological differences between musician's dystonia and writer's cramp. Brain 128: 918–931.
3. BENGTSSON, S.L. & F. ULLEN. 2005. Dissociation between melodic and rhythmic processing during piano performance from musical scores. Neuroimage. In press.
4. BAADER, A.P., O. KAZENNIKOV & M. WIESENDANGER. 2005. Coordination of bowing and fingering in violin playing. Cogn. Brain Res. 23: 436–443.

Enhanced P1-N1 Auditory Evoked Potential in Patients with Musicians' Cramp

VANESSA K. LIM,[a,d] JOHN L. BRADSHAW,[b] MICHAEL E. R. NICHOLLS,[c] AND ECKART ALTENMÜLLER[d]

[a]Department of Psychology, University of Auckland, New Zealand

[b]Neuropsychology Research Unit, Psychology Department, Monash University, Clayton 3168, Victoria, Australia

[c]Department of Psychology, University of Melbourne, 3010 Parkville, Victoria, Australia

[d]University of Music and Drama, D-30161 Hannover, Germany

ABSTRACT: Auditory evoked potentials (AEPs) were examined in patients with musician's cramp (focal dystonia) in order to determine whether these patients have electrophysiological changes in a sensory system that is not usually associated with symptoms. All participants were professional guitarists and were required to listen to 2000 monaurally presented stimuli (middle C, with duration of 7 ms). During one block, 250 stimuli were presented to one ear. Once a block was finished, another block was presented in the other ear; in total there were eight blocks of stimuli. During this task, EEGs from 10 scalp electrodes and one bipolar eye channel were continuously recorded. There were no significant latency or topographical differences in the electrophysiological recordings. However, there was a significant group difference in the peak-to-peak amplitude of the P1-N1a component. The patients had a larger peak-to-peak difference than controls (1.63 vs. 0.62 μV). The P1 and N1a are cortically generated potentials. Patients with focal dystonia had an increase in activity compared to controls when processing simple auditory stimuli. Such changes in electrophysiological responses may be a result of increases in excitation or lack of inhibition; alternatively the changes may represent cross-modal maladaptive plasticity from the somatosensory modality to the auditory modality. Thus, this study provides further evidence that patients with focal dystonia have alterations of the central nervous system that are not limited to their symptomatic sensory domain.

KEYWORDS: focal dystonia; musicians; auditory evoked potentials; electroencephalography

INTRODUCTION

With regard to the pathophysiology of focal dystonia, it appears that the somatosensory and motor circuits on several levels of the central nervous system are

Address for correspondence: Prof. Dr. Eckart Altenmüller, University for Music and Drama Hannover, Institute of Music Physiology and Musicians' Medicine, Hohenzollernstr. 47, D-30161 Hannover, Germany. Voice: +49-511-310-552; fax: +49-511-3100-557.
altenmueller@hmt-hannover.de

Ann. N.Y. Acad. Sci. 1060: 349–359 (2005). © 2005 New York Academy of Sciences.
doi: 10.1196/annals.1360.029

affected.[1,2] Compared to controls, patients with focal dystonia have clear changes in the somatosensory (symptomatic) domain. For example, the N30 component of the somatosensory evoked potential is enhanced.[3] There are alterations in the cortical representation of fingers in the somatosensory cortex[4] and increases in task-related power (10–12 Hz) during simple and complex finger movement.[5] Changes occur in electrophysiological activity prior to movement,[6–9] and such changes in physiology have also been found in the nonsymptomatic limb.[10,11]

Not only are there changes in physiology, but patients with focal dystonia also have changes in perception.[12,13] Interestingly, evidence suggests that there are differences in temporal discrimination across different modalities—particularly the tactile and visual domains.[14–16]

Such cross-modal interactions between tactile and visual modalities are important for other forms of focal dystonia. Therefore, for the musician, the sensory modality typically does not appear to exhibit symptoms, but what is functionally important to musicians is the auditory modality. Audition is of upmost importance for musicians because their skills rely crucially on accurate auditory feedback, and musicians must have good functional links or couplings between the auditory and somatosensory modalities.[17–20] In professional musicians, such multimodal sensorimotor and auditory integration has been demonstrated.[21–23]

If the somatosensory domain has functional changes in patients with focal dystonia and there are cross-modal interactions, as evidence has suggested above, then it is possible that another highly coupled/linked area, such as the auditory cortex, will also demonstrate electrophysiological changes. Thus, an investigation involving a sensory system not traditionally linked to symptoms, but that has clear importance for a musician, is warranted.

In this study, an auditory stimulus was used to investigate whether there were electrophysiological changes in patients with musicians' cramp compared to control musicians. It was hypothesized that there would be no group differences in the auditory (asymptomatic) modality if focal dystonia did not involve generalized multimodal sensory problems; conversely, differences between groups in the auditory domain would be expected if the disorder involved generalized sensory problems.

METHOD

Participants

All participants played the guitar professionally (TABLES 1 and 2). There were six patients (40 ± 6 years; ± 1 SD) with involuntary flexion of their fingers while playing (TABLE 1). The average duration of dystonia was 8 ± 5 years. Five of the patients had had previous injections of Botulinum-Toxin A (Dysport®; Ipsen Pharma); however, the last injection was approximately 3–4 months prior to the study. None of the six patients had any family history of movement disorders. Seven controls (35 ± 10 years old) participated in this study (TABLE 2). Two of the six patients were left-handed.[24] The patients had relatively similar symptoms and were classified as having mild dystonia; the average severity score (in percent) was $84 \pm 2\%$.[25] Informed consent was obtained from each participant. Furthermore, none of these patients had any obvious lesions along the auditory pathway and the cortex, and all had normal

TABLE 1. Clinical profiles for patients with musicians' cramp in this study

Patient	Age	Number of years playing	Age of entrance into music academy	Handedness and side of symptom	Length of problem (years)	Symptoms (fingers, all flexion problems)[a]	Severity (%)*	Task specific[b]	Previous medication	Type of music played[c]
1	41	30	19	left	6	3	86	yes	**40 U	3
2	35	24	22	left	1	3, 4	86	no (w)	**50 U	2
3	36	14	21	right	11	2	81	no (w)	No	3
4	42	28	21	right	11.5	4, 5	76	no (w)	No	1
5	49	38	17	right	15	3	85	yes	**150 U	3
6	34	25	23	right	4	2, 4	81	no (c)	**75 U	3

[a]Symptoms: (fingers), 2 = index; 3 = middle; 4 = ring; 5 = small.
[b]w = writing; c = computer.
[c]Music played: 1 = classical; 2 = jazz and rock/pop; 3 = classical, jazz, and rock/pop.
* Severity (Fahn[25]).
** Dysport; U = units.

TABLE 2. Profiles for the guitarist controls in this study

Control	Age	Number of years playing	Age of entrance into music academy	Handedness	Type of music played[a]
1	49	40	23	left	3
2	23	16	22	left	1
3	33	22	22	right	1
4	40	26	20	right	3
5	43	33	17	right	3
6	34	24	21	right	2
7	20	10	15	right	3

[a]Music played: 1 = classical; 2 = jazz and rock/pop; 3 = classical, jazz, and rock/pop.

hearing (all participants were examined by EA, a neurologist who specializes in examining and treating musicians). All procedures were approved by a local ethics committee (University of Hannover, Germany).

Electrophysiology: Auditory Evoked Potentials

Electroencephalography (EEG) was recorded continuously from 10 scalp electrodes (Fz, FC3, FCz, FC4, C3, Cz, C4, CP3, CPz, CP4) mounted on an elastic cap (EasyCap, FMS, Herrsching-Breitbrunn, Germany). Linked mastoids were used as a reference, and the electrode impedances were kept below 5 kOhms for the EEG. EEG signals were amplified (Synamps Amplifiers, Neuroscan Inc., USA) between DC (high pass) and 200 Hz (low pass) and were digitized with a sampling frequency of 1000 Hz.

Stimuli

The auditory stimuli were 7 ms in duration and consisted of a single stimulus (261.6 Hz/middle C, 49 ± 3 dB). A computer sound card generated the auditory stimuli, and the output was measured with a Lutron, SL-4001 sound level meter. The stimuli were presented monaurally; earphones were kept in the ears at all times, irrespective of the side of stimulation. The earphones were stereo EEG earphones (EAR Tone™).

Procedure

During the trials, participants looked at a fixation cross in the middle of a computer monitor that was 2 meters away. The task was to listen to 2000 monaurally presented stimuli at a frequency close to 1 Hz (± 0.01–0.5 s; this stimulus presentation allowed any 50 Hz interference to be minimized). That is, during one block, 250 stimuli were presented to one ear; there were, in total, eight blocks of stimuli. The starting ear was counterbalanced between participants, and subsequently, the stimulated side was alternated between blocks. Alternating between left and right sides continued until all eight blocks were run. Participants were encouraged not to move during this experiment, which took approximately 45 minutes to complete, including breaks (approximately two minutes).

Data Processing

Eye-movement artifacts were mathematically removed using the commercial software algorithms (Neuroscan Inc., USA). Custom written software (Delphi 5) was employed to obtain the amplitudes and latencies of the auditory evoked potentials (AEPs), and these were subsequently visually inspected to ensure that the computer captured the correct wave forms. This study limited assessment of AEPs to the late-latency components because surface electrodes over a limited number of sites on the cortex were employed; also the sampling rate was limited to 1000 Hz rather than the usual 4000 Hz required for accurate detection of earlier components. The components identified were P1 (43–53 ms), N1a (62–72 ms), N1 (85–100 ms), N1b (110–140 ms) and the P2 (140–180 ms). After the AEPs were identified, the mean amplitudes across the above-mentioned times were averaged. Then the peak-to-peak amplitudes were calculated for the components (P1-N1a, N1a-N1, N1-N1b, and N1b-P2).

Statistical Analyses

The peak-to-peak amplitudes were investigated using nonnormalized data to investigate amplitude differences. For investigations of differences in scalp distributions, the data were normalized.[26] Here, the maximum and minimum values for the grand averaged waveforms were identified for each condition and group. The minimum from the corresponding condition or group was subtracted from each data point, and these values were subsequently divided by the range. Three mixed-repeated measures analysis of variance (MANOVAs) were conducted with a between-group (patients and controls) factor and within-group factors side (left and right) and electrodes (Fz, FC3, FCz, FC4, C3, Cz, C4, CP3, CPz, CP4) (1) on latencies of the components (P1, N1a, N1, N1b, and P2); (2) for the amplitudes; and (3) on the scalp distribution of the peak-to-peak times (P1-N1a, N1a-N1, N1-N1b, and N1b-P2). Due to the multiple electrodes and violations of sphericity, a conservative correction was adopted for the degrees of freedom for the MANOVAs.[27]

RESULTS

Latencies

There were no significant main effects or interactions involving the factor "group" for the latencies of the seven components (all F statistics$_{(1,11)}$ were less than 3.15; n.s.). The five components followed standard AEP latencies.[28]

Scalp Distributions of AEPs

The standard EEG normalization of amplitude to investigate scalp distributions was performed for the four peak-to-peak amplitudes.[26] There were no significant main effects or interactions (all F statistics$_{(1,11)}$ were less than 1.3; n.s.).

Amplitude

There was a significant group difference for the peak-to-peak P1-N1a ($F_{(1,11)} = 5.7$; $P < .05$). The patients had a greater P1-N1a difference (1.63 μV) than controls

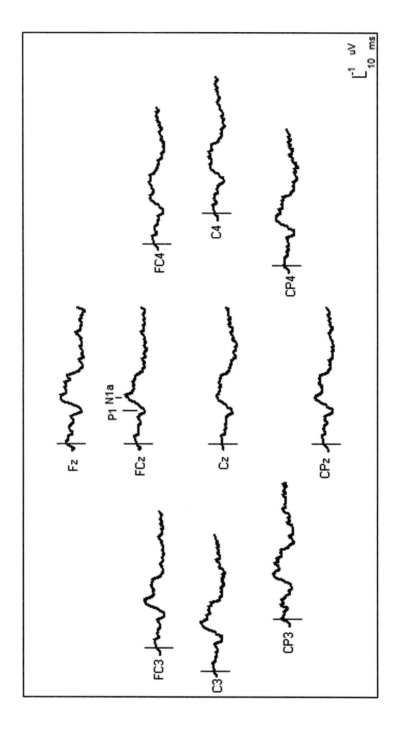

FIGURE 1. The grand averaged AEP difference wave is shown. The controls were subtracted from the patients, and the P1 and N1a components are shown on electrode FCz. The negative voltages are presented on the upper half of the y axis.

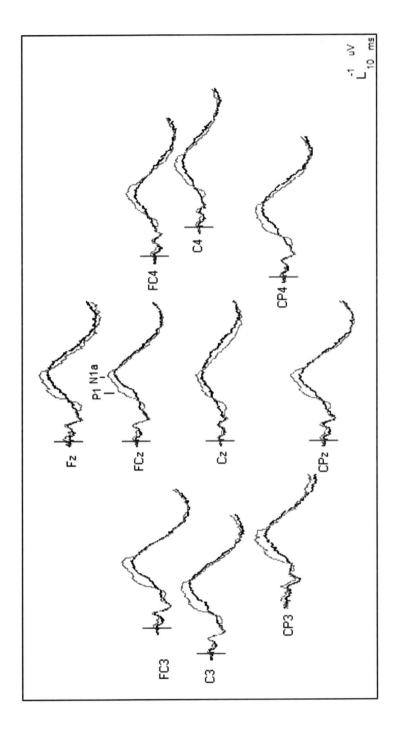

FIGURE 2. Grand average of the AEPs for controls (*black*) and patients (*gray*). Fz is a frontal electrode close to the forehead. The components P1 and N1a are marked on the electrode: FCz.

(0.62 µV); this difference is clearly represented in FIGURE 1. FIGURE 2 shows the grand averaged AEPs for stimulation from both sides for the controls in the black (on average, 887 trials were artifact free for each participant) and patients in the gray (on average, 947).

DISCUSSION

The current study investigated the late-latency AEPs occurring between 50 ms to 200 ms in two groups of musicians (with and without musicians' cramp). There were no significant latency differences between groups. This suggests that patients with musicians' cramp do not have deficiencies in the initial generation of the late-latency AEPs. In contrast, this study revealed that the patients with musician's cramp had a larger peak-to-peak difference in the P1 and N1a components than controls (1.63 vs. 0.62 µV).

The intracerebral recordings suggest that the generator of the P1 is lateral and widely distributed in the primary auditory cortex (Heschl's gyrus).[29] Evidence suggests that the generator for the N1a, the broad negativity over the frontocentral regions of the scalp, is the planum temporale.[30]

The group difference in the peak-to-peak amplitude of the P1-N1a component without changes in scalp distribution is consistent with altered generator strength. In the case of the patients, it appears that the generators of the P1 and N1a are increased compared to those of the controls. It is possible that the increase in overall activation of the primary and secondary auditory cortices is another example of changes in the balance of inhibitory and excitatory networks in patients with focal dystonia. Evidence suggests that patients with focal dystonia lack inhibition on several levels of sensorimotor pathways.[31–34] An imbalance of the firing rates along the basal ganglia-thalamocortical pathway (most probably excess activity in the internal globus pallidus) may cause an excess of activity and/or a lack of inhibition. In this case, if patients lack inhibition or have increases in excitation, these pathways are not regulated as well as in controls, leading to an overall increase in activity. Thus, the current results provide evidence suggesting that patients with musicians' cramp have changes at the cortical level, which are not limited to the symptomatic sensory modality. The results suggest that there are changes in the auditory-motor loops in patients with focal dystonia.

Although this study did not directly test across modalities, the increased peak-to-peak amplitude of the cortical P1 and N1a may reflect cross-modal maladaptive plasticity. Such cortical changes may have occurred from the interaction between the aberrant sensory afferents from the fingers and hands with the processing of auditory stimuli. There is clear evidence that experience changes the representation in the brain. Both advantageous and maladaptive changes have been reported. Increased brain representations in musicians can be considered advantageous,[4,35,36–40] while decreased spatial acuity in the representation of digits in both monkeys and humans may lead to maladaptive muscle contractions that are not advantageous.[4,41,42]

Evidence suggests[a] that patients with focal dystonia have aberrant sensorimotor feedback.[1,2,5] It is possible that continued aberrant feedback may "spill" over from

[a]It should be noted that five of the six patients in this study were also in the study reported by Lim et al.[5]

the somatosensory domain into another heavily associated sensory (i.e., auditory) domain. The continued aberrant feedback may subsequently alter functional coupling between the two sensory areas, such that the damage or changes in one domain alter another. The changes in electrophysiological activity for the P1 and N1a may therefore be a result of aberrant feedback or feedforward signals in the somatosensory domain that subsequently alter associated areas.

The results of the current study suggest that compared to controls with similar playing experiences, patients with focal dystonia have changes in the electrophysiology of two cortical potentials while listening to simple auditory stimuli. It can be argued that although the patients with dystonia have more disrupted practice due to their symptoms, compared to their nonaffected controls, the amplitude difference is unlikely to be explained by the change in practice. It would be expected that reduced playing would decrease the generator strength rather than increase it.

In conclusion, patients with musicians' cramp did not differ from controls in the overall latencies of the AEPs, nor were the scalp distributions different between groups. However, there was a significant difference in the size of the peak-to-peak AEP amplitude for the P1-N1a component. Patients had a larger difference compared to controls, and such effects are likely to be the result of changes in generator strength. Whether this amplitude difference represents a general dysfunction of sensory processing across multiple sensory modalities, and whether the increase in activity is due to structural or functional changes, or both, remains to be examined in further studies.

This study provided evidence to support the idea that patients with musicians' cramp have multiple sensory deficiencies, as demonstrated by the group difference in the peak-to-peak amplitude of the AEPs. It is recommended for future studies that the electrode coverage be larger; increasing the electrode coverage may reveal the extent of the changes in the patients with musicians' cramp, and source localization of the components would thus be possible. This result provides further evidence that patients with focal dystonia have changes in the central nervous system that occur at multiple levels.

ACKNOWLEDGMENTS

The first author is supported by the Deutscher Akademischer Austauschdienst (DAAD) Short-term Research Scholarship and a New Zealand Neurological Foundation Grant—0225WF (Philip Wrightson Postdoctoral Fellowship).

[Competing interests: The authors declare that they have no competing financial interests.]

REFERENCES

1. LIM, V.K., E. ALTENMÜLLER & J.L. BRADSHAW. 2001. Focal dystonia: current theories. Hum. Mov. Sci. **20:** 875–914.
2. TINAZZI, M., T. ROSSO & A. FIASCHI. 2003. Role of the somatosensory system in primary dystonia. Mov. Disord. **18:** 605–622.

3. REILLY, J.A. *et al.* 1992. The N30 component of somatosensory evoked potentials in patients with dystonia. Electroencephalogr. Clin. Neurophysiol. **84:** 243–247.
4. ELBERT, T. *et al.* 1998. Alteration of digital representations in somatosensory cortex in focal hand dystonia. Neuroreport **9:** 3571–3575.
5. LIM, V.K. *et al.* 2003. Aberrant sensorimotor integration in patients with musicians' cramp. J. Psychophysiol. **17:** 195–202.
6. HAMANO, T. *et al.* 1999. Increased contingent negative variation in writer's cramp. Clin. Neurophysiol. **110:** 508–515.
7. IKEDA, A. *et al.* 1996. Abnormal sensorimotor integration in writer's cramp: study of contingent negative variation. Mov. Disord. **11:** 683–690.
8. KAJI, R. *et al.* 1995. Physiological study of cervical dystonia: task-specific abnormality in contingent negative variation. Brain **118:** 511–522.
9. LIM, V.K., J.L. BRADSHAW, M.E.R. NICHOLLS & E. ALTENMÜLLER. 2004. Abnormal sensorimotor processing of pianists with focal dystonia. *In* S. Fahn, M. Hallett & M. R. DeLong, Eds.: 367–373. Annals of Neurology, Vol. 94. Lippincott, Williams, and Williams. Philadelphia.
10. MEUNIER, S. *et al.* 2001. Human brain mapping in dystonia reveals both endophenotypic traits and adaptive reorganization. Ann. Neurol. **50:** 521–527.
11. SOMMER, M. *et al.* 2002. Intracortical excitability in the hand motor representation in hand dystonia and blepharospasm. Mov. Disord. **17:** 1017–1025.
12. BARA-JIMENEZ, W., P. SHELTON & M. HALLETT. 2000. Spatial discrimination is abnormal in focal hand dystonia. Neurology **55:** 1869–1873.
13. SANGER, T.D., D. TARSY & A. PASCUAL-LEONE. 2001. Abnormalities of spatial and temporal sensory discrimination in writer's cramp. Mov. Disord. **16:** 94–99.
14. AGLIOTI, S.M., M. FIORIO, B. FORSTER & M. TINAZZI. 2003. Temporal discrimination of cross-modal and unimodal stimuli in generalized dystonia. Neurology **60:** 782–785.
15. FIORIO, M., M. TINAZZI, L. BERTOLASI & S.M. AGLIOTI. 2003. Temporal processing of visuotactile and tactile stimuli in writer's cramp. Ann. Neurol. **53:** 630–635.
16. TINAZZI, M., M. FIORIO, L. BERTOLASI & S.M. AGLIOTI. 2004. Timing of tactile and visuotactile events is impaired in patients with cervical dystonia. J. Neurol. **251:** 85–90.
17. EIMER, M., D. COCKBURN, B. SMEDLEY & J. DRIVER. 2001. Cross-modal links in endogenous spatial attention are mediated by common external locations: evidence from event-related brain potentials. Exp. Brain Res. **139:** 398–411.
18. LEWALD, J., W.H. EHRENSTEIN & R. GUSKI. 2001. Spatio-temporal constraints for auditory-visual integration. Behav. Brain Res. **121:** 69–79.
19. LUTKENHONER, B., C. LAMMERTMANN, C. SIMOES & R. HARI. 2002. Magnetoencephalographic correlates of audiotactile interaction. Neuroimage **15:** 509–522.
20. OKAJIMA, Y., N. CHINO, E. SAITOH & A. KIMURA. 1991. Recovery functions of somatosensory vertex potentials in man: interaction of evoked responses from right and left fingers. Electroencephalogr. Clin. Neurophysiol. **80:** 531–535.
21. BANGERT, M. & E. ALTENMÜLLER. 2003. Mapping perception to action in piano practice: a longitudinal DC-EEG-study. BMC Neuroscience **4:** 26–36.
22. BANGERT, M., U. HAEUSLER & E. ALTENMÜLLER. 2001. On practice: how the brain connects piano keys and piano sounds. *In* Biological Foundations of Music. Annals of the New York Academy of Sciences, Vol. 930. pp. 425–428. New York Academy of Sciences. New York.
23. SCHLAUG, G. 2001. The brain of musicians: a model for functional and structural adaptation. *In* Biological Foundations of Music. Annals of the New York Academy of Sciences, Vol. 930. pp. 281–299. New York Academy of Sciences. New York.
24. OLDFIELD, R.C. 1971. The assessment and analysis of handedness: the Edinburgh inventory. Neuropsychologia **9:** 77–113.
25. FAHN, S. 1989. Assessment of the primary dystonias. *In* Quantification of Neurologic Deficit. T. L. Munsat, Ed.: 241–270. Butterworth Publishers. Stoneham, MA.
26. McCARTHY, G. & C.C. WOOD. 1985. Scalp distributions of event-related potentials: an ambiguity associated with analysis of variance models. Electroencephalogr. Clin. Neurophysiol. **62:** 203–208.
27. GREENHOUSE, S. & S. GEISSER. 1959. On methods in analysis of profile data. Psychometrika **24:** 95–112.

28. CELESIA, G.G. & M.G. BRIGELL. 1999. Auditory evoked potentials. *In* Electroencephalography: Basic Principles, Clinical Applications, and Related Fields. E. Niedermeyer & F. Lopes Da Silva, Eds. Williams & Wilkins. Baltimore.

29. LIÉGEOIS-CHAUVEL, C. *et al.* 1994. Evoked-potentials recorded from the auditory-cortex in man: evaluation and topography of the middle latency components. Electroencephalography Clin. Neurophysiol. **92:** 204–214.

30. WOODS, D.L. 1995. The component structure of the N1 wave of the human auditory evoked potential. Electroencephalography & Clin. Neurophysiol. **44:** 102–109.

31. HALLET, M. 2004. Dystonia: abnormal movements result from loss of inhibition. *In* Advances in Neurology, Vol. 94. S. Fahn, M. Hallet & M.R. DeLong, Eds.: 1–9. Lippincott, Williams, and Williams. Philadelphia.

32. HUMMEL, F. *et al.* 2002. Inhibitory control of acquired motor programmes in the human brain. Brain **125:** 404–420.

33. RIDDING, M.C. *et al.* 1995. Changes in the balance between motor cortical excitation and inhibition in focal, task-specific dystonia. J. Neurol. Neurosurg. Psychiatry **59:** 493–498.

34. RONA, S. *et al.* 1998. Alterations of motor cortical inhibition in patients with dystonia. Mov. Disord. **13:** 118–124.

35. ALTENMÜLLER, E., W. GRUHN, D. PARLITZ & J. KAHRS. 1997. Music learning produces changes in brain activation patterns: a longitudinal DC-EEG study. Int. J. Arts Med. **5:** 28–33.

36. ELBERT, T. *et al.* 1995. Increased cortical representation of the fingers of the left hand in string players. Science **270:** 305–307.

37. HIRATA, Y., S. KURIKI & C. PANTEV. 1999. Musicians with absolute pitch show distinct neural activities in the auditory cortex. NeuroReport **10:** 999–1002.

38. KOELSCH, S., E. SCHRÖGER & M. TERVANIEMI. 1999. Superior preattentive auditory processing in musicians. NeuroReport **10:** 1309–1313.

39. PANTEV, C. *et al.* 1998. Increased auditory cortical representation in musicians. Nature **392:** 811–814.

40. PANTEV, C. *et al.* 2001. Timbre-specific enhancement of auditory cortical representations in musicians. NeuroReport **12:** 169–174.

41. BYL, N.N., M.M. MERZENICH & W.M. JENKINS. 1996. A primate genesis model of focal dystonia and repetitive strain injury: I. Learning-induced dedifferentiation of the representation of the hand in the primary somatosensory cortex in adult monkeys. Neurology **47:** 508–520.

42. WANG, X., M.M. MERZENICH, K. SAMESHIMA & W.M. JENKINS. 1995. Remodeling of hand representation in adult cortex determined by timing of tactile stimulation. Nature **378:** 71–75.

Time Course of Retrieval and Movement Preparation in Music Performance

CAROLINE PALMER

Psychology Department, McGill University, Montreal, Quebec, H3A 1B1, Canada

ABSTRACT: Music performance requires that musicians represent many different kinds of sequence structure: musicians must remember which pitch to produce, when to produce it, and how to produce it (with what movements). The time course of item retrieval and movement preparation processes during music performance are considered. Serially ordered stage models of retrieval, in which item retrieval ends before movement preparation begins, are compared with interactive cascade models, in which the time course of both processes overlap, permitting interaction. Evidence from transfer of learning paradigms, production errors, and anticipatory movements, as measured in motion capture, are described. This early evidence suggests different time courses for item retrieval (slower, earlier) than for movement preparation (faster, later) with significant temporal overlap during music performance.

KEYWORDS: performance; motor planning; memory retrieval; stage models; cascade models

Most research on the brain functions that support music focuses on perception. Although live music performances are the stimulus for most music perception, our understanding of the neural mechanisms underlying music performance is just beginning. The research reported here addresses two lines of study on the time course of brain processes during music performance: one on memory retrieval, and one on movement preparation. Music performance requires that musicians represent many different kinds of sequence structure: musicians must remember which pitch to produce, when to produce it, and how to produce it (with what movements). Music contains pitches that repeat within a melody in different contexts; the mapping of musical pitches to the movements that produce them is not one-to-one. FIGURE 1 shows the pitch sequence of a simple melody and the right-hand finger sequence that a pianist might use to produce it. In this example, the pitch E is produced with finger 3 as well as with finger 1; likewise, finger 3 produces the pitch G as well as the pitch E. When one adds rhythmic and timbral complexity to the process, music performance becomes a complex sequencing task.[1] The objective of this paper is to discuss how memory retrieval functions and movement preparation combine in music performance.

Address for correspondence: Caroline Palmer, Psychology Dept., McGill University, 1205 Dr. Penfield Ave, Montreal QC H3A 1B1, Canada. Voice: 514-398-6128; fax: 514-398-4896.
caroline.palmer@mcgill.ca

Ann. N.Y. Acad. Sci. 1060: 360–367 (2005). © 2005 New York Academy of Sciences.
doi: 10.1196/annals.1360.030

Pitch: C − **E** − D − **E** − F − **E** − A − G − F
Finger: 1 − **3** − 2 − **3** − 4 − 1 − 4 − **3** − 2

FIGURE 1. Sequencing demands of piano performance in pitch memory and finger movements: a piano melody with right hand fingering.

One important sequencing issue is the time course of memory retrieval and movement processes. Although musical events unfold in time, the brain processes that support them do not necessarily occur in the same order. Theoretical models of the time course of mental processes in such domains as object recognition, speech production, and word recognition have indicated several general classes of models. One possibility is that item retrieval and movement preparation are independent stages of processing and are serially ordered. Stage models assume that multiple processes are accessed in a serial manner, one after the other, and that processing at an early stage must be completed before a subsequent stage can begin, permitting no feedback from the later to the earlier stages.[2] A schematic for this class of model is shown in Figure 2a; process B (movement preparation) cannot affect process A (item retrieval) in this view, because B is serially ordered after A. In Sternberg's[3] original description of reaction times in psychological tasks that appear to be additive combinations of various factors, each factor reflects the operation of discrete stages; each processing stage must finish before the next begins. Stage models of production processes in speech propose that phonological (sound) processes associated with words cannot begin until the word to be spoken is done being selected from memory.[4,5] A stage model would predict that, as in FIGURE 2a, pitches are retrieved from memory before movement preparation can begin in music performance, and movement preparation cannot affect pitch retrieval.

Another possibility is that item retrieval and movement preparation are distinct processes that overlap temporally and thus offer the possibility for interaction. In this view, depicted in FIGURE 2b, process B (movement preparation) begins after process A (item retrieval), but overlaps in time with process A. As a result, some information from process B may affect process A. Cascade models of word recognition propose that multiple processes are activated at least partially in parallel, and later and earlier stages of processing can interact.[6,7] McClelland's[6] cascade model accounts for factor additivity in reaction times in terms of a series of N stages that operate at the same time (not sequentially), with the output of one stage serving as input to another stage. The cascade class of models allows that, as depicted in FIGURE 2b, movement preparation would overlap and allow interaction with item retrieval in music performance.

The distinction between serially ordered and cascade models is debated in speech production, with regard to how words are selected from memory and when phonological processes (including information that defines sound production) begin. One major view holds that speech production is strictly serial, with phonological encoding beginning only after a lexical item has been selected.[4,5,8] Latencies to name pictures support the temporal separation of lexical selection and phonological

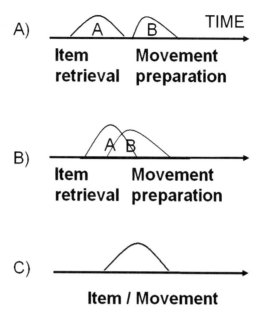

FIGURE 2. Time course of item retrieval and movement preparation. (**A**) Serial stage models; (**B**) cascade models; (**C**) single model.

processes. Another widely held view proposes that although phonological forms are activated after lexical nodes, activation from the lexical level can serve as input to the phonological level before lexical selection has taken place.[9–11] Speech production errors provide evidence that phonological similarity influences lexical retrieval, supporting temporally overlapping processes that interact.

A third possibility is that retrieval and execution are the same process. This single-process model, depicted in FIGURE 2c, might predict that the movements to be prepared are integrated with the items being retrieved. Evidence from transfer of learning studies suggests, however, that item retrieval and movement preparation are not integrated. Palmer and Meyer[12] showed that pianists who learn to perform a novel melody can generalize that knowledge to another novel melody independently for movement sequences and pitch sequences. Pianists performed a melody with one set of finger/hand movements. When asked to perform a second novel melody, their ability to perform was facilitated when either the finger sequence or the pitch sequence (or both) were the same as in the first novel melody. The degree of facilitation was independent for the movements and for the pitch items, inconsistent with an integrated process. Furthermore, child beginners showed less facilitation of the same finger movements across melodies than adults, indicating that the motor and pitch retrieval processes may progress at different learning rates.[12] Further studies indicated that movement dimensions and temporal dimensions (rhythm and meter) showed independent amounts of transfer from one novel melody to another in performance.[13] These studies suggest that movements and pitches are not retrieved in an integrated fashion.

This paper considers whether item retrieval and movement preparation in music performance can best be described by serially ordered stages or by cascaded processes. Is the time course of item retrieval and movement preparation sufficient to allow information from one process to influence the other? Item retrieval during music performance is considered in terms of when pitch information is available in memory, as evidenced in the occasional production errors that musicians make during well-learned performances. Palmer and Pfordresher's[14] model of memory retrieval during music performance describes specific predictions for the time course of retrieval. Then the time course of movement preparation is examined in terms of pianists' anticipatory finger movements, based on motion capture information. The final section returns to the classes of models that address how memory retrieval and movement preparation might combine to support sequencing in music performance.

TIME COURSE OF ITEM RETRIEVAL

Item retrieval during production is often examined in analyses of serial ordering errors: errors in which an event intended for elsewhere in the sequence is produced.[4] These errors are informative because they indicate what information from the sequence is available from memory at a given moment during production. Serial ordering errors are often caused by confusion among similar sequence events.[15,16] Serial ordering errors in speech and music reveal *distance*-based constraints on the accessibility of events during planning: an event is more or less accessible based on its separation (in number of events) from the current event.[14,17,18]

Palmer and Pfordresher[14] proposed a model of memory processes during sequence production, which predicts the time course of item retrieval during music performance. Events are considered accessible from memory, spanning from the current event being produced to nearby sequence events in the past and future, forming

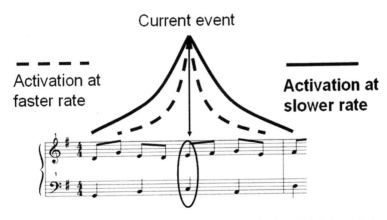

FIGURE 3. Range model predictions for temporal activation of pitch items during performance. (Adapted from Palmer and Pfordresher.[14])

a gradient of event activations across the sequence. Memory activations are determined in terms of the tempo at which the performance is produced and each event's distance from the current event produced. FIGURE 3 shows time-based predictions of the range model for item retrieval at the time the performer is producing the circled event in the melody; the gradient of activation above the score indicates how active sequence events are in memory during a slow tempo performance (the solid line) and a fast tempo performance (the dashed line). The range model predicts that items nearby in serial order will be more accessible during memory retrieval at slower tempi than at faster tempi; also, items that are metrically similar will be more accessible during retrieval.

Analyses of serial ordering errors in piano performances supported the model's predictions; pitch errors were more likely to arise from nearby sequence positions.[14] Most important, the gradient of activation was steeper for fast tempo performances than for slow tempo performances of the same music by the same performers; errors tended to reflect sequence events from a larger range at slow tempi (four events) than at fast tempi (three events), consistent with the model's predictions. This range of three to four events is consistent with models of working memory that posit an attentional constraint on elements that must be retained temporarily during ongoing tasks[19] (Cowan, 1998). The effects of production tempo on item retrieval suggested that pitches were available 400–700 ms prior to their production.

Further studies extended these findings to novice child performers, who showed a smaller serial range of planning than advanced child performers (three to four events).[20] The role of auditory feedback in performance also supports a memory retrieval constraint of three to four items. Altered feedback presented to pianists over headphones during a performance tended to disrupt a pianist's memory when that feedback corresponded to pitches within three events from the past or future; pianists tended to produce pitches that compensated for that feedback (i.e., when they heard an event from the past, they produced an event from the future, and vice versa).[21] Overall, these experiments suggest that items from three to four events around the current event are accessible in memory during performance, and that performance tempo stretches or shrinks that range.

TIME COURSE OF MOVEMENT PREPARATION

When is the next movement prepared in music performance? Do the same serial or temporal constraints apply as during item retrieval? Recordings of finger movements during piano performance, made with motion capture techniques, provide information to address the time course of movement preparation. Anticipatory movements can be measured in terms of how early fingers begin their trajectories toward keypresses. Using a Vicon motion capture system with passive reflective markers on pianists' fingers and the piano keyboard, Dalla Bella and Palmer[22] captured the finger motion of performances of melodies at different tempi. Fourteen cameras positioned around the pianist recorded light reflected from each 3 mm marker, placed on the joints of pianists' fingers and on the edge of the electronic piano keys. From memory, pianists performed simple melodies at different tempi that required little or no hand repositioning. The goal was to determine whether finger trajectories toward keypresses began at different times when the same performer played the same melody at fast or at slow tempi.

Analyses of the finger motion in the height plane (upwards from the piano keys) showed the most change in motion across the tempo conditions. Measures of variance in each finger's velocity and acceleration one, two, and three events prior to the finger's arrival on a key indicated that finger trajectories began to change one to three events prior to a keystroke. By four events prior, the finger motion was equivalent to that of other fingers that were not producing keystrokes (a control condition). Thus, the serial distance over which finger trajectories showed movement preparation was within that of the item retrieval (three to four events prior). The movement preparation was constrained by the tempo: finger accelerations changed in variance only one event prior (150 ms) to the keypress at the fastest tempi, and up to 3 events prior (500 ms) at the slower tempi.[23]

CONCLUSIONS

Evidence from production errors and motion capture in music performance indicated temporal overlap in the time course of item retrieval and movement preparation, with item retrieval (three to four events, 400–700 ms prior to an event) preceding movement preparation (one to three events, 150–500 ms prior). Although more evidence is necessary to address the roles of important performance factors, such as individual differences in working memory, performer expertise, and instrument differences, this early evidence is most consistent with the class of cascade models. At the least, the evidence argues against serially ordered processes in which movement preparation does not begin until item retrieval is complete.

Cascade models are popular in modeling many psychological tasks. Behavioral evidence consistent with cascade models suggests that stimulus identification does not need to be complete before response preparation can begin,[24,25] and that perceptual object identification is not completed before semantic representations can be formed.[26] Cascade models are also prominent in neurophysiological modeling,[27,28] based on evidence from single-cell studies that indicates significant temporal overlap of target-related spike activity at different neural sites.[29,30]

What are the advantages of a cascade model? One advantage may be to accommodate different timescales of neural processes. Item retrieval in word recognition, speech production, and music performance tasks is relatively slow (300 ms) compared with movement preparation (150 ms); reflexive muscle stiffening (30 ms) is even faster. If faster processes have to wait for slower processes to finish, as in stage models, then fast and accurate performance typified by music performance might not be possible. Cascade models also make it possible that more easily executed items may be more easily remembered, if information from process B informs process A.

Study of the time course of mental processes in music performance is in its infancy, in part because of the recent development of time-sensitive measurement techniques that do not disrupt performance. This early evidence suggests that item retrieval is an early, slow process, whereas movement preparation is later and faster. The time course of these processes suggests significant overlap. Further evidence is necessary to examine possible feedback among the cascaded processes. Interaction among processes may be evidenced in production errors and the movements that accompany them. When the wrong item is retrieved, how early does the finger producing that event alter its trajectory, compared with production of the correct event? Can incorrect trajectories be adjusted after item retrieval? Techniques, such as motion capture and imaging with musical instruments, make it possible to address memory retrieval and movement preparation in music performance, a domain in which brain processes are pushed to their temporal limits. Yet performers excel.

ACKNOWLEDGMENTS

I wish to thank my colleagues and students who contributed to the research described in this paper. Funding has been provided by grants from NSERC and the Canada Research Chairs program.

[Competing interests: The author declares that she has no competing financial interests.]

REFERENCES

1. PALMER, C. 2005. Sequence memory in music performance. Curr. Dir. Psychol. Sci. **14:** 247–250.
2. GLASER, W.R. 1992. Picture naming. Cognition **42:** 61–105.
3. STERNBERG, S. 1969. The discovery of processing stages: extensions of Donder's method. Acta Psychol. **30:** 276–315.
4. GARRETT, M.F. 1980. Levels of processing in sentence production. *In* Language Production: Speech and Talk. B. Butterworth, Ed: 177–220. Academic Press. London.
5. LEVELT, W.J.M., A. ROELOFS & A. MEYER. 1999. A theory of lexical access in speech production. Behav. Brain Sci. **22:** 1–75.
6. MCCLELLAND, J.L. 1979. On the time relations of mental processes: an examination of systems of processes in cascade. Psychol. Rev. **86:** 287–330.
7. HUMPHREYS, G.W., M.J. RIDDOCH & P.T. QUINLAN. 1988. Cascade processes in picture identification. Cogn. Neuropsychol. **5:** 67–103.
8. SCHRIEFERS, H., A.S. MEYER & W.J.M. LEVELT. 1990. Exploring the time-course of lexical access in production: picture–word interference studies. J. Mem. Lang. **29:** 86–102.
9. DELL, G.S. 1986. A spreading–activation theory of retrieval in sentence production. Psychol. Rev. **93:** 283–321.
10. HARLEY, T.A. 1993. Phonological activation of semantic competitors during lexical access in speech production. Lang. Cognit. Processes **8:** 291–309.
11. MACKAY, D.G. 1987. The Organization of Perception and Action: A Theory for Language and Other Cognitive Skills. Springer-Verlag. New York.
12. PALMER, C. & R.K. MEYER. 2000. Conceptual and motor learning in music performance. Psychol. Sci. **11:** 63–68.
13. MEYER, R.K. & C. PALMER. 2003. Temporal and motor transfer in music performance. Mus. Percept. **21:** 81–104.
14. PALMER, C. & P.Q. PFORDRESHER. 2003. Incremental planning in sequence production. Psychol. Rev. **110:** 683–712.
15. CONRAD, R. 1965. Order error in immediate recall of sequences. J. Verb. Learn. Verb. Behav. **4:** 161–169.
16. HEALY, A.F. 1974. Separating item from order information in short-term memory. J. Verb. Learn. Verb. Behav. **13:** 644–655.
17. GARCIA-ALBEA, J.E., S. DEL VISO & J.M. IGOA. 1989. Movement errors and levels of processing in sentence production. J. Psycholing. Res. **18:** 145–161.

18. PALMER, C. & C. VAN DE SANDE. 1995. Range of planning in skilled music performance. J. Exp. Psychol. Hum. Percept. Perform. **21:** 947–962.
19. COWAN, N. 2000. The magical number four in short-term memory: reconsideration of mental storage capacity. Behav. Brain Sci. **24:** 87–185.
20. DRAKE C. & C. PALMER. 2000. Skill acquisition in music performance: relations between planning and temporal control. Cognition **74:** 1–33.
21. PFORDRESHER, P.Q. & C. PALMER. 2005. Effects of hearing the past, present, or future during music performance. Percept. Psychophys. In press.
22. DALLA BELLA, S. & C. PALMER. 2004. Tempo and dynamics in piano performance: the role of movement amplitude. *In* Proceedings of the International Conference on Music Perception and Cognition. S.D. Lipscomb, R. Ashley, R.O. Gjerdingen & P. Webster, Eds: 256–257. Causal Productions. Adelaide, Australia.
23. PALMER, C. & S. DALLA BELLA. 2004. Movement amplitude and tempo change in piano performance. J. Acoust. Soc. Am. **115:** 2590.
24. MILLER, J.O. & R. ULRICH. 1998. Locus of the effect of the number of alternative responses: evidence from the lateralized readiness potential. J. Exp. Psychol. Hum. Percept. Perform. **24:** 1215–1231.
25. MILLER, J.O., R. ULRICH & G. RINKENAUER. 1999. Effects of stimulus intensity on the lateralized readiness potential. J. Exp. Psychol. Hum. Percept. Perform. **25:** 1454–1471.
26. PARASURA, R. & A. MARTIN. 2001. Interaction of semantic and perceptual processes in repetition blindness. Visual Cogn. **8:** 103–118.
27. FOURTES, M.G.F & P.M. O'BRYAN. 1972. Responses to single photons. *In* Handbook of Sensory Physiology: VII/2. Physiology of photoreceptor organs. M.G.F. Fuortes, Ed.: 321–338. Springer. Berlin.
28. LEVINSON, J. 1966. One-stage model of visual temporal integration. J. Opt. Soc. Am. **56:** 95–97.
29. SCHALL, J.D., D.G. HANES & T.L. TAYLOR. 2000. Neural control of behavior: countermanding eye movements. Psychol. Res. **63:** 299–307.
30. SCHALL, J.D. & K.G. THOMPSON. 1999. Neural selection and control of visually guided eye movements. Annu. Rev. Neurosci. **22:** 241–259.

Neural Control of Rhythmic Sequences

FREDRIK ULLÉN, SARA L. BENGTSSON, H. HENRIK EHRSSON,
AND HANS FORSSBERG

*Neuropediatric Research Unit, Department of Woman and Child Health,
Karolinska Institute, SE-171 76 Stockholm, Sweden*

ABSTRACT: We investigated whether the temporal structure of movement
sequences can be represented and learned independently of their ordinal structure, and whether some brain regions are particularly important for temporal
sequence performance. Using a learning transfer design, we found evidence for
independent temporal representations: learning a spatiotemporal sequence facilitated learning its temporal and ordinal structure alone; learning a temporal
and an ordinal structure facilitated learning of a sequence where the two were
coupled. Second, learning of temporal structures was found during reproduction of sequential stimuli with random ordinal structure, suggesting independent mechanisms for temporal learning. We then used functional magnetic
resonance imaging to investigate the neural control of sequences during well-
learned performance. The temporal and ordinal structures of the sequences
were varied in a 2×2 factorial design. A dissociation was found between brain
regions involved in ordinal and temporal control, the latter mainly involving
the presupplementary motor area, the inferior frontal gyrus and precentral
sulcus, and the superior temporal gyri. Finally, in a second fMRI experiment,
well-learned temporal sequences were performed with the left or right index
fingers, or using rhythmic speech. The overlap in brain activity during performance with the different effectors included a similar set of brain regions as that
found in the first fMRI experiment: the supplementary motor area (SMA), the
superior temporal gyrus, and the inferior frontal cortex. We thus suggest that
this set of regions is important for abstract, movement-independent, temporal
sequence control. This organization may be important for increased flexibility
in voluntarily timed motor tasks.

KEYWORDS: fMRI; music; timing; supplementary motor area; superior
temporal gyrus; Broca's area

INTRODUCTION

An influential view in motor control theory has been that movement sequences
are controlled by generalized motor programs, where the temporal structure (relative
timing) is invariant, once the sequence is well learned.[1] This theory received support
from findings of a fixed relative timing at different speeds and over long periods of
time for such skills as typing and handwriting.[2–6]

Address for correspondence: Neuropediatric Research Unit, Department of Woman and Child
Health, Karolinska Institute, SE-171 76 Stockholm, Sweden. Voice: +46-8-517-773-55; fax:
+46-8-517-773-49.
Fredrik.Ullen@ki.se

Ann. N.Y. Acad. Sci. 1060: 368–376 (2005). © 2005 New York Academy of Sciences.
doi: 10.1196/annals.1360.031

More recent studies have examined temporal sequence learning using the serial reaction time paradigm.[7–9] These results have been less clear-cut, which may reflect certain disadvantages of this paradigm for the study of temporal learning.[10] Lee[8] thus found indications of an independent representation of temporal structures, when studying spatial sequences of hand movements, with a temporal structure of alternating long and short time intervals. Phase-shifting this temporal structure in relation to the ordinal structure gave small deficits in performance. However, Shin and Ivry[9] suggested that temporal and ordinal learning, rather, may be integrated processes. In that study, temporal sequence learning was found only when a temporal sequence was coupled to a particular ordinal sequence.

In an attempt to resolve this question, we performed two behavioral experiments to directly investigate the representation and learning of temporal sequential structures.[10] First (experiment 1), we used a learning transfer paradigm to test whether independent representations of the temporal and ordinal structure of movement sequences can indeed be formed. Second (experiment 2), we investigated whether independent processes for learning of temporal and ordinal information exist. These experiments gave support for the existence of independent mechanisms for both learning and representation of temporal structures.

We therefore used functional magnetic resonance imaging (fMRI) to test the hypothesis that there is a dissociation of brain regions preferentially involved in temporal and ordinal control of spatiotemporal sequence performance (experiment 3).[11] Several earlier studies have investigated learning, encoding, and performance of temporal sequences.[12–18] New in this study was to ask the same subjects to perform well-learned sequences, the ordinal and temporal structures of which were varied independently, so that increases in activity specifically related to temporal processing could be revealed. Finally (experiment 4), we used fMRI to investigate which brain regions are involved in effector-independent aspects of temporal sequence control. The participants performed a well-learned temporal sequence, using tapping movements with the left or right index finger, or silent rhythmic speech (imagery) on the syllable *pa*. Sequence performance was contrasted with rest for each effector, and a conjunction analysis of these contrasts was employed to reveal overlap in activity.

METHODS

Participants

In all four experiments, the participants were right-handed[19] nonmusicians. No subject participated in more than one experiment. The experimental procedures were approved by the Ethical Committee of the Karolinska Institute. The number of subjects in the different experiments were as follows: experiment 1, $n = 24$ (12 males); experiment 2, $n = 12$ (6 males); experiment 3, $n = 7$ (5 males); and experiment 4, $n = 7$ (4 males).

Experimental Procedures and Analysis

Experiment 1

A detailed description of the procedures in experiments 1 and 2 has been presented elsewhere.[10] The tasks consisted of rhythmic sequences of key presses, performed with the right index finger on the numerical keypad of a personal computer. All subjects learned three different sequences: *Combined* had an ordinal structure of nine key presses and a corresponding temporal structure of eight temporal intervals, with durations of 375, 750, or 1125 milliseconds. *Temporal* had the same temporal structure as Combined but was performed on the central key only. *Ordinal* had the same ordinal structure as Combined but a temporal structure of one element, that is, a regular rhythm. In addition, 14 of the subjects learned a control sequence, *Control*, that had the same length of the temporal and ordinal structure as Combined.

Each sequence was learned in a single continuous session. The session started with one audiovisual presentation of the sequence. The subject tried to reproduce the sequence on the keyboard. If the response contained an ordinal or temporal error (relative error > 30%), the sequence was presented again. If the sequence had been produced without any errors, temporal or ordinal, the participant was instructed to reproduce the sequence again, without a new sequence presentation. The learning session was terminated when the subject had successfully performed twelve successful productions in a row.

To quantify transfer effects, subjects were divided into two groups, with 12 subjects in each group. Group I started by learning Combined, and thereafter learned Temporal and Ordinal; group II started by learning Temporal and Ordinal, and thereafter learned Combined. Of the 14 subjects that learned the sequence Control, 7 learned it before the experimental sequences, and 7 afterwards. Learning transfer was estimated by between-group comparisons of the total number of failed reproductions before a particular sequence was learned (Mann-Whitney U Test). Within-subject correlations of the duration of the same temporal interval, as performed in the Temporal and Combined tasks, were analyzed by calculating the Pearson product-moment correlation coefficient.

Experiment 2

Sequence presentation and data collection were performed as in experiment 1, but here all sequences consisted of seven key presses separated by six temporal intervals. Three different tasks were included in the experiment. For each task, the participant performed a session of 50 trials without a break. Each trial consisted of one sequence presentation followed by a reproduction. No feedback was provided. The only instruction to the subject before each session was to repeat both the temporal pattern and the key sequence as accurately as possible in each trial. In the first task, *Fix Temporal*, the sequences in all 50 trials of the session had a constant temporal structure (750–375–375–1125–375–750; all in ms), while the ordinal structure was varied randomly. *Fix Ordinal* had a constant ordinal structure (left–right–center–right–left–center–left) but varying temporal structure. *Random*, the third task, had both varying temporal and varying ordinal structure.

The median number of correct key presses and correct temporal intervals were calculated for each trial in the different tasks, using data from all subjects. Correlations between trial number and number of correct responses were analyzed by calculating the Pearson product-moment correlation coefficient. To evaluate if subjects had explicit awareness of the structure of the different sequences, subjects filled out a brief questionnaire after each session, where they were asked if they noticed any pattern in the sequences and, if so, what pattern. This information was used to

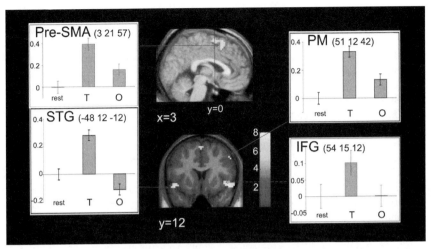

FIGURE 1. Direct comparisons between long temporal structure (*left*) and long ordinal structure (*right*). Activity maps of brain regions with significantly increased BOLD contrast signals ($P < .05$ corrected) are shown. Activated brain regions included the pre-SMA ($x = 3$), the bilateral superior temporal gyri (STG) ($y = 12$), the inferior frontal gyrus pars opercularis (IFG) ($y = 12$), and the right inferior part of the precentral sulcus (PM) ($y = 12$). The *bars* illustrate the adjusted relative hemodynamic response for each task in one local peak voxel (C = *Combined*, T = *Temporal*, O = *Ordinal*, E1 = *Even One Key*, E2 = *Even Two Keys*). The *color scale* (*online version only*) shows Z values. (Modified from Bengtsson *et al.*[11] Reproduced with kind permission of Blackwell Publishing.)

investigate if learning took place in those subjects that reported that they detected no regularities in the stimuli.

Experiment 3

A full description of this experiment is given elsewhere.[11] As in experiment 1, the tasks consisted of rhythmic sequences, where the length of the temporal and ordinal structures were varied to allow analysis with a 2 × 2 factorial model. The ordinal structure of a sequence had either eight or one element. The temporal structure of a sequence also had either eight or one element (durations of 375, 750, 1125, or 1500; all in ms), that is, an isochronous pulse with the period 750 milliseconds. The tasks are illustrated schematically in FIGURE 1. Combined, Temporal, and Ordinal correspond to the tasks with the same names in experiment 1. Two control sequences were used: *Even One Key* consisted of a regular pulse performed on the middle key, and *Even Two Keys* had a isochronous rhythm performed on the two outer keys, alternatingly, and was included to control for lateral wrist movements in Ordinal and Combined. A rest condition (*Rest*) was also included, where the subject was completely relaxed. All tasks were practiced for about one hour in one training session one or two days before the fMRI experiment and for 30 min immediately before the scanning. In the experiment the tasks were performed in epochs lasting 40 seconds. During the first 8 s of each epoch, the subjects were given a verbal instruction, followed by five beats of an auditory metronome at 80 beats per minute providing the correct tempo. We analyzed the brain scans obtained during the 32-s periods when the metronome was turned off and the subjects repetitively generated the movement sequences.

fMRI was conducted on a 1.5 T scanner (Signa Horizon Echospeed, General Electric Medical Systems, Milwaukee, WI, USA). Functional imaging data was then recorded as gradient-echo echo-planar (EPI) T2*–weighted images with blood oxygenation level–dependent (BOLD) contrast.[20–22] The fMRI data was analyzed using the SPM-99 software package (Wellcome Department of Imaging Neuroscience, London, UK), using a standard linear regression model, with six conditions of interest corresponding to the periods when the subjects performed the tasks. The significance of the effects was assessed using t statistics for every voxel from the brain to create statistical parametric maps (SPMs), which were subsequently transformed into Z statistics, pooling data from all subjects in a fixed-effects group analysis. Reported activations were significant at $P < .05$, after correction for multiple comparisons, on the basis of a test of peak height.[23]

Here we will discuss brain regions active in the main effect contrast for the long temporal structure, as well as regions that showed a significantly higher activity during performance of long temporal structures than during performance of long ordinal structures, as revealed by the contrast Temporal–Ordinal.

Experiment 4

A full description of this study will be given elsewhere.[24] Six experimental conditions and one baseline rest condition were used. Here we will discuss results pertaining to three experimental conditions, *SeqRight*, *SeqLeft*, and *SeqSpeech*, where the subjects performed a rhythmic sequence of six temporal intervals, with the right index finger, the left index finger, and using rhythmic speech, respectively. The rhythmic sequence had the temporal pattern 375–375–750–1125–375–1500 (all in ms). Task training was performed as in experiment 3. During the MR scans, the tasks were performed repetitively in epochs of 40 seconds. fMRI data was processed using the SPM-99 software package, as described for experiment 3. The focus of the experiment was to investigate brain regions involved in effector-independent control of motor timing. For this purpose, we used conjunction analyses,[25] which can be interpreted as logical AND-operations, giving only those activations that are significant in all individual contrasts of the conjunction. Here we will discuss the results of the conjunction between the three contrasts, SeqRight–Rest, SeqLeft–Rest, and SeqSpeech–Rest.

RESULTS

Experiment 1

Strong transfer effects were seen on all tasks. Learning of Combined in group II, which before that had learned Temporal and Ordinal, was significantly quicker ($M = 4.5$ trials; $P = .0003$) than in group I ($M = 12.5$ trials). Likewise, a quicker learning of both Temporal ($M = 1$ trial; $P = .015$) and Ordinal ($M = 2$ trials; $P = .0002$) was seen in group I subjects than in group II subjects ($M = 12$ and 13 trials, respectively). Learning of Control was significantly faster ($M = 3$ trials;

$P = .01$) in participants that trained this task at the beginning of the experiment, than in the group that trained at the end ($M = 16$ trials). The transfer effects were thus specific to the experimental sequences and not due to an increase in general ability to learn this type of task.

Further evidence for independent representations of temporal structures were found when comparing the relative errors in the temporal pattern produced in Temporal and Combined. Pooling data from all subjects, high positive within-subject correlations were found for the duration of all individual temporal intervals of the sequence during performance of Temporal and Combined (r values: .65–.86; $P < .001$).

Experiment 2

In the three conditions Fix Temporal, Fix Ordinal, and Random, learning—as indexed by a significant correlation between trial number and number of correct responses—was observed only in two cases: for number of correct temporal intervals in Fix Temporal ($P < .00001$; $r = .73$) and for number of correct key presses in Fix Ordinal ($P < .01$; $r = .35$). The weaker correlation in the latter case is due to the fact that the ordinal structure of the sequence was learned very rapidly, so that the learning curve reached an early plateau, and the relationship with trial number was curvilinear. Independent mechanisms for learning of temporal and ordinal information thus appear to exist. Interestingly, learning of temporal information ($P < .00001$; $r = .65$) was also seen in a subgroup of subjects ($n = 9$) that reported having detected no regularities in the stimuli of Fix Temporal, suggesting that temporal sequences can be learned implicitly.

Experiment 3

The main effect contrast for long temporal structure revealed a large cluster of active voxels located in the presupplementary motor area and the supplementary motor area (SMA). Activity was also seen in the bilateral superior temporal gyri (STG), extending into the bilateral insula and the right inferior frontal gyrus (IFG). Bilateral activity in the inferior part of the precentral sulcus was also seen (PMV/BA 44). Subcortical activations were found in the cerebellum: lobule V of the right anterior vermis and the left lobule VI of the hemisphere and in the right globus pallidus.

A similar pattern of activity was seen in the contrast Temporal–Ordinal (FIG. 1). Activations were seen in the pre-SMA, extending from the rostral pre-SMA to the SMA/pre-SMA border; the right IFG; the right inferior part of the precentral sulcus (PMV/BA 44); the right anterior part of the STG; the left posterior part of the STG; and the right insula.

Experiment 4

The conjunction analysis among SeqRight–Rest, SeqLeft–Rest, and SeqSpeech–Rest revealed a similar set of brain regions to that found in experiment 3. On the medial wall of the frontal lobe, activations were seen in the left SMA, extending into the caudal cingulate motor area and rostrally to the SMA/preSMA border. Further, activity was seen in the left IFG and in the left anterior STG, as well as the posterior left STG, close to the temporoparietal junction. Subcortical activations included the left lobule V/VI of the vermis and lobule VI of the cerebellar hemispheres.

DISCUSSION

A main conclusion from these studies is that the processing of temporal sequences in voluntarily timed motor tasks is largely independent from the processing of ordinal information. Temporal sequences can be learned and represented independently, and performing well-learned rhythmic sequences predominantly involves a set of brain regions, different from those controlling ordinal structures. These brain regions appear to be involved in rhythm production independently of both the effector and the ordinal structure of the movements, suggesting that they represent rhythmic patterns in a more abstract, movement-independent manner. We propose that this organization gives larger flexibility and faster learning of alternative behavioral strategies to reach a certain goal. A detailed discussion of the results is given elsewhere.[10,11,24] We will here add some comments on the findings in relationship to recent studies.

Garraux and coworkers[26] manipulated ordinal and temporal structure in a sequence reproduction task. In the temporal condition, participants were asked to produce random temporal intervals in the range 1.3–3.3 seconds. In this study, timing-related increases in brain activity were found only in the right putamen. Interestingly, this suggests different neural mechanisms for the voluntary production of specific temporal patterns[11,16,27] and production of sequences of random durations.

The supplementary motor area and the adjacent nonprimary motor areas on the medial wall are active in a large variety of tasks that involve voluntary timing,[28] including the sequential tasks discussed here. The specific contributions of the different areas for temporal sequence production are not well understood, however. Recent work suggests that the pre-SMA may be of particular importance for such superordinate functions as initiation of repetitive rhythm performance[29] and metric organization of temporal sequences.[30]

In studies of self-paced rhythm production, the movement rate is often established by the use of a metronome, which is switched off after a while. Recent studies have shown that the neural activity during the self-paced phase is dependent both on the modality of the metronome[31,32] and on the coordination between movements and metronome beats[33] in the initial phase, before the metronome is turned off. The activity in auditory association as seen in the present experiments could therefore reflect either that the temporal sequences were presented in the auditory modality during training or that an auditory metronome was used to set the movement rate during scanning. Further experiments will be needed to elucidate this question.

The studies discussed here were performed on nonmusicians. We have recently found a similar dissociation between brain regions predominantly involved in rhythmic and melodic processing during piano playing from musical scores by professional pianists.[34] A new finding in that study was that the dorsal visual stream was more involved in the processing of melodic information (i.e., which key to press), whereas the ventral visual stream was more engaged in rhythmic processing. This presumably reflects that in standard Western musical notation melodic information is encoded spatially (the location of a note head determines which piano key to strike), whereas various features of the note, such as the stem, dots, and flags, determine the rhythmic structure.

ACKNOWLEDGMENTS

This work was supported by grants from the Swedish Research Council, the Royal Swedish Academy of Sciences, the Swedish Foundation for Strategic Research, Sällskapet Barnavård, the Ax:son Johnsson Foundation, and the Magnus Bergwall Foundation.

[Competing interests: The authors declare that they have no competing financial interests.]

REFERENCES

1. SCHMIDT, R.A. & T.D. LEE. 1999. Motor Control and Learning. Human Kinetics. Champaign, IL.
2. SHAFFER, L.H. 1980. Analyzing piano performance: a study of concert pianists. *In* Tutorials in Motor Behavior. G. E. Stelmach & J. Requin, Eds.: 443–455. Elsevier. Amsterdam.
3. VIVIANI, P. & C. TERZUOLO. 1980. Space–time invariance in learned motor skills. *In* Tutorials in Motor Behavior. G.E. Stelmach & J. Requin, Eds. North-Holland Publishing Company.
4. SHAPIRO, D.C. 1977. A preliminary attempt to determine the duration of a motor program. *In* Psychology of Motor Behavior and Sport. D.M. Landers & R.W. Christina, Eds.: 17–24. Human Kinetics. Champaign, IL.
5. ARMSTRONG, T.R. 1970. Training for the production of memorized movement patterns.
6. SCHMIDT, R.A. 1999. Central contributions to motor control. *In* Motor Control and Learning: A Behavioral Emphasis.: 131–170. Human Kinetics. Champaign, IL.
7. NISSEN, M.J. & P. BULLEMER. 1987. Attentional requirements of learning: evidence from performance measures. Cogn. Psychol. **19:** 1–32.
8. LEE, D. 2000. Learning of spatial and temporal patterns in sequential hand movements. Cogn. Brain Res. **9:** 35–39.
9. SHIN, J.C. & R.B. IVRY. 2002. Concurrent learning of temporal and spatial sequences. J. Exp. Psychol. Learn Mem. Cogn. **28:** 445–457.
10. ULLÉN, F. & S. BENGTSSON. 2003. Independent processing of the temporal and ordinal structure of movement sequences. J. Neurophysiol. **90:** 3725–3735.
11. BENGTSSON, S. *et al.* 2004. Dissociating brain regions controlling the temporal and ordinal structure of learned movement sequences. Eur. J. Neurosci. **19:** 2591–2602.
12. HALSBAND, U. *et al.* 1993. The role of premotor and the supplementary motor area in the temporal control of movement in man. Brain **116:** 243–266.
13. PENHUNE, V.B., R.J. ZATORRE & A.C. EVANS. 1998. Cerebellar contributions to motor timing: a PET study of auditory and visual rhythm reproduction. J. Cogn. Neurosci. **10:** 752–765.
14. SAKAI, K. *et al.* 1999. Neural representation of a rhythm depends on its interval ratio. J. Neurosci. **19:** 10074–10081.
15. RAMNANI, N. & R.E. PASSINGHAM. 2001. Changes in the human brain during rhythm learning. J. Cogn. Neurosci. **13:** 952–966.
16. SCHUBOTZ, R.I. & D.Y. VON CRAMON. 2001. Interval and ordinal properties of sequences are associated with disctinct premotor areas. Cereb. Cortex **11:** 210–222.
17. PENHUNE, V.B. & J. DOYON. 2002. Dynamic cortical and subcortical networks in learning and delayed recall of timed motor sequences. J. Neurosci. **22:** 1397–1406.

18. ULLÉN, F., H. FORSSBERG & H.H. EHRSSON. 2003. Neural networks coordinating the hands in time. J. Neurophysiol. **89:** 1126–1135.
19. OLDFIELD, R.C. 1971. The assessment and analysis of handedness: the Edinburgh inventory. Neuropsychologia **9:** 97–113.
20. KWONG, K.K. *et al.* 1992. Dynamic magnetic resonance imaging of the human brain activity during primary sensory stimulation. Proc. Natl. Acad. Sci. USA **89:** 5675–5679.
21. OGAWA, S. *et al.* 1992. Intrinsic signal changes accompanying sensory stimulation: functional brain mapping with magnetic resonance imaging. Proc. Natl. Acad. Sci. USA **89:** 5951–5955.
22. LOGOTHETIS, N.K. 2002. The neural basis of the blood-oxygen-level-dependent functional magnetic resonance imaging signal. Philos. Trans. R. Soc. Lond. Ser. B Biol. Sci. **357:** 1003–1037.
23. FRISTON, K.J. *et al.* 1995. Statistical parametric maps in functional imaging: a general linear approach. Hum. Brain Mapp. **2:** 189–210.
24. BENGTSSON, S.L. *et al.* 2005. Neural correlates of effector-independent voluntary timing. In preparation.
25. NICHOLS, T., M. BRETT, J. ANDERSSON, *et al.* 2005. Valid conjunction inference with the minimum statistic. NeuroImage **25:** 653–660.
26. GARRAUX, G. *et al.* 2005. Shared brain areas but not functional connections controlling movement timing and order. J. Neurosci. **25:** 5290–5297.
27. SAKAI, K., N. RAMNANI & R.E. PASSINGHAM. 2002. Learning of sequences of finger movements and timing: frontal lobe and action-oriented representation. J. Neurophysiol. **88:** 2035–2046.
28. MACAR, F. *et al.* 2002. Activation of the supplementary motor area and of attentional networks during temporal processing. Exp. Brain Res. **142:** 475–485.
29. LEWIS, P.A. *et al.* 2004. Brain activity correlates differentially with increasing temporal complexity of rhythms during initialisation, synchronisation, and continuation phases of paced finger tapping. Neuropsychologia. **42:** 1301–1312.
30. CHEN, J.L., V.B. PENHUNE & R.J. ZATORRE. 2005. Tapping in synchrony to auditory rhythms: effect of temporal structure on behavior and neural activity. Ann. N. Y. Acad. Sci. **1060:** 400–403. [This volume].
31. JANTZEN, K.J., F.L. STEINBERG & J.A.S. KELSO. 2005. Functional MRI reveals the existence of modality and coordination-dependent timing networks. NeuroImage **25:** 1031–1042.
32. JÄNCKE, L. *et al.* 2000. Cortical activations during paced finger-tapping applying visual and auditory pacing stimuli. Cogn. Brain Res. **10:** 51–66.
33. JANTZEN, K.J., F.L. STEINBERG & J.A.S. KELSO. 2004. Brain networks underlying human timing behavior are influenced by prior context. Proc. Natl. Acad. Sci. USA **101:** 6815–6820.
34. BENGTSSON, S.L. & F. ULLÉN. 2005. Dissociation between melodic and rhythmic processing during piano performance from musical scores. NeuroImage. In press. Oct: 20 [e-pub]; PMID:16246591.

A Neurocognitive Approach to Music Reading

LAUREN STEWART

Wellcome Department of Imaging Neuroscience,
University College, London, United Kingdom, and
University of Newcastle upon Tyne, NE1 7RU, United Kingdom

ABSTRACT: Music reading offers a unique perspective on the acquisition of a notational system. Many people cannot read music, but a large proportion are motivated to learn. Musical literacy is therefore amenable to studies of acquisition in a way that language literacy is not. The studies reviewed here investigate how musical symbols on the page are decoded into a musical response. The studies address the nature of the mental representations used in music reading, as well as their instantiation within the brain. The results of a musical Stroop paradigm are described, in which musical notation was present but irrelevant for task performance. The presence of musical notation produced systematic effects on reaction time, demonstrating that reading of the written note, like the written word, is obligatory for those who are musically literate. Spatial interference tasks are also described that suggest that music reading, at least for the pianist, can be characterized as a set of vertical to horizontal mappings. These behavioral findings are mirrored by the results of an fMRI training study in which musically untrained adults were taught to read music and play piano keyboard over a period of three months. Specific learning-related changes were seen in the superior parietal cortex and fusiform gyrus, for melody reading and rhythm reading, respectively. These changes are suggested to correspond to the acquisition of processes that deal with the extraction of spatial and featural properties of notation, respectively.

KEYWORDS: music reading; learning; fMRI; automaticity; melody; rhythm

INTRODUCTION

While language notation primarily functions to carry semantic meaning, music notation carries instructions for the production of a musical performance. Music reading thus lies at the interface between perception (encoding of visual symbols on the page) and action (production of the musical response). Although it is undeniably the case that music reading is "more than a visuo-motor task,"[1] it is also the case that without a theory of music reading at this level of transcription, studies of musical expertise and the role of higher cognitive processes in music reading will necessarily be limited in their scope.

Address for correspondence: Lauren Stewart, Institute of Cognitive Neuroscience, Alexandra House, 17 Queen Square, London WC1N 3AR. Voice: +44-20-7679-1160; fax: +44-20-7813-2835.
l.stewart@ucl.ac.uk

Ann. N.Y. Acad. Sci. 1060: 377–386 (2005). © 2005 New York Academy of Sciences.
doi: 10.1196/annals.1360.032

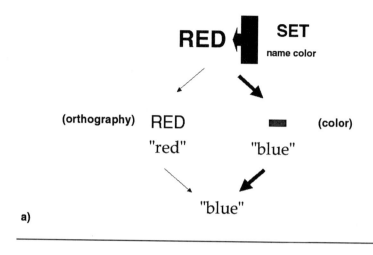

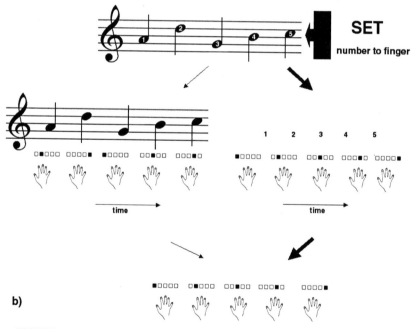

FIGURE 1. Schematic of the language Stroop task and the musical stroop task. In the language Stroop task (**a**), the participant is required to name the ink color of a color word. In the incongruent situation, the color name and the color word do not correspond (as above). In order to make the appropriate response (naming the color), the participant must inhibit a tendency to read the color word. In the musical Stroop task (**b**), the participant is required to make a sequence of keypresses, on the basis of the numbers, superimposed on musical notes. In the incongruent situation, the numbers and the notes do not correspond (as above), and the participant must inhibit a tendency to play the notes. Participants who cannot read music have only one mapping, from the numbers to the fingers, and are therefore unaffected by the note/number congruence of the pairing.

AUTOMATICITY IN MUSIC READING

In investigating how musical symbols are decoded into a musical response, one question that can be asked is whether this process occurs automatically. Does musical literacy result in obligatory processing of musical notation, in much the same way that standard literacy results in the automatic processing of the written word?[2] Reports from accomplished musicians suggest that this may be the case:

> I have come to the conviction that sight-reading has something to do with the speed at which the visual image is converted into a muscular act; and with some people this is a very rapid transformation. As a matter of fact, it is so rapid, that the transformation of a visual image into its pianistic result is so rapid that a great deal of what happens escapes awareness—it just happens.[3]

The classic Stroop task[2] demonstrates the obligatory nature of word reading. Naming the ink color of a color word is faster when then the ink color and the color word correspond, for example, the word *red*, written in red ink, compared to the word *red*, written in blue ink, because reading words is more automatic than naming colors. In order to test whether this automaticity extends to music reading, a musical version of the classic Stroop task[4] was used, as illustrated in FIGURE 1. Pianists and nonmusicians placed their right hand over the keys G to D of a piano keyboard and made keypresses according to the appearance of numbers, between 1 and 5, on the screen. If the number 1 was seen, they pressed the key beneath their thumb as quickly as possible. If the number 2 was seen, they pressed the key beneath their index finger as quickly as possible, and so on. The numbers were superimposed onto musical notes, which were either congruent or incongruent with respect to the numbers. In a congruent trial, the number 1 would be superimposed onto the note G, the number 2 superimposed onto the note A, and so on. In an incongruent trial, the number 1 may be superimposed onto the note A, the number 2 on the note D, and so on, as shown in FIGURE 1. Baseline trials were also included in which numbers were presented in a straight line and in the absence of musical notes.

Both musicians and nonmusicians were easily able to perform the number-to-finger mapping. However, the pianists took several hundreds of milliseconds longer to make a sequence of keypresses when the notes and numbers were incongruent compared with trials in which only numbers were present and were significantly facilitated when notes and numbers were congruent. The nonmusicians showed no

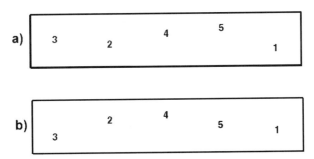

FIGURE 2. Musical Stroop task in a nonmusical context. Example of a motorically matched pair of stimuli: congruent (**a**) and incongruent (**b**).

effect of congruence, performing at the same speed regardless of the note/number correspondence. A similar effect was also observed when the musical notes were removed so that the numbers were presented in vertical locations that were either musically congruent or incongruent (FIG. 2).

The experiments previously described used a fixed association between particular fingers and particular piano keys (e.g., thumb above G, index finger above A). However, pianists do not form absolute associations between individual fingers and individual notes; rather they can play any note with any finger, and the choice of fingering will depend on the musical context. In order to show that the effect we observed was independent of the previous association between fingers and notes, the musical Stroop experiment was repeated with the small difference that the numbers now referred to particular keys rather than particular fingers, and participants responded using a single finger. The systematic effects on response time in pianists were still seen. Equally, when the participants responded by pressing keys on a computer keyboard, rather than a piano keyboard, the effect was still observed. Thus the effect does not depend upon which effectors are used; nor does it depend on a musical context for the response. Instead, the critical locus of interference seems to be the spatial location at which the response is made. Overall, these experiments provide evidence that musical notation is automatically read by those who are musically literate, even when the process of reading music is detrimental to task performance.

SPATIAL MAPPINGS IN PIANISTS

On the basis of these findings, it was hypothesized that music reading involves a set of spatial mappings from a note's position on the musical staff to the specified location of the instrumental response. Notes on the musical staff are organized along the vertical dimension so that, for pianists, notes that are vertically lower on the staff correspond to keys that are further to the left on the keyboard, while notes that are vertically higher correspond to keys that are further to the right on the keyboard. Musicians and nonmusicians were compared on a nonmusical version of the musical Stroop task (FIG. 3), which again required them to place their right hand over five adjacent keys and to think about their fingers in terms of the numbers 1 to 5.

Numbers appeared in different vertical locations, but the spatial location at which they appeared was always irrelevant. The subjects were required to perform the same number-to-finger mapping as before (1 = thumb, 2 = index finger, and so on) using a computer keyboard. A congruent trial was one in which the number 1 (specifying a keypress beneath the left-most effector) would appear in the vertically lowest box, the number 2 in the second lowest box, and so on (a piano-like vertical to horizontal relationship between visual stimulus and its associated response).

There were two kinds of incongruent trials: random and systematic. An incongruent (random) trial was one in which the number 1 would appear in a box other than the vertically lowest one, the number 2 would appear in a box other than the second to lowest one, and so on. An incongruent (systematic) trial was one in which the number 1 would appear in the uppermost box, the number two in the box second from top, and so on (opposite to the mapping used for congruent trials). Such stimulus–response mappings would be incongruent with respect to those used by pianists when reading from a musical score. Pianists took longer to make the same sequence

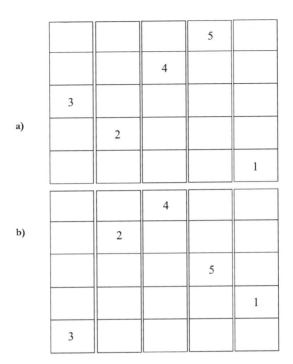

FIGURE 3. Nonmusical Stroop task. In congruent trials (**a**), the number-to-finger mapping corresponds to the spatial stimulus–response mapping used by pianists when playing from a musical score. In incongruent trials (**b**), numbers were presented in spatial locations that did not correspond to these stimulus–response mappings. Each trial consisted of 5 stimuli, specifying a sequence of 5 different keypresses. Stimuli appeared consecutively, triggered by a keypress response to the previous stimulus.

of keypresses when the numbers appeared in incongruent locations (whether random or systematic), compared to congruent locations. Even when neither the stimulus nor the response elements of the task were in any way musical, the spatial mappings inherent to the reading and playing keyboard music could still be seen. The demands of constantly mapping from vertically organized visual stimuli (musical notes on the staff) to a set of horizontally organized response elements (piano keys) thus appear to forge a set of spatial mappings that are in evidence even outside of a musical context.

NEURAL CORRELATES OF MUSICAL LITERACY ACQUISITION

These cognitive studies of music reading have shed light on the nature of the representations that are involved in reading music. However, it is also pertinent to ask which brain areas are involved in this transcriptional process. Music reading has the advantage of being a real-world skill that can, to some extent, be fractionated into behavioral components that are already well understood. By studying the neural basis of music reading at a basic transcriptional level, we can ask how these components are integrated into a complex skill. A previously conducted study scanned professional pianists as they read, played, and listened to music within the confines of a positron emission tomography (PET) scanner.[5] When playing music from a score, with auditory feedback, was contrasted with reading from a score and hearing it played, the superior parietal cortex was activated, leading the authors to suggest that this area is involved in mediating the "sensori-motor transformations for visually guided skilled actions and finger positioning," although this study lacked a nonmusical sensorimotor transformation for comparison, so the musical specificity of these results was open to debate.

However, a subsequent study[6] contrasted the pattern of brain activity involved in playing a note, based on a musical note on the stave, with playing a note based on a verbal label corresponding to that note or based on a number corresponding to the required finger. When musical transcoding was compared to either verbal or numerical transcoding, activation of right superior parietal cortex was found. The major difference between the musical transcoding and the verbal and numerical transcoding is the reliance of the transformation on the spatial information contained within the musical stimulus. Thus we can infer that the activation of the superior parietal cortex appears specific to the kind of sensorimotor transformations required for reading music.

Converging evidence for the role of the superior parietal cortex in music reading comes from an fMRI study that measured changes in the brains of individuals as they learned to read music and play keyboard over a period of three months.[7] Adult par-

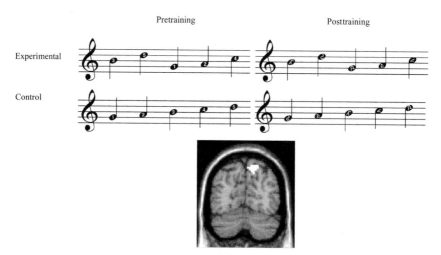

FIGURE 4. fMRI study: reading music for melody. *Upper panel*: Example of stimuli. Pretraining: participants performed a number-to-finger mapping. Experimental trials were unpredictable in their specified motor response compared to control trials, which specified an ascending or descending sequence of responses. Posttraining: the same stimuli were used, but nonsense symbols replaced the numbers, so that participants now performed a note-to-finger mapping. *Lower panel*: Statistical parametric map, rendered onto a normalized structural image, showing a learning-related change in superior parietal cortex (coordinates: 18, −72, 57).

ticipants were scanned twice, once before training and once after training. During both scanning sessions, the participants saw musical notes that varied in pitch (between G and D) and responded by making the appropriate keypresses on an MRI-compatible keyboard. Before training, when musical notes held no meaning for the participants, they simply made their responses based on numbers that were superimposed onto the notes (notes and numbers were always congruent). After training, these numbers were replaced by visually similar but meaningless symbols, so that stimuli were visually similar across the pretraining and posttraining scanning sessions (FIG. 4). Now participants made their responses based on their newly acquired musical literacy skills. A training effect was seen in the right superior parietal cortex, less than half a centimetre from the area identified in the musical transcoding study.

The neuroimaging studies that have so far been mentioned have concentrated on delineating the processes involved in reading musical notation for pitch. However, apart from knowing what to play, it is equally important to know when to play it. The rhythmic instructions in musical notation are conveyed not by spatial information contained within the notation but rather by its visual features: whether a notehead is filled or open, or whether there is a stem or a dot. In the same training study men-

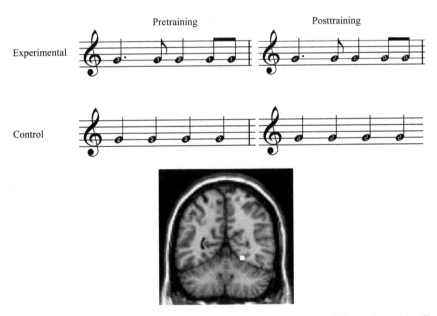

FIGURE 5. fMRI study: reading music for rhythm. *Upper panel*: Example of stimuli. Participants produced rhythms, using only a single key. Pretraining: the numbers indicated the duration of each keypress. An auditory beep was used to give the duration of the single beat. Experimental trials always comprised three different durations, whereas control trials specified a sequence of keypresses of the same duration. In order to keep the number of keypresses constant across control and experimental trials, control trials were either a bar of eight short duration keypresses, a bar of four medium duration keypresses, or a bar of two long duration keypresses. *Lower panel*: Statistical parametric map, rendered onto a normalized structural image, showing a learning-related change in the fusiform gyrus (coordinates: 21, −57, −12).

tioned above, one condition required participants to tap out an equitonal rhythm from notation. Before training, numbers that were superimposed onto the noteheads conveyed note durations, but after training, these symbols were once again replaced with visually similar meaningless symbols so that rhythmic literacy was required to successfully decode the notation (FIG. 5).

A pre-/posttraining comparison of the brain's response to rhythmic notation revealed learning-related changes in the fusiform gyrus.[8] This region and adjacent regions in the ventral occipital cortex have been shown to be important for visually discriminating between exemplars of a particular category, for instance, faces,[9] buildings/scenes,[10] letters,[11] and now, rhythmic notation. It is striking that reading music for pitch and for rhythm, respectively, involves the dorsal and ventral visual processing streams.[12,13] Of course, it is usual for musical notation to contain both pitch and rhythmic information, and future studies will be necessary to investigate how these two aspects of music reading draw upon processing in both of these streams to produce a response that is integrated in space and time.

The melody and rhythm reading conditions previously described required participants to attend to these different dimensions of musical notation and respond accordingly, based on their explicit knowledge concerning the meaning of musical notation. In order to investigate whether the brain automatically processes musical

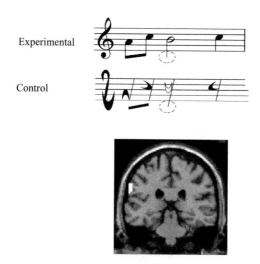

FIGURE 6. fMRI study: implicit music reading. *Upper panel*: Example of stimuli. Participants performed a visual feature detection task in which the target (a vertical stem that protruded above or below the horizontal stave) could form part of a musical stimulus (musical trial, above) or a nonmusical stimulus (nonmusical trial, below), constituting experimental and control trials, respectively. Participants indicated whether the target was ascending or descending, using an arbitrary up/down mapping to the index and middle fingers. Musical trials were visually and motorically matched with nonmusical trials, but only the musical trials were musically interpretable. *Lower panel*: Statistical parametric map, rendered onto a normalized structural image, showing a learning-related change in the left supramarginal gyrus (coordinates: −63, −30, 27).

notation, participants engaged in a third condition, in which notation was present but irrelevant. The task was a visual one that required participants to indicate whether a visual ascender or descender (vertical stem extending above or below the five horizontal lines) was present (see FIG. 6). Control trials were similar, but the ascenders and descenders were embedded within a kind of notation that was visually similar to musical notation but could not be musically deciphered. The prediction was that musical notation and the visually matched nonmusical notation would both be meaningless before training but that after training the musical notation alone would automatically activate elements of a newly acquired musical lexicon.

A comparison of the brain's activity before and after training revealed a learning-related effect in the left supramarginal gyrus. As predicted, this region did not distinguish between the musical notation and the nonmusical notation before training but was more active for musical notation than for nonmusical notation after training. Neuroimaging studies, transcranial magnetic stimulation (TMS) studies, and patient studies have highlighted the importance of this area in the processing of "motor intention."[14–18] A plausible interpretation of the activation change seen in the supramarginal gyrus during the implicit music reading condition, therefore, is that during the course of training, participants learned to make specific keypresses in response to particular musical notes. After training, the visual appearance of musical notes was automatically and unconsciously interpreted as an instruction to act, though processing was terminated before the response was executed. Support for this interpretation comes from the demonstration that the musical novices showed a musical Stroop effect after training,[7] suggesting that the mere presence of musical notation is sufficient to result in musical response preparation.

CONCLUSIONS

The studies described above reveal that music reading, at least for keyboard performance, requires coordinated sensorimotor translation between a set of vertically organized stimuli and a horizontally organized set of responses. These spatial mappings, which are likely to relate to the decoding of pitch from notation, develop in the early stages of skill acquisition and are associated with functional changes in the superior parietal cortex. The ability to decode the rhythmic elements of notation, by contrast, appears to depend upon a visual discrimination process subserved by the fusiform gyrus. These "what" and "when" aspects of music reading seem to map onto the what and when occipitoparietal and occipitotemporal streams, respectively.

ACKNOWLEDGMENTS

The work referred to in this article was supported by the Medical Research Council and the Wellcome Trust. The behavioral experiments also involved Uta Frith and Vincent Walsh. The neuroimaging work was conducted in collaboration with Rik Henson, Knut Kampe, Robert Turner, Uta Frith, and Vincent Walsh.

[Competing interests: The author declares that she has no competing financial interests.]

REFERENCES

1. SLOBODA, J.A. 1984. Experimental studies of musical reading: a review. Mus. Percept. **2:** 222–236.
2. STROOP, J.R. 1935. Studies of interference in serial verbal reactions. J. Exp. Psychol. 643–662.
3. WOLF, T. 1976. A cognitive model of sight-reading. J. Psycholing. Res. **5:** 143–171.
4. STEWART, L., V. WALSH & U. FRITH. 2004. Reading music modifies spatial mapping in pianists. Percept. Psychophys. **66:** 183–195.
5. SERGENT, J., E. ZUCK, S. TERRIAH & B. MACDONALD. 1992. Distributed neural network underlying musical sight-reading and keyboard performance. Science **257:** 106–109.
6. SCHÖN, D., J.L. ANTON, M. ROTH & M. BESSON. 2002. An fMRI study of music sight-reading. Neuroreport **13:** 2285–2289.
7. STEWART, L., R. HENSON, K. KAMPE, et al. 2003. Becoming a pianist. An fMRI study of musical literacy acquisition. Ann. N. Y. Acad. Sci. **999:** 204–208.
8. STEWART, L. 2003. Neurocognitive Studies of Music Reading. PhD Thesis. University College. London.
9. KANWISHER, N., J. MCDERMOTT & M.M. CHUN. 1997. The fusiform face area: a module in human extrastriate cortex specialized for face perception. J. Neurosci. **17:** 4302–4311.
10. EPSTEIN, R. & N. KANWISHER. 1998. A cortical representation of the local visual environment. Nature **392:** 598–601.
11. COHEN, L., S. DEHAENE, L. NACCACHE, et al. 2000. The visual word form area: spatial and temporal characterization of an initial stage of reading in normal subjects and posterior split-brain patients. Brain **123:** 291–307.
12. MISHKIN, M. & L.G. UNGERLEIDER. 1982. Contribution of striate inputs to the visuospatial functions of parieto-preoccipital cortex in monkeys. Behav. Brain Res. **6:** 57–77.
13. GOODALE, M.A. & A.D. MILNER. 1992. Separate visual pathways for perception and action. Trends Neurosci. **15:** 20–25.
14. RUSHWORTH, M.F., T. PAUS & P.K. SIPILA. 2001. Attention systems and the organization of the human parietal cortex. J. Neurosci. **21:** 5262–5271.
15. RUSHWORTH, M.F., A. ELLISON & V. WALSH. 2001. Complementary localization and lateralization of orienting and motor attention. Nat. Neurosci. **4:** 656–661.
16. HARRINGTON, D.L. & K.Y. HAALAND. 1992. Motor sequencing with left hemisphere damage. Are some cognitive deficits specific to limb apraxia? Brain **115:** 857–874.
17. KIMURA, D. 1993. Neuromotor Mechanisms in Human Communication. Oxford University Press. New York.
18. SIRIGU, A., E. DAPRATI, P. PRADAT-DIEHL, et al. 1999. Perception of self-generated movement following left parietal lesion. Brain **122:** 1867–1874.

Structural, Functional, and Perceptual Differences in Heschl's Gyrus and Musical Instrument Preference

PETER SCHNEIDER,[a] VANESSA SLUMING,[b,c] NEIL ROBERTS,[c] STEFAN BLEECK,[d] AND ANDRÉ RUPP[a]

[a]Department of Neurology, University Hospital Heidelberg, Heidelberg, Germany

[b]School of Health Sciences, Division of Medical Imaging, University of Liverpool, United Kingdom

[c]Magnetic Resonance and Image Analysis Research Centre (MARIARC), University of Liverpool, United Kingdom

[d]Institute of Sound and Vibration Research, University of Southampton, United Kingdom

ABSTRACT: The musical pitch of harmonic complex sounds, such as instrumental sounds, is perceived primarily by decoding either the fundamental pitch (keynote) or spectral aspects of the stimuli, for example, single harmonics. We divided 334 professional musicians, including symphony orchestra musicians, 75 amateur musicians, and 54 nonmusicians, into either fundamental pitch listeners or spectral pitch listeners. We observed a strong correlation between pitch perception preference and asymmetry of brain structure and function in the pitch-sensitive lateral areas of Heschl's gyrus (HG), irrespective of musical ability. In particular, fundamental pitch listeners exhibited both larger gray matter volume measured using magnetic resonance imaging (MRI) and enhanced P50m activity measured using magnetoencephalography (MEG) in the left lateral HG, which is sensitive to rapid temporal processing. Their chosen instruments were percussive or high-pitched instruments that produce short, sharp, or impulsive tones (e.g., drums, guitar, piano, trumpet, or flute). By contrast, spectral pitch listeners exhibited a dominant right lateral HG, which is known to be sensitive to slower temporal and spectral processing. Their chosen instruments were lower-pitched melodic instruments that produce rather sustained tones with characteristic changes in timbre (e.g., bassoon, saxophone, french horn, violoncello, or organ). Singers also belonged to the spectral pitch listeners. Furthermore, the absolute size of the neural HG substrate depended strongly on musical ability. Overall, it is likely that both magnitude and asymmetry of lateral HG, and the related perceptual mode, may have an impact on preference for particular musical instruments and on musical performance.

KEYWORDS: pitch perception; Heschl's gyrus; asymmetry; musical instrument preference; harmonic complex tones; orchestra

Address for correspondence: Peter Schneider, Department of Neurology, Section of Biomagnetism, INF 400, 69120 Heidelberg, Germany. Fax: +49-6221-565258.
Peter.Schneider@med.uni-heidelberg.de.

Ann. N.Y. Acad. Sci. 1060: 387–394 (2005). © 2005 New York Academy of Sciences.
doi: 10.1196/annals.1360.033

Harmonic complex tones, such as instrumental sounds, may be described by objective aspects as the time period of the sound pressure curve or the envelope of the frequency spectrum. However, the perceived musical pitch differs largely by up to three or four octaves, when the same sound is presented to different individuals, even for professional musicians. For example, if the 5th, 6th, and 7th harmonic of 500 Hz is played, the perceived pitch ranged between one-line octave B and four-line octave F sharp. Some subjects recognized predominantly the fundamental pitch, that is, the keynote of a sound. Others predominantly perceived single harmonics of the complex sounds.[1–4] Such subjective aspects have been described earlier by Hermann von Helmholtz[5,6] who introduced a "synthetical mode," based on fundamental pitch perception, and an "analytical mode," based on the perception of single harmonics of the complex sounds. To psychoacoustically quantify the large perceptual differences, we performed a new pitch test[4] using tone pairs of complex tones. Participants were asked to identify the dominant direction of pitch shift in a large sample of 144 tone pairs. For each subject, an index of pitch perception preference ($\delta_p = (f_{sp} - f_0)/(f_{sp} + f_0)$) was measured by identifying the percentage of fundamental (f_0) versus spectral (f_{sp}) pitch perception. We measured 334 professional musicians, including symphony orchestra musicians from the Royal Liverpool Philharmonic Orchestra (RLPO),[7] 75 amateurs, and 54 nonmusicians and observed a large bimodal distribution that enabled the classification in fundamental and spectral pitch listeners.

How are these perceptual differences related to structural and functional attributes of the human auditory cortex (AC)? In a subgroup of 87 subjects, magnetic resonance imaging (MRI) and magnetoencephalography (MEG) studies demonstrated a strong neural basis for both types of pitch perception irrespective of musical ability.[4] The fundamental pitch listeners exhibited a pronounced leftward asymmetry of gray matter volume and functional P50m activity in the pitch-sensitive[8,9] areas of lateral Heschl's gyrus (lHG) whereas, in contrast, spectral pitch listeners exhibited a corresponding pronounced rightward asymmetry.[4]

The relative hemispheric lateralization was observed to correlate with pitch preference irrespective of musical ability. In FIGURE 1 we depict the strong correlation between pitch perception asymmetry (δ_p) and either gray matter aymmetry of lHG structurally (δ_s) on the left panel (A) or pitch asymmetry and auditory-evoked P50m asymmetry functionally (δ_f) on the right panel (B). The correlation is equally strong for nonmusicians and musicians. Furthermore, there is a link to musical instrument preference. The left AC is known to be sensitive to rapid temporal processing.[10–12] Therefore, fundamental pitch listeners may make use of short, sharp, and impulsive tones. Indeed, as shown in FIGURE 1, they play mainly percussive instruments (drums or guitar) or higher-pitched solo instruments (trumpet, flute, or piccolo). By contrast, the right AC is sensitive to slower temporal and spectral processing.[10,12] Therefore, spectral pitch listeners may make use of sustained tones with characteristic formants in their frequency spectrum. Consistently, we observed that the musical instruments of their choice were melodic instruments producing sustained tones (strings, woodwind, or brass, without trumpet, organ, or the singing voice). However, most professional musicians simultaneously perceive both the keynote and single harmonics from an ambiguous tone, and the subjective differences are relative rather than absolute.

To quantify in more detail the relationship between pitch perception and musical instrument preference, all 463 psychometrically tested musicians were subdivided

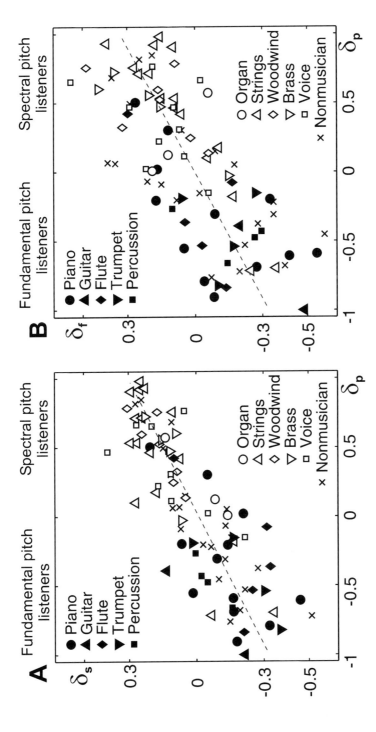

FIGURE 1. Correlation of pitch perception preference (δ_p) versus (**A**) structural asymmetry (δ_s) and (**B**) functional asymmetry (δ_f) of the neural substrate in lateral HG. The correlation is strong for both nonmusicians (x symbols) and musicians (symbols indicate their main instrument).

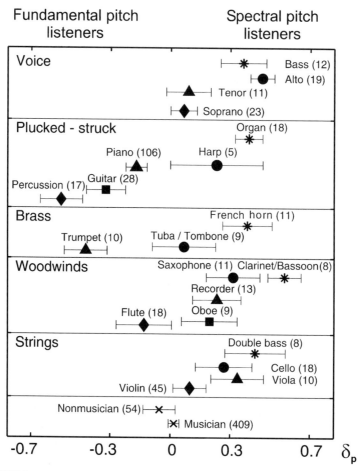

FIGURE 2. The index of pitch perception preference (δ_p) was classified for musical instrument families (mean ± SEM), nonmusicians, and all musicians (*bottom*). Fundamental pitch listeners played predominantly percussive or high-pitched instruments, whereas spectral pitch listeners prefered lower-pitched melodic instruments and singing.

according to their main instrument (FIG. 2). The index of pitch perception preference (δ_p) was averaged over instrumental groups, and the psychometric results were analyzed for different instrument families. Percussionists showed the most pronounced fundamental pitch percept, followed by trumpeters, guitarists, and flutists. On the other hand, players of lower-pitched melody instruments (bassoon, double bass, organ, or basses and altos in a choir) were found to show the most pronounced spectral pitch percept. Regarding the distribution of one single group, 65% of the pianists were fundamental pitch listeners, and 35% were spectral pitch listeners. We believe that this pronounced variation may be explained by an additional influence of the perceptual and neuronal asymmetry on musical performance style. For example, a pianist predominantly perceiving the fundamental pitch may prefer to perform with virtuosity and enjoy playing complex rhythmic patterns, whereas a pianist predominantly perceiving harmonics may prefer slower music and may concentrate more on timbral or melodic aspects of the music.

Within singers, sopranos showed enhanced fundamental pitch perception as compared to altos, women in the female bass register ($F_{1,41} = 12.2$, $P < .0001$), and male basses ($F_{1,34} = 3.7$, $P < .05$). Pianists exhibited strongly enhanced fundamental pitch perception as compared to organists ($F_{1,123} = 13.5$, $P < .0001$). Within the brass family, trumpet players demonstrated an almost opposite perception mode as compared to french horn and tuba players ($F_{1,29} = 21.3$, $P < .0001$). Within the woodwinds, flute and piccolo players were completely separated from the saxophonists ($F_{1,28} = 5.7$, $P < .01$). Last, but not least, percussionists frequently perceived the missing fundamental pitch, but string players did only relatively rarely ($F_{1,97} = 39.2$, $P < .0001$). Consistently, the players of higher pitched descant instruments of each instrument family (trumpet, flute, or violin) demonstrated enhanced fundamental pitch perception as compared to the lower-pitched instruments. However, the pitch perception preference was not significantly different between musicians and nonmusicians ($F_{1,462} = 2.6$, n.s.).

Musical sounds from percussive and high-pitched instruments all exhibit a short attack time[13,14] (i.e., raise time of the amplitude envelope) in the time range of 20–50 ms, whereas lower-pitched instruments typically show attack times of 100–300 milliseconds. These time ranges fit excellently with the timescales of temporal sensitivity in lHG and may also be reflected in the separation of transient and sustained AC activity.[15] As a consequence, the pronounced left lHG of percussionists and guitarists may lead to the predilection of impulsive short tones with rapid decay. By contrast, the large right lHG of singers or string players may facilitate the processing of slower sustained tones, including characteristic spectral information in the form of natural resonances like formants or vowels in the singing voice.[16] Accordingly, the rightward lateralization of voice processing has been demonstrated by functional imaging.[17]

In FIGURE 3 we show a large collection of individual 3D-reconstructed auditory cortices[4] of both musicians and nonmusicians (bottom). Fundamental pitch listeners were grouped left of the dashed line and spectral pitch listeners right of the dashed line. This grouping is already consistent with the standard seating arrangement of a modern symphony orchestra. In an orchestra, the first violins are positioned to the conductor's left, the lower strings to the right. The woodwinds form a center square directly behind the strings, followed by brass and percussion. Historically, trumpets and timpani have been placed together to play fanfares, even before the time of Beethoven.[18] Conductors were predominantly fundamental pitch listeners. Special arrangements may include a choir or solo instruments (e.g., piano, harp, or organ). Overall, percussive or higher-pitched solo instruments are placed more to the left in the orchestra and lower-pitched melody instruments more to the right.

Detailed interviews regarding musical qualifications and the intensity of musical practice revealed that musical performance depends on pitch perception preference. Thus, listeners with a different type of pitch perception who prefer the same instrument may differ in musical performance style.

Our data demonstrated, furthermore, that the absolute size of the neural HG substrate depended strongly on musical ability.[4,19] First, individual segmentation and

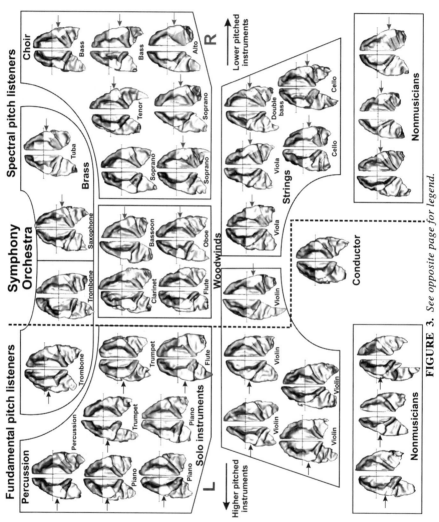

FIGURE 3. See opposite page for legend.

3D-reconstruction of the AC revealed that the gray matter volume of HG and anterior supratemporal gyrus was twofold larger in professional musicians as compared to nonmusicians.[4,19] Second, the magnitude of the auditory-evoked P50m activity in response to harmonic complex tones as measured by MEG was, on average, fivefold larger in professional musicians as compared to nonmusicians. However, this effect vanished completely for the subsequent prominent N100m activation. To understand these large group–specific differences in relation to musical ability, we collected in detail from all 463 subjects the individual histories of musical training, and performed the Advanced Measure of Music Audiation (AMMA) test of Edwin E. Gordon.[20] According to Gordon, musical aptitude represents the potential to learn music and stabilizes prior to the outset of intensive musical education at the age of about nine years. Furthermore, musical aptitude was related to *audiation*, a term coined by Gordon, which means "to hear and comprehend music for which the sound is not physically present." Thus, the result of the AMMA test may be linked to auditory imagery.

The volume of gray matter in both the HG and supratemporal gyrus correlated strongly with the tonal score of the AMMA test. Thus, the neural substrate subserving musical aptitude and audiation may reside in the HG and may not be largely modified by long-term musical performance throughout life. By contrast, the auditory-evoked P50m magnitude in lHG only correlated strongly with long-term musical practice. Therefore, our data suggest a functional–structural dissociation between long-term musical training and musical aptitude in the AC of musicians and nonmusicians.

In conclusion, our psychoacoustic data provide evidence that the perceptual mode of listening may have an impact on preference of timbre, tone, and size of particular musical instruments and in particular on musical performance. Interestingly, the perceptual mode may be predicted by MRI and MEG studies of the asymmetry and absolute magnitude of structure and function of the auditory cortex.

ACKNOWLEDGMENTS

We thank K. Sartor for providing the 3D-MRI in Heidelberg; the radiographic staff at MARIARC for assistance with MRI data acquisition from members of the Royal Liverpool Philharmonic Orchestra (RLPO); and E. Hofmann (Music Academy, Basel), D. Geller, R. Schmitt, and T. van der Geld (University of Music and Performing Arts, Mannheim), C. Klein (Institute of Music Pedagogy, Halle), and D. Schmidt (Conservatory of Music and Performing Arts, Stuttgart) for assistance with collecting the psychometric data.

FIGURE 3. Individual HG morphometry of orchestra musicians. The classification in fundamental pitch and spectral pitch listeners (left/right to the *dashed line*) is consistent with a typical symphony orchestra seating arrangement. Fundamental pitch listeners exhibit a larger left lHG and spectral pitch listeners a larger right lateral Heschl's gyrus in most cases (*highlighted, arrows*). Percussive and high-pitched instruments are placed more to the left and low-pitched instruments more to the right. Furthermore, professional musicians showed significantly increased gray matter volume within the anterior part of the auditory cortex (*gray shaded part, top–bottom*) irrespective of asymmetry (*left–right*).

[Competing interests: The authors declare that they have no competing financial interests.]

REFERENCES

1. SMOORENBURG, G.F. 1970. Pitch perception of two-frequency stimuli. J. Acoust. So. Am. **48:** 924–942.
2. LAGUITTON, V., L. DEMANY, C. SEMAL & C. LIÉGEOIS-CHAUVEL. 1998. Pitch perception: a difference between right- and left-handed listeners. Neuropsychologia **36:** 201–207.
3. PATEL, A.D. & E. BALABAN. 2001. Human pitch perception is reflected in the timing of stimulus-related cortical activity. Nat. Neurosci. **4:** 839–844.
4. SCHNEIDER, P. et al. 2005. Structural and functional asymmetry of lateral Heschl's gyrus reflects pitch perception preference. Nat. Neurosci. **8:** 1241–1247.
5. VON HELMHOLTZ, H.L.F. 1885. On the Sensations of Tone. Longmans. London.
6. TERHARDT, E. Pitch, consonance, and harmony. 1974. J. Acoust. Soc. Am. **55:** 1061–1069.
7. SLUMING, V. et al. 2002. Voxel-based morphometry reveals increased gray matter density in Broca's area in male symphony orchestra musicians. Neuroimage **17:** 1613–1622.
8. GRIFFITHS, T.D. 2003. Functional imaging of pitch analysis. 2003. Ann. N. Y. Acad. Sci. **999:** 40–49.
9. PENAGOS, H., J.R. MELCHER & A.J. OXENHAM. 2004. A neural representation of pitch salience in nonprimary human auditory cortex revealed with functional magnetic resonance imaging. J. Neurosci. **24:** 6810–6815.
10. BOEMIO, A., S. FROMM & D. POEPPEL. 2005. Hierarchical and asymmetric temporal sensitivity in human auditory cortices. Nat. Neurosci. **8:** 389–395.
11. RUPP, A., A. GUTSCHALK, S. UPPENKAMP & M. SCHERG. 2004. Middle latency auditory-evoked fields reflect psychoacoustic gap detection thresholds in human listeners. J. Neurophysiol. **92:** 2239–2247.
12. ZATORRE, R. & P. BELIN. 2001. Spectral and temporal processing in human auditory cortex. Cereb. Cortex **11:** 946–953.
13. GISELER, W., L. LOMBARDI & R.D. WEYER. 1985. Instrumentation in der Musik des 20. Jahrhunderts. Moeck. Celle.
14. REUTER, C. 1995. Der Einschwingvorgang nichtperkussiver Musikinstrumente. Lang. Frankfurt.
15. SEIFRITZ, E. et al. 2002. Spatiotemporal pattern of neural processing in the human auditory cortex. Science **297:** 1706–1708.
16. JOLIVEAU, E., J. SMITH & J. WOLFE. 2004. Tuning of vocal tract resonance by sopranos. Nature **427:** 116.
17. BELIN, P., R.J. ZATORRE & P. LAFAILLE. 2000. Voice-selective areas in human auditory cortex. Nature **403:** 309–312.
18. BERLIOZ, H. 1844. Grand traité d'instrumentation et d'orchestration modernes. Translated 1882. M.C. Clarke, Ed. Novello. London.
19. SCHNEIDER, P. et al. 2002. Morphology of Heschl's gyrus reflects enhanced activation in the auditory cortex of musicians. Nat. Neurosci. **5:** 688–694.
20. GORDON, E.E. 1998. Introduction to research and the psychology of music. GIA Publications. Chicago.

Differences in Gray Matter between Musicians and Nonmusicians

PATRICK BERMUDEZ AND ROBERT J. ZATORRE

Montreal Neurological Institute, McGill University, Montreal, Quebec, Canada H3A 2B4

ABSTRACT: Voxel-based morphometry is used to examine differences in cerebral morphology between musicians and nonmusicians. Principal results show differences in gray matter concentration in the right auditory cortex.

KEYWORDS: auditory cortex; musicians; VBM

INTRODUCTION

There exists a controversial literature regarding gross morphological differences in the brain structure of musicians as compared to nonmusicians. Several authors have shown changes in diverse cerebral regions associated with musical training, notably in the motor and auditory cortices (e.g., Refs. 1 and 2), but these have not always been consistent across studies. Here we present new data bearing on such differences with the use of voxel-based morphometry (VBM) of magnetic resonance images (MRIs).

METHODS

Subject groups were made up of both males and females, 51 nonmusicians and 43 musicians (10 years or more of musical experience, 22 with absolute pitch). T1 MRIs were linearly registered to the symmetric MNI152 template with a 12-parameter cost-minimization fit[3] and then RF inhomogeneity corrected[4] (FIG. 1a) and tissue classified (FIG. 1b).[5] The gray matter class was extracted and blurred using an 8-mm Gaussian kernel (FIG. 1c). Musician and nonmusician images were then contrasted and evaluated according to the general linear model and random field theory correction for multiple comparisons.[6]

Address for correspondence: Patrick Bermudez, Cognitive Neuroscience Unit, Room 276, Montreal Neurological Institute, 3801 University Street, Montréal, Québec, H3A 2B4. Voice: 514-398-2579; fax: 514-398-1338.

pat@ego.psych.mcgill.ca

Ann. N.Y. Acad. Sci. 1060: 395–399 (2005). © 2005 New York Academy of Sciences.
doi: 10.1196/annals.1360.057

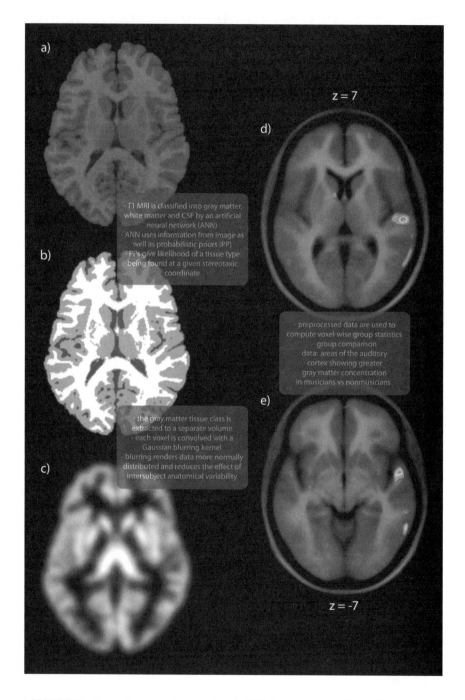

FIGURE 1. Illustration of methods and main VBM results in the right auditory cortex.

RESULTS AND DISCUSSION

The main result is a greater gray matter (GM) concentration in musicians as compared to nonmusicians in the right lateral surface of the superior temporal gyrus, posterior to Heschl's gyrus (planum temporale or caudal-lateral auditory belt cortex, FIG. 1d). A region of interest was defined using this t-statistic peak for the purpose of extracting GM values for each subject. A plot of these values by group (musicians with absolute pitch, musicians with relative pitch, and nonmusicians) shows that absolute pitch and relative pitch musicians did not differ from one another in GM concentration in this region, although both differed from nonmusicians (FIG. 2).

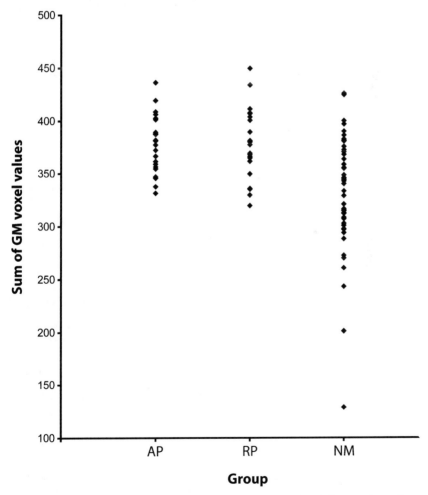

FIGURE 2. The t-statistic peak illustrated in FIG. 1d is used to define a region of interest from which the GM values are plotted for absolute pitch (AP), relative pitch (RP), and nonmusician (NM) subject groups separately.

Another peak of greater GM concentration in musicians lies anterior to Heschl's gyrus in the planum polare (FIG. 1e).

VBM results derived from images that are linearly transformed to a stereotaxic space concurrently communicate information about size, position, and morphology. We must therefore be cautious in our interpretation. These and adjacent areas of the superior temporal gyrus have been previously implicated in a number of functional imaging studies, as well as in older brain-lesion studies, as being important for the processing of pitch (see Ref. 7 for a review). They have also been implicated as differential morphological markers in volumetric studies using manual segmentation (e.g., Ref. 8 and 9), and neurophysiological data indicate neuronal pitch sensitivity in an analogous region of the marmoset monkey brain.[10] VBM is a data-driven technique that does not rely upon apriori definitions of anatomical circumscription and, in this way, is free from systematic errors that can arise from such definitions. Our results suggest an experience-dependent difference between musicians and nonmusicians in areas of the right hemisphere known to be important in pitch processing and that seem to preferentially subserve spectral and pitch resolution. However, this anatomical effect seems unrelated to absolute pitch, which may depend more on interactions between pitch-sensitive regions and the dorsolateral frontal cortex.[11]

ACKNOWLEDGMENTS

We thank Joyce Chen, Jennifer Johnson, Karine Delhommeau, and the staff of the McConnell Brain Imaging Centre. This work was funded by the International Foundation for Music Research (IFMR) and the Canadian Institutes of Health Research (CIHR).

[Competing interests: The authors declare that they have no competing financial interests.]

REFERENCES

1. GASER, C. & G. SCHLAUG. 2003. Brain structures differ between musicians and nonmusicians. J. Neurosci. **23:** 9240–9245.
2. SCHNEIDER, P., M. SCHERG, H.G. DOSCH, et al. 2002.. Morphology of Heschl's gyrus reflects enhanced activation in the auditory cortex of musicians. Nat. Neurosci. **23:** 688–694.
3. COLLINS, D.L., P. NEELIN, T.M. PETERS & A.C. EVANS. 1994. Automatic 3D intersubject registration of MR volumetric data in standardized Talairach space. J. Comput. Assist. Tomogr. **281:** 567–585.
4. SLED, J.G., A.P. ZIJDENBOS & A.C. EVANS. 1998. A nonparametric method for automatic correction of intensity nonuniformity in MRI data. IEEE Trans. Med. Imag. **17:** 87–97.
5. ZIJDENBOS A.P. & B.M. DAWANT. 1994. Brain segmentation and white matter lesion detection in MR images. Crit. Rev. Biomed. Eng. **22:** 401–465.
6. WORSLEY, K.J., C.H. LIAO, J. ASTON, et al. 2002. A general statistical analysis for fMRI data. Neuroimage **15:** 1–15.
7. ZATORRE, R.J., P. BELIN & V.B. PENHUNE. 2002. Structure and function of auditory cortex: music and speech. Trends Cogn. Sci. **6:** 37–46.
8. ZATORRE, R.J., D.W. PERRY, C.A. BECKETT, et al. 1998. Functional anatomy of musical processing in listeners with absolute pitch and relative pitch. Proc. Natl. Acad. Sci. USA **95:** 3172–3177.

9. KEENAN, J.P., V. THANGARAJ, A.R. HALPERN & G. SCHLAUG. 2001. Absolute pitch and planum temporale. Neuroimage **14:** 1402–1408.
10. BENDOR, D. & X. WANG. 2005. The neuronal representation of pitch in primate auditory cortex. Nature **436:** 1161–1165.
11. BERMUDEZ, P. & R.J. ZATORRE. 2005. Conditional associative memory for musical stimuli in nonmusicians: implications for absolute pitch. J. Neurosci. **25:** 7718–7723.

Tapping in Synchrony to Auditory Rhythms

Effect of Temporal Structure on Behavior and Neural Activity

JOYCE L. CHEN,[a,b] VIRGINIA B. PENHUNE,[c] AND ROBERT J. ZATORRE[a,b]

[a]Montreal Neurological Institute, Montreal, Quebec, Canada

[b]McGill University, Montreal, Quebec, Canada

[c]Concordia University, Montreal, Quebec, Canada

ABSTRACT: This study examines how synchronizing movements to auditory rhythms affects behavioral performance and neural activity as assessed using fMRI. Subjects perform worse as they tap along with increasingly complex musical rhythms, with responses becoming progressively asynchronous. This behavioral change also correlates with increasing neural activity in several motor-related brain regions, such as the cerebellum and premotor cortex (medial and lateral), highlighting the role of these areas in auditory-motor interactions during temporal processing.

KEYWORDS: fMRI; rhythm; temporal; motor timing; metricality

INTRODUCTION

When we listen to music, we may synchronize our body movements, for example, by tapping our feet to the rhythm's beat. Features of rhythm, such as its temporal structure, may aid beat detection and thus synchronization of movements. Those rhythms whose total number of beats can be equally subdivided are known as metric (or regular) sequences; these are better reproduced by finger tapping than nonmetric (or irregular) rhythms that cannot be subdivided into equal partitions.[1,2] The present study examines how tapping performance during auditory-motor synchronization is affected by manipulations of metricality and investigates the neural correlates of this behavior using functional magnetic resonance imaging (fMRI). We hypothesized that as rhythm synchronization becomes more difficult, brain regions mediating auditory-motor temporal processing will become progressively more engaged.

METHODS

Twelve right-handed (six male) healthy volunteers with no musical training were recruited for this study. Subjects listened and tapped in synchrony to three auditory

Address for correspondence: Joyce L. Chen, MNI, 3801 University Street, Rm. 276, Montreal, QC H3A 2B4. Voice: 514-398-6644, ext. 00268; fax: 514-398-1338.

joyce.chen@mail.mcgill.ca

Ann. N.Y. Acad. Sci. 1060: 400–403 (2005). © 2005 New York Academy of Sciences.
doi: 10.1196/annals.1360.044

rhythms (metric simple, metric complex, and nonmetric) using the index finger of the right hand on a mouse button. These rhythms differed only in their temporal structure (temporal intervals between musical notes), resulting in progressively more complex and less metrically structured rhythms as confirmed by pilot data. All rhythms were played with a woodblock sound at 75dB SPL (sound pressure level) and contained the same number of musical notes of the same duration but were re-arranged to produce the three levels of metrical complexity; they were equal in total duration (6 s). There were a total of seven conditions; subjects listened and tapped in synchrony to each of the three rhythms, and a silent condition was included as a baseline. On each trial (two fMRI acquisition frames), subjects were presented with a rhythm twice: in the first frame they listened passively, and in the second they tapped synchronously to the same rhythm. Trials were randomized for order and tap responses (mouse button on and off time) were recorded online. The dependent variable of interest was tap asynchrony, a measure of how accurately subjects were able to synchronize their motor response with each note of the rhythms. Subjects learned to tap in synchrony with these rhythms 1 day before the fMRI session to avoid con-founds of motor learning. On the day of the fMRI session, subjects first underwent a high-resolution T1-weighted anatomical scan (voxel size: 1 mm^3, FOV: 256 mm^2). Next, two functional runs using T2*-weighted gradient-echo planar imaging were performed with a sparse-sampling paradigm (25 whole-head interleaved slices; TE = 50 ms; TR = 10,000 ms; voxel size = 5 mm^3; matrix size = 64 mm^3). The brain images then were motion corrected, smoothed with a 12-mm filter, transformed into stereotaxic MNI space, and analyzed using fMRISTAT.[3] The contrasts of interest were (1) blood oxygenation level dependent (BOLD) signal covariation as a function of increasing rhythm complexity; (2) subtraction of the average of all tap conditions minus silence; and (3) subtraction of the average of all listen conditions minus silence. A region-of-interest analysis was performed on the peaks obtained from contrast (1) to characterize the function of the covariation analysis. A conjunction analysis was performed on the two subtraction contrasts to identify regions similarly active in both tapping and listening conditions.

RESULTS

The behavioral results demonstrated that as subjects tapped to progressively more complex and less metrically structured rhythms, their tapping became more asyn-chronous (one-way repeated-measures ANOVA, $F[2,22] = 46.23$, $P < .0005$; FIG. 1). Pairwise comparisons confirmed that performance was worse on the metric complex than metric simple rhythms ($F[1,11] = 34.39$, $P < .0001$) and worse for nonmetric than metric complex rhythms ($F[1,11] = 11.32$, $P < .006$). The fMRI analysis iden-tified regions of BOLD signal covariation with parametric changes in rhythm com-plexity. Tapping synchronously to progressively complex rhythms resulted in increased BOLD signal at bilateral cerebellum lobule VI, presupplementary motor area, dorsal premotor cortex, thalamus, and parietal Brodmann area 7. The region-of-interest analysis confirmed that the signal changes in these regions increased linearly across the parametric conditions. The conjunction of the two subtraction contrasts revealed that motor regions active during the tapping conditions were also active during the listening conditions. These areas included bilateral cerebellum lob-

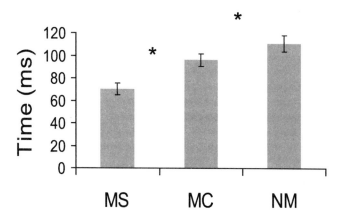

FIGURE 1. Asynchrony (ms) across the three rhythms: metric simple (MS), metric complex (MC), and nonmetric (NM). Data reported as means ± SE.

ules VI and VIIIa, bilateral putamen, dorsal and ventral premotor cortex, and the supplementary motor area.

DISCUSSION

Synchronizing movements with musical rhythms may depend on how successfully one deconstructs the temporal structure. Manipulating the temporal intervals between notes such that the rhythms became progressively complex and less metrically structured resulted in greater tap asynchrony. Beats that fall on time are readily predicted and thus synchronization can be more accurately reproduced during performance on the metric rhythms. As the organized structure degrades, expectation of when to synchronize movements is more variable resulting in poorer performance. Previous studies in which subjects reproduced metrical and nonmetrical rhythms have demonstrated similar behavioral findings.[1,2,4] As tap asynchrony increased due to rhythm complexity, percentage of BOLD signal change in brain regions involved in auditory-motor temporal processing also increased. The lateral cerebellum and pre-supplementary motor area may be jointly engaged during rhythm synchronization, each region processing a different cognitive component involved in performing the task, such as motor timing[5] and rhythm sequencing,[6] respectively. The dorsal premotor cortex on the other hand may have a more general role in integrating information from the auditory modality with motor regions of the brain.[7] Additional findings show that whether one is tapping or just listening to these rhythms, an overlapping network of motor areas are recruited, providing further support to existing literature that motor execution and "imagery" share similar neural substrates.[8] This latter finding is consistent with the idea that motor rehearsal occurs before actual execution.

ACKNOWLEDGMENTS

We thank Marc Bouffard and the staff of the McConnell Brain Imaging Centre. This work was supported by the Canadian Institutes of Health Research.

[Competing interests: The authors declare that they have no competing financial interests.]

REFERENCES

1. POVEL, D.J. 1984. A theoretical framework for rhythm perception. Psychol. Res. **45:** 315–337.
2. POVEL, D.J. & P.J. ESSENS. 1985. Perception of temporal patterns. Mus. Percept. **2:** 411–440.
3. WORSLEY, K.J., C.H. LIAO, J. ASTON, *et al.* 2002. A general statistical analysis for fMRI data. Neuroimage **15:** 1–15.
4. SAKAI, K., O. HIKOSAKA, S. MIYAUCHI, *et al.* 1999. Neural representation of a rhythm depends on its interval ratio. J. Neurosci. **19:** 10074–10081.
5. IVRY, R.B., S.W. KEELE & H.C. DIENER. 1988. Dissociation of the lateral and medial cerebellum in movement timing and movement execution. Exp. Brain Res. **73:** 167–180.
6. KENNERLEY, S.W., K. SAKAI & M.F. RUSHWORTH. 2004. Organization of action sequences and the role of the pre-SMA. J. Neurophysiol. **91:** 978–993.
7. CHEN, J.L., V.B. PENHUNE & R.J. ZATORRE. Interactions between auditory and dorsal premotor cortex during synchronization to musical rhythms. Submitted for publication.
8. JEANNEROD, M. & V. FRAK. 1999. Mental imaging of motor activity in humans. Curr. Opin. Neurobiol. **9:** 735–739.

Neural Substrates Governing Audiovocal Integration for Vocal Pitch Regulation in Singing

JEAN MARY ZARATE AND ROBERT J. ZATORRE

*Cognitive Neuroscience Unit and Department of Neurology and Neurosurgery,
Montréal Neurological Institute, McGill University, Montréal, Québec H3A 2B4*

ABSTRACT: The integration of auditory feedback with the vocal motor system is essential for singing, because precise control of vocal pitch is necessary to produce notes and melodies accurately. We used singing tasks with fMRI to determine the neural substrates involved in audiovocal integration. In tasks that specifically required monitoring of auditory feedback and pitch control, neural activity increased in the anterior cingulate cortex and insula, which suggests that these regions may play a special role in audiovocal integration.

KEYWORDS: auditory feedback; nonmusicians; pitch-shifted feedback; singers; training

INTRODUCTION

Vocal pitch regulation in singing requires the integration between auditory feedback and a stable vocal motor system; however, the neural correlates for audiovocal integration are not well understood. We manipulated auditory feedback during singing tasks to investigate which brain regions may integrate auditory processing with vocal pitch control.

Within the functional network for singing, three cortical regions may be candidates for audiovocal integration: (1) the anterior cingulate cortex (ACC), a vocal motor region that elicits vocalization when electrically stimulated;[1] (2) the superior temporal gyrus (STG), which contains auditory processing areas and whose activity is altered by the ACC during vocalization;[1] and (3) the insula, an area that shares connections with both the ACC and STG and is active during speaking and singing aloud, when compared to silently rehearsing speech and song.[2] We hypothesized that the ACC, STG, and insula may be involved in audiovocal integration due to their roles in the network for singing and their connections with each other. Among the different singing tasks, we further hypothesized that singers would exhibit greater modulation of cortical activity in the ACC, STG, and insula than nonmusicians, since singers have more experience fine-tuning their vocal pitch under different feedback conditions.

Address for correspondence: Jean Mary Zarate, Montréal Neurological Institute, Cognitive Neuroscience Unit, 3801 University St., Room 276, Montréal, Québec, Canada H3A 2B4. Voice: 514-398-2579; fax: 514-398-1338.

jean.zarate@mail.mcgill.ca

Ann. N.Y. Acad. Sci. 1060: 404–408 (2005). © 2005 New York Academy of Sciences.
doi: 10.1196/annals.1360.058

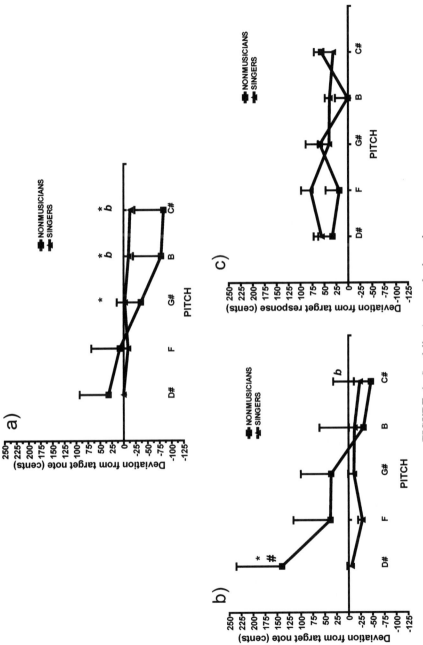

FIGURE 1. *See following page for legend.*

METHODS

Twenty-four subjects were divided into two gender-balanced groups; nonmusicians were classified as persons with less than three years of musical experience, while persons with at least three years of singing experience (mean: 9.5 years) and currently practicing were considered singers. Prior to scanning, subjects were trained with tasks that required singing with either normal or pitch-shifted auditory feedback. For each task, subjects heard one of five target notes ($D^{\#}3$, F3, $G^{\#}3$, B3, $C^{\#}4$ for males; the same notes an octave higher for females) and were visually cued to sing the note back on the syllable "ah." In the "normal" task, subjects sang for four seconds with unaltered auditory feedback after hearing the target. In the "ignore" task, as subjects sang, we shifted their auditory feedback by 200 cents up or down (on different trials); subjects were told to disregard the shifted feedback and continue singing the original target note. In the "compensate" task, subjects were instructed to correct for the pitch shift in order to make the feedback sound like the original note.

In a 1.5-Tesla magnetic resonance (MR) scanner, subjects were placed comfortably on the scanner bed and given an MR-compatible headset with microphone. After a T1-weighted anatomical scan, subjects went through two functional runs that contained the same tasks used in the training session. A perception condition, which entailed hearing a target note without singing the note back, was included as an auditory control for all singing tasks. In each run, we used temporally sparse acquisitions (TE = 2.125 ms, TR = 10 s) to obtain T2*-weighted echo planar images of the entire head.

RESULTS AND DISCUSSION

For normal singing, singers were more accurate across all notes than nonmusicians (FIG. 1a). In the ignore task, singers maintained all pitches in the presence of shifted feedback better than nonmusicians (FIG. 1b). Because of the difficulty of the compensate task, vocal training had no significant effect on performance—both groups performed equally on the task (FIG. 1c).

When contrasting normal singing with the perception task, both groups recruited the same functional network for singing as reported in a previous singing study from our lab,[3] which includes bilateral auditory cortices, bilateral primary motor cortices, the supplementary motor area (SMA), ACC, thalamus, insula, and cerebellum. For both pitch-shifted tasks, nonmusicians recruited only the ACC (an area more anterior

---

FIGURE 1. Performance for all singing tasks across five notes in nonmusicians and singers. Data points represent mean vocal accuracy + SEM. (**a**) Behavioral results for normal singing. Nonmusicians were less accurate overall than singers—the three highest notes were significantly flatter than the lowest note (marked by $*$, $P < .05$), while the two highest notes were flatter than the second lowest note (marked by b, $P < .01$). (**b**) Behavioral results for the ignore task. After the pitch shift, nonmusicians were less accurate than singers on the lowest note (marked by $*$, $P < .05$), and they sang the lowest note sharper than the other four notes (marked by $\#$, $P < .01$). Nonmusicians also sang the highest note flatter than the lowest two notes (marked by b, $P < .05$). (**c**) Behavioral results for the compensate task. After the pitch shift, both groups performed similarly on this task.

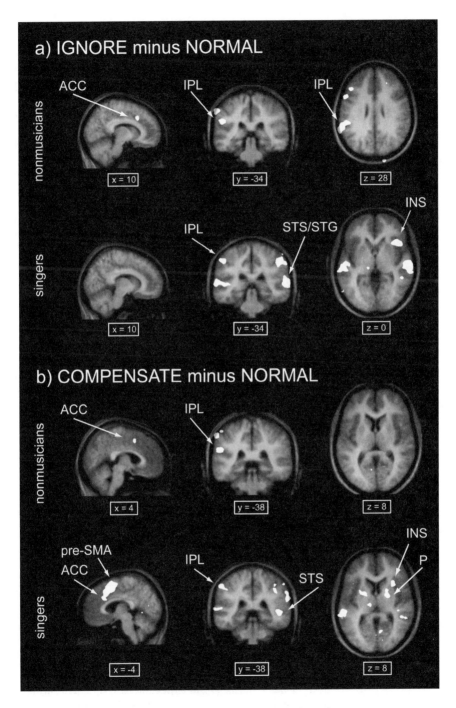

FIGURE 2. *See following page for legend.*

to the region recruited in normal singing) and the inferior parietal lobule (IPL) when compared to normal singing (FIG. 2). The ACC and IPL, which have been associated with response selection[4] and error processing,[5] may have been recruited as nonmusicians perceived the shifted feedback and attempted to produce the appropriate vocal response, depending on the given instruction. By contrast, singers exhibited different patterns of cortical activity between the two pitch-shifted singing tasks. During the ignore task, singers showed increased activity in the IPL, STG, superior temporal sulcus (STS), and right insula, relative to normal singing (FIG. 2a). As with the nonmusicians, the IPL may have been used for error processing. Recruitment of the STG and STS, which is involved in processing voice characteristics,[6,7] may reflect increased perceptual analysis of the incoming signal. For the compensate task, which specifically required audiovocal integration to be performed successfully, singers displayed enhanced activity in the ACC, STS, insula, putamen, pre-SMA, and IPL, compared to normal singing (FIG. 2b). Since enhanced activity was seen in the ACC and insula (areas that were also recruited for normal singing), we suggest that these regions play a special role in audiovocal integration. Since this pattern was only seen in singers, we propose that their training enables them to use the audiovocal integration system more effectively to control vocal pitch.

[Competing interests: The authors declare that they have no competing financial interests.]

REFERENCES

1. MULLER-PREUSS, P., J.D. NEWMAN & U. JURGENS. 1980. Anatomical and physiological evidence for a relationship between the "cingular" vocalization area and the auditory cortex in the squirrel monkey. Brain Res. **202:** 307–315.
2. RIECKER, A., H. ACKERMANN, D. WILDGRUBER, *et al.* 2000. Opposite hemispheric lateralization effects during speaking and singing at motor cortex, insula, and cerebellum. NeuroReport **11:** 1997–2000.
3. PERRY, D.W., R.J. ZATORRE, M. PETRIDES, *et al.* 1999. Localization of cerebral activity during simple singing. NeuroReport **10:** 3979–3984.
4. PICARD, N. & P.L. STRICK. 2001. Imaging the premotor areas. Curr. Opin. Neurobiol. **11:** 663–672.
5. FASSBENDER, C., K. MURPHY, J.J. FOXE, *et al.* 2004. A topography of executive functions and their interactions revealed by functional magnetic resonance imaging. Brain Res. Cogn. Brain Res. **20:** 132–143.
6. BELIN, P., R.J. ZATORRE, P. LAFAILLE, *et al.* 2000. Voice-selective areas in human auditory cortex. Nature **403:** 309–312.
7. KRIEGSTEIN, K.V. & A.-L. GIRAUD. 2004. Distinct functional substrates along the right superior temporal sulcus for the processing of voices. Neuroimage **22:** 948–955.

FIGURE 2. fMRI results for pitch-shifted singing tasks. Images were directly contrasted with images obtained during normal singing to determine which areas were specifically involved in each task. ACC, anterior cingulate cortex; INS, insula; IPL, inferior parietal lobule; P, putamen; pre-SMA, presupplementary motor area; STG, superior temporal gyrus; STS, superior temporal sulcus.

Part VII: Music and the Emotional Brain

Introduction

ISABELLE PERETZ[a] AND JOHN SLOBODA[b]

[a]Department of Psychology, University of Montreal, Montreal, Quebec, H3C 3J7, Canada
[b]Keele University, Keele, United Kingdom

Music experience is intimately related to its emotional appeal.[1] This emotional power is something of a mystery or paradox because of the abstract, nonrepresentational nature of much music. Musical sequences do not refer to specific events in the external world in the way that verbal utterances do.[2] Nevertheless, for most of us, participating in musical activities or simply listening to music produces exquisite pleasure. The question is, Where does the emotional power of musical activities come from?

Many psychologists would argue that we treat music as intentional and personal, emanating from a person (persona, agent) with communicative intentions. One possible origin of the emotional power is the highly emotional vocal communication that is directed to the infant. Juslin and Laukka[3] have come to the conclusion on the basis of a recent metanalysis of behavioral and acoustical studies that listeners perceive music as expressive of emotions because music mimics vocal expression in speech. However, infants prefer singing to speaking,[4] suggesting that music is more powerful for expressing emotions. Interestingly, maternal singing is not only emotionally engaging, it is also highly individuated, suggesting that singing may act as a vocal signature of the mother.[5] From this perspective, musical effects on emotions may be universal and also unique, depending on individual experience.

Indeed, emotional appraisal of music is modulated by prior experience. The "mere exposure effect" on musical taste is a well-established phenomenon. As listeners, we tend to like what we already know. We prefer familiar over unfamiliar music, even after a single prior exposure and often without awareness.[6] This unconscious effect of prior exposure to music may account for a vast array of phenomena, such as the preference for consonance over dissonance[7] and the association of the major and minor modes with happy and sad emotions (e.g., Refs. 8 and 9). Here, Samson and Peretz[10] examine this relationship between music liking and prior exposure in patients with damage to brain structures known to be involved in both memory (e.g., the hippocampus) and emotions (e.g., the amygdala). In doing so, they find that memory and preferences are tightly connected, making music memorable despite the presence of severe memory losses in other domains.

Address for correspondence: Isabelle Peretz, Department of Psychology, University of Montreal, C.P. 6128 succ. Centre-ville, Montreal, Quebec, H3C 3J7, Canada. Voice: 514-343-5840; fax: 514-343-5787.
Isabelle.Peretz@umontreal.ca

Ann. N.Y. Acad. Sci. 1060: 409–411 (2005). © 2005 New York Academy of Sciences.
doi: 10.1196/annals.1360.067

Although our emotional responses to music may be conditioned by memory, the ability to recognize the emotional character of musical excerpts is remarkably consistent across listeners, independently of music education, as reported by Bigand, Filipic, and Lalitte.[11] Furthermore, as these authors point out, the emotional appraisal of subtle structural aspects of music can be immediate. Less than a quarter of a second of music is sufficient to elicit reliable emotional judgments (see also Ref. 12 for converging evidence). Extremely fast-acting processes are typically observed in response to biologically important stimuli. The fact that short musical extracts can arouse emotional responses similarly in every human being with rapid onset and with little awareness qualify such experiences as *reflexes*.

Translated in neural terms, the characterization of musical emotions as reflexes need not be assimilated to subcortical or primitive reflexes. As reviewed by Koelsch,[13] all available neuroimaging data point to the involvement of the neocortex (e.g., the orbitofrontal cortex, an evolutionary recent brain structure), in addition to the classical limbic structures (e.g., the amygdala) in the emotional appraisal of music. As developed further by Koelsch,[13] the contribution of this frontal area is highlighted by the temporal dynamics of music that offer departure points from expectancies, thereby creating tension and relaxation, an emotionally appealing feature of music.[14] The time course of the emotions experienced is an important aspect that can be easily monitored with music, as emphasized in articles in Bigand *et al.*[11] and Koelsch.[13] Yet, dynamic stimuli are rarely used in studies of the emotional brain. Emotions are typically triggered by static stimuli, like the widely used set of images of the International Affective Picture System.[15] Hence, music provides a unique window for studying the operation of the emotional brain at both the cortical and subcortical levels.

The fact that musical emotions can act at the subcortical level, by triggering the limbic system, an evolutionary ancient brain structure, is also of primary importance. Under certain circumstances, music can access neural substrates that are associated with either primary reinforcers, such as food and sex,[16] or with anticipation of danger.[17] Thus, with limbic mediation, consistency, and precociousness, musical emotions resemble other important classes of biological stimuli. Whether music is unique in this respect remains to be seen; it may be one of a class of human constructs that elicits pleasure by co-opting ancient neural systems via inputs from the neocortex. In this respect, music may serve as an excellent paradigm to explore the interactions between neocortically mediated cognitive processes and subcortically mediated affective responses.

The study of musical emotion is still in its infancy. Some of the challenges that still remain include (a) developing reliable methods of distinguishing appraisal of emotion ("this music sounds sad") from experience of emotion ("I feel sad");[18,19] (b) exploring emotional responses in complete musical works, where the nature and intensity of emotion may vary significantly over many minutes;[20] and (c) resolving the dilemmas posed by the fact that many emotional responses to music are specific to the situation in which they are heard, and to the degree of control that the listener has over the music he or she hears.[21] Studies of the musical brain must generally take place in highly controlled and somewhat culturally impoverished environments, limiting the generality of their findings. The challenges that music poses to brain researchers are likely to tax the best brains in the discipline for many years to come.

REFERENCES

1. JUSLIN, P. & J. SLOBODA. 2001. Music and Emotion: Theory and Research. Oxford University Press. Oxford.
2. SLOBODA, J.A. 1998. Does music mean anything? Mus. Sci. **2:** 21–32.
3. JUSLIN, P.N. & P. LAUKKA. 2003. Communication of emotions in vocal expression and music performance: different channels, same code? Psychol. Bull. **129:** 770–814.
4. NAKATA, T. & S. TREHUB. 2004. Infants' responsiveness to maternal speech and singing. Inf. Behav. Dev. **27:** 455–464.
5. BERGESON, T.R. & S.E. TREHUB. 2002. Absolute pitch and tempo in mother's songs to infants. Psychol. Sci. **13:** 72–75.
6. PERETZ, I., D. GAUDREAU & A.-M. BONNEL. 1998. Exposure effects on music preference and recognition. Mem. Cogn. **26:** 884–902.
7. ZENTNER, M.R. & J. KAGAN. 1996. Perception of music by infants. Nature **383:** 29.
8. HEVNER, K. 1935. The affective character of the major and minor modes in music. Am. J. Psychol. **47:** 103–118.
9. DALLA BELLA, S., I. PERETZ, L. ROUSSEAU & N. GOSSELIN. 2001. A developmental study of the affective value of tempo and mode in music. Cognition **80:** B1–10.
10. SAMSON, S. & I. PERETZ. Effects of prior exposure on music liking and recognition in patients with temporal lobe lesions. Ann. N. Y. Acad. Sci. **1060:** 419–428. [This issue.]
11. BIGAND, E., S. FILIPIC & P. LALITTE. The time course of emotional responses to music. Ann. N. Y. Acad. Sci. **1060:** 429–437. [This issue.]
12. PERETZ, I., L. GAGNON & B. BOUCHARD. 1998. Music and emotion: perceptual determinants, immediacy, and isolation after brain damage. Cognition **68:** 111–141.
13. KOELSCH, S. Investigating emotion with music: neuroscientific approaches. Ann. N. Y. Acad. Sci. **1060:** 412–418. [This issue.]
14. MEYER, L.B. 1956. Emotion and Meaning in Music. University of Chicago Press. Chicago.
15. LANG, P.J., M.M. BRADLEY & B.N. CUTHBERT. 1997. International Affective Picture System (IAPS): Technical Manual and Affective Ratings.
16. BLOOD, A.J. & R. ZATORRE. 2001. Intensely pleasurable responses to music correlates with activity in brain regions implicated in reward and emotion. Proc. Natl. Acad. Sci. USA **98:** 11818–11823.
17. GOSSELIN, N., I. PERETZ, M. NOULHIANE, et al. 2005. Impaired recognition of scary music following unilateral temporal lobe excision. Brain **128:** 628–640.
18. GABRIELSSON, A. 2002. Perceived emotion and felt emotion: same or different? Mus. Sci. [special issue 2001–2002: 123–148]
19. SCHERER, K. & M. ZENTNER. 2001. Emotional effects of music: production rules. In Music and Emotion: Theory and Research. P.N. Juslin & J.A. Sloboda, Eds.: 361–392. Oxford University Press. Oxford.
20. SCHUBERT, E. 2001. Continuous measurement of self-report emotional response to music. In Music and Emotion: Theory and Research. P.N. Juslin & J.A. Sloboda, Eds.: 393–414. Oxford University Press. Oxford.
21. SLOBODA, J.A. & S.A. O'NEILL. 2001. Emotions in everday listening to music. In Music and Emotion: Theory and Research. P.N. Juslin & J.A. Sloboda, Eds.: 415–430. Oxford University Press. Oxford.

Investigating Emotion with Music

Neuroscientific Approaches

STEFAN KOELSCH

Max Planck Institute for Human Cognitive and Brain Sciences, Leipzig, Germany

ABSTRACT: This article briefly reviews the few functional imaging studies conducted so far on the investigation of emotion with music. Basically, these studies showed involvement of limbic and paralimbic cerebral structures (such as amygdala, hippocampus, parahippocampal gyrus, temporal poles, insula, ventral striatum, orbitofronal, as well as cingulate cortex) during the processing of music with emotional valence (such as pleasant or unpleasant). The second part of this article highlights the role of unexpected musical events for the elicitation of emotional responses. Recent studies suggest that music-syntactically irregular chords elicit changes in electrodermal activity, and that such chords activate orbital frontolateral cortex, as well as the amygdala (that is, brain structures that have been implicated in emotion processing). The third part of this article mentions findings on the temporal dynamics of emotion (that is, changes in the physiological correlates of emotion processing over time). This issue has so far been mainly neglected in the functional imaging (and psychophysiological) literature.

KEYWORDS: music; emotion; fMRI; PET

NEUROSCIENCE STUDIES ON MUSIC AND EMOTION

So far, the majority of neuroscience studies on human emotion has used static visual images as experimental stimuli. However, during the past years, the neurosciences have discovered that music is also a valuable tool to investigate emotion. Important advantages of music are (1) that music is capable of inducing emotions with a fairly strong intensity, (2) that such emotions can usually be induced quite consistently across subjects,[1,2] and (3) that music can induce not only unpleasant, but also pleasant emotions (which are rather difficult to induce by static images). Neuroscience studies on the investigation of emotion with music basically indicate that networks of limbic and paralimbic structures (such as amygdala, hippocampus, parahippocampal gyrus, insula, temporal poles, ventral striatum, orbitofrontal cortex, and cingulate cortex) are involved in the emotional processing of music. These structures have previously been implicated in emotion, but the functional significance of each of these structures is still not well understood.

Address for correspondence: Stefan Koelsch, Max Planck Institute for Human Cognitive and Brain Sciences. Stephanstr. 1a, 04103 Leipzig, Germany. Voice: +49-341-35521711; fax: +49-341-9940113.
koelsch@cbs.mpg.de

Ann. N.Y. Acad. Sci. 1060: 412–418 (2005). © 2005 New York Academy of Sciences.
doi: 10.1196/annals.1360.034

Using PET, Blood et al.[3] investigated the emotional dimension of pleasantness/ unpleasantness with sequences of harmonized melodies. The stimuli varied in their degree of (permanent) dissonance and were accordingly perceived as less or more unpleasant (stimuli with highest permanent dissonance were rated as the most unpleasant). Stimuli were presented under computerized control without musical expression. This paradigm was not intended to induce the full range of (pleasant) musical mood, yet it allowed Blood and colleagues to examine emotional processing with music while simultaneously excluding effects of musical preference on the perception of the emotional valence of the stimuli. Variations in pleasantness/unpleasantness affected activity in a number of paralimbic structures: increasing unpleasantness of the stimuli correlated with activations of the (right) parahippocampal gyrus, while decreasing unpleasantness of the stimuli correlated with activations of frontopolar, orbitofrontal, and subcallosal cingulate cortex.

A PET study by Brown et al.[4] investigated activations elicited by unfamiliar (pleasant) music (instrumental songs in the rembetika style; only two songs were used, which were, unfortunately, composed by the same Greek composer). Contrasted to a rest condition, pleasant music activated limbic and paralimbic structures, including subcallosal cingulate cortex, anterior insula, the posterior part of the hippocampus, and part of the ventral striatum (possibly nucleus accumbens). Similarly, another PET experiment by Blood and Zatorre[5] measured changes in regional cerebral blood flow (rCBF) during "chills" when participants were presented with a piece of their own favorite music (using normal CD recordings; as a control condition, participants listened to the favorite piece of another subject). Increasing chills intensity correlated with increases in rCBF in brain regions thought to be involved in reward and emotion, including the insula, orbitofrontal cortex, the ventral medial prefrontal cortex, and the ventral striatum. Decreases in rCBF (with increasing chills intensity) were observed in the amygdala and the hippocampus.

Reminiscent of the study from Blood et al.,[3] a recent fMRI study from Koelsch et al.[6] also used pleasant and unpleasant musical stimuli. In contrast to the study from Blood et al.,[3] the pleasant musical excerpts were not computerized sounds, but natural musical stimuli (joyful instrumental dance tunes, recorded from normal CDs). Unpleasant stimuli were electronically manipulated, permanently dissonant counterparts of the original musical excerpts. The stimuli of that study were, thus, not only intended to induce unpleasantness, but also pleasantness in response to the joyful, naturalistic music. The use of identical stimuli across subjects enabled the investigation of emotion independent of personal preferences of listeners.

Unpleasant music elicited increases in blood oxygen level–dependent (BOLD) signals in the amygdala, the hippocampus, the parahippocampal gyrus, and the temporal poles (strong decreases in BOLD signal were observed in these structures in response to the pleasant music). During the presentation of the pleasant music, increases in BOLD signal were observed in the ventral striatum and the insula (and in cortical structures not belonging to limbic- or paralimbic circuits; this activity will not be reported further here).

Involvement of the amygdala in the emotional processing of music has also been reported by a recent lesion study from Gosselin et al.,[7] in which patients with medial temporal lobe resections (including the amygdala) showed impaired recognition of fearful music. Involvement of the ventral striatum (nucleus accumbens) in response to normal music (contrasted to scrambled music) is also reported by a study from Levitin and Menon.[8]

The mentioned studies show that listening to music can elicit activity changes in limbic and paralimbic structures that have previously been implicated in emotion (amygdala, hippocampus, parahippocampal gyrus, insula, temporal poles, cingulate cortex, orbitofrontal cortex, and ventral striatum), and that, thus, functional brain imaging using musical stimuli can potentially contribute to the investigation of emotion.

MUSICAL EXPECTATIONS AND EMOTIONAL RESPONSES

The studies reported in the previous section used experimental paradigms with "pleasant," "unpleasant," "scary," "happy," or "peaceful" tunes. This section deals with the possible role of music-structural aspects for the processing of emotion. Meyer[9] proposed a theory of musical emotions on the basis of fulfilled or suspended musical expectations. He proposed that the confirmation and violation of musical expectations produces emotions in the listener. According to this proposal, Sloboda[10] found that specific musical structures lead to specific psychophysiological reactions, and he showed that new or unexpected harmonies can evoke shivers.

FIGURE 1 shows two chord sequences, ending on a structurally regular, and thus expected, chord (left) and ending on a structurally irregular chord, which is, thus, perceived as unexpected (right). These chord sequences have so far been used in a number of studies (for an overview, see Koelsch and Friederici),[11] which originally aimed at investigating the processing of musical structure (and not the processing of emotion). Interestingly, recent functional imaging experiments using such chord sequences[12] have shown activations of orbital frontolateral cortex (OFLC) in response to unexpected chords. The OFLC (lateral orbital gyrus of BA11, as well as medial inferior frontal gyrus, BA47; for an overview, see Ref. 13) is a paralimbic structure that plays an important role in the processing of emotion: the OFLC has been implicated in the evaluation of the emotional significance of a sensory stimulus[13] and is considered as a gateway for preprocessed sensory information into the (medial) orbitofrontal paralimbic division.[13]

The unexpected chords (which sound odd to listeners familiar with major–minor tonal regularities) violate the sensory expectancies of listeners.[14] As mentioned above, the violation of musical expectancies has been regarded as an important aspect of generating emotions when listening to music.[1,9] Moreover, the perception of irregular chord functions has been shown to lead to an increase of perceived tension,[15] and the perception of tension has been linked to emotional experience dur-

FIGURE 1. Chord sequences ending on a structurally regular chord (*left*) and ending on a structurally irregular chord (*right*) The irregular chord is indicated by the *arrow.*

ing music listening.[1] Thus, the activation of the OFLC points to the possibility that unexpected chords generate emotional responses.

The activation of the OFLC in response to unexpected chords has been replicated in a recent fMRI study from Barbara Tillmann and colleagues (personal communication with BT), in which less related harmonies (subdominant chords presented at the end of eight-chord sequences) elicited an activation of the OFLC (contrasted to regular tonic chords). Similar activations have also been reported by Levitin and Menon[16] in response to normal music contrasted to scrambled music, and it is likely that this activation is due to differences in the emotional valence of natural music on the one side, and of scrambled music on the other.

With a lower statistical threshold, the data obtained in the study from Koelsch et al.[12] also showed bilateral activations of the amygdala in response to the unexpected chords, underlining the assumption that unexpected, or irregular chords, can elicit emotional responses.

A recent study from Steinbeis et al.[17] further tested the hypothesis that unexpected chords generate emotional responses. In that study, physiological measures, including EEG, electrodermal activity (EDA), and heart rate, were recorded while subjects listened to three versions of Bach chorales: one version was the original version composed by Bach with a harmonic sequence that ended on an irregular chord function. The same chord was rendered expected, and very unexpected. The EDA to these three different chord types showed clear differences between the expected and the (very) unexpected chords. Because the EDA reflects activity of the sympathetic nervous system, and because this system is intimately linked to emotional experiences, these data clearly corroborate the assumption that unexpected harmonies elicit emotional responses in normal listeners.

The reported findings show that unexpected musical events often elicit emotional responses, and not only responses related to the processing of the structure of the music (or of other stimulus features that may systematically be perceived as more or less expected). It is, thus, important to be aware of this. Research using stimuli that are systematically more or less expected should ideally assess the emotional valence (and emotional arousal) of the experimental stimuli (even if an experiment is not originally designed to investigate emotion), so that these variables can potentially be used to explain variance in the data.

TIME COURSE OF EMOTION

The intensity of emotions usually changes over time (even if the emotion itself might be the same). Intuitively, it seems plausible that aversive sounds elicit fast emotional responses (although longer durations of such sounds might even increase the degree of unpleasantness), and that especially tender emotions might take a while to unfold. To date, only little is known about the time course of emotional processing and the underlying neural mechanisms.

One of the very few psychophysiological studies that has investigated the time course of emotion was conducted by Krumhansl.[1] In this study, several physiological measures (including cardiac, vascular, electrodermal, and respiratory functions) were recorded while listeners heard musical excerpts chosen to represent one of three emotions (sadness, fear, and happiness). Significant correlations were found between most of the recorded physiological responses and time (measured in one-second intervals from the beginning of the presentation of each musical excerpt). The strongest physiological effects for each emotion type generally tended to increase over time, suggesting that the intensity of an emotional experience is likely to increase over time during the perception of a musical excerpt.

Activity changes of physiological correlates of emotion were also observed in the previously mentioned fMRI study from Koelsch et al.[6] In that study, the (pleasant and unpleasant) excerpts had a duration of about 1 min, and data were not only modeled for the entire excerpts, but also separately for the first 30 s, and for the remaining 30 s to investigate if there are differences in brain activity over time. As described above, activity changes during the entire pieces were observed in response to the (un)pleasant music in a number of limbic and paralimbic structures (amygdala, hippocampus, parahippocampal gyrus, temporal poles, insula, and ventral striatum). When looking at activation differences between the first 30 s and the remaining 30 s, activations of all of these structures, except the hippocampus, were stronger during the second block of the musical excerpts, presumably because the intensity of listeners' emotional experiences increased during the perception of both the pleasant and the unpleasant musical excerpts. This finding corroborates the notion that emotion processing has a temporal dynamic,[1] especially when listening to music (that unfolds over time; see also the study from Ref. 5, in which musical stimuli selected to evoke chills had a duration of 90 s). It remains to be specified why no clear differences between both blocks were observed in the hippocampus. One possibility is that the signal-to-noise ratio was not high enough to yield a substantial difference, and future studies should investigate if this finding can be replicated.

As mentioned before, it is intersting to note that increases of BOLD signals were observed in the amygdala (as well as in the hippocampus, the parahippocampal gyurs, and the temporal poles) in response to the unpleasant music, but clear decreases of BOLD signals were additionally observed in response to the pleasant music. Note that signal changes observed with fMRI (and often also with PET) do not necessarily have to be due to excitatory (post)synaptic processes, as they can as well be due to inhibitory synaptic processes. It is, for example, possible that the BOLD responses observed in the contrast *unpleasant > pleasant* (in the study from Koelsch et al.[6]) originate from inhibitory, rather than from excitatory synaptic activity,[18] and that, thus, the unpleasant stimuli inhibited emotional activity in limbic regions (compared to activity as present during a positive emotional state), rather than activating excitatory processes in those regions.

With respect to this, it is interesting to note that the hippocampus is presumably one of the most sensitive cerebral structures, because it appears to be the only brain structure that can be damaged by traumatic stressors (such as extreme violence; see, e.g., Ref. 19). Thus, inhibition of neural pathways projecting to the hippocampus during the perception of unpleasant stimuli could represent a neural mechanism that serves the prevention from potential damage of hippocampal neurons. In other words, in a number of studies on the investigation of emotion it is well possible that activity changes observed in the amygdala were not due to the generation of fear (or other unpleasant emotions), but rather due to inhibitory processes that reflect a mechanism activated to prevent the hippocampus from traumatization during the exposure to potentially harmful stimuli. In the mentioned experiment from Koelsch et al.[6] it is, thus, possible that the signal increase ("activation") of the amygdala (and the hippocampus) during the unpleasant music actually reflects inhibitory processes. This issue remains to be specified.

With respect to the amygdala, it is also worth mentioning that the amygdala is not a single nucleus but composed of several distinct groups of cells, usually referred to as the lateral, basal, and accessory basal nuclei (which are often collectively termed the *basolateral amygdala*), as well as of several surrounding structures, including the central, medial, and cortical nuclei. These surrounding structures, together with the basolateral amygdala, are often referred to as the *amygdala*, although including the basolateral nucleus with surrounding nuclei into a single entity does not make anatomical sense.[20] Because the spatial resolution of PET and 3-T fMRI is not high enough to differentiate signal changes originating from different nuclei of the amygdala, it can often be problematic to compare, and relate, signal changes within the amygdala between different studies.

In summary, the results of the studies presented in this section suggest that processing of emotion may have a temporal dynamic. Thus, temporal dynamics of emotion (and the underlying neural correlates) could be taken into account for an appropriate description of emotional processes. Therefore, the length of musical stimuli for experiments investigating emotion should be sufficient. However, the details about the time course of emotion (for example, details about how pleasant emotions unfold over time) remain to be specified. One way to do this is to conduct more fine-grained investigations of the activity of the structures involved in emotional processing over time. Information about activity changes over time of the structures implicated in emotion (e.g., information about how activity in one structure affects activity in another) could provide insight into the functional significance of these structures. Note that, so far, most neuroscience studies on emotion have used static visual images when investigating emotion; thus, with respect to the time course of emotion, music is particularly appropriate to investigate this issue.

[Competing interests: The author declares that he has no competing financial interests.]

REFERENCES

1. KRUMHANSL, C.L. 1997. An exploratory study of musical emotions and psychophysiology. Can. J. Exp. Psychol. **51:** 336–353.
2. PANKSEPP, J. 1995. The emotional sources of "chills" induced by music. Music Percept. **13:** 171–207.
3. BLOOD, A.J., R.J. ZATORRE, P. BERMUDEZ & A.C. EVANS. 1999. Emotional responses to pleasant and unpleasant music correlate with activity in paralimbic brain regions. Nat. Neurosci. **2:** 382–387.
4. BROWN, S., M. MARTINEZ & L.M. PARSONS. 2004. Passive music listening spontaneously engages limbic and paralimbic systems. Neuroreport **15:** 2033–2037.
5. BLOOD, A. & R.J. ZATORRE. 2001. Intensely pleasurable responses to music correlate with activity in brain regions implicated in reward and emotion. Proc. Natl. Acad. Sci. **98:** 11818–11823.
6. KOELSCH, S., T. FRITZ, D.Y. VON CRAMON, *et al.* 2005. Investigating emotion with music: an fMRI study. Hum. Brain Mapp. [Epub ahead of print].
7. GOSSELIN, N., I. PERETZ, M. NOULHIANE, *et al.* 2005. Impaired recognition of scary music following unilateral temporal lobe excision. Brain **128:** 628–640.
8. MENON, V. & D.J. LEVITIN. 2005. The rewards of music listening: response and physiological connectivity of the mesolimbic system. Neuroimage **28:** 175–184.

9. MEYER, L.B. 1956. Emotion and meaning in music. University of Chicago Press. Chicago.
10. SLOBODA, J.A. 1991. Music structure and emotional response: some empirical findings. Psychol. Music 19: 110–120.
11. KOELSCH, S. & A.D. FRIEDERICI. 2003. Toward the neural basis of processing structure in music: comparative results of different neurophysiological investigation methods. Ann. N. Y. Acad. Sci. 999: 15–28.
12. KOELSCH, S., T. FRITZ, K. SCHULZE, et al. 2005. Adults and children processing music: an fMRI study. Neuroimage 25: 1068–1076.
13. MEGA, M.S., J.L. CUMMINGS, S. SALLOWAY & P. MALLOY. 1997. The limbic system: an anatomic, phylogenetic, and clinical perspective. In The Neuropsychiatry of Limbic and Subcortical Disorders. S. Salloway, P. Malloy & J.L. Cummings, Eds.: 3–18. American Psychiatric Press. Washington, D.C./London.
14. KOELSCH, S., T.C. GUNTER, A.D. FRIEDERICI & E. SCHRÖGER. 2000. Brain indices of music processing: "nonmusicians" are musical. J. Cogn. Neurosci. 12: 520–541.
15. Bigand, E., R. Parncutt & J. Lerdahl. 1996. Perception of musical tension in short chord sequences: the influence of harmonic function, sensory dissonance, horizontal motion, and musical training. Perc. Psychophys. 58: 125–141.
16. LEVITIN, D.J. & V. MENON. 2003. Musical structure is processed in "language" areas of the brain: a possible role for Brodmann Area 47 in temporal coherence. Neuroimage 20: 2142–2152.
17. STEINBEIS, N., S. KOELSCH & J.A. SLOBODA. Emotional processing of harmonic expectancy violations. Ann. N. Y. Acad. Sci. 1060: 458–462. [This volume.]
18. BUXTON, R.B. 2002. An Introduction to Functional Magnetic Resonance Imaging: Principles and Techniques. Cambridge University Press. Cambridge.
19. BREMNER, J.D. 1999. Does stress damage the brain? Biol. Psychiatry 45: 797–805.
20. DAVIS, M. & P.J. WHALEN. 2001. The amygdala: vigilance and emotion. Mol. Psych. 6: 13–34.

Effects of Prior Exposure on Music Liking and Recognition in Patients with Temporal Lobe Lesions

SÉVERINE SAMSON[a,b] AND ISABELLE PERETZ[c]

[a]Department of Psychology, University of Lille 3, Villeneuve d'Ascq, France

[b]Department of Epilepsy, la Salpêtrière Hospital, Paris, France

[c]Department of Psychology, University of Montreal, Montreal, Canada

ABSTRACT: Prior exposure to music typically increases liking. This manifestation of implicit memory can be dissociated from explicit memory recognition. To examine the contribution of the medial temporal lobe to musical preference and recognition, we tested patients with either left (LTL) or right (RTL) temporal lobe lesions as well as normal control (NC) participants using the procedure of Peretz *et al.* The results in the affect task showed that NC and LTL participants preferred the studied over nonstudied melodies, thereby demonstrating an implicit exposure effect on liking judgments, whereas RTL patients failed to exhibit this effect. Explicit recognition was impaired in both LTL and RTL patients as compared to NC participants. On the basis of these findings, we suggest that RTL structures play a critical role in the formation of melody representations that support both priming and memory recognition, whereas LTL structures are more involved in the explicit retrieval of melodies. Furthermore, we were able to test an amnesic patient (PC) with bilateral lesions of the temporal lobe. In this case, the exposure effect on liking was also absent. However, repeated exposure to melodies was found to enhance both liking and recognition judgments. This remarkable sparing of memory observed through melody repetition suggests that extensive exposure may assist both implicit and explicit memory in the presence of global amnesia.

KEYWORDS: exposure effect; musical liking; temporal lobe; amnesia

INTRODUCTION

Exposure to music may induce various behavioral changes. Thus, it can result in an increase of positive affect toward the presented music. This reliable effect has been observed by social psychologists for more than a century. By using Asian-like music, Meyer in 1903[3] demonstrated the effect of repeated exposure on affect ratings of music, revealing more positive ratings with increasing familiarity (see also Refs. 4 and 5 for consistent findings). Although considerable empirical data have been

Address for correspondence: Séverine Samson, Department of psychology, Université de Lille 3, BP 60 149, 59653 Villeneuve d'Ascq Cedex, France. Voice: +33-3-20-41-64-43.
severine.samson@univ-lille3.fr

Ann. N.Y. Acad. Sci. 1060: 419–428 (2005). © 2005 New York Academy of Sciences.
doi: 10.1196/annals.1360.035

reported in the literature (see Ref. 6 for a review), this phenomenon remains poorly understood, and its cerebral substrate is still unknown.

In Zajonc,[7] the *mere exposure effect* refers to the positive effect of prior exposure on music liking. In the classical paradigm, exposure to unfamiliar material during the study phase is followed by a preference task in which studied and nonstudied stimuli are presented. Typically, participants prefer the studied over nonstudied stimuli, although they may not recognize the studied stimuli. Such dissociation between affect and recognition suggests that the two tasks tap different cognitive processes. Zajonc[7,8] proposed that the mere exposure effect is an expression of a precognitive emotional system that is distinct from memory. However, this interpretation has been questioned by Mandler *et al.*,[9] who demonstrated that the exposure effect is not limited to affect judgments but could also bias other types of nonemotional evaluation (e.g., brightness). This phenomenon may therefore result from perceptual fluency due to repetition. A previous exposure to a stimulus may facilitate its subsequent processing without awareness, as indicated by data obtained in various musical[1,10] and nonmusical domains.[11–16]

MERE EXPOSURE EFFECT IN MUSIC

In music, the positive effect of prior exposure on liking is a robust finding that can be considered an implicit memory phenomenon, as suggested by Peretz, Gaudreau, and Bonnel.[1] In this study, the exposure effects to melodic excerpts on subsequent liking and recognition judgments were studied in a series of experiments. In each experiment, participants incidentally encoded 20 familiar and 20 unfamiliar musical excerpts in the study phase. They were required to judge the familiarity of each musical excerpt. They were subsequently presented with the same material, half of which corresponded to the stimuli that were presented before. In the affect task, the subjects were asked to rate their liking. In the recognition task, they had to rate their recognition. The results showed that familiarity produced memory effects in opposite directions from the affect and the recognition tasks. Memory effects were maximal on affect ratings for unfamiliar melodies, even after a single exposure, whereas recognition memory was better for familiar melodies. The authors also explored the longevity of these memory effects. By varying time delay between study and test, the data demonstrated that affect and recognition judgments decline differently over time. Finally, inducing subjects to attend to surface attributes or to structural features at encoding had little influence on the preference judgments, whereas it had a marked impact on recognition. All these findings provide support for the idea that exposure effects on liking and recognition judgments may reflect implicit and explicit memory mechanisms, respectively.

In a subsequent study, Gaudreau and Peretz[17] investigated the effect of aging on liking and recognition using the exact same paradigm. The results showed that exposure effects on liking did not differ between the old and young participants, the effect being relatively immune to aging. By contrast, the analysis revealed recognition difficulties in the elderly. These data support the notion that the aging process spares implicit memory but impairs explicit recognition, illustrating once more a dissociation between the two memory tasks in neurologically intact individuals. Consistent results were also reported by Halpern and O'Connor[18] in a task composed of a small set of short unfamiliar melodies presented twice. Both young and older adults exhibited exposure effects on liking, whereas recognition was impaired in the aged.

MERE EXPOSURE IN NEUROPSYCHOLOGY

Very few neuropsychological studies have investigated the effects of prior exposure on liking in music.[10,18,19] In 1985, Johnson, Kim, and Risse[10] used Asian music to examine patients suffering from severe explicit memory deficits. For this purpose, they tested amnesic patients with alcoholic Korsakoff syndrome as well as nonalcoholic control participants. In the study phase, subjects listened to six melodies played either 1, 5, or 10 times and tried to categorize each stimulus according to its style, for example, Chinese, American or neither. Following a 5-min retention interval, an equal number of old and new melodies was presented. Participants determined to what extent they liked each melody on a 5-point rating scale. Korsakoff patients as well as normal participants preferred studied over nonstudied musical excerpts, hence showing a robust effect of exposure on liking judgments. As expected, the patients presented a deficit in music recognition. Therefore, these patients demonstrated a memory-biased performance in liking with little recognition, revealing a dissociation between affect and recognition judgments in music.

Convergent results were obtained in Alzheimer disease patients in which explicit memory was also severely impaired.[19] In the latter study, a slightly different procedure was used (i.e., unfamiliar occidental instead of oriental music), and elderly depressed patients were also examined. Unlike Alzheimer patients, depressed patients were unable to develop a positive affective bias in liking judgments for previously heard melodies, although they were able to adequately recognize these melodies. This double dissociation provides further support to the notion that exposure effect on affect can be independent from exposure effect on recognition. However, the results obtained in Alzheimer disease patients were different in the Halpern and O'Connor's[18] study, which found that Alzheimer patients and their age- and education-matched controls had the same the level of recognition performance while only the patients failed to demonstrate an exposure effect on liking. It seems plausible that melodic excerpts presented twice in this latter study were not sufficiently repeated to enhance liking judgments.

MERE EXPOSURE IN PATIENTS WITH UNILATERAL
TEMPORAL LOBE LESIONS

The goal of our study was to test the neuropsychological dissociation between liking and recognition judgments for melodies. To this aim, we tested patients with unilateral medial temporal lobe dysfunction with the paradigm designed by Peretz et al.[1] As previously described, this paradigm offers the opportunity to compare preference and recognition of familiar and unfamiliar melodic excerpts. The experimental paradigm consisted of an initial study phase in which 20 familiar and 20 unfamiliar excerpts were presented. The participants had to determine if the melody was familiar or not. Following a 5-min retention interval, 80 melodic excerpts, including the 40 studied melodies mixed with 40 nonstudied ones, were presented to the participants. The subject's task was to rate how much they liked each melody on a 10-point scale, in which 1 meant "I don't like it," and 10 "I like it a lot." Then, a second task, consisting of 80 melodic excerpts corresponding to the 40 studied stimuli mixed with 40 nonstudied (or new) ones, was presented. The subject's task was to decide whether they had heard the melodic excerpt before using a 10-point scale, in which 1 meant "No, I certainly did not hear this melody in the prior test" and 10 "Yes, I certainly heard it in the prior test." To control for individual differences in the use of response scales, each rating was converted to a z score using the subject's own mean and standard deviation for all the judgments obtained in the liking and the recognition tasks.

Thirty-seven patients with right (RTL = 19) or left (LRL = 18) medial temporal lobe lesions participated in this study. They were tested before or after a surgical treatment of medically intractable epilepsy at La Salpêtrière Hospital (Paris). A group of 16 normal control (NC) participants matched for age, sex, and education were also tested. Patients assessed before surgery presented medial temporal lobe epilepsy associated with lateralized hippocampal atrophy as identified by magnetic resonance imaging (MRI). Patients tested after surgery had undergone a medial temporal lobe resection, including the hippocampal structures and surrounding cortex, and in some patients the excision involved the temporal pole as well. None of the patients presented language disorders or suffered from extratemporal lesions. Language was lateralized in the left hemisphere for all subjects.

According to the literature, we hypothesized a dissociation between the memory effects on liking and recognition judgments in patients with unilateral temporal lobe lesions. Since these patients usually present explicit memory disorders, whereas implicit memory abilities remain relatively spared,[20,21] we predicted a deficit in melody recognition but not in liking judgments. If memory effects on liking were perceptual, a deficit in the affect task was predicted in patients with right temporal lobe lesions, considering the predominant contribution of the right temporal lobe structures in melodic perception.[22–24] Finally, prior knowledge of the melodies was expected to facilitate the recognition of the studied items without increasing their relative preferences, whereas the opposite effect was predicted for unfamiliar melodies. That is, studied unfamiliar melodies should be preferred over nonstudied ones.

As can be seen in FIGURE 1, we found that the liking judgments of the NC, RTL, and LTL groups for the studied and the nonstudied musical excerpts differ $[F (2.50) = 3.96; P < .05]$. As predicted, the NC participants and the patients with LTL lesions preferred the studied over nonstudied material $(P < .005)$, hence exhibiting a mere exposure effect of music on liking judgments. Interestingly, the patients with RTL lesions failed to demonstrate such an effect $(P > .05)$. This lack of exposure effect on liking may indicate that implicit recovery of melodies depends on the integrity of right mesial temporal lobe structures.

In the recognition task, the 10 response categories were dichotomized in two classes. Each rating above 5 was considered an *old* tune response, and each rating below 6 was considered a *new* tune response. The individual recognition ratings were analyzed by computing proportions of correct old tune responses given to studied items (hit rates) minus the proportion of erroneous old tune responses in responses to nonstudied items (false-alarm rates). The results revealed a significant effect of group $[F (2.50) = 3.35; P < .05]$ and of familiarity $[F (1.50) = 6.70; P = .01]$ on the

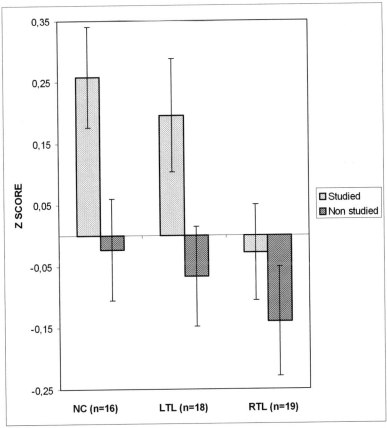

FIGURE 1. Mean z score transform data of the subjects' ratings in the affect task obtained by the three groups of participants (NC, normal control patients; LTL, left temporal lobe patients; RTL, right temporal lobe patients) for the studied and nonstudied musical excerpts. The *bars* correspond to the standard error of the mean.

proportion of hit minus false alarms, although we found no interaction between the two factors [$F (2.50) = 0.22$; ns]. Patients with both RTL and LTL lesions were impaired in melody recognition as compared to NC participants. Thus, both the right and left mesial temporal lobe structures seem involved in explicit retrieval of melodies, as observed in prior studies.[23,25,26] Our results also show that familiar melodies are better recognized than unfamiliar ones. Preexisting material is retained better than novel material, which is in agreement with previously published results obtained in musical[27–29] and visual domains.[30,31] Finally, the lack of correlation between liking and recognition judgments suggests that performance on the implicit task was not accounted for by performance on the explicit task.

Altogether, the results are consistent with the existence of different processes for liking and recognition judgments. Although the affect and the recognition tests employed the same study and test materials, hence providing the same external cues to subjects in both tasks, the manipulation of instructions at the test affected the performance of patients differently with RTL and LTL lesions. Implicit retrieval of melodies can still be preserved despite disrupted explicit retrieval abilities in patients with LTL lesions, whereas both implicit and explicit tasks were impaired in patients with RTL patients.

MERE EXPOSURE EFFECT IN AMNESIA

To further explore the mere exposure effects on liking and recognition in a neuropsychological setting, we tested an amnesic patient with bilateral lesions of the mesial temporal lobes structures. We then pursued the investigation by using another procedure, as did Johnson and collaborators,[10] to allow comparison with the literature.

Case Description

PC is a 43-year-old right-handed man with 7 years of education who suffered from an amnesic syndrome caused by a streptococcus pneumoniae meningitis. He was tested 4 months after the onset of the amnesic syndrome in La Salpêtrière Hospital. The MRI taken at the time of testing revealed a bilateral atrophy of the temporal poles along the anterior part of the first and second temporal gyri, as well as a symmetrical amygdala and hippocampal atrophy involving mainly the head of both hippocampi. A summary of PC's neuropsychological functioning is presented in TABLE 1. Although his full-scale IQ was relatively low, his logical reasoning and executive functions were good in relation to his low level of education. The overall results generally indicate that cognitive function outside the domain of episodic memory was largely, although not completely, preserved. More specifically, PC presented with anterograde amnesia, as illustrated by the very low scores obtained in most learning tasks, and suffered from retrograde amnesia limited to the last couple of years.

Mere Exposure Effect in PC

Since we obtained a deficit in both liking and recognition judgments in patients with RTL lesions, we predicted a similar but more drastic dysfunction in PC. It is important to mention that PC was able to recognize familiar melodies with 90% accuracy (NC = 92%) in the study phase of the test. This result attests to PC's familiarity with the musical material and to preserved perceptual abilities. Yet, PC did not exhibit a mere exposure effect on liking judgments. His rating of studied (z score transform = −0.11) and nonstudied (z score transform = −0.20) unfamiliar melodies did not differ (nonparametric sign test: $z = 0.63$, n.s.). Similar results were obtained with the familiar melodies (with z score transform = 0.86 and 0.75 for the studied and nonstudied melodies; nonparametric sign test: $z = 0.90$, n.s.). In general, PC preferred the familiar (z score transform = 0.81) over the unfamiliar melodies (z score transform = −0.16; nonparametric sign test $z = 3.80$; $P < .005$). Thus, PC failed to exhibit an exposure effect on liking judgments but displayed the general tendency to prefer the highly familiar melodies over the unfamiliar ones, as reported in normal participants.[1]

TABLE 1. Neuropsychological results of the PC case

Tests	PC	Control mean (SD)
Wais-III (standard score)		
Full Scale IQ	74	
Verbal IQ	75	
Performance IQ	77	
Logical reasoning		
PM38 (percentile)	> to 90	
MEM III (standard score)		
General memory score	66	
BEM 144 (battery for memory evaluation)		
Global memory score/144	51	119.72 (12.39)
Auditory verbal score/72	22.5	58.44 (6.93)
Visual nonverbal score/72	28.5	61.28 (7.18)
Verbal memory tests		
Digit span		
Forward	4 digits	
Backward	4 digits	
Rey Auditory Verbal Learning Test		
Recall 1 to 5/15	4, 6, 6, 4, 6	8–14 (1.5)
Interference list recall/15	5	8 (1.8)
6th recall/15	0	13.5 (1.8)
Delayed recall (30 min)/15	1	13.5 (1.9)
Recognition hits/15	8	15 (0.3)
False recognition/35	12	
Visuospatial memory tests		
Visual span		
Forward	6 blocs	
Backward	4 blocs	
Rey-Osterrieth Complex Figure Test		
Copy/36	26	32 (2)
Immediate recall/36	6	
Delayed recall (40 min)/36	6	22 (5)
Executive functions		
Wisconsin Card Sorting Test		
Categories/6	6 in 90 cards	
Trail making test		
Part A	26 seconds	28.7 (8.4)
Part B	91 seconds	58.8 (16.6)
Language		
Denomination (DO80)/80	79	
Word fluency (words/min)		
Semantic	12	19.4 (3.1)
Phonemic	6	14.6 (3.4)
Recognition of emotion		
Ekman face (% correct recognition)	83	91.42 (0.97)

NOTE: WAIS-III, Wechsler Adult Intelligence Scale-Form III; WMS-III, Wechsler Memory Scale–Form III; BEM144, Batterie d'efficience mnésique (Signoret).

In the recognition test, PC produced two different profiles of responses for familiar and unfamiliar musical excerpts. PC recognized the familiar excerpts very poorly (proportion of hit minus false alarm = 0.00), unlike control participants (mean = 0.50; s.d = 0.18), responding that all familiar melodies had been presented earlier. This difficulty to differentiate the studied from the nonstudied items seems to result from a confusion between preexisting and intraexperimental familiarity of the items. Such a difficulty in distinguishing the sources of global familiarity in melody recognition confirms and extends previous findings reported in Alzheimer patients.[27] Conversely, and surprisingly, recognition of unfamiliar melodies was relatively spared in PC (proportion of hit minus false alarm = 0.35), as compared to normal control values (mean = 0.43; SD = 0.26). This result is highly unexpected, given PC's global amnesia syndrome.

PC's dissociation between impaired implicit and spared explicit memory for unfamiliar melodies resembles the pattern of results observed in elderly depressed patients.[19] However, in that study, a different procedure involving repeated exposure of few unfamiliar items was used.

In order to further document this unusual pattern, we subsequently tested PC with the same procedure as the one used with depressed patients. In that particular procedure,[19] only six unfamiliar melodies were presented, two being played one time, two being played five times, and the other two being played ten times. This study phase was followed by two separate tests using different sets of 12 melodic excerpts for the affect and recognition judgments, respectively. With this procedure, PC exhibited a mere exposure effect, with liking ratings being higher for the studied (z score transform = 0.64) than for the nonstudied (z score = −0.06) melodies. This effect was already present after a single exposure (z score = 0.53) and was larger after five (z score = 0.88) and ten (z score = 0.53) exposures. The obtainment of implicit memory effects in PC with this procedure, employing few repeated melodies, suggests that interference resulting from a large number of melodies (40 items), of varying level of familiarity, may have contributed to the lack of exposure effect observed with the previous procedure.[1] However, it remains difficult to understand why such manipulations have affected implicit rather than explicit memory.

Another remarkable finding of this study is that PC, who suffers from severe amnesia, exhibits spared recognition, with a proportion of hit minus false responses of 0.50. Thus, formation of new representations that support explicit recognition seems possible in PC. It remains to be determined to what extend this remarkable sparing is related to the use of musical material.

CONCLUSION

The results obtained in patients with unilateral temporal lobe lesions in the affect task showed that NC and LTL participants preferred the studied over nonstudied melodies, thereby demonstrating an implicit exposure effect on liking judgments, whereas RTL patients failed to exhibit this effect. Explicit recognition was impaired in both LTL and RTL patients, as compared to NC participants. On the basis of these findings, we suggest that RTL structures play a critical role in the formation of melody representations that support both priming and memory recognition, whereas LTL structures are more involved in the explicit retrieval of melodies. The evaluation of an amnesic patient (PC) with bilateral lesions of the temporal lobe provided additional results. Using the same procedure, we found that the exposure effect on liking was also absent, whereas recognition of unfamiliar melodies was preserved. However, repeated exposure of melodies was found to enhance both liking and recognition judgments in a subsequent study. This remarkable sparing of memory observed through melody repetition suggests that extensive exposure of study items in the context of little interfering material may assist both implicit and explicit memory in the presence of global amnesia, emphasizing an idea already proposed by Tulving and his collaborators.[2,32] This finding as well as previous data published in the literature may indicate that exposure effect on liking depends not only on the etiology of the brain anomaly but also on the conditions under which this effect occurs. The frequency of exposure and the interference resulting from other study items and pre-existing familiarity of the melodies should be independently manipulated in future neuropsychological studies in order to better understand the nature of the memory representations tapped by preference tasks.

ACKNOWLEDGMENTS

We are indebted to Marisa Denos, Sophie Dufour, and Nathalie Gosselin for testing the patients, and to Michel Baulac from la Salpêtrière Hospital (epilepsy unit) for facilitating access to the patients. This work was supported by a grant from the Natural Science and Engineering Research Council of Canada to Isabelle Peretz.

[Competing interests: The authors declare that they have no competing financial interests.]

REFERENCES

1. PERETZ, I., D. GAUDREAU & A.M. BONNEL. 1998. Exposure effects on music preference and recognition. Mem. Cogn. **26:** 884–902.
2. TULVING, E., C.A. HAYMAN & C.A. MACDONALD. 1991. Long-lasting perceptual priming and semantic learning in amnesia: a case experiment. J. Exp. Psychol. Learn. Mem. Cogn. **17:** 595–617.
3. MEYER, M. 1903. Experimental studies in the psychology of music. Am. J. Psychol. **14:** 456–478.
4. MOORE, H.T. & A.R. GILLILAND. 1924. The immediate and long term effects of classical and popular phonograph selections. J. Appl. Psychol. **8:** 309–323.
5. WASHBURN, M.F., M.S. CHILD & T.M. ABEL. 1927. The effects of immediate repetition on the pleasantness or unpleasantness of music. *In* The Effects of Music. M. Schoen, Ed.: 199–210. Harcourt, Brace. New York.
6. BORNSTEIN, R.F. 1989. Exposure and affect: overview and meta-analysis of research, 1968–1987. Psychol. Bull. **106:** 265–289.
7. ZAJONC, R. 1968. Attitudinal effects of mere exposure. J. Personality Soc. Psychol. Monographs. **9:** 1–28.
8. ZAJONC, R. 1980. Feeling and thinking: preferences need no inferences. Am. Psychol. **35:**151–175.
9. MANDLER, G., Y. NAKAMURA & B. VAN ZANDT. 1987. Nonspecific effects of exposure on stimuli that cannot be recognized. J. Exp. Psychol. Learn. Mem. Cogn. **13:** 646–648.
10. JOHNSON, M., J.K. KIM & G. RISSE. 1985. Do alcoholic Korsakoff's syndrome patients acquire affective reactions? J. Exp. Psychol. Learn. Mem. Cogn. **11:** 22–36.

11. GREVE, K.W. & R.M. BAUER. 1990. Implicit learning of new faces in prosopagnosia: an application of the mere-exposure paradigm. Neuropsychologia 28: 1035–1041.
12. SCHACTER, D.L. 1987. Implicit expressions of memory in organic amnesia: learning of new facts and associations. Hum. Neurobiol. 6: 107–118.
13. SEAMON, J.G., N. BRODY & D.M. KAUFF. 1983. Affective discrimination of stimuli that are not recognized: II. Effect of delay between study and test. Bull. Psychol. Soc. 21: 187–189.
14. SEAMON, J.G. et al. 1997. A mere exposure effect for transformed three-dimensional objects: effects of reflection, size, or color changes on affect and recognition. Mem. Cogn. 25: 367–374.
15. SQUIRE, L.R. 1992. Memory and the hippocampus: a synthesis from findings with rats, monkeys, and humans. Psychol. Rev. 99: 195–231.
16. TOBIAS, B., J. KIHLSTROM & D.L. SCHACTER. 1992. Emotional and implicit memory. In The Handbook of Emotion and Memory: Research and Theory. S. Christanson, Ed.: 67–92. Erlbaum. New Jersey.
17. GAUDREAU, D. & I. PERETZ. 1999. Implicit and explicit memory for music in old and young adults. Brain Cogn. 40: 126–129.
18. HALPERN, A.R. & M.G. O'CONNOR. 2000. Implicit memory for music in Alzheimer's disease. Neuropsychology 14: 391–397.
19. QUONIAM, N. et al. 2003. Implicit and explicit emotional memory for melodies in Alzheimer's disease and depression. Ann. N. Y. Acad. Sci. 999: 381–384.
20. BLAXTON, T.A. 1992. Dissociations among memory measures in memory-impaired subjects: evidence for a processing account of memory. Mem. Cogn. 20: 549–562.
21. ZAIDEL, D.W., S.M. OXBURY. & J.M. OXBURY. 1994. Effects of surgery in unilateral medial temporal lobe regions on verbal explicit and implicit memory. Neuropsych. Behav. Neurol. 7: 104–108.
22. MILNER, B. 1962. Laterality effects in audition. In Interhemispheric Relations and Cerebral Dominance. V. Mountcastle, Ed.: 173–201. John Hopkins University Press. Baltimore.
23. SAMSON, S. & R.J. ZATORRE. 1988. Melodic and harmonic discrimination following unilateral cerebral excision. Brain Cogn. 7: 348-360.
24. ZATORRE, R.J. 1985. Discrimination and recognition of melodies after unilateral cerebral excisions. Neuropsychologia 23: 31–41.
25. SAMSON, S. 1999. Musical function and temporal lobe structures: a review of brain lesion studies. J. New Music Res. 28: 217–228.
26. SAMSON, S. & R.J. ZATORRE. 1992. Learning and retention of musical and verbal information after unilateral temporal lobectomy. Neuropsychologia 30: 815–826.
27. BARTLETT, J.C., A.R. HALPERN & W.J. DOWLING. 1995. Recognition of familiar and unfamiliar melodies in normal aging and Alzheimer's disease. Mem. Cogn. 23: 531–546.
28. JAVA, R., Z. KAMINSKA & J. GARDINER. 1995. Recognition memory and awareness for famous and obscure musical themes. Eur. J. Cogn. Psychol. 7: 41–53.
29. PERETZ, I. 1996. Can we lose memories for music? The case of music agnosia in a non-musician. J. Cogn. Neurosc. 8: 481–496.
30. ELLIS, H., J. SHEPERD & G. DAVIS. 1979. Identification of familiar and unfamiliar faces from internal and external features: some implications for theories of face recognition. Perception. 8: 431–439.
31. KLATZKY, R.L & F.H. FORREST. 1984. Recognizing familiar and unfamiliar faces. Mem. Cogn. 12: 60–70.
32. ROSENBAUM, R. et al. 2005. The case of K.C.: contributions of a memory-impaired person to memory theory. Neuropsychologia 43: 989–1021.

The Time Course of Emotional Responses to Music

EMMANUEL BIGAND, SUZANNE FILIPIC, AND PHILIPPE LALITTE

LEAD-CNRS, Université de Bourgogne, Dijon

ABSTRACT: Two empirical studies investigate the time course of emotional responses to music. In the first one, musically trained and untrained listeners were required to listen to 27 musical excerpts and to group those that conveyed a similar emotional meaning. In one condition, the excerpts were 25 seconds long on average. In the other condition, excerpts were as short as 1 second. The groupings were then transformed into a matrix of emotional dissimilarity that was analyzed with multidimensional scaling methods (MDS). We compared the outcome of these analyses for the 25-s and 1-s duration conditions. In the second study, we presented musical excerpts of increasing duration, varying from 250 to 20 seconds. Participants were requested to evaluate on a subjective scale how "moving" each excerpt was. On the basis of the responses given for the longer duration, excerpts were then sorted into two groups: highly moving and weakly (or less) moving. The main purpose of the analysis was to identify the point in time where these two categories of excerpts started to be differentiated by participants. Both studies provide consistent findings that less than 1 s of music is enough to instill elaborated emotional responses in listeners.

KEYWORDS: emotional responses; Mozart; acoustic structure; cultural stimuli

INTRODUCTION

Music is a complex acoustic and temporal structure that induces a large variety of emotional responses in listeners. The adaptive value of these emotional responses to music remains a matter of debate,[1] and several arguments support the idea that emotional responses to music rest on a complex psychological and neurophysiological architecture that could be specifically devoted to music processing. If music was a simple by-product of evolution (some kind of "cheesecake" as suggested by Pinker[2]), emotional responses to music would certainly not be as elaborated as they are in human communities. Understanding the psychological and neurophysiological foundation of these responses is an issue of importance for the cognitive neurosciences, as is shown by the increasing number of studies investigating emotional responses to music at both the behavioral level[3–8] and neurophysiological level.[9–11]

Address for correspondence: E. Bigand, LEAD-CNRS, 2, Esplanade Erasme, Pole AAFE, Université de Bourgogne. F-21000 Dijon, France. Voice: +33-380-395-781; fax: +33-380-395-757.

bigand@u-bourgogne.fr.

Ann. N.Y. Acad. Sci. 1060: 429–437 (2005). © 2005 New York Academy of Sciences.
doi: 10.1196/annals.1360.036

FOUR ISSUES FOR THE PSYCHOLOGICAL
STUDY OF MUSICAL EMOTION

Psychological approaches should contribute to, at least, four main issues. First of all, the nature of emotions induced by music should be described in detail. Preliminary empirical investigations have demonstrated that basic emotions, such as happiness, anger, fear, and sadness, can be recognized in, and induced by, musical stimuli in adults and in young children. These studies converge to demonstrate a strong consistency among participants, as long as musical excerpts are chosen to convey very basic emotions.

The conclusion that music induces three or four basic emotions is, however, far from compelling for music theorists, composers, and music lovers. Indeed, they are likely to underestimate the richness of the emotional reactions to music that may be experienced in real life. An alternative approach is to stipulate that musical emotions evolve in a continuous way along two or three major psychological dimensions. A lot of previous research has established a bidimensional structure of emotion for facial expressions,[12,13] voice perception,[14] and affect words.[15,16] In the music domain, an attempt to model the perception of expressive content of Western music in multidimensional space has been developed,[16-21] and specific neurophysiological reactions associated with the arousal and valence dimensions have also been reported.[22,23] In Wedin[24-26] musical excerpts were rated on a number of semantic scales defined by emotionally colored adjectives. Factor analysis was then used to extract the dimensions accounting for the evaluation.

More recently, Canazza, De Poli, Vidolin, and Zanon[27] presented participants with an excerpt from Mozart's Concerto for Clarinet, K622, that was interpreted and recorded with different levels of expressiveness. Participants were asked to use a set of sensorial adjectives (e.g., light, soft, dark, hard) to express the emotion they felt the excerpt possessed. Using multidimensional and factorial analyses, they obtained a semantic bidimensional space of musical expressiveness in which two dimensions appeared: one called *kinematics* that corresponded to the tempo parameter, and another called *energy* related to the intensity parameter. Such a kinematics–energy space provides useful information on how expressiveness is organized in the listener's mind. Interestingly, this study indicates that kinematics aspects could be a main determinant of participant judgments on musical expressiveness.

A common feature of these studies was to investigate emotional responses to music using linguistic labels. The use of verbal labels is potentially problematic since it can encourage participants to simplify what they actually experience.[28] Some philosophers even argued that musical emotions are definitely ineffable.[29,30] As nicely coined by the composer Felix Mendelssohn, the emotion induced by music may be so rich that even language may not be able to account for it. If this is the case, using linguistic labels in empirical research could result in missing important aspects of musical emotion. We will report below a method to investigate the emotions conveyed by musical pieces without using linguistic responses.

A second important issue of the psychological approach is to characterize the factors that contribute to musical emotion. A broad distinction of extra- and intramusical factors has been proposed. Extramusical factors designate the features inherent to both the inner state of the subject and the social context in which he/she stands that contribute to emotional responses. Listening to music alone, with partners, or within an experimental setting, for example, may have considerable influence on the perceived emotion. Similarly, being in a positive or in a negative mood when going to a concert may drastically change the emotional experience. By contrast, inner musical factors designate the features in the auditory signal that induce emotions. Some psychoacoustical parameters, such as loudness or roughness, obviously have an quasi-immediate impact on the nature of perceived emotion (with negative valence of emotion being generally correlated with higher roughness). Other more cultural factors linked to the compositional processes also play an important role in modulating the variety of emotions we may experience. These factors deal with harmonic processes (modulation, ornamentation, tonal tension, and relaxation), or rhetoric processes (theme elaboration and variations, violation of expectancies).[31,32] It is likely that emotions induced by a given piece are derived from the way both type of factors (psychoacoustics and cultural factors) are combined in the performance. That is to say, a sudden change in loudness (or in timbre) per se is not crucial for emotion: the way this change occurs in the musical process is, however, of central importance.

A third important issue for the psychological approach would be to describe the timing of emotional response to music. Emotion is a dynamic process that evolves through musical time. Tracking these changes would illuminate how sophisticated the processes governing emotions are. Recently, provocative findings relative to this issue have been reported by authors showing that emotional responses to music may occur very fast. A previous study by Watt and Ash[33] showed that 3- to 5-second extracts of orchestral interludes from Wagner operas were quite sufficient to generate consistent, and emotionally relevant, categorizations in listeners. Peretz *et al.*[32] went one step further by showing that emotional responses rest on very fast-acting mechanisms, so that 250 ms of music may be enough to distinguish happy from sad excerpts. This suggests that some basic emotions may function without cortical mediation. Independence between the affective and cognitive systems has been demonstrated in the visual domain.[34] Cerebral lesions have been found to alter the recognition of facial expressions while the ability to identify faces was spared. In the field of music, I.R., a brain-damaged patient who suffers from considerable deficits in music cognition, was shown to be able to discriminate between happy and sad excerpts while being unable to determine whether the music was familiar or not (Peretz *et al.*[32]) Moreover, I.R. performed similarly to controls when differentiating sad from happy excerpts of extremely short duration (500 ms). This suggests that severe deficits in the cognitive processing of music may leave emotional responses unimpaired.

Finally, the fourth issue of the psychological approach to musical emotion is to delineate the potential influence of musical expertise. There is no doubt that emotion is the core of musical experience, and it is what motivates people to listen to music and to produce it. As such, it is of crucial importance to investigate how emotional experience may change with musical expertise. Surprisingly, there is no clear *a priori* predication about this issue. On the one hand, we may assume that musical expertise results in a greater ability for musically trained listeners to process very subtle changes in auditory musical signals that may be highly relevant for emotion. Research on auditory plasticity in musicians provided a lot of empirical arguments showing that musicians' brains differ from the brains of nonmusicians and that this difference reveals the greater ability of the former group to process subtle changes in pitch or timbre (see papers on plasticity, this volume). Accordingly, important differences in the emotional experience to music should be reported. On the other hand, several studies pointed out that musicians and nonmusicians do not differ a lot when processing musical structures of critical importance for Western music.[35–37] According to these studies, musically untrained listeners implicitly process subtle changes in musical structure, as trained listeners do. This finding challenges the idea that the emotional experience of both groups should differ importantly. Moreover, if we accept the possibility that music serves important evolutionary functions, such as regulating emotion in human communities, then we should also assume that musical competence is largely shared by human beings and does not require a complex explicit training. Ethnomusicological studies provide a large set of evidence along this line.

TRACKING THE TIME COURSE OF EMOTIONAL RESPONSES TO MUSIC

Our two studies focus more specifically on the timing of emotional experience in music but also deal more or less directly with the other issues. The first study involved the multidimensional scaling of emotional responses to musical excerpts that were 25 seconds and 1 second long. A complete account of this study will appear in Bigand et al.[38] The second study adapts a gating paradigm to the study of musical emotion.[39]

In the former study, participants were encouraged to focus entirely on their emotional experience of the musical excerpts. A sample of 27 musical excerpts of serious nonvocal music was selected by music theorists and psychologists according to several constraints.[a] All excerpts were expected to convey a strong emotional experience. They were chosen to illustrate a large variety of emotions and to be representative of key musical periods of Western classical music (baroque, classic, romantic, and modern) as well as of the most important instrumental groups (solo, chamber music, orchestra). This final constraint is of methodological importance, since it helps to neutralize any confounding effect between the structural surface similarity and the emotional similarity of excerpts.

Participants were presented with a random visual pattern of 27 loudspeakers, representing the 27 excerpts. They were first required to listen to all the excerpts and to focus their attention on their emotional experience. They were then asked to look for excerpts that induced similar emotional experiences (whatever they might be) and to drag the corresponding icons to group these excerpts. This task encouraged participants to focus on induced emotions and not to recognize the emotional information encoded in the music. They were allowed to listen to the excerpts as many times as they wished, and to group together as many excerpts as they wished. A simple rule then transformed the partition obtained into a dissimilarity matrix: stimuli grouped together were set with a dissimilarity of 0, while everything else was set to 1. The multidimensional scaling (MDS) method was then used to analyze the psychological dimensions underlying this matrix. The critical point of this study was to

---

[a]Musical stimuli are available at http://www.u-bourgogne.fr/LEAD/people/bigand.html.

compare two experimental conditions: in one condition, excerpts lasted 25 seconds on average. In the other condition, excerpts lasted 1 second. We expected that this drastic length change would have a profound impact on the outcome of the MDS. Moreover, the experiment was run with musically trained and untrained listeners in order to assess the potential influence of training on emotional experience.

Here are the main important outcomes. First, emotional response was found to be very consistent within participants, as well as between participants. Second, we found that a three-dimensional solution accounted for a substantial part of the emotional responses: the vertical axis separates musical excerpts that vary by their arousal level. The horizontal axis presumably separates musical excerpts that differ by their emotional valence. The third axis tends to separate pieces with broad and regular melodic contours from those that proceed harmonically or by broken arpeggios. A possible interpretation is that this dimension expresses the influence body gestures evoked by musical excerpts might have on perceived emotions. The link between music and movement is well established, and several authors have emphasized that musical affects arise in large part from their relationship to physical patterns of posture and gesture.[40] Given that a considerable amount of emotional experience is presumably embedded in cortical sensory motor maps,[41] the evocation of gestures by music could have relevant influence on listeners' emotional experiences.

The geometrical solution found for musicians and nonmusicians was not distinguishable. Both groups of musicians and nonmusicians produced an equal number of clusters, and their matrices of emotional similarity were highly correlated. This weak difference is all the more surprising given the complexity of the musical stimuli used. For the present purpose, the critical point of the study was to find that the outcome of the MDS is weakly influenced by the duration of the excepts: that is to say, the structure of the emotional experience, as revealed by MDS, did not apparently differ between the 1-s and the 25-s conditions. Reducing considerably the length of the musical excerpts (25 seconds to 1 second) only had a weak effect on emotional responses. The present data thus extends the findings of Peretz *et al.*[42] to a larger set of complex musical stimuli. At first glance this finding may sound extremely surprising. In half of the cases, the first second of our excerpts contained a single chord or pitch interval (excerpts 1, 3, 4, 7, 8, 9, 10, 17, 21) and sometimes even a single tone (excerpts 5, 6, 19). What factors may contribute to the emotional experience in this case? A cautious analysis of these one-tone excerpts suggests that performance cues are enough to induce emotions in Western listeners, even at this extremely short duration. The importance of performance for musical expressiveness is well established.[8,18] The way the first tone of a piece of a given expression should be played is highly constrained. Expert performers shape the amplitude and the spectral envelope of the first tone of a piece in a way that prefigures the main mood of the piece. Acculturated listeners have probably internalized these regularities of Western performances (implicitly for nonmusicians). Thanks to this knowledge, they may experience an emotion as soon as the first tone of a piece is played (at least when the performer is highly skilled). In short, we suggest that cognitive appraisal occurs with these extremely short excerpts and that it immediately induces emotional feelings in listeners.

It may be argued that participants in our experiments may have identified the musical expression of the piece without being really moved by the piece. Our experimental paradigm encouraged them to focus on what they experienced, but this did not insure that their response actually reflected this instruction, notably for very short excerpts. Several control experiments were done to address this issue. In one of them, we slightly modified the task and asked participants to evaluate on a subjective scale how much they were moved by the musical excerpts. The critical manipulation consisted of presenting musical excerpts by slices of increasing duration. That is to say, participants started the experiment by listening to 250-ms excerpts and then continued the task with excerpts of 500 ms, 1 s, 2 s, 5 s, and 20 s (on average). One group of musically trained and untrained participants worked with excerpts from classical (serious) music, another group with a sample of excerpts for pop/rock music. Excerpts were initially chosen so that half of them were likely to be perceived as strongly moving and the other half as being less moving. In addition, half of the musical excerpts had high dynamic contrast, and the other half had low dynamic contrast. In order to explain the task to participants, we encouraged them to imagine that they were looking for moving (emotional) music on a radio. Shifting from one radio station to another would be an ecological situation that was similar to the experiment. The critical question of the experiment was to specify how many milliseconds they needed to decide whether the excerpts they were listening to were highly moving.

To analyze the data, we first considered the responses given to the longest excerpts. On the basis of these responses, we sorted the excerpts into two groups: those that were perceived as highly moving by the participants, and the other with those that were perceived as less moving. The critical point was to assess the point in time when participants became able to differentiate significantly between these two groups of excerpts. We were wondering whether this minimum time to feel an emotion would vary as a function of musical style (classic versus pop/rock), as function of dynamics (low versus high), or as a function of musical expertise (musically trained versus untrained listeners). The outcome was highly consistent for all groups of listeners and for both styles: statistically significant differences between these two groups of excerpts were found by 250 ms and notably for the pieces with a low dynamic contrast. However, there was neither an effect for musical style nor for musical expertise. This finding suggests that 250 ms of music may be enough to induce strong or weak feelings of emotion in listeners, whatever the musical style being played.

CONCLUSION

The present data lead to several conclusions. First, they demonstrated that refined emotional responses to music occur from the very beginning of music listening. This finding is consistent with a number of others in the domain of emotion: it has been shown that responses to emotional stimuli such as human faces, human body gestures, or other stimuli of biological importance occur extremely fast. It is remarkable, however, that a highly cultural stimulus such as music, which seems to have no clear adaptive value (at least for some authors), can trigger emotions in such a fast-acting way. Further research is needed to settle the issue of whether the speed of emotional responses is a general characteristic of human emotions that could be confirmed irrespective of the type of stimuli used, or whether it is restricted to stimuli of evolutionary importance for human species.

Second, the present findings suggest that emotional responses to very short musical stimuli presumably involved cortical mediation. An analysis of the musical and psychoacoustical structures of very short excerpts (in both experiments) suggests that emotions are likely to be governed by features in both compositional structure (harmony) and performances that all are highly cultural. Musical emotions induced by very short excerpts are too refined to be simply derived from basic emotional properties of sound. We argue that these responses required a cognitive appraisal. From this point of view, our conclusion differs from Peretz *et al.*,[43] who considered that fast emotional responses to music may be viewed as the product of some subcortical reflexes.

Taken in combination with other findings reported on music cognition that show that cognitive processing of subtle musical structure occurs extremely fast,[44] the present findings provide evidence that both cognitive and emotional processes are very fast-acting processes that seem to occur automatically in acculturated listeners. The fact that these findings were obtained for the musically trained as well as for the musically untrained provided further evidence that music is a highly relevant sound structure of the environment, and that processing it does not require an intensive explicit training.

Finally, our data suggest that emotional responses are quasi-immediate as soon as music is played. Of course this does not mean that musical emotion does not change as music goes by. On the contrary, it is likely that emotional experiences accumulated from the very beginning of the piece contribute to color and to intensify the emotions experienced later.

[Competing interests: The authors declare that they have no competing financial interests.]

REFERENCES

1. CROSS, I. 2003. Music cognition, culture, and evolution. *In* The Cognitive Neuroscience of Music. I. Peretz & R. Zatorre, Eds.: 42–56. Oxford University Press. New York.
2. PINKER, S. 1999. How the Mind Works. W. W. Norton & Company. New York.
3. GABRIELSON, A. & P.N. JUSLIN. 1996. Emotional expression in music performances: between the performer's intention and the listener's experience. Psychol. Mus. **24:** 68–91.
4. GABRIELSON, A. & P.N. JUSLIN. 2003. Emotional expression in music. *In* Handbook of affective sciences. R. J. Davidson, K. R. Scherer & H. H. Goldsmith, Eds.: 503–535. Oxford University Press. New York.
5. ROBINSON, J. 1997. Music and Meaning. Cornell University Press. Ithaca, NY.
6. SLOBODA, J.A. & P.N. JUSLIN. 2001. Music and Emotion: Theory and Research. Oxford University Press. New York.
7. GABRIELSSON, A. 2001. Emotions in strong experiences with music. *In* Music and Emotion: Theory and Research. J.A. Sloboda & P. N. Juslin, Eds.: 431–449. Oxford University Press. New York.
8. GABRIELSON, A. & E. LINDSTROM. 2001. The influence of musical structure on emotional expression. *In* Music and Emotion: Theory and Research. J. A. Sloboda & P. N. Juslin, Eds.: 223–249. Oxford University Press. New York.
9. BARTLETT, D. 1999. Physiological responses to music and sound stimuli. *In* Handbook of Music Psychology. D. Hodges, Ed.: 343–385. IMR Press. San Antonio, Texas.

10. PERETZ, I. 2001. Listen to the brain: a biological perspective on music and emotion. *In* Music and Emotion: Theory and Research. J.A. Sloboda & P.N. Juslin, Eds.: 105–135. Oxford University Press. New York.
11. SCHERER, K.R. & M.R. ZENTNER. 2001. Emotional effects of music productions rules. *In* Music and Emotion: Theory and Research. J.A. Sloboda & P.N. Juslin, Eds.: 361–392. Oxford University Press. New York.
12. ABELSON, R.P. & V. SERMAT. 1962. Multidimensional scaling of facial expressions. J. Exp. Psychol. **63:** 546–554.
13. OSGOOD, C.E. 1966. Dimensionality of the semantic space for communication via facial expressions. Scand. J. Psychol. **7:** 1–30.
14. GREEN, R.S. & N. CLIFF. 1975. Multidimensional comparisons of structures of vocally and facially expressed emotions. Percept. Psychophys. **17:** 429–438.
15. RUSSEL, J.A. 1978. Evidence of convergent validity on the dimensions of affect. J. Pers. Soc. Psychol. **36:** 1152–1168.
16. RUSSELL, J.A. 1980. A circumplex model of affect. J. Pers. Soc. Psychol. **39:** 1161–1178.
17. HEVNER, K. 1935. The affective character of the major and minor modes in music. Am. J. Psychol. **47:** 103–118.
18. JUSLIN, P.N. 2001. Communicating emotion in music performance: a review and a theoretical framework. *In* Music and Emotion: Theory and Research. J.A. Sloboda & P.N. Juslin, Eds.: 309–337. Oxford University Press. New York.
19. SCHUBERT, E. 1996. Enjoyment of negative emotions in music: an associative network explanation. Psychol. Mus. **24:** 18–28.
20. SCHUBERT, E. 2004. Modeling perceived emotion with continuous musical features. Mus. Percept. **21:** 561–585.
21. MADSEN, C.K. 1997. Emotional response to music as measured by the two-dimensional CRDI. J. Mus. Ther. **34:** 187–199.
22. SCHMIDT, L.A. & L.J. TRAINOR. 2001. Frontal brain electrical activity distinguishes valence and intensity of musical emotions. Cogn. Emot. **25:** 487–500.
23. TSANG, C., L. TRAINOR, D. SANTESSO, *et al.* 2001. Frontal EEG responses as a function of affective musical features. *In* The Biological Foundations of Music. R. Zatorre & I. Peretz, Eds.: 930: 439–442. Annals of the New York Academy of Science. New York.
24. WEDIN, L. 1969. Dimension analysis of emotional expression in music. Swed. J. Musicol. **51:** 119–140.
25. WEDIN, L. 1972a. Multidimensional scaling of emotional expression in music. Swed. J. Musicol. **51:** 1–43.
26. WEDIN, L. 1972b. A multidimensional study of perceptual-emotional qualities in music. Scand. J. Psychol. **13:** 241–257.
27. CANAZZA S., G. DE POLI, A. RODÀ, *et al.* 2001. Kinematics-energy space for expressive interaction in music performance. In Proc. of MOSART; workshop on current research directions in Computer Music. 35–40. November 15–17. Barcelona.
28. SCHERER, K.R. 1994. Affect bursts. *In* Emotions: Essays on Emotion Theory. S. van Goozen, N. E. van de Poll & J. A. Sergeant, Eds.: 161–196. Erlbaum. Hillsdale, NJ.
29. JANKÉLÉVITCH, V. 1974. Fauré et l'inexprimable, Paris, Plon.
30. JANKÉLÉVITCH, V. 1983. La musique et l'ineffable. Paris, Seuil. English translation by C. Abbate, Music and the Ineffable. 2003. Princeton University Press.
31. SLOBODA, J.A. 1991. Music structure and emotional response: some empirical findings. Psychol. Mus. **19:** 110–120.
32. PERETZ, I., L.GAGNON & B. BOUCHARD. 1998. Music and emotion: perceptual determinants, immediacy, and isolation after brain damage. Cognition **68:** 111–141.
33. WATT, R.J. & R.L. ASSH. 1998. A psychological investigation of meaning in music. Mus. Sci. **2:** 33–54.
34. LEDOUX, J.E. 2000. Emotion circuits in the brain. Annu. Rev. Neurosci. **23:** 155–184.
35. BIGAND, E. & B. POULIN-CHARRONNAT. Submitted. Are we all "experienced listeners"? Cognition
36. BIGAND, E. 2003. More about the musical expertise of musically untrained listeners. Ann. N. Y. Acad. Sci. **999:** 304–312.

37. TILLMANN, B., J. BHARUCHA & E. BIGAND. 2000. Implicit learning of tonality: a self-organizing approach. Psychol. Rev. **107:** 885–913.
38. BIGAND, E., S. VIEILLARD, F. MADURELL, *et al.* Multidimensional scaling of emotional responses to music: the effect of musical expertise and of the duration of the excerpts. Cogn. Emot. In press.
39. FILIPIC, S., B. TILLMANN & E. BIGAND. The time-course of the emotional response to music. Submitted.
40. FRANCES, R. 1958. La Perception de la Musique. Vrin. Paris.
41. DAMASIO, A. 1995. Descartes' Error: Emotion, Reason, and the Human Brain. Avon Books. New York.
42. PERETZ, I., A. J. BLOOD, V. PENHUNE & R. ZATORRE. 2001. Cortical deafness to dissonance. Brain **124:** 928–940.
43. PERETZ, I., A. J. BLOOD, V. PENHUNE & R. ZATORRE. 2001. A developmental study of the affective value of tempo and mode in music. Cognition **80:** 1–10.
44. BIGAND, E., B. POULIN, B. TILLMANN, *et al.* 2003. Sensory versus cognitive components in harmonic priming. J. Exp. Psycol. Hum. **29:** 159–171.

Remember Bach

An Investigation in Episodic Memory for Music

SUSANN ESCHRICH[a,b] THOMAS F. MÜNTE,[c] AND ECKART O. ALTENMÜLLER[a]

[a]Institute of Music Physiology and Musicians' Medicine, Hannover D-30161, Germany

[b]Center for Systems Neurosciences, Hannover, Germany

[c]Department of Neuropsychology, Magdeburg University, Magdeburg D-39106, Germany

ABSTRACT: Emotional events are remembered better than nonemotional ones, especially after a long period of time. In this study, we investigated whether emotional music is kept better in episodic long-term memory than less emotional music and to which extent musical structure is important.

KEYWORDS: episodic memory; musical memory; emotions

INTRODUCTION

The long-term memory for music and its representation in the brain have been investigated very little (but see Refs. 1 and 2). It is still under debate whether a separate (episodic) memory for music exists in contrast with verbal memory. However, some evidence has been found in favor of this.[1,3] Music can elicit strong emotions[4,5] and can be remembered, possibly in connection with these emotions, even years later. Evidently, strong emotions related to the musical experience facilitate memory formation and retrieval. However, until now, the episodic memory for highly emotional music compared with less emotional music had not yet been examined. Episodic memory is defined as the kind of memory that allows one to remember past events of one's own life, whereas semantic memory is defined as memory for general facts of the world. Tulving[6] introduced the "remember/know paradigm," a test in which subjects, whenever they judge an item as "known," indicate whether they "remember" the item's occurrence on the studied list or whether they have some other reasons for "knowing" the item on the list.

In this study, we investigated whether emotional music is kept better in episodic long-term memory than less emotional music and examined the emotional rating of short piano pieces of J. S. Bach.

Address for correspondence: Susann Eschrich, Dipl.-Psych., Institute of Music Physiology and Musicians' Medicine, University of Music and Drama Hannover, Hohenzollernstrasse 47, D-30161 Hannover, Germany. Voice: +49-511-3100-576; fax: +49-511-3100-557.

eschrich@hmt-hannover.de

Ann. N.Y. Acad. Sci. 1060: 438–442 (2005). © 2005 New York Academy of Sciences.
doi: 10.1196/annals.1360.045

TABLE 1. Recognition rates and d' measures per subject

Subject no.	Total correct (%)	Hit (%)	False alarm (%)	d'
2	49	80	83	−0.11
3	60	66	47	0.52
4	59	59	40	0.43
5	36	50	77	−0.64
6	57	63	50	0.33
7	50	40	40	0
8	66	66	33	0.88
9	48	43	47	−0.08
10	70	90	53	1.2
11	57	66	53	0.36

METHODS

As target pieces, 30 J. S. Bach piano pieces (French and English suites, inventions and sinfonias, and the *Well-Tempered Clavier*) were cut off after 30 to 60 seconds. In a preassessment, these pieces were categorized by valence and arousal ratings. Pieces rated as less emotional were contrasted with pieces of high emotional content. For the second session, 30 other piano pieces (distractors) were chosen from the same piece collection, cut off, and matched by valence and arousal with the targets. Ten nonmusicians (graduate and undergraduate students with less than two years of instrumental or choral expertise) participated in this experiment. Five subjects were male, and five subjects were female. The mean age of the subjects was 33.9 years.

In the first session, after each presentation of a target piece, subjects had to rate the emotions induced by this music according to the two-dimensional valence–arousal model of emotions by Russel.[7] According to this model, *arousal* is defined as the amount of excitation elicited by the music (from "very pacifying" to "very arousing") and for *valence* as the emotional value of the music and the degree to which a subject wants to continue listening to the music ("positive") or to turn it off ("negative"). Additionally, we asked for the strength of emotion the music elicited in the subject ("emotional strength"), ranging from "no emotions" to "very strong emotions." We also asked for the feeling of control a subject had over the feelings elicited by the music ("control"), ranging from "very low control/overwhelming feelings" to "very high control."

Approximately two weeks later, target and distractor pieces were presented together in random order. Subjects were asked to indicate whether they had heard the piece of music before according to the remember/know paradigm.[7]

RESULTS

Emotional ratings of the musical pieces were highly subjective. Because the recognition task was rather difficult, recognition performance was not high, however, above chance level. Four subjects with the best recognition rate and the best hits to

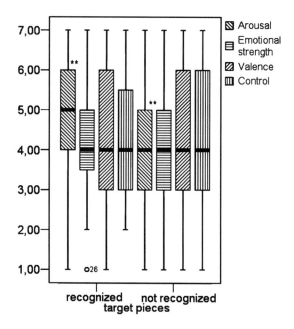

FIGURE 1. Differences of emotional ratings between target musical pieces recognized by most of the subjects and badly recognized pieces.

false alarm ratio were considered separately. For the hits to false alarm ratio, d′ measures were calculated. All of the four subjects showed a d′ greater than 0.5 (see TABLE 1). The d′ of these four subjects differed significantly from the other six ($P < .01$).

Contrasting well-recognized (wr) pieces of music (recognized by at least 80% of the subjects) with badly recognized (br) pieces, we found that arousal ratings proved to be significantly higher for well-recognized pieces ($M_{wr} = 1.70$ and $M_{br} = 1.30$, $P < .001$; FIG. 1). The four subjects who remembered best were contrasted with the other subjects. They showed significantly higher ratings in emotional strength ($M_{wr} = 1.75$ and $M_{br} = 1.29$, $P < .00$) and more positive valence ratings ($M_{wr} = 1.71$ and $M_{br} = 1.34$, $P < .00$) as well as almost significantly higher arousal ratings ($M_{wr} = 1.59$ and $M_{br} = 1.43$, $P < .1$; FIG. 2). Pieces that were well recognized by the four subjects were rated significantly more positive in valence ($M_{wr} = 1.32$ and $M_b = 1.68$, $P < .012$) and almost significantly higher in arousal ($M_{wr} = 1.36$ and $M_b = 1.64$, $P < .07$).

CONCLUSIONS

The current study shows that the emotional dimension arousal seems to be more important for episodic long-term storage and retrieval of music than emotional valence. However, in the subgroup of the four subjects who remembered best, emo-

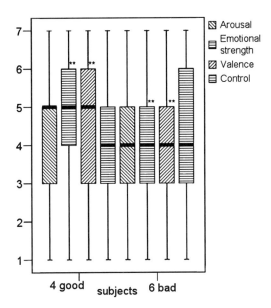

FIGURE 2. Rating differences between the four subjects who remembered best and the other six subjects.

tional content of music (positive valence) seems to be a major characteristic. Thus, a predisposition for emotional resonance seems to facilitate encoding and retrieval of music, given that a high state of arousal is achieved. The results of this study, however, are preliminary. It should be considered that these data are based on a small sample of subjects and recognition rates, especially the remembrance rates that were generally rather low. Additionally, the pieces of music did not elicit very strong emotions and did not differ very much emotionally. Further experiments on a larger group of subjects and with more emotionally "extreme" musical stimuli are planned.

[Competing interests: The authors declare that they have no competing financial interests.]

REFERENCES

1. PLATEL, H., J.-C. BARON, B. DESGRANGES, *et al.* 2003. Semantic and episodic memory of music are subserved by distinct neural networks. NeuroImage **20:** 244–256.
2. HALPERN, A.R. & J.C. BARTLETT. 2002. Aging and memory for music: a review. Psychomusicology **18:** 10–27.
3. PERETZ, I. 1996. Can we lose memories for music? The case of music agnosia in a non-musician. J. Cogn. Neurosci. **8:** 481–496.

4. KRUMHANSL, C.L. 1997. An exploratory study of musical emotions and psychophysiology. Can. J. Exp. Psychol. **51:** 336–353.
5. PANKSEPP, J. & G. BERNATZKY. 2002. Emotional sounds and the brain: the neuro-affective foundations of musical appreciation. Behav. Processes **60:** 133–155.
6. TULVING, E. 1985. How many memory systems are there? Am. Psychol. **40:** 385–398.
7. RUSSEL, J.A. 1980. A circumplex model of affect. J. Pers. Soc. Psychol. **39:** 1161–1178.

Key Processing Precedes Emotional Categorization of Western Music

SUZANNE FILIPIC AND EMMANUEL BIGAND

LEAD-CNRS, Université de Bourgogne, Dijon, France

ABSTRACT: To investigate whether key processing precedes the appraisal of valence in music, participants listened to pairs of clips of same or different valence, played either in the same key or one semitone apart. They judged whether the second clip expressed the same emotion as the first one. Our predictions were confirmed: the response times obtained were shorter when both clips were played in the same key than when they were played one semitone apart.

KEYWORDS: key; appraisal; valence

Most theorists agree that emotions are fast processes that can be elicited in the absence of conscious cognitive mediation, but that some cognitive processing is required for most emotion.[1] According to Ekman and Davidson, the challenge now is to specify the types of cognitive operations that are critical to the emotion-generation process. The goal of this study was to show the importance of key processing for emotions generated by music.

Indeed, using five experiments testing an amusic patient (I.R.), Peretz and colleagues[2] challenged this idea by showing that there may be separate affective and cognitive pathways.[3] The pathway for emotional evaluation would be quick, precocious, and present in everyone. The route for nonemotional judgments would probably be slower, appear somewhat later in development, and be sensitive to musical expertise. This conclusion would be very much in line with Zajonc.[4,5]

Our study was an attempt to show whether key processing may be a "stimulus antecedent"[6] involved in appraisal processes of valence in music. Over the years,[7,8] studies on the categorization of the emotional valence in music have shown that the two most important factors were tempo and mode. To identify the mode of a piece of music, identifying its key is a necessary step because music theory defines modes according to the nature of the intervals that exist between the different scale notes and the tonic.[9] Therefore, we predicted that participants would process the key of a piece before categorizing its emotional valence.

Sixty-two participants took part in the experiment: 38 students in psychology with little or no music education (referred to as nonmusicians) and 24 musicians,

Address for correspondence: Suzanne Filipic, LEAD-CNRS, Pôle AAFE, Université de Bourgogne, 2 Esplanade Erasme, BP 26513, 21065 Dijon cedex, France. Voice: +33-03-80-39-57-85; fax: +33-03-80-39-57-67.
suzanne.filipic@leadserv.u-bourgogne.fr

Ann. N.Y. Acad. Sci. 1060: 443–445 (2005). © 2005 New York Academy of Sciences.
doi: 10.1196/annals.1360.039

TABLE 1. Average correct response times (msec [SDs]) as a function of key, valence similarity, and musical expertise

Emotion	Musicians		Nonmusicians	
	Same	Different	Same	Different
Same key	2,947 (1,225)	2,541 (995)	2,953 (1,393)	2,898 (1,365)
Different key	3,198 (1,535)	2,960 (1,248)	3,058 (1,460)	3,010 (1,494)

including music students and professional musicians. They were presented with 24 musical excerpts from Gosselin and colleagues:[10] 12 peaceful and 12 sad clips. The stimuli were played on a Midi keyboard by a professional pianist in order to convey the clearest possible expression. To prevent peaceful and sad clips from being distinguished on the basis of small differences in range (the peaceful melodies being, on average, higher in pitch than the sad ones) or small tempo differences (the peaceful clips being slightly faster than the sad ones), we modified the clips slightly to be aligned to similar ranges and tempi. The key of the clips was also manipulated so that the tonic of the two clips in the pair were either identical or one semitone apart. The 24 clips thus were combined to create 48 pairs of clips: 24 "same emotion" pairs (peaceful-peaceful, or sad-sad, but the two melodies of a pair were never identical) and 24 "different emotion" pairs (peaceful-sad, or sad-peaceful). Each series of 12 pairs was made of six pairs in the same key, and six pairs in different keys.

The experimental procedure was split into two sessions, between which participants had to take a short break. During the experiment, participants worked at their own pace, starting each excerpt by clicking on the space bar. For each pair of clips, participants listened to the first melody entirely and had to indicate if it expressed peacefulness or sadness, by clicking on "peaceful" or "sad" on the screen. The main task was performed on the second clip of the pair. As soon as they started the second clip, participants had to decide as quickly as possible whether it expressed the same emotion as the previous clip. They were informed by a feedback signal if they gave an incorrect response. Note that participants were not asked to identify the emotion (category) verbally in the second task, when their response times were measured. They pressed a key labeled "same" or "different."

The first analysis, run on the emotional judgments made on the first clip of each pair, showed that participants correctly identified the emotional valence of the clips, on average, 90.86% of the time and thus replicated the findings of Gosselin and colleagues.[10] Musicians were slightly more accurate (93.58%) than nonmusicians (89.14%), but the difference was not significant.

Correct response times for the responses on the second clip on each pair are displayed in TABLE 1. We kept in the analysis the correct response times only when the first clip of the pair had been correctly categorized. A 2 (key relationship) × 2 (valence similarity) × 2 (sessions) × 2 (musical expertise) ANOVA was performed on correct response times. There was a main effect of key relationship, with longer response times for pairs of melodies played in different keys (3,057 ms) than for pairs played in the same key (2,835 ms), $F(1,60) = 9.41$, $P < .004$, MSE = .614E+06.

The present finding thus suggests that participants may have processed the key of each piece before appraising its valence. Further research would be necessary to document whether key processing also occurs before distinguishing between musical excerpts of a high arousal level (e.g., happy or fearful music).

[Competing interests: The authors declare that they have no competing financial interests.]

REFERENCES

1. DAVIDSON, R.J. & P. EKMAN. 1994. Afterword to question five: what are the minima cognitive prerequisites for emotion? *In* The Nature of Emotion. Series in Affective Science. P. Ekman & R.J. Davidson, Eds: 232–234. Oxford University Press. New York.
2. PERETZ, I. *et al.* 1998. Music and emotion: perceptual determinants, immediacy, and isolation after brain damage. Cognition **68:** 111–141.
3. PERETZ, I. & L. GAGNON. 1999. Dissociation between recognition and emotional judgment for melodies. Neurocase **5:** 21–30.
4. ZAJONC, R.B. 1980. Feeling and thinking: preferences need no inferences. Am. Psychol. **35:** 151–175.
5. ZAJONC, R.B. 1984. On the primacy of affect. Am. Psychol. **39:** 117–123.
6. SCHERER, K. & M. ZENTNER. 2001. Emotional effects of music: production rules. *In* Music and Emotion. P.N. Juslin & J.A. Sloboda, Eds.: 361–392. Oxford University Press. New York.
7. HEVNER, K. 1933. The affective character of the major and minor modes in music. Am. J. Psychol. **47:** 103–118.
8. GAGNON, L. *et al.* 2003. Mode and tempo relative contributions to "happy-sad" judgments in equitone melodies. Cogn. Emot. **17:** 25–40.
9. DOWLING, W. & D. HARWOOD. 1986. Music Cognition. Academic Press. Orlando, FL.
10. GOSSELIN, N. *et al.* 2005. Impaired recognition of scary music following unilateral temporal lobe excision. Brain **128:** 628–640.

How Does Music Arouse "Chills"?

Investigating Strong Emotions, Combining Psychological, Physiological, and Psychoacoustical Methods

OLIVER GREWE,[a] FREDERIK NAGEL,[a] REINHARD KOPIEZ,[b]
AND ECKART ALTENMÜLLER[a]

[a]Institut für Musikphysiologie und Musikermedizin,
[b]Institut für Musikpädagogische Forschung,
Hochschule für Musik und Theater Hannover, D-30161 Hannover, Germany

ABSTRACT: Music can arouse ecstatic "chill" experiences defined as "goose pimples" and as "shivers down the spine." We recorded chills both via subjects' self-reports and physiological reactions, finding that they do not occur in a reflex-like manner, but as a result of attentive, experienced, and conscious musical enjoyment.

KEYWORDS: emotion; chill; music

Music can arouse extraordinarily strong emotional responses, even up to ecstatic "chill" experiences.[1–3] Such strong psychological reactions are often accompanied by measurable bodily reactions, such as goose pimples or shivers. Because emotional states can change over the course of a piece of music, it is necessary to measure psychological and bodily reactions continuously. To investigate distinct musical events related to chill reactions, we combined psychological, psychoacoustical, and physiological methods.

METHODS

We asked 38 subjects to press a mouse button whenever they experienced a chill while listening to music. A selection of seven pieces from different musical styles was used for all subjects. Additionally, subjects were asked to bring 5 to 10 "personal" pieces of music that regularly induced strong emotions. All kinds of musical styles were accepted. The subjects' mean age was 38 (SD, 16), with a range of 11 to 72 years. Subjects had different musical experiences and education.

Pressing of the mouse button was recorded simultaneously to the music. The mere pressing of the mouse button did not influence skin conductance response.

Address for correspondence: Oliver Grewe, Institut für Musikphysiologie und Musikermedizin, Hochschule für Musik und Theater, Hohenzollernstrasse 47, 30161 Hannover, Germany. Voice: +49-511-3100-561; fax: +49-511-3100-557.
oliver_grewe@gmx.de

Ann. N.Y. Acad. Sci. 1060: 446–449 (2005). © 2005 New York Academy of Sciences.
doi: 10.1196/annals.1360.041

Within the whole piece, musical events that triggered chills could thereby be identified with an accuracy range of ~5 seconds. After each piece of music, subjects completed a questionnaire regarding their knowledge of the piece and perceived bodily reactions.

Chills had to fulfill three criteria: (1) pressing of the mouse button; (2) a measurable skin conductance response; and (3) report of goose pimples or shivers down the spine indicated on the questionnaire.

After the experimental session, subjects answered further questions concerning their musical taste and experience, and completed three standardized personality inventories.[4-6] Psychoacoustical parameters of the stimuli were analyzed using dBSonic software.

To check for the reproducibility of our results, we repeated the same experiment with one female musician (a 24-year-old soprano) on seven subsequent days. Conditions were kept as stable as possible.

RESULTS

Chills are rare events. A maximum of 8 out of 37 subjects had chills within the same piece of music. Chills do not occur interindividually as a deterministic response to a single musical event. Chills occurred mainly during defined semantic musical structures. Subjects with an extremely high number of chills exhibited differences in character and musical experience as compared with subjects with no chill reactions. Changes in loudness seem to have an influence on chill events.

Here, we present an example from the retest experiment. FIGURE 1 shows all chill reactions to the fourth movement ("Urlicht") of Gustav Mahler's Symphony No. 2.[7]

Three musical excerpts are presented for the three parts of the piece where chills occurred repeatedly on 5 to 6 days. Each shows a highly similar musical motive repeated throughout the movement in different voices of the orchestral score (trumpets, alto [human voice]). This motive does not occur at other times during the movement. Mahler indicated "espressivo" in the score when the motive is performed in the alto voice. Additionally, he specified a small crescendo, that is, an increase in loudness, in two of the three excerpts. The change in loudness can also be seen in the psychoacoustical loudness analysis.

The three excerpts are very similar because of their musical structure. They are three harmonic variances of the same motive, all being a V-I or v-I sequence. The first two excerpts finish in G-flat major, the dominant; the last example is the final resolution [closing] on the tonic, D-flat major.

The psychoacoustical time series analysis reveals strong changes in loudness for all three excerpts, but also at points in time that stimulate less stable or no chills. In the questionnaires, the subject mentioned that excerpts 1 and 3 were extraordinarily pleasant.

DISCUSSION

The preliminary results presented here reveal evidence that strong emotions in response to music are related to structural musical elements. These can partially

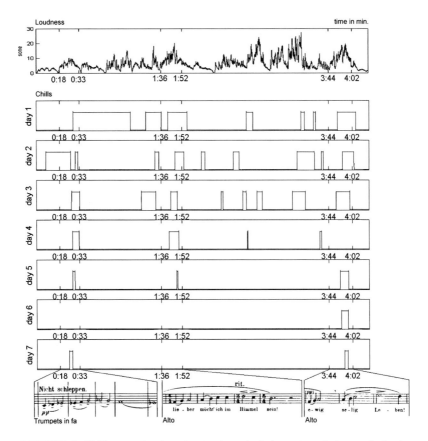

FIGURE 1. Chills are related to structural musical elements and changes in loudness. Example: Mahler, Symphony No. 2, "Urlicht." The experiment was repeated for 7 days with one subject. Chills are presented for each day as time series for the duration of the piece. Chills are presented as bars, and the length reveals how long the mouse button was pressed. The loudness time series analysis is shown in the *upper panel*. All data are synchronized. The musical parts related to chills occurring on 5 or more days are presented as excerpts of single voices from the score. The excerpts last from 0:18 to 0:33, 1:36 to 1:52, and 3:44 to 4:02, respectively. The whole piece lasts 4:17 minutes.

be described psychoacoustically; a distinct chill-triggering acoustical pattern could not be found, though. Important musical factors seem to be harmonic sequences,[3] the entrance of a voice, and the beginning of a new part, that is, a violation of expectancies.[8,9] Even if chills do not interindividually occur as a deterministic response to a single acoustical trigger, in single individuals, chills can be stable responses to distinct musical structures. Subjects often report the musical structures related to chills as being extraordinarily pleasant. On the basis of these results, we hypothesize that strong emotions in response to music do not occur in a reflex-like manner, but as a result of attentive, experienced, and conscious musical enjoyment.

ACKNOWLEDGMENTS

This work was supported by the Deutsche Forschungsgemeinschaft (grant no. AL 269-6) and the Center for Systemic Neurosciences Hannover.

[Competing interests: The authors declare that they have no competing financial interests.]

REFERENCES

1. PANKSEPP, J. 1995. The emotional sources of "chills" induced by music. Mus. Percept. **13:** 171–207.
2. PANKSEPP, J. & G. BERNATZKY. 2002. Emotional sounds and the brain: the neuro-affective foundations of musical appreciation. Behav. Processes **60:** 133–155.
3. SLOBODA, J.A. 1991. Music structure and emotional response: some empirical findings. Psychol. Mus. **19:** 110–120.
4. CLONINGER, R.C. *et al.* 1999. Das Temperament- und Charakter-Inventar. Swets & Zeitlinger B.V. Frankfurt.
5. DAVIS, K.L., J. PANKSEPP & L. NORMANSELL. 2003. The affective neuroscience personality scales: normative data and implications. Neuro-Psychoanalysis **5:** 57–69.
6. LITLE, P. & M. ZUCKERMAN. 1986. Sensation seeking and music preferences. Pers. Indiv. Diff. **7:** 575–577.
7. SOLTI, G. 1996. Mahler Symphonies 1 and 2. Music recording. Decca (Universal) ASIN: B000025RIM.
8. MEYER, L.B. 1956. Emotions and Meaning in Music, 1961 ed. University of Chicago. Chicago.
9. MEYER, L.B. 2001. Music and emotion: distinctions and uncertainties. *In* Music and Emotion: Theory and Research. J.A. Sloboda and P.N.S. Juslin, Eds.: 341–360. Oxford University Press. Oxford.

Emotion Processing of Major, Minor, and Dissonant Chords

A Functional Magnetic Resonance Imaging Study

KAREN JOHANNE PALLESEN,[a,b] ELVIRA BRATTICO,[c]
CHRISTOPHER BAILEY,[a] ANTTI KORVENOJA,[d] JUHA KOIVISTO,[b]
ALBERT GJEDDE,[a] AND SYNNÖVE CARLSON[b]

[a]Center of Functionally Integrative Neuroscience and PET Centre,
Aarhus University Hospital, 8000 Aarhus C, Denmark

[b]Neuroscience Unit, Helsinki Brain Research Center, Institute of Biomedicine/
Physiology, [c]Cognitive Brain Research Unit, Helsinki Brain Research Center,
University of Helsinki, 00014 Helsinki, Finland

[d]Functional Brain Imaging Unit, Helsinki Brain Research Center,
00029 Helsinki University Central Hospital, Helsinki, Finland

ABSTRACT: Musicians and nonmusicians listened to major, minor, and dissonant musical chords while their BOLD brain responses were registered with
functional magnetic resonance imaging. In both groups of listeners, minor and
dissonant chords, compared with major chords, elicited enhanced responses in
several brain areas, including the amygdala, retrosplenial cortex, brain stem,
and cerebellum, during passive listening but not during memorization of the
chords. The results indicate that (1) neural processing in emotion-related brain
areas is activated even by single chords, (2) emotion processing is enhanced in
the absence of cognitive requirements, and (3) musicians and nonmusicians do
not differ in their neural responses to single musical chords during passive
listening.

KEYWORDS: emotions; music; musical competence; working memory;
emotion–cognition interaction

INTRODUCTION

Major, minor, and dissonance in music are commonly said to cause happy, sad,
and unpleasant experiences, respectively. Behavioral studies have shown that these
emotional effects may be elicited by brief melodic fragments and even by isolated
chords in musicians as well as nonmusicians.[1] Dissonance presented in melodies
was previously related to activity in the right parahippocampal gyrus and right

Address for correspondence: Karen Johanne Pallesen, Center for Functionally Integrative
Neuroscience, Aarhus University Hospital, Nørrebrogade 44, 8000 Aarhus C, Denmark. Voice:
+45-89494095; fax: +45-89494400.
karenjohanne@pet.auh.dk

Ann. N.Y. Acad. Sci. 1060: 450–453 (2005). © 2005 New York Academy of Sciences.
doi: 10.1196/annals.1360.047

precuneus brain areas.[2] Moreover, the downregulating effects of appraisal and cognition on emotion processes were demonstrated in terms of amygdalar responses to aversive visual stimuli.[3] In spite of the well-proven power of music to elicit positive or negative emotional experiences, the mechanisms for emotional regulation have not so far been studied in this domain. Because musical competence, reflected in the neurophysiological auditory responses, implies a more analytical approach to musical sounds, this may also reflect in the neural emotion processes. We studied whether (1) simple musical chords activate brain areas previously associated with emotion analysis, (2) cognitive evaluation has an influence on these responses, and (3) musical competence influences the emotional responses.

METHODS

Twenty-one right-handed individuals (mean age 26; 14 females), 11 subjects with a classical music education (musicians), and 10 subjects with no musical training (nonmusicians), were subjected to nine piano chords belonging to three pitch classes (major, minor, dissonant), each spanning three octaves from A3 to A5. The subjects either listened passively to the chords or performed an n-back working memory task with respect to pitch. After the brain scanning, the subjects were asked to rate the emotional connotations of each chord on two 11-point scales (as unpleasant-pleasant and sad-happy, respectively). In all conditions, subjects pressed a button after each chord to maintain motor-related brain activity constant across conditions. Magnetic resonance images were acquired with a 1.5-T Siemens Sonata scanner. Analysis was performed using FMRIB Software Library (FSL, Oxford, UK). Each subject's structural and functional data were coregistered to the MNI152 standard template.

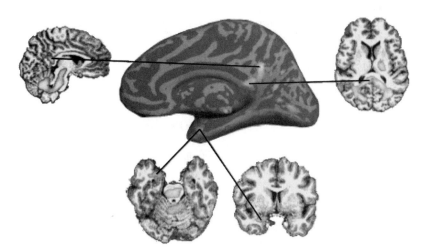

FIGURE 1. BOLD responses that were larger during passive listening to minor chords than during passive listening to major chords, including the amygdala, retrosplenial cortex, brain stem, and cerebellum. The responses in this contrast were not present during working memory. Group results are overlayed on transverse sections and inflated cortex of a single individual (corrected for multiple comparisons: Z > 3, P < .05).

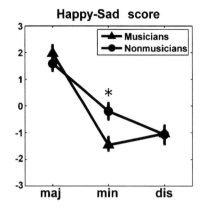
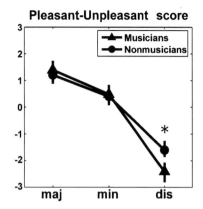

FIGURE 2. Emotional ratings (mean ± SE) of major, minor, and dissonant chords. Musicians rated dissonant chords as significantly more unpleasant, and minor chords as significantly more sad than nonmusicians did (*$P < .05$).

RESULTS

During passive listening, minor and dissonant chords elicited larger BOLD responses than did major chords in several brain areas, including the amygdala, retrosplenial cortex, brain stem, and cerebellum (FIG. 1 represents the minor versus major chords contrast). Together with the thalamus and brain stem, the amygdala has been implicated in the evolution of an adaptive "alarm system."[4] These differential responses to minor and dissonant chords, compared with major chords, were not present during the pitch working memory task, requiring cognitive evaluation of the chords. Although musicians rated minor chords as sadder and dissonant chords as more unpleasant, than did nonmusicians (FIG. 2), there were no significant differences between the two groups of subjects in the neural responses to the chords during passive listening.

CONCLUSION

The results provide evidence for a role of emotional reactions to isolated musical sound units in the musical experience. Moreover, they confirm that cognitive evaluation leads to decreased emotional responsiveness. We suggest that the amygdala–brain stem responses, during passive listening, to minor and dissonant chords, compared with major chords, reflect a mechanism that automatically interprets these chords as being potentially alarming stimuli. The group difference in the emotional ratings of the chords, but not in the neural responses, may reflect a musician's ability to recognize and categorize the chords in terms of the conventional emotional connotations, rather than an effective enhanced emotional experience.

[Competing interests: The authors declare that they have no competing financial interests.]

REFERENCES

1. PALLESEN, K.J., E. BRATTICO & S. CARLSON. 2003. Emotional connotations of major and minor musical chords in musically untrained listeners. Brain Cogn. **51:** 188–190.
2. BLOOD, A. *et al.* 1999. Emotional responses to pleasant and unpleasant music correlate with activity in paralimbic brain regions. Nat. Neurosci. **2:** 382–387.
3. HARIRI, A.R., S.Y. BOOKHEIMER & J.C. MAZZIOTTA. 2000. Modulating emotional responses: effects of a neocortical network on the limbic system. Neuroreport **11:** 43–48.
4. LIDDELL, B.J. *et al.* 2005. A direct brainstem-amygdala-cortical "alarm" system for subliminal signals of fear. Neuroimage **24:** 235–243.

Left Auditory Cortex Specialization for Vertical Harmonic Structure of Chords

NATALIA PASSYNKOVA, KERSTIN SANDER, AND HENNING SCHEICH

Leibniz Institute for Neurobiology, Magdeburg, Germany

ABSTRACT: The representation of consonant and dissonant chords in the auditory cortex was investigated using low-noise functional magnetic resonance imaging and different experimental paradigms to separate the effects of vertical harmony from those of other musical features. The results revealed higher activation by consonant compared with dissonant chords in the left posterior auditory cortex, suggesting contributions of mechanisms of encoding the acoustical chord structure rather than mechanisms based on sequential integration of chords.

KEYWORDS: consonance; dissonance; harmony; fMRI

INTRODUCTION

The present study was designed to explore the representation of the vertical harmony of music, specifically the consonance–dissonance dimension, within the human auditory cortex (AC). There is some possibility of confounding this representation with those of other principal properties of music. For example, harmonically related chords in a sequence, that is, in the horizontal dimension, are perceptually grouped and form an auditory Gestalt. To avoid such confounding factors, we conducted four experiments in a functional magnetic resonance imaging (fMRI) block design, in which we modified step-by-step the individual stimuli and their presentation order, destroying the effect of auditory integration in the horizontal dimension and promoting the perception of single chords as isolated auditory objects.

MATERIALS AND METHODS

The auditory cortex of nonmusicians was scanned using low-noise fMRI[1,2] while subjects listened (1) to four-note consonant chords in major and minor mode, (2) to four-note dissonant chords created from the consonant chords by shifting the tone in the bass position upward by one semitone, and (3) to the individual harmonic tones, of which the chords were made up. All stimuli were created to be of piano timbre using the "Cubasis VST 4.0" software and were presented in a block design.

Address for correspondence: Dr. Natalia Passynkova, Leibniz Institute for Neurobiology (IfN), Special Lab Non-Invasive Brain Imaging, Brenneckestrasse 6, 39118 Magdeburg, Germany. Voice: +49-391-62-63-124; fax: +49-391-62-63-328.
  natalia.passynkova@ifn-magdeburg.de [*or*] passynkova@mail.ru

Ann. N.Y. Acad. Sci. 1060: 454–456 (2005). © 2005 New York Academy of Sciences.
doi: 10.1196/annals.1360.048

In the first experiment (14 subjects), consonant and dissonant chords as well as individual tones were presented at a rate of one stimulus per second in ascending and descending orders of pitch. Thus, the chords were harmonically related by three attributes: melody, temporal structure, and tonality, which might have bound successive chords together in the horizontal dimension.

In the second experiment (4 subjects), the order of presentation of chords and individual tones was randomized. This destroyed all the melodic lines that were present in experiment I.

To destroy the temporal relationship between stimuli, we randomly jittered the interstimulus intervals, from 700 to 1,400 msec, in the third experiment (4 subjects).

In the fourth experiment (7 subjects), in addition to a randomized order of presentation and jittering of stimuli, we created individual harmonic chords, which were played out of any tonality, that is, chords were mistuned relative to each other in the horizontal dimension. We produced such chords by increasing or decreasing the frequencies of each of the four tones of a chord by 1%, 2%, 3%, or 4% of the original. Thus, the different chords could have originated from different pianos, each tuned with a different tuning fork resonating at a different frequency (e.g., 440 Hz and 443 Hz).

Data were analyzed using the multisubject general linear model approach implemented in BrainVoyager 2000™ and the individual regions of interest approach implemented in custom-made software[1–4] and followed by ANOVAs and Scheffé test.

We contrasted three experimental conditions directly and also each with silence blocks (no stimulus presentation). In experiments I to III, we contrasted mutually (1) consonant chords, (2) dissonant chords, and (3) single note conditions; in experiment IV these were (1) consonant chords out of tonality, (2) dissonant chords out of tonality, and (3) consonant chords in tonality (as in experiment III).

RESULTS AND DISCUSSION

The results revealed significantly stronger BOLD response to consonant than to dissonant chords in the left posterior auditory cortex (AC), mainly the planum temporale (Brodmann area 22). This left-hemispheric effect was not altered by variations of presentation order or interstimulus interval. Furthermore, it was present regardless of whether consonant or dissonant chords were played "in a scale" or whether they were mistuned in the horizontal dimension. Only a trend for significant effects was found between individual note conditions and consonance as well as dissonance conditions. The BOLD response to single notes was lower than to consonance, but higher than to dissonance. The comparison of the two consonant conditions "out of tonality" and "in tonality" (experiment IV) did not show any differences. Thus, the results indicate that the sequential integration of chords does not provide an explanation for the left-hemispheric effect of the consonance–dissonance contrast. Rather, this left-hemispheric preference may be associated with mechanisms of encoding the acoustical structure of the two types of chords. Taking into account studies of specialization of left and right AC,[1,2,4,5] we suggest that the left lateralization effect in the AC can be explained by the contribution of the temporal coding mechanism.

[Competing interests: The authors declare that they have no competing financial interests.]

ACKNOWLEDGMENTS

This study was supported in part by SFB 426, Deutsche Forschungsgemeinschaft.

REFERENCES

1. BRECHMANN, A. & H. SCHEICH. 2005. Hemispheric shifts of sound representation in auditory cortex with conceptual listening. Cerebr. Cortex **15**: 578–587.
2. DEIKE, S., B. GASCHLER-MARKEFSKI, A. BRECHMANN, *et al.* 2004. Auditory stream segregation relying on timbre involves left auditory cortex. Neuroreport **15**: 1511–1514.
3. GASCHLER, B., F. SCHINDLER & H. SCHEICH. 1996. KHORFu: a KHOROS-based functional image post processing system. A statistical software package for functional magnetic resonance imaging and other neuroimage data sets. COMPSTAT 1996: Proceedings in Computational Statistics. 12th symposium, Barcelona. pp. 57–58. Physica-Verlag. Heidelberg.
4. SANDER, K. & H. SCHEICH. 2005. Left auditory cortex and amygdala, but right insula dominance for human laughing and crying. J. Cogn. Neurosci. **17**: 1519–1531.
5. ZATORRE, R.J., P. BELIN, & V. PENHUNE. 2002. Structure and function of auditory cortex: music and speech. Trends Cogn. Sci. **6**: 37–46.

Emotional Processing of
Harmonic Expectancy Violations

NIKOLAUS STEINBEIS,[a,b] STEFAN KOELSCH,[a] AND JOHN A. SLOBODA[b]

[a]Max-Planck-Institute for Human Cognitive and Brain Sciences,
04103 Leipzig, Germany

[b]University of Keele, Keele, United Kingdom

ABSTRACT: This study is about whether harmonic expectancy violations can trigger emotional processes, as indexed by physiological and subjective measures.

KEYWORDS: harmonic expectancy; emotion; Bach chorales; EDA; EEG

INTRODUCTION

Music is capable of eliciting strong emotions in listeners.[1] It has been argued that a reason for this is the suspension and fulfilment of implicit musical expectations of the listener and that music becomes increasingly emotional with the suspension of musical expectations.[2] The validity of this claim was directly tested in this study.

It has been shown that listeners have harmonic expectations given a salient harmonic context.[3] These expectations mirror the relationships of harmonic keys in Western music. Previous electroencephalogram (EEG) experiments have indicated event-related potentials (ERPs) to be sensitive to the harmonic context and its violation, as reflected by an early right anterior negativity (ERAN).[4] In turn, it has been demonstrated that listeners reported heart racing when a prominent event occurred earlier than expected and shivers in response to a new or unprepared harmony.[1] However, there is yet no direct evidence that the systematic manipulation of harmonies leads to an increase in emotions.

The purpose of the present study was to investigate the influence of three different degrees of harmonically expected harmonizations of the same pieces of music on the neural processing of these events, as well as on subjective and psychophysiological indicators of emotion.

Address for correspondence: Nikolaus Steinbeis, Max-Planck-Institute for Human Cognitive and Brain Sciences, Junior Research Group, Neurocognition of Music, Stephanstr. 1A, 04103 Leipzig, Germany. Voice: +49-341-355-217-10.
steinb@cbs.mpg.de

Ann. N.Y. Acad. Sci. 1060: 457–461 (2005). © 2005 New York Academy of Sciences.
doi: 10.1196/annals.1360.055

METHOD

Participants

Twelve musicians (mean years of musical training: 13.8 years) and 12 nonmusicians participated in the study. Mean age was 25.15 years. Twelve of the participants were male.

Stimuli

Six Bach chorales were chosen explicitly for sections containing unexpected harmonies. This unexpectedness was measured using the circle of fifths, which demonstrates the distance between the 12 major and minor keys of Western music. Harmonically expected and very unexpected alternatives to a single cadential ending in each chorale were composed, by varying the distance of the target chord along the circle of fifths. Eventually there were three versions of each chorale, producing 18 stimuli in total.

Materials

Emotional Self-Report

This was the first part of the study, for which subjects heard chorales in their entirety. At the end of each chorale, participants stated how emotional they felt in response to the music.

Electrodermal Activity

Several studies have shown that emotions in response to music can be captured by psychophysiological measures.[1,5] Electrodermal activity (EDA) has been suggested to be a good indicator of emotions, reflecting its arousal dimension,[6] also while listening to music.[7] Two electrodes were applied to the medial phalanx of the middle and index fingers of the nondominant hand.

Electroencephalogram

An EEG was recorded from 30 locations distributed over the scalp to study the neural processing of expectancy violations. EDA and EEG were recorded simultaneously for the second part of the study, in which shortened versions of the chorales were presented randomly.

Procedure

For the second part, participants had to perform a judgement task to show that they were paying attention to the music. The length of each stimulus had to be compared with the previously heard one and judged if it was shorter or longer. There was an 8-s pause between each response and the start of a new stimulus to allow the EDA to return to a baseline.

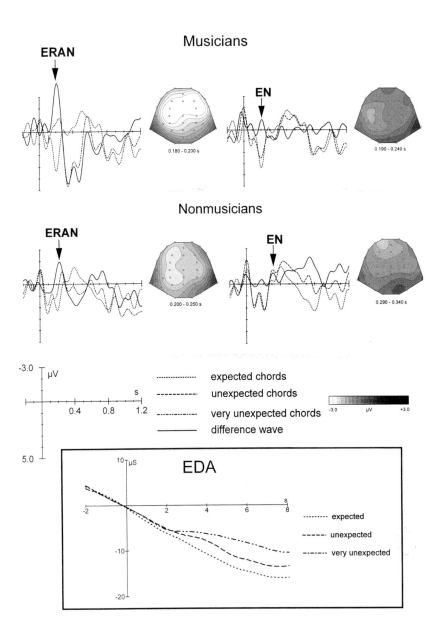

FIGURE 1. ERPs to harmonic expectancy violations. The difference waves indicate the processing of the violations (very unexpected–expected = ERAN; unexpected–expected = EN). The time course of the ERPs are shown from the F4 (right-anterior) electrode only, whereas adjacent scalp distributions display their strengths over all electrodes. Because of group differences in peak latency, the data for musicians and nonmusicians are displayed separately. The EDA data (*bottom*) are shown for all subjects.

RESULTS

Event-Related Potentials

The difference between ERP responses to the very unexpected and the expected harmonic events displays an ERAN for both groups (statistic from frontal electrodes: $F\{1,22\} = 11.43$; $P < .005$), which occurred slightly earlier for musicians than for nonmusicians ($t\{22\} = 2,048$; $P = .053$).

The difference between responses to the unexpected and the expected harmonic events also reveals a negativity for both groups ($F\{1,22\} = 4.99$; $P < .05$), which occurred considerably earlier for musicians than nonmusicians ($t\{22\} = -3.585$; $P < .005$). Because it is neither lateralized to the right, nor stronger over anterior leads, this negativity will be referred to as an early negativity (EN). It is assumed to reflect the same neural processes as the ERAN. This view is supported by a marginally significant difference in negativity size ($F\{1,22\} = 3.99$; $P = .058$), suggesting a sensitivity to increasing harmonic unexpectedness.

EDA and Subjective Response

The EDA increased in response to harmonic unexpectedness over time ($F\{2,46\} = 6.15$; $P = .01$). Equally, listeners reported an increased emotionality at the end of chorales containing increased harmonic unexpectedness ($F\{2,141\} = 17.591$; $P = .0001$). Because this increase was systematic with harmonic unexpectedness, this suggests that heightened harmonic unexpectedness is increasingly emotional (FIG. 1).

CONCLUSION

The data suggest that harmonically unexpected events do elicit emotional effects. Whereas the EEG data demonstrate that early negativities are sensitive to the degree of harmonic expectancy violation, the subjective and physiological data suggest that these violations predispose subjects to an increased emotionality. It can be argued that the violations lead to an increase in physiological "arousal," which in turn leads to a heightened emotional experience of the musical pieces. These data cannot be explained by manipulated versus original Bach, because the unexpected harmonies are the original compositions.

The role of the early negativities remains to be specified. It seems most likely that the detection of a harmonic expectancy violation will trigger a variety of different processes, including emotional ones. Recent neuroimaging work reports the activation of areas involved in emotional processing, such as the orbital frontolateral cortex, in response to harmonic expectancy violations.[8] In conjunction with the present data, this provides compelling support for the idea that suspensions of musical expectations are an important pathway to generating emotions in the listener.

ACKNOWLEDGMENT

This research was kindly supported by an award from the Arts and Humanities Research Board to Nikolaus Steinbeis.

[Competing interests: The authors declare that they have no competing financial interests.]

REFERENCES

1. SLOBODA, J.A. 1991. Music structure and emotional response: some empirical findings. Psychol. Mus. **19:** 110–120.
2. MEYER, L.B. 1956. Emotion and Meaning in Music. University of Chicago Press. Chicago.
3. BHARUCHA, J.J. & K. STOECKIG. 1986. Reaction time and musical expectancy: priming of chords. J. Exp. Psychol. Hum. Percept. Perform. **12:** 403–410.
4. KOELSCH, S., T. GUNTER, A.D. FRIEDERICI, *et al.* 2000. Brain indices of musical processing: "nonmusicians" are musical. J. Cogn. Neurosci. **12:** 520–541.
5. KRUMHANSL, C.L. 1997. An exploratory study of musical emotions and psychophysiology. Can. J. Exp. Psychol. **51:** 336–352.
6. BRADLEY, M.M. & P.J. LANG. 2000. Affective reaction to acoustic stimuli. Psychophysiology **37:** 204–215.
7. KHALFA, S., I. PERETZ, J. BLONDIN, *et al.* 2002. Event-related skin conductance responses to musical emotions in humans. Neurosci. Lett. **328:** 145–149.
8. KOELSCH, S., T. FRITZ, K. SCHULZE, *et al.* 2005. Adults and children processing music. An fMRI study. NeuroImage **25:** 1068–1076.

Dimensions of Emotion in Expressive Musical Performance

BRADLEY W. VINES,[a,b] CAROL L. KRUMHANSL,[c] MARCELO M. WANDERLEY,[b] IOANA M. DALCA,[a] AND DANIEL J. LEVITIN[a,b]

[a]Psychology Department, McGill University, Montreal, Quebec, Canada

[b]Faculty of Music and Center for Interdisciplinary Research in Music Media and Technology (CIRMMT), McGill University, Montreal, Quebec, Canada

[c]Psychology Department, Cornell University, Ithaca, New York, USA

ABSTRACT: This paper explores the dimensions of emotion conveyed by music. Participants rated emotion terms after seeing and/or hearing recordings of clarinet performances that varied in expressive content. A factor analysis revealed four independent dimensions of emotion. Changes to the clarinetists' expressive intentions did not significantly affect emotions conveyed by sound. It was largely through the visual modality that expressive intentions influenced the experience for observers.

KEYWORDS: crossmodal interactions; music cognition; emotion; performance

BACKGROUND

Multisensory Integration in Music

Musical performance involves both auditory and visual communication between musicians and observers. Body movements and postures convey a person's mental states and emotions.[1,2] Past research revealed that musicians' physical gestures convey their expressive intentions more accurately than the musical sound,[3] and that seeing a musician has a significant impact on the temporal dynamics of the experience for observers.[4] Additionally, music and dance can induce similar experiences of emotion and structural form.[5] Thus, the visual aspect of musical performance contributes information that is relevant to the audience's emotional response.

Address for correspondence: Daniel J. Levitin, Department of Psychology, McGill University, 1205 Avenue Penfield, Montreal, Quebec, Canada H2V 2A3. Voice: 514-398-8263; fax: 514-398-4896.

dlevitin@psych.mcgill.ca

Ann. N.Y. Acad. Sci. 1060: 462–466 (2005). © 2005 New York Academy of Sciences.
doi: 10.1196/annals.1360.052

TABLE 1. Factor analysis solution

Factor no. and name	Variance accounted for	Emotion terms
I. Active positive	24%	expressivity (.86), intensity (.81), movement (.81), quality (.75), surprise (.70), interest (.68), amusement (.59)
II. Active negative	16%	disgust (.76), anxiety (.76), anger (.75), contempt (.75), fear (.66)
III. Passive positive	14%	contentedness (.77), pleasantness (.67), relief (.62), happiness (.61), familiarity (.56)
IV. Passive negative	8%	embarrassment (.80), sadness (.66)

NOTE: The numbers in parentheses are correlation values for the emotion terms and their most closely related factor dimension.

METHODS

Stimuli

Stimuli were audio–video recordings of two professional clarinetists playing a Stravinsky piece for solo clarinet in three different manners: immobile, standard, and exaggerated.[6]

Participants

Thirty musically trained members of the McGill University community were divided randomly into three treatment groups: auditory only, visual only, and auditory + visual.

Task

For each stimulus, participants rated how much they experienced 19 emotions, using a five-point Likert scale, ranging from 1 ("not at all") to 5 ("very much"). The emotion terms (see TABLE 1) were drawn from the emotion and music cognition literatures.[5,7,8]

RESULTS AND DISCUSSION

Factor Analysis

We conducted a factor analysis to identify major independent dimensions of emotion. A principal component analysis extraction method with varimax rotation produced a four-factor solution, accounting for 62% of the total variance.

TABLE 1 shows the factors (with names based on past research[9]), total variance accounted for, and the associated emotion terms. Note that the dimensions corresponding to positive and negative emotions are orthogonal to each other, not bipolar.

These data provide preliminary evidence that musical emotions are like complex real-life experiences that can involve ambivalent states with both positive and negative emotions.[10,11]

Repeated Measures ANOVA Results

We conducted repeated-measures ANOVAs to the factor loadings (Z-scores for each participant on each emotion dimension). We focus here on the significant effects of the independent variables.

Active Positive Valence

The immobile condition induced significantly lower ratings overall ($P < .001$, $\eta^2 = .46$), primarily because of participants who could see the performances (see FIG. 1). Thus, the musicians' restriction of body movement tended to dampen active positive emotions when performances could be seen.

Auditory-only ratings changed very little across performance manners. Variation in the musicians' expressive intentions did not have an emotional impact by means of sound.

Only the auditory + visual group registered a linear increase in emotion from the standard manner to exaggerated. This is evidence of an emergent emotional intensity when performances can be both seen and heard (FIG. 1).

Passive Positive Valence

In contrast with ratings for performer R, passive positive emotion increased for performer W's immobile performance. There was a significant interaction between

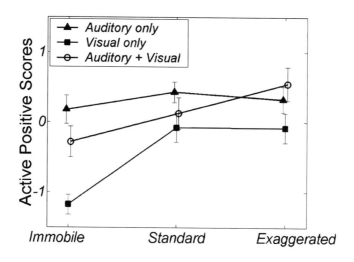

FIGURE 1. Factor loadings for the active positive dimension, with the mean and standard error of the mean shown for each presentation group and performance manner.

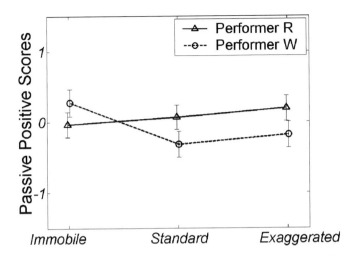

FIGURE 2. Factor loadings for the passive positive dimension, showing the mean and standard error of the mean for each performer and performance manner.

the performer and performance manner variables ($P = .015$, $\eta^2 = .14$). The relationship between induced emotion and performance intention, therefore, might be non-linear (i.e., a decrease in movement does not necessarily induce a decrease in amount of emotion conveyed), and it depends on the movement pattern of each performer (FIG. 2).

Active and Passive Negative Valence

Analyses did not reveal significant effects for the negative emotions. We posit that negative valence depends largely on the musical composition itself, and not on variations in the manipulated variables.

CONCLUSIONS

Visual experience was the primary channel through which variation in the clarinetists' performance intentions influenced the emotions of observers. Depending on the performer, a restriction in movement can actually increase positive emotions for the audience. These data suggest that, like complex real-life emotions, musical emotions can involve ambivalent states for both positive and negative emotions at once.

ACKNOWLEDGMENTS

Thanks go to the clarinetists and to research assistants in the Levitin Laboratory: Nora Hussein, Sawsan M'Birku, Hadiya Roderique, and Laura Morris. This work was supported by the J.W. McConnell Foundation and the CIRMMT Doctoral Fellowship and travel award (B.W.V.); a Guggenheim Fellowship and the National Science Foundation (C.L.K.); FQRNT Strategic Professor award (M.M.W.); and Valorisation Recherche Québec, the Natural Science and Engineering Research Council of Canada, and the Canadian Foundation for Innovation (D.J.L.).

[Competing interests: The authors declare that they have no competing financial interests.]

REFERENCES

1. DITTRICH, W.H. *et al.* 1996. Perception of emotion from dynamic point-light displays represented in dance. Perception **25:** 727–738.
2. RUNESON, S. & G. FRYKHOLM. 1983. Kinematic specification of dynamics as an informational basis for person-and-action perception. J. Exp. Psychol. Gen. **112:** 585–615.
3. DAVIDSON, J.W. 1993. Visual perception of performance manner in the movements of solo musicians. Psychol. Mus. **21:** 103–113.
4. VINES, B.W. *et al.* 2005. Cross-modal interactions in the perception of musical performance. Cognition. In press.
5. KRUMHANSL, C.L. & D.L. SCHENCK. 1997. Can dance reflect the structural and expressive qualities of music? A perceptual experiment on Balanchine's choreography of Mozart's Divertimento No. 15. Mus. Sci. **1:** 63–85.
6. WANDERLEY, M.M. 2002. Quantitative analysis of non-obvious performer gestures. *In* Gesture and Sign Language in Human-Computer Interaction. I. Wachsmuth & T. Sowa, Eds.: 241–253. Springer Verlag. Berlin.
7. ORTONY, A. & T.J. TURNER. 1990. What's basic about basic emotions? Psychol. Rev. **97:** 315–331.
8. RUSSELL, J.A. 1979. Affective space is bipolar. J. Pers. Soc. Psychol. **37:** 345–356.
9. TELLEGEN, A. *et al.* 1999. On the dimensional and hierarchical structure of affect. Psychol. Sci. **10:** 297–303.
10. LARSEN, J.T. *et al.* 2001. Can people feel happy and sad at the same time? J. Pers. Soc. Psychol. **81:** 684–696.
11. CACIOPPO, J.T. *et al.* 1997. Beyond bipolar conceptualizations and measures: the case of attitudes and evaluative space. Pers. Soc. Psychol. Rev. **1:** 3–25.

Complete List of Poster Papers Presented at the Conference

Part I. Ethology/Evolution: Do Animals Have Music or Something Else?

Entrainment, evolution, and ethology: together in time, sound, and action

J. Bispham and I. Cross

Department of Biological Anthropology, and Centre for Music and Science, Cambridge University, Cambridge, United Kingdom
e-mail address: jcb59@cam.ac.uk

Neuronal mechanisms underlying the perception of pitch and harmony

G. Langner

Department of Neuroacoustics, TU-Darmstadt, Germany
e-mail address: gl@bio.tu-darmstadt.de

Tonality and nonlinear resonance

E.W. Large and A.E. Tretakis

Center for Complex Systems and Brain Sciences, and Department of Psychology, Florida Atlantic University, Boca Raton, Florida, USA
e-mail address: large@ccs.fau.edu

Rhythmicity is a key feature of memory-enhancing auditory stimuli in an animal model

N.S. Rickard, S.R. Toukhsati, and S.E. Field

Psychology Department, Monash University, Victoria, Australia
e-mail address: nikki.rickard@med.monash.edu.au

Part II. Music and Language

Music: *Lingua Mundi*

A. Agnetta

Department of Linguistics, University of Palermo, Palermo, Italy
e-mail address: agnetta.alessandro@lettere.unipa.it

Ann. N.Y. Acad. Sci. 1060: 467–487 (2005). © 2005 New York Academy of Sciences.
doi: 10.1196/annals.1360.100

Processing of sung vowels under extreme pitch conditions
T. Andreas and T. Pechmann

Graduate Program "Universality and Diversity,"
University of Leipzig, Leipzig, Germany
e-mail address: andreas@uni-leipzig.de

Musicality as a sign for language in autism and SLI
L. Brandi, C. Simonetti, and B. Salvadori

Department of Linguistics, University of Florence, Florence, Italy
e-mail address: brandi@unifi.it

Ataxic dysarthria in children with congenital cerebellar dysmorphology is related to rhythmic tapping but not to perceptual or motor timing
M. Dennis and J.H. Okrainec

Brain and Behaviour Program, The Hospital for Sick Children,
Toronto, Ontario, Canada
e-mail address: maureen.dennis@sickkids.ca

Neural correlates of rapid spectrotemporal processing in musicians and nonmusicians
N. Gaab, P. Tallal, C.H. Kim, K. Lakshminarayanan, J.J. Archie, G.H. Glover, and J.D.E. Gabrieli

Department of Psychology, Stanford University,
Stanford, California, USA
e-mail address: gaab@MIT.EDU

"Traduttore/traditore." Communication challenges in musical interpretation
V. Gaspar

National University of Music, Bucharest, Romania
e-mail address: veronicagaspar@hotmail.com

Differences in musicians and nonmusicians in rapid auditory temporospectral processing of nonlinguistic and linguistic stimuli
H. Kim, N. Gaab, K. Lakshminarayanan, P. Tallal, and J.D.E. Gabrieli

Department of Psychology, Stanford University,
Stanford, California, USA
e-mail address: heesoo@mit.edu

Interactions between lyrics and tunes: vowels sing but consonants speak
P. Lidji, R. Kolinsky, I. Peretz, C. Astesano, M. Besson, and J. Morais

Université Libre de Bruxelles, Brussels;
Université de Montréal, Montréal, Quebec, Canada
e-mail address: plidji@ulb.ac.be

The role of selective attention in the processing of syntactical violations in music and speech

C. Maidhof, S. Koelsch, and U. Seifert

Institute of Musicology, University of Cologne, and the Max Planck Institute for Human Cognitive and Brain Sciences, Leipzig, Germany
e-mail address: maidhof@cbs.mpg.de

The perception of music phrase structure: from a cross-cultural perspective

Y. Nan, T.R. Knösche, and A.D. Friederici

Max Planck Institute for Human Cognitive and Brain Sciences, Leipzig, Germany
e-mail address: nanyun@cbs.mpg.de

Musical expertise and local acoustic cues affect phrase perception in music

C. Neuhaus, T.R. Knösche, and A.D. Friederici

Max Planck Institute for Human Cognitive and Brain Sciences, Leipzig, Germany
e-mail address: neuhaus@cbs.mpg.de

Auditory chimaeras: distinct and shared networks for speech and music perception

T. Overath, C. Narain, S.K. Scott, and P.M. Matthews

Wellcome Department of Imaging Neuroscience, University College, London, United Kingdom
e-mail address: t.overath@fil.ion.ucl.ac.uk

The effects of melodic intonation therapy on the neural basis of speech in a patient with severe nonfluent aphasia

K. Overy, E. Ozdemir, A. Norton, N. Helm-Estabrooks, and G. Schlaug

Institute for Music in Human and Social Development, School of Arts, Culture and Environments, University of Edinburgh, Edinburgh, Scotland
e-mail address: k.overy@ed.ac.uk

Music and language intersections in 20th century posttonal music: overlooked opportunities for research in cognitive neuroscience

B. Ruviaro and R. Segnino

CCRMA Center for Computer Research in Music and Acoustics, Department of Music, Stanford University, Stanford, California, USA
e-mail address: ruviaro@stanford.edu

Working memory for music and language in five-year-old children with normal versus impaired language development

S. Sallat and F.J. Stachowiak

Department of Special Education Speech and Language Pathology, Justus-Liebig-University of Gießen, Gießen, Germany
e-mail address: sallat@foepaed.net

Spectral and temporal processing in the human auditory cortex—revisited
M. Schönwiesner, R. Rübsamen, and D.Y. von Cramon

Institute of Zoology, Faculty of Biology II, University of Leipzig;
Max Planck Institute of Human Cognitive and Brain Sciences,
Leipzig, Germany
e-mail address: marcs@uni-leipzig.de

Preattentive auditory processing of vowel duration changes in monosyllabic Thai words: comparison between native and nonspeakers of Thai
W. Sittiprapaporn, W. Khampan, T.-C. Wang, P. Ananchaipatthana,
C. Chindaduangratn, and N. Kotchabhakdi

Neuro-Behavioural Biology Center, Mahidol University,
Salaya, Nakhonpathom, Thailand
e-mail address: stwst@mahidol.ac.th

Influence of musical training on pitch processing: event-related brain potential studies of adults and children
S. Moreno and M. Besson

INCM Institut de Neurosciences Cognitives de la Méditerranée,
Marseille, France
e-mail address: moreno@incm.cnr_mrs.fr

Words versus pitch: differences in the perception of speech and song
R. van Besouw, D.M. Howard, and S. Ternström

Department of Electronics, University of York, York, United Kingdom
e-mail address: rmvb101@ohm.york.ac.uk

To musicians, the message is in the meter
P. Vuust, K.J. Pallesen, C. Bailey, T.L. van Zuijen, A. Gjedde,
A. Roepstorff, and L. Østergaard

Centre for Functionally Integrative Neuroscience, University Hospital of
Aarhus and Royal Academy of Music, Aarhus, Denmark
e-mail address: pv@pet.auh.dk

Part III. Mental Representations

How scrambling chord progressions with secondary dominant chords affects global expectations
N.B. Atalay and H.G. Tekman

Cognitive Science/Informatics Institute, Ankara, Turkey
e-mail address: bedin@ii.metu.edu.tr

Modeling intrinsic oscillations in the auditory system: a neuronal mechanism for quantal pitch shifts and absolute pitch?
A. Bahmer and G. Langner

Institute for Zoology, Department of Neuroacoustics,
Technical University of Darmstadt, Darmstadt, Germany
e-mail address: a_bahmer@gmx.de

A network for sensory-motor integration: what happens in the auditory cortex during piano playing without acoustic feedback?
S. Baumann, S. Koeneke, M. Meyer, K. Lutz, and L. Jäncke

Division of Neuropsychology, University of Zurich, Zurich, Switzerland
e-mail address: s.baumann@psychologie.unizh.ch

The neuronal representation of pitch in primate auditory cortex
D. Bendor and X. Wang

Biomedical Engineering, School of Medicine,
Johns Hopkins University, Baltimore, Maryland, USA
e-mail address: dbendor@yahoo.com

Culture-specific tonal knowledge drives judgments about an unfamiliar tonality: a probe-tone study
M. Curtis and J. Bharucha

Department of Psychology, Tufts University, Boston, Massachusetts, USA
e-mail address: mcurtis@dartmouth.edu

The role of culture in musical memory
S.M. Demorest, S.J. Morrison, M. Beken, and D. Jungbluth

School of Music, University of Washington, Seattle, Washington, USA
e-mail address: demorest@u.washington.edu

Similarity rating and JND measurements on timbre perception in hearing-impaired and normal listeners
S. Emiroglu and B. Kollmeier

Institute of Physics, Oldenburg University, Oldenburg, Germany
e-mail address: suzan.emiroglu@mail.uni-oldenburg.de

Cannabis and music perception in a pre/post qEEG brain mapping investigation
J. Fachner

Chair for Qualitative Research in Medicine, Faculty for Medicine,
University of Witten/Herdecke, Germany
e-mail address: joergf@uni-wh.de

Neural correlation underlying the hierarchical ordering of chords in tonal context: a magnetoencephalographic study
D. Fukuda, A. Otsuka, S. Kuriki, Y. Takei, K. Sumie,
M. Fukuda, and N. Murata

Department of Science and Engineering, University of Waseda,
Tokyo, Japan
e-mail address: flying_duck58@toki.waseda.jp

Music error detection by absolute and relative pitch possessors
G.P. Hoplaros, E. Costa-Giomi, and B. Pennycook

School of Music, University of Texas, Austin, Texas, USA
e-mail address: georgehoplaros@mail.utexas.edu

Factors influencing the perception of rhythm in music

K.M. Howland

The Boston Conservatory, Lesley University,
Cambridge, Massachusetts, USA
e-mail address: kmhowland@freshvinyl.com

Modulation of brain responses to rhythmic sequences by subjective metrical interpretation

J.R. Iversen, B.H. Repp, and A.D. Patel

The Neurosciences Institute, San Diego, California, USA
e-mail address: iversen@nsi.edu

Dividing attention between music and vision: an fMRI study

J.A. Johnson and R.J. Zatorre

Neuropsychology, Montreal Neurological Institute, McGill University,
Montreal, Quebec, Canada
e-mail address: JennyJ@ego.psych.mcgill.ca

When the tempo changes: different brain mechanisms for detecting speeding up or slowing down

M.L.A. Jongsma, T. Eichele, and P.G. Vos

NICI—Nijmegen Instiute of Cognition and Information,
Department of Biological Psychology, Radboud University of Nijmegen,
Nijmegen, The Netherlands
e-mail address: jongsma@nici.ru.nl

Neural representations of tonal categories

K. Krohn, E. Brattico, and M. Tervaniemi

Cognitive Brain Research Unit, University of Helsinki,
Helsinki, Finland
e-mail address: kaisu.krohn@helsinki.fi

The prominent role of voice in melody recognition

O. Ladinig and O. Vitouch

Cognitive Psychology Unit, Department of Psychology,
University of Klagenfurt, Klagenfurt, Austria
e-mail address: olivia.ladinig@uni-klu.ac.at

The power of listening: auditory-motor interactions in musical training

A. Lahav, A. Boulanger, G. Schlaug, and E. Saltzman

The Music, Mind and Motion Lab, Department of Rehabilitation Science,
Sargent College of Health, Boston University,
Boston, Massachusetts, USA
e-mail address: lahav@bu.edu

A piano in the head? Spatial associations of response codes for musical stimuli

P. Lidji, R. Kolinsky, A. Lochy, and J. Morais

Université Libre de Bruxelles, Brussels, Belgium;
Université de Montréal, Montréal, Quebec, Canada
e-mail address: plidji@ulb.ac.be

Harmonic expectation, selective attention, and affect in music

P. Loui and D. Wessel

Department of Psychology, Center for New Music and Audio
Technologies, University of California at Berkeley,
Berkeley, California, USA
e-mail address: psyche@berkeley.edu

Cognitive priming in melody perception

F. Marmel and B. Tillmann

CNRS-UCBL UMR 5020 "Neurosciences et Systèmes sensoriels,"
Université de Lyon, Lyon, France
e-mail address: frederic.marmel@olfac.univ-lyon1.fr

Brain-computer music interfacing: assistive music technology for severe disability

E. Miranda, A. Brouse, B. Boskamp, and H. Mullaney

Computer Music Research, University of Plymouth,
Plymouth, United Kingdom
e-mail address: eduardo.miranda@plymouth.ac.uk

Auditory gestalt formation in musicians and nonmusicians—an ERP study

C. Neuhaus and T.R. Knösche

Max Planck Institute for Human Cognitive and Brain Sciences,
Leipzig, Germany
e-mail address: neuhaus@cbs.mpg.de

Analysis of timbral microvariations of noisy musical sounds using computational models of the auditory periphery and multidimensional scaling

G. Papadelis and K. Pastiadis

Department of Music Studies, Aristotles University of Thessaloniki,
Thessaloniki, Greece
e-mail address: papadelis@mus.auth.gr

Contributions of motor and vestibular sensory systems to musical rhythm perception

J. Phillips-Silver and L.J. Trainor

Department of Psychology, Neuroscience and Behaviour,
McMaster University, Hamilton, Ontario, Canada
e-mail address: phillij@mcmaster.ca

Attention effects on the processing of musical syntax in nonmusicians
B. Poulin-Charronnat, S. Koelsch, and E. Bigand

Max Planck Institute for Human Cognitive and Brain Sciences,
Leipzig, Germany
e-mail address: poulin@cbs.mpg.de

Tapping on and off a self-imposed beat: metrical interpretation affects coordination
B.H. Repp

Haskins Laboratories, New Haven, Connecticut, USA
e-mail address: repp@haskins.yale.edu

Absolute pitch confers an enhanced ability to distinguish between properly and improperly tuned stimuli
D.A. Ross and L.E. Marks

Yale University, New Haven, Connecticut, USA
e-mail address:

Absolute pitch confers enhanced memory for musical sequences
D.A. Ross and L.E. Marks

Yale University, New Haven, Connecticut, USA
e-mail address: david.a.ross@yale.edu

The mental space of pitch height
E. Rusconi, B. Kwan, B. Giordano, C. Umiltà, and B. Butterworth

Institute of Cognitive Neuroscience, University College,
London, United Kingdom
e-mail address: e.rusconi@ucl.ac.uk

The relative influence of pitch and timbre on perceived interval size
F.A. Russo and W.F. Thompson

Department of Psychology, University of Toronto at Mississauga,
Mississauga, Ontario, Canada
e-mail address: frusso@utm.utoronto.ca

Musicians' dreams
P. Salzarulo, V. Uga, M.C. Lemut, C. Zampi, and I. Zilli

Sleep Lab, Department of Psychology,
University of Florence, Florence, Italy
e-mail address: salzarulo@psico.unifi.it

Working memory in musicians and nonmusicians: an fMRI study
K. Schulze and S. Koelsch

Max Planck Institute for Human Cognitive and Brain Sciences,
Leipzig, Germany
e-mail address: kschulze@cbs.mpg.de

Structural brain correlates of visuospatial memory in musicians
V. Sluming, D. Page, C. Denby, J. Downes, A. Mayes, and N. Roberts

Department of Medical Imaging, University of Liverpool,
Liverpool, United Kingdom
e-mail address: vanessa.sluming@liv.ac.uk

Jeannie, Kirk, and John Boy
S. Strauß S. and O. Vitouch

Cognitive Psychology Unit (CPU), Department of Psychology,
University of Klagenfurt, Klagenfurt, Austria
e-mail address: s2straus@edu.uni-klu.ac.at

A universal map of major–minor relationships
A. Tangian

Hans Böckler Stiftung, Düsseldorf, Germany
e-mail address: Andranik-Tangian@BOECKLER.DE

Absolute pitch and Greek culture
M. Vraka and G. Welch

Institute of Education, University of London,
London, United Kingdom
e-mail address: mariavraka@hotmail.com

Matching symbols and sounds: evidence from oscillatory gamma activity
A. Widmann, T. Gruber, T. Kujala, M. Tervaniemi, and E. Schröger

Institut für Psychologie I, University of Leipzig, Leipzig, Germany
e-mail address: widmann@uni-leipzig.de

Interval cycles and tonal hierarchies
M. Woolhouse and I. Cross

Centre for Music and Science, Faculty of Music, University of Cambridge,
Cambridge, United Kingdom
e-mail address: mhw31@cam.ac.uk

Effects of basic musical structure on the neural responses: a magnetoencephalographic study
Asuka Otsuka, Toshikazu Hasegawa, Yoshihiro Hirata, and Shinya Kuriki

Research Institute for Electronic Science, University of Hokkaido, Japan
e-mail address: otsuka@ecs.c.u-tokyo.ac.jp

Part IV. Developmental Aspects and Impact of Music on Education

An Indian musical counting game evaluated for use in speech therapy for stutterers
S. Bull

Escola Superior de Música, Lisbon, Portugal
e-mail address: sgb@ip.pt

Does music instruction improve fine motor abilities?

E. Costa-Giomi

Center for Music Learning, University of Texas, Austin, Texas, USA
e-mail address: costagiomi@mail.utexas.edu

A review of the practical uses of music to improve human performance, function, and health

D. Craig

Department of Occupational Therapy, Department of Music,
Tufts University, Boston, Massachusetts, USA
e-mail address: Craiger91@hotmail.com

The role of rhythm, syntax, and musical training in similarity judgments of series of words and letters

M.M. Marin and R. Parncutt

Language Development and Cognitive Science Unit,
Department of Musicology, Graz University, Graz, Austria
e-mail address: marinmanuela@hotmail.com

Neural and behavioral discrimination of music attributes in children involved in music specialization programs

R. Milovanov, M. Tervaniemi, and V. Välimäki

Department of English/Centre for Cognitive Neuroscience,
University of Turku, Turku, Finland
e-mail address: riheino@utu.fi

The development of short-term memory for melodies in children and adolescents: the effects of intervals, tonal context, and musical training

A.E. Paltoglou

Medical Research Council, Institute of Hearing Research,
Nottingham, United Kingdom
e-mail address: Aspa@ihr.mrc.ac.uk

The effect of early musical training on adult motor performance: evidence for a sensitive period in motor learning

V. Penhune, D. Watanabe, and T. Savion-Lemieux

Department of Psychology, Concordia University,
Montreal, Quebec, Canada
e-mail address: vpenhune@vax2.concordia.ca

Melody perception in two-month-old infants

J. Plantinga and L.J. Trainor

Department of Psychology, Neuroscience and Behaviour,
McMaster University, Hamilton, Ontario, Canada
e-mail address: plantija@mcmaster.ca

Long-range synchrony in the gamma band and musical styles
D. Spada and M. Poli

Department of Psychology, Medical School,
University of Milan, Milan, Italy
e-mail address: danilo.spada@unimi.it

The effects of education and career path on the categorization of trumpet articulations
S. Spiegelberg

School of Music, DePauw University, Greencastle, Indiana, USA
e-mail address: Spiegelberg@depauw.edu

Children's songs effectively teach reading to kindergarten children
P. Walton

Thompson Rivers University, British Columbia, Canada
e-mail address: pwalton@cariboo.bc.ca

Roundtable on Music Therapy

Music therapy in tinnitus: a prospective pilot study
H. Argstatter, S. Hoth, G. Dyckhoff, A.K. Nickel, H. Weidauer, and H.V. Bolay

German Center for Music Therapy, Research Institute of the University of Applied Sciences, Heidelberg, Heidelberg, Germany
e-mail address: heike.argstatter@fh-heidelberg.de

Evaluating the effects of a structured music training on adult autistic patients: first follow-up (6 months)
F. Barale, M. Boso, P.L. Politi, A. Pace, L. Delfino, P. Aliata, and S. Ucelli

Department of Applied Health and Psychobehavioural Sciences, University of Pavia, Pavia, Italy
e-mail address: francesco.barale@unipv.it

Music and sensorial response of the fetus in the uterus
F. Di Stefano and N. Cantarella

II Unità Operativa Complessa, Hospital "San Bambino," Catania, Sicily
e-mail address: nellycantarella@virgilio.it

Phenomenon of a sound–color simultaneous representation (SCSR)
I. Miroshnik and E. Gavrilin

The Centre for Personality-Oriented Computerized Psychotherapy (Poc-Therapy), Ukraine
e-mail address: pocther@ok.ru

Sound to sense, sense to sound: the S2S^2 project
D. Cirotteau and N. Bernardini

Media Innovation Unit, Firenze Tecnologia, Florence, Italy
e-mail address: d.cirotteau@miu-ft.org

Twenty years of research in music/medicine sponsored by ISMM
R. Spintge and R. Droh

ISMM, International Society for Music in Medicine, Department of Algesiology Sportkrankenhaus Hellersen, Lüdenscheid, Germany
e-mail address: Ralph.Spintge@sportkrankenhaus.de

Effectiveness of music therapy in the treatment of children with migraine headache
A.K. Nickel, R. Oelkers-Ax, P. Parzer, T.K. Hillecke, H. Argstatter, F. Resch, and H.V. Bolay

German Center for Music Therapy Research, Heidelberg Child and Adolescent Psychiatry of the University Heidelberg, Heidelberg, Germany
e-mail address: anne.nickel@fh-heidelberg.de

Undertaking and improving the effectiveness of musical interventions in clinical settings
C. Preti

Music Department, Institute of Education, University of London, London, United Kingdom
e-mail address: cpreti@ioe.ac.uk

Music performance and expressiveness in music therapy against chronic pain disease
V. Busch, T. Hillecke, H.V. Bolay, and W. Auhagen

Department of Musicology, Martin Luther University, Halle-Wittenberg, Germany
e-mail address: veronika@busch.net

Musical behavior during music therapy in children with Williams syndrome
A. Colletti and M. Rodocanachi

Don Calabria Insitute, Peppino Vismara Center, Milan, Italy
e-mail address: angeloraffaele.colletti@fastwebnet.it

Part V. Neurological Disorders and Music

The effect of music on dopaminergic function in the brain
K. Akiyama and D. Sutoo

Institute of Medical Science, University of Tsukuba, Tsukuba, Japan
e-mail address: akiyamak@md.tsukuba.ac.jp

Music perception in children with right or left hemispherectomy in infancy
M. Dennis

Brain and Behaviour Program, The Hospital for Sick Children, Toronto, Ontario, Canada
e-mail address: maureen.dennis@sickkids.ca

Perception of metrical structure is related to cerebellar volumes in children with spina bifida

M. Dennis, T. Misakyan, and G.E. Schellenberg

Brain and Behaviour Program, The Hospital for Sick Children, Toronto, Ontario, Canada
e-mail address: maureen.dennis@sickkids.ca

Semeiological features of auditory auras in patients with IPEAF

I. Florindo, F. Pittau, F. Bisulli, I. Naldi, R. Michelucci, P. Striano, S. Striano, P. Avoni, G. D'Orsi, C.A. Tassinari, A. Baruzzi, and P. Tinuper

Department of Neurology, University of Bologna, Bologna, Italy
e-mail address: ireneflo@libero.it

Memory impairment and preservation of episodic musical memory in Alzheimer disease

L. Fornazzari, T. Castle, S. Nadkarni, M. Dielle, N. Apanasiewicz, M. Ambrose, and F. Phillips

Multicultural Multilingual Memory Clinic, Geriatric Mental Health Centre for Addiction and Mental Health, University of Toronto, Toronto, Ontario, Canada
e-mail address: fornazzari@sympatico.ca

Rhythm analysis in tone deafness

J.M. Foxton, R.K. Nandy, and T.D. Griffiths

Auditory Group, University of Newcastle upon Tyne, United Kingdom
e-mail address: j.m.foxton@ncl.ac.uk

Rhythm processing in patients with basal ganglia dysfunction

J. Grahn and M. Brett

MRC Cognition and Brain Sciences Unit, Cambridge, United Kingdom
e-mail address: jessica.grahn@mrc-cbu.cam.ac.uk

Morphometry of the amusic brain: a two-site study

K.L. Hyde, R.J. Zatorre, T.D. Griffiths, J. Lerch, and I. Peretz

Department of Psychology, University of Montreal, Montreal, Quebec, Canada
e-mail address: krista.hyde@umontreal.ca

Focal dystonia in pianists: objective treatment monitoring reveals beneficial effects of medical treatment and retraining therapy

H.C. Jabusch, L. Boullet, and E. Altenmüller

Institute of Music Physiology and Musicians' Medicine, Hannover University of Music and Drama, Hannover, Germany
e-mail address: jabusch@hmt-hannover.de

The effect of external rhythmic stimulation on auditory syntactic processing: ERP evidence from patients with basal ganglia lesions

S.A. Kotz, T.C. Gunter, and S. Wonneberger

Max Planck Institute of Human Cognitive and Brain Sciences, Leipzig, Germany
e-mail address: kotz@cbs.mpg.de

Musicians with tic disorders

R.J. Lederman

Department of Neurology and Medical Center for Performing Artists, Cleveland Clinic Foundation, Cleveland, Ohio, USA
e-mail address: ledermr@ccf.org

Enhanced P1-N1 auditory evoked potential in patients with musicians' cramp

V.K. Lim, J.L. Bradshaw, M.E.R. Nicholls, and E. Altenmüller

University of Auckland, New Zealand, and Hochschule für Musik und Theater, Hannover, Germany
e-mail address: v.lim@auckland.ac.nz

Music to electric ears: pitch and timbre perception by cochlear implant patients

D. Pressnitzer, J. Bestel, and B. Fraysse

Equipe Audition, LPE-CNRS UMR 8581, Département d'Etudes Cognitives, Ecole Normale Supérieure, Paris, France
e-mail address: Daniel.Pressnitzer@ens.fr

Processing of music-syntactic irregularities and basic auditory features in patients with left hemisphere lesions—an ERP study

D. Sammler and S. Koelsch

Max Planck Institute for Human Cognitive and Brain Sciences, Leipzig, Germany
e-mail address: sammler@cbs.mpg.de

Neural correlates of congenital amusia

G. Schlaug, K. Schulze, and J. Mandell

Department of Neurology, Music and Neuroimaging Laboratory, Beth Israel Deaconess Medical Center and Harvard Medical School, Boston, Massachusetts, USA
e-mail address: gschlaug@bidmc.harvard.edu

Facilitating recovery of expressive language functions through melodic intonation therapy

G. Schlaug, A. Norton, E. Ozdemir, K. Overy, and N. Helm-Estabrooks

Department of Neurology, Music and Neuroimaging Laboratory, Beth Israel Deaconess Medical Center and Harvard Medical School, Boston, Massachusetts, USA
e-mail address: gschlaug@bidmc.harvard.edu

Music performance in autism: the special case of a patient with "musical stereotypies"
F. Barale, M. Boso, A. Pace, P.L. Politi, L. Delfino, P. Aliata, and S. Ucelli

Department of Applied Health and Psychobehavioural Sciences, University of Pavia, Pavia, Italy
e-mail address: francesco.barale@unipv.it

Part VI. Music Performance

Adoption of a new neurobiofeedback approach to the evaluation, education, and correction of musical performing skills: advanced diagnosis technologies, treatment planning, and follow-up
O.M. Bazanova and M.B. Shtark

Molecular Biology and Biophysics, Siberian Branch, Russian Academy of Medical Science
e-mail address: bazanova@soramn.ru

Music performance training and learnability: an EEG investigation
O.M. Bazanova and A.V. Kondratenko

Molecular Biology and Biophysics, Siberian Branch, Russian Academy of Medical Science
e-mail address: bazanova@soramn.ru

Differences in gray matter between musicians and nonmusicians
P. Bermudez and R.J. Zatorre

Cognitive Neuroscience Unit, Montreal Neurological Institute, McGill University, Montreal, Quebec, Canada
e-mail address: pat@ego.psych.mcgill.ca

Notational audiation: perceiving the mind's voice
W. Brodsky, A. Henik, and B-S. Rubinstein

Department of the Arts, Ben-Gurion University of the Negev, Beer-Sheva, Israel
e-mail address: wbrodsky@bgumail.bgu.ac.il

Tapping in synchrony to auditory rhythms: effect of temporal structure on behavior and neural activity
J. L. Chen, V.B. Penhune, and R.J. Zatorre

Department of Neuropsychology, Montreal Neurological Institute, McGill University, and Concordia University, Montreal, Quebec, Canada
e-mail address: joyce.chen@mail.mcgill.ca

Singing is stable and accurate in adult nonmusicians
S. Dalla Bella, J.F. Giguère, and I. Peretz

Department of Cognitive Psychology, University of Finance and Management, Warsaw, Poland
e-mail address: sdallabella@vrja.pldesain@nici.ru.nl

Motor behavior in drumming with different striking tools: effects of tempo on control of rebound

S. Dahl and R. Huys

KTH Speech, Music and Hearing, Royal Institute of Technology,
Stockholm, Sweden
e-mail address: sofia@speech.kth.se

Using multichannel near infrared spectroscopy for the study of rhythm processing

P. Desain, H. Katayose, and M. Hashida

MMM Group, NICI, Radboud University Nijmegen,
Nijmegen, The Netherlands
e-mail address: desain@nici.ru.nl

Kinematics and kinetics of a simple octave keystroke movement in the piano

S. Furuya, T. Aoki, H. Tsuda, and H. Kinoshita

Graduate School of Medicine, University of Osaka, Osaka, Japan
e-mail address: furuya@moted.hss.osaka-u.ac.jp

An event-related brain potential study of the responses to human voices versus musical instruments

R.Y. Granot, D. Abecasis, D. Levy, and S. Bentin

Department of Musicology, The Hebrew University of Jerusalem,
Jerusalem, Israel
e-mail address: rgranot@huji.013.net.il

Electrocortical activity as a function of expertise in execution and perception of expressive tonal music

C. James, M.P. Deiber, R. Caldara, E. Bigand, C. Michel, P. Vuilleumier, and C.A. Hauert

University of Geneva, Geneva, Switzerland
e-mail address: clara.james@pse.unige.ch

Polyrhythm production is affected by cognitive as well as motor constraints

P.E. Keller, E. Saltzman, and B.H. Repp

Department of Cognitive Psychology, University of Finance and
Management, Warsaw, Poland
e-mail address: keller@vizja.pl

Bow-arm kinematics in trained and untrained violin playing

J. Kiggins, J. James, T. Ingalls, M. Santello, and S.H. Tillery

Harrington Department of Bioengineering, Arts, Media and Engineering
Program, Department of Kinesiology, Arizona State University,
Tempe, Arizona, USA
e-mail address: justin.kiggins@asu.edu

Keystroke force in the piano
H. Kinoshita, S. Furuya, T. Aoki, and H. Tsuda

Graduate School of Medicine and School of Health and Sports Sciences,
University of Osaka, Osaka, Japan
e-mail address: hkinoschita@moted.hss.osaka-u.ac.jp

Non-right-handers do it better: reduced lateralization results in better sight-reading achievement
R. Kopiez and N. Galley

Hannover University of Music and Drama, Institute for Research in Music
Education, Hannover, Germany
e-mail address: kopiez@hmt-hannover.de

Listen to hearing—On David Tudor's composition "Neural Synthesis"
J. Kursell

Max Planck Institute for the History of Science, Berlin, Germany
e-mail address: kursell@mpiwg-berlin.mpg.de

Changes in corticospinal excitability while listening to newly learned melodies: a TMS-study
I. Mutschler, A. Schulze-Bonhage, U. Halsband, S. Gräf, R. Martmüller,
S. Rummler, and T. Ball

Department of Psychology, Neuropsychology, University of Freiburg,
Freiburg, Germany
e-mail address: isabella.mutschler@psychologie.uni-freiburg.de

Following changes in tempo
S.K. Rankin, E.W. Large, P. Fink, and S. Houlton

Center for Complex Systems and Brain Sciences,
Florida Atlantic University, Boca Raton, Florida, USA
e-mail address: rankin@ccs.fau.edu

Cortical imaging of imagery and motor components of piano playing
M. Richter, O. Vitouch, F.Ph.S Fischmeister, and H. Bauer

Fakultät für Psychologie, Institute for Clinical, Biological and Differential
Psychology, Vienna, Austria
e-mail address: maria.richter@univie.ac.at

Metaphoric and conventional conceptions of musical expression
R.K. Rosenthal

VanderCook College of Music, Chicago, Illinois, USA
e-mail address: rrosenthal@vandercook.edu

Facial expression of pitch relations
W.F. Thompson and F.A. Russo

Communication, Culture and Information Technology,
University of Toronto at Mississauga, Mississauga, Ontario, Canada
e-mail address: b.thompson@utoronto.ca

Neural substrates governing audiovocal integration for vocal pitch regulation in singing
J.M. Zarate and R.J. Zatorre

Cognitive Neuroscience Unit and Department of Neurology and Neurosurgery, Montreal Neurological Institute, McGill University, Montreal, Quebec, Canada
e-mail address: jean.zarate@mcgill.ca

Local and global processing of pitch patterns: an fMRI study
L. Stewart, J. Warren, J. Foxton, and T. Griffiths

UCL and University of Newcastle, Newcastle, United Kingdom
e-mail address: l.stewart@ucl.ac.uk

Part VII. Emotion in Music

Rating expressive music performance with colors
R. Bresin

Speech Music and Hearing Department, KTH Royal Institute of Technology, Stockholm, Sweden
e-mail address: roberto@speech.kth.se

Music influences absorption in motion pictures: genre interactions
A.J. Cohen, K.A. MacMillan, and R.N. Drew

Department of Psychology, University of Prince Edward Island, Charlottetown, Prince Edward Island, Canada
e-mail address: acohen@upei.ca

Physiological changes during chills induced by music
D. Craig

Department of Neuroscience/Physiology, Department of Music, University of Connecticut, Storrs, Connecticut, USA
e-mail address: Craiger91@hotmail.com

Meaningfulness of music as a function of emotional response
D. Craig

Department of Occupational Therapy, Department of Music, Tufts University, Boston, Massachusetts, USA
e-mail address: Craiger91@hotmail.com

Recognition of emotion expressed by music in patients with medial temporal lobe epilepsy
D. Dellacherie, N. Ehrlé, C. Lancelot, M. Baulac, and S. Samson

Research Unit on Behavioural and Learning Development, Psychology Unit, University of Lille 3/Epilepsy Unit, La Salpêtrière Hospital, Paris, France
e-mail address: delphine.dellacherie1@libertysurf.fr

Remember Bach: an investigation in episodic memory for music

S. Eschrich, T.F. Münte, and E.O. Altenmüller

Institute of Music Physiology and Musicians' Medicine,
University of Music and Drama, Hannover, Germany
e-mail address: eschrich@hmt-hannover.de

Key processing precedes emotional categorization of Western music

S. Filipic and E. Bigand

LEAD-CNRS 5022, University of Bourgogne, Dijon, France
e-mail address: Suzanne.Filipic@leadserv.u-bourgogne.fr

Investigating the initial response to pleasant and unpleasant music with fMRI

T. Fritz and S. Koelsch

Neurocognition of Music Perception, Max Planck Institute for Human
Cognitive and Brain Sciences, Leipzig, Germany
e-mail address: fritz@cbs.mpg.de

The effects of music listening on endocrine hormones from the viewpoint of interval changes and musical taste

H. Fukui, K. Toyoshima, and K. Kuda

Department of Education, Nara University of Education, Japan
e-mail address:

Why does this sound so sad? Searching for the psychophysical parameters underlying affective categorization of expressive musical sounds

K.N. Goydke, H. Colonius, T.F. Münte, and E. Altenmüller

Institute of Music Physiology and Musicians' Medicine,
University of Music and Drama, Hannover, Germany
e-mail address: fukuih@nara-edu.ac.jp

How does music arouse "chills"? Investigating strong emotions, combining psychological, physiological, and psychoacoustical methods

O. Grewe, F. Nagel, R. Kopiez, and E. Altenmüller

Institute of Music Physiology and Musicians' Medicine,
University of Music and Drama, Hannover, Germany
e-mail address: Oliver_Grewe@gmx.de

Frontal brain activation during listening to pleasant versus unpleasant and high- versus low-arousal music: the moderating effects of habitual listening to classical music, and extrovert (impulsivity) and introvert (neuroticism-anxiety) dimensions of personality

K. Kallinen and N. Ravaja

Helsinki School of Economics, Helsinki, Finland
e-mail address: kari.kallinen@maxinetti.fi

Affective valence of music as a context for word processing: the effect of musical expertise
J. Kantor-Martynuska, E. Bigand, and C. Delbé

LEAD-CNRS, University of Burgundy, Dijon, France
e-mail address: joanna.kantor@leadserv.u-bourgogne.fr

Brain regions involved in recognition of happiness and sadness in music
S. Khalfa, D. Schön, and C. Liégeois-Chauvel

Neurophysiology and Neuropsychology Laboratory, Inserm EMI-U 9926 Medical School, Université de la Méditerranée La Timone, Marseille, France
e-mail address: skhalfa@skhalfa.com

The role of musical features in perceiving and inducing emotions
I. Mutschler, L. Schiel, A. Schulze-Bonhage, C. Berger, S. Rotter, and U. Halsband

Department of Psychology, Neuropsychology, University of Freiburg, Freiburg, Germany
e-mail address: isabella.mutschler@psychologie.uni-freiburg.de

The impact of emotional music on heart rate and skin conductivity
F. Nagel, O. Grewe, R. Kopiez, and E. Altenmüller

Institute of Music Physiology and Musicians' Medicine, University of Music and Drama, Hannover, Germany
e-mail address: frederik.nagel@hmt-hannover.de

Emotion processing of major, minor, and dissonant chords: a functional magnetic resonance study
K.J. Pallesen, E. Brattico, C. Bailey, A. Korvenoja, J. Koivisto, A. Gjedde, and S. Carlson

Center of Functionally Integrative Neuroscience and PET Center, University of Aarhus, Aarhus, Denmark
e-mail address: karenjohanne@pet.auh.dk

Left auditory cortex specialization for vertical harmonic structure of chords
N. Passynkova, K. Sander, and H. Scheich

Special Laboratory of Non-Invasive Brain Imaging, Leibniz Institute for Neurobiology, Magdeburg, Germany
e-mail address: npassynk@ifn-magdeburg.de

Pleasant music reduces pain independently from its relaxing effect
M. Roy, J. Gauthier, P. Rainville, and I. Peretz

Department of Psychology, Université de Montréal, Montréal, Quebec, Canada
e-mail address: mathieu.roy.2@UMontreal.CA

Neurophysiological effects of excitative and sedative music—a transcranial magnetic stimulation study
A. Sahal

Department of Psychology, University of Sheffield, Sheffield, United Kingdom
e-mail address: anil777sahal@hotmail.com

Converging cognitive, physiological, and subjective aspects of acoustically induced stress
V. Simoens, S. Puttonen, M. Kivimäki, A. Hirvonen, R. Näätänen, and M. Tervaniemi

Cognitive Brain Research Unit, Helsinki University, Helsinki, Finland
e-mail address: veerle.simoens@cbru.helsinki.fi

Emotional processing of harmonic expectancy violations
N. Steinbeis, S. Koelsch, and J.A. Sloboda

Max Planck Institute for Human Cognitive and Brain Sciences, Leipzig, Germany
e-mail address: Stein@mail.cbs.mpg.de

Dimensions of emotion in expressive musical performance
B.W. Vines, C.L. Krumhansl, M.M. Wanderley, I. Dalca, and D.J. Levitin

Department of Psychology, McGill University, Montreal, Quebec, Canada
e-mail address: bradley.vines@mcgill.ca

Emotional communication through music performance played by hearing-impaired people
T. Yamasaki, R. Hiraga, and N. Kato

Faculty of Human Science, Shoin Women's University, Osaka, Japan
e-mail address: yamasak@gamelans.org

Music and instinctive behavior
I. Yevin and A. Koblyakov

Department of Biomechanics, Mechanical Engineering Institute, Russian Academy of Sciences
e-mail address: yevin@online.ru

Index of Contributors